AF847154

Nonlinear Physical Science

Series Editors

Albert C. J. Luo, Department of Mechanical and Mechatronics Engineering, Southern Illinois University Edwardsville, Edwardsville, IL, USA

Dimitri Volchenkov, Department of Mathematics and Statistics, Texas Tech University, Lubbock, TX, USA

Advisory Editors

Eugenio Aulisa, Department of Mathematics and Statistics, Texas Tech University, Lubbock, TX, USA

Jan Awrejcewicz, Department of Automation, Biomechanics and Mechatronics, Lodz University of Technology, Lodz, Poland

Eugene Benilov, Department of Mathematics, University of Limerick, Limerick, Limerick, Ireland

Maurice Courbage, CNRS UMR 7057, Universite Paris Diderot, Paris 7, Paris, France

Dmitry V. Kovalevsky, Climate Service Center Germany (GERICS), Helmholtz-Zentrum Geesthacht, Hamburg, Germany

Nikolay V. Kuznetsov, Faculty of Mathematics and Mechanics, Saint Petersburg State University, Saint Petersburg, Russia

Stefano Lenci, Department of Civil and Building Engineering and Architecture (DICEA), Polytechnic University of Marche, Ancona, Italy

Xavier Leoncini, Case 321, Centre de Physique Théorique, MARSEILLE CEDEX 09, France

Edson Denis Leonel, Departmamento de Física, Sao Paulo State University, Rio Claro, São Paulo, Brazil

Marc Leonetti, Laboratoire Rhéologie et Procédés, Grenoble Cedex 9, Isère, France

Shijun Liao, School of Naval Architecture, Ocean and Civil Engineering, Shanghai Jiao Tong University, Shanghai, China

Josep J. Masdemont, Department of Mathematics, Universitat Politècnica de Catalunya, Barcelona, Spain

Dmitry E. Pelinovsky, Department of Mathematics and Statistics, McMaster University, Hamilton, ON, Canada

Sergey V. Prants, Pacific Oceanological Institute of the RAS, Laboratory of Nonlinear Dynamical System, Vladivostok, Russia

Laurent Raymond, Centre de Physique Théorique, Aix-Marseille University, Marseille, France

Victor I. Shrira, School of Computing and Maths, Keele University, Keele, Staffordshire, UK

C. Steve Suh, Department of Mechanical Engineering, Texas A&M University, College Station, TX, USA

Jian-Qiao Sun, School of Engineering, University of California, Merced, Merced, CA, USA

Simon Villain-Guillot, Laboratoire Ondes et Matière d'Aquitaine, Université de Bordeaux, Talence, France

Michael Zaks, Institute of Physics, Humboldt University of Berlin, Berlin, Germany

Nonlinear Physical Science focuses on recent advances of fundamental theories and principles, analytical and symbolic approaches, as well as computational techniques in nonlinear physical science and nonlinear mathematics with engineering applications.

Topics of interest in *Nonlinear Physical Science* include but are not limited to:

- New findings and discoveries in nonlinear physics and mathematics
- Nonlinearity, complexity and mathematical structures in nonlinear physics
- Nonlinear phenomena and observations in nature and engineering
- Computational methods and theories in complex systems
- Lie group analysis, new theories and principles in mathematical modeling
- Stability, bifurcation, chaos and fractals in physical science and engineering
- Discontinuity, synchronization and natural complexity in physical sciences
- Nonlinear chemical and biological physics

This book series is indexed by the SCOPUS database.

To submit a proposal or request further information, please contact Dr. Mengchu Huang(Email: mengchu.huang@springer.com).

Vladimir V. Uchaikin

Electron-Photon Cascades

A Probabilistic Point of View

Higher Education Press

Springer

Vladimir V. Uchaikin
Department of Theoretical Physics
Ulyanovsk State University
Ulyanovsk, Russia

ISSN 1867-8440 ISSN 1867-8459 (electronic)
Nonlinear Physical Science
ISBN 978-981-99-7523-5 ISBN 978-981-99-7524-2 (eBook)
https://doi.org/10.1007/978-981-99-7524-2

Jointly published with Higher Education Press Limited Company
The print edition is not for sale in the mainland of China. Customers from the mainland of China please order the print book from: Higher Education Press Limited Company.

© Higher Education Press Limited Company 2024

This work is subject to copyright. All rights are solely and exclusively licensed by the Publisher, whether the whole or part of the material is concerned, specifically the rights of reprinting, reuse of illustrations, recitation, broadcasting, reproduction on microfilms or in any other physical way, and transmission or information storage and retrieval, electronic adaptation, computer software, or by similar or dissimilar methodology now known or hereafter developed.
The use of general descriptive names, registered names, trademarks, service marks, etc. in this publication does not imply, even in the absence of a specific statement, that such names are exempt from the relevant protective laws and regulations and therefore free for general use.
The publishers, the authors, and the editors are safe to assume that the advice and information in this book are believed to be true and accurate at the date of publication. Neither the publishers nor the authors or the editors give a warranty, express or implied, with respect to the material contained herein or for any errors or omissions that may have been made. The publishers remain neutral with regard to jurisdictional claims in published maps and institutional affiliations.

This Springer imprint is published by the registered company Springer Nature Singapore Pte Ltd.
The registered company address is: 152 Beach Road, #21-01/04 Gateway East, Singapore 189721, Singapore

Paper in this product is recyclable.

This book is dedicated to memory of Professor Igor Ivanenko (Moscow State University) 1929–1993.

Preface

On July 4, 2012, the physical community learned the long-awaited news: a new particle corresponding to the hypothetic Higgs boson was discovered. After a 40-year search, the subatomic particle with the expected properties was discovered in the ATLAS and CMS groups at the Large Hadron Collider (LHC) at CERN. The new particle was subsequently confirmed to match the expected properties of a Higgs boson, which promises to provide an extraordinary insight into our Universe.

At the core of the CMS detector sits a high-magnetic-field and large-bore superconducting solenoid surrounding an all-silicon pixel and strip tracker, a lead tungstate scintillating-crystal electromagnetic calorimeter, and a brass-scintillator sampling hadron calorimeter.

ATLAS is a detector of elementary particles, having the largest sensitive volume ever constructed for a particle collider. It looks like a cylinder 46 m long, 25 m in diameter, located in a cave at a depth of 100 m and weighs 7,000 tons, almost the same as the Eiffel Tower. The detector itself is a many-layered instrument designed to detect some of the tiniest yet most energetic particles ever created on earth. It consists of six different detecting subsystems wrapped concentrically in layers around the collision point to record the trajectory, momentum, and energy of particles, allowing them to be individually identified and measured. Huge magnet system bends the paths of the charged particles so that their momenta can be measured as precisely as possible. Most of the particles are electrons, positrons, and accompanying them are always photons. Being related to them by origin, they form many-particle families called the *electron–photon cascades* (EPC). The processes of cascade multiplication in these experiments play the role of photomultipliers customary in nuclear physics, differing from them only in their by larger sizes.

I dedicate this book to the blessed memory of Igor Pavlovich Ivanenko, who led in the 70s–80s the computational laboratory of the Scientific Research Institute of Nuclear Physics (Lomonosov Moscow State University), with which I was lucky to cooperate in those unforgettable young years, full of strength and self-confidence.

The author expresses deep gratitude to the professor of Moscow State University Tatyana Roganova for numerous consultations and materials provided, and Elena Kozhemyakina for assistance in preparing the manuscript for publication.

Ulyanovsk, Russia
November 2023

Vladimir V. Uchaikin

Contents

Chapter 1
Introduction

The entire Universe is a scene of continuous creation and development of electromagnetic cascades.

Felix Aharonian

The outer space surrounding us and the entire Universe as a whole are filled with high-energy particles of matter—photons, electrons, protons, hydrogen atoms, helium, heavier atoms, the nuclei of which can consist of several tens of nucleons. All of them perform, interacting with each other, as well as with gravitational and electromagnetic fields. In most cases, the particle–particle interactions are treated in terms of collision events, between which the particles move along smooth continuous trajectories. It is in such collisions particles can vanish, being absorbed by other particles, change their momentum being scattered by other particles, and produce new particles, sometimes in rather large numbers (multiple production). Often, particles disappear as a result of their absorption by other particles or decay due to their own instability. As a result, the passage of a high-energy particle through the matter is accompanied by secondary particles, which then generate particles of the next generation, and so on. Such a process is called the *cascade process* (for short, a *cascade*) and denoted by CP. The graphic representation of CP resembles a tree; for this reason, its mathematical model is called the *branching process*. However, it should be remembered that in this model the branches of the process are assumed to be mutually independent, and the history of each of them represents a Markov random process with continuous time.

It is the process of neutron multiplication in a nuclear reactor, which can serve as a typical examples of CP [1]. The topic of this book, however, is other cascades that play a major role in cosmic rays and high-energy physics. These are electron–photon cascades (EPC), as well as extensive air showers (EAS), containing, in addition to EPC, nucleon and nuclear cascades playing role of the basic carriers of information in astrophysics and accelerator physics.

Probabilistic characteristics of interaction of a fast particle with matter (say, cross section, momentum distribution, or so) are determined by its dynamical state at a

© Higher Education Press Limited Company 2024

V. V. Uchaikin, *Electron-Photon Cascades*, Nonlinear Physical Science,
https://doi.org/10.1007/978-981-99-7524-2_1

given time and the medium property at a given place; the other CP particles are wherein ignored. Because of this, cascade processes are usually treated within the framework of the branching processes theory, assuming independence of the future from the past with a fixed present and the mutual independence of different branches from each other.

In principle, these assumptions, taken as postulates, are sufficient to derive the kinetic equations of the cascade, formulate the Cauchy problem, and find the space-time distributions of particles of different types, following the model of solving similar ones in statistical mechanics of gas-dynamical kind [2, 3].

Meanwhile, there is a fundamental difference between the astrophysical theory of transport and, say, the routine theory of transport in liquids and gases. This difference is precisely due to the presence of cascade phenomena. Considering cosmic rays as a system of a very large number of particles filling the volume of our galaxy, we could calculate the concentrations and correlations of particles of different types from general kinetic consideration (as it is done in gas dynamics). But as soon as it comes to observation, we are faced with completely different conditions. A single particle with fantastically high energy entering the Earth's atmosphere generates in it an *extensive air shower* with fantastically high energies invading the Earth's atmosphere generate in it a wide—a thin disk consisting of secondary particles, with a near-light speed, crossing the atmosphere and plunging some of its components (muons, neutrinos) to a great depth in the ground. The charged part of this disc emits electromagnetic (Cherenkov) radiation, having managed to illuminate large areas by dark nights. The short flash coming to the mirror system brings the information equivalent those obtained by kilometer-sized telescopes. The events send a large amount of information to the array's server for subsequent processing. The purpose of this is to extract new information about the laws of particle interaction at such energies, about the origin of particles with such energies, about the properties of the interstellar medium, which is crossed by the particles that generate the cascades.

To some extent, one can talk about the analogy with an important task of physical kinetics: based on the cluster analysis of the observed inhomogeneous structures results, establish the properties of constituents the laws of their interaction. But the astrophysical aspect of the similar problem is much more complicated.

Having adopted the above philosophy, it seems natural to start discussing the problem as a whole with the concept of a detector—a device for measuring some property of a cascade. We imagine a cascade as a space-time cluster, the constituents of which are linked by correlations. The reading of a detector that measures some of its characteristics is a random variable. By repeating this experiment on an accelerator, where it is easy to keep the initial characteristics of the process constant, one can make sure that this value is subject to statistical fluctuations.

If the detector reading q fluctuations are small enough, it is sufficient to calculate only its mathematical expectation (m. e.) $\bar{q} \equiv \mathsf{M}q$. The calculation algorithm and the required set of characteristics describing the properties of the source and environment, in which the cascade develops, depend on the type of detector. The simplest class is formed by *additive detectors* [4–6], the readings of which are presented as the sum of the contributions of the elements of the volume of the medium (or individual elements of the trajectory of the cascade). This class includes detectors measuring particle number, energy absorbed, number of some elementary events, and other cascade characteristics.

Calculation of m.e. of additive characteristics can be carried out in two ways: to solve the basic kinetic equation for the particle density (containing the source term in its right-hand side) and find the detector reading as some functional of this solution, or to solve the adjoint (in the Lagrange sense) equation for the detector reading as a function of the primary particle initial position and find the functional of it depending on the source. Both of these approaches form the traditional basis of transport theory, and are widely applied to many problems in reactor physics, dosimetry, radiation shielding, experimental physics of high energies, etc.

The simplest characteristic of the measured quantity fluctuations (statistical error) is its second central moment (variance). Calculation of the second moment of the indication of the additive the detector can also be performed in two ways. One of them is based on the classical method of statistical mechanics—solving equations for two-particle distribution functions with subsequent double integration with a weighting function depending on the measured value. In the cascade theory, it was first used in the works of Baba , Ramakrishnan (see, for details, [7, 8]). The disadvantage of this approach (especially in terms of numerical calculations) is that the sought functions depend on two sets of phase variables.

Alternative, more effective and popular approach is based on considering the characteristics of the cascade as a function of one primary particle phase position. The time-asymmetry of the branching process (particles may be born in groups, but "die in alone" independently of each other) allows you to write equations for any moments of the additive functional, and each of them will have only one "active" variable—state of primary particle—and include information only about lower moments. A. Ramakrishnan named this approach the *regeneration point method* [8].

The concept of *neutron importance* as an average neutrons produced in the reactor by one primary neutron with given coordinates was introduced in the physics of nuclear reactors [9, 10]. A more general definition of the importance function as a linear functional from a transport Green function as a function of the source variable was given in [10], where the function was called the *influence function*. A connection between the equations for the particle flux density and for the importance function, the terms *adjoint function*, variational principles, perturbation theory, optimization methods, methods for solving inverse problems, etc. has been developed. The concept of importance has become a very popular tool in the theory of the Monte Carlo method and the computational methods in general [12]. The importance function is used to formulate a theory of transport in the framework of the canonical form field theory [11–13].

The further generalization of this concept developed due to stochastic nature of the most transfer process, whereby the detector reading may essentially be different from its mean value. In case of an additive detector, the mathematical expectation (m.e.) of this quantity as a function of the coordinates of the primary particle satisfies the adjoint equation, i.e., coincides with the above deterministic definition. The very same random variable is naturally to be called *linear stochastic importance* (the term was introduced in [14]).

Particular problems of fluctuations in the number of particles in a cascade arose and were solved long before the formation of a general concept. The pioneering work of Janossy [15], who discovered the renewal points method and obtained an equation for the generating function of a branching process in a continuous phase space (Janossy G-equation), gave impetus to a wide range of works devoted to calculating fluctuations in ionization [16], nuclear [18, 27], electron–photon [19–21] cascades. Pal, and independently of him Bell extended this approach to the theory of neutron transport [22, 23], Williams improved it in relation to cascades particles of the same type with a deterministic multiplicity of secondary particles equal to two [24].

Common to all these works is the particular type of the linear stochastic importance: the number of particles in this area of the phase space at the time of observation. Generalization to the case of an arbitrary additive functional and a general type of cascade process are given in [6], where the equations are derived for the distribution function of linear stochastic importance and its moments. So far we have been talking about additive detectors and fluctuations in their readings. In relation to this range of problems, Bell has used the term "stochastic transport theory" [1].

However, the recovery point method has found the most fertile field of application in the field of branching cascade processes (electronic, electron-photon, neutron, nuclear cascades). The reason for the preference for inverse equations, when the active variable is the phase position of one (primary) particle, over direct equations, depending on many positions of secondary particles, lies in the asymmetry between the past and the future: trajectories can branch but not merge again. The newly formed branches become mutually independent. This is why backward equations are preferable to forward ones.

The Chaps. 3 and 4 describe both approaches: the multipoint approach using the direct (forward) equation, and the Ramakrishnan's method of regeneration points using the backward equation. Comparing them in applications to cascade problems, one can conclude, that the second one has found the most fertile field of application in the branching cascade processes (electron, electron-photon, neutron, nuclear cascades) [15–26].

The reason for the backward preference for inverse equations, when the active variable relates to one (primary) particle, over the forward equations, depending on many positions of secondary particles, lies in the asymmetry between the past and the future: trajectories can branch but not merge again. The newly formed branches become mutually independent. This is why backward equations are preferable to forward ones.

By using generating functionals technique, the dual representation for its m.e. is obtained. Forward and backward equations are derived for cascade processes. The corresponding many-particle functions characterizing the process are obtained. Considered different ways of describing cascade process: using multiparticle probability densities and factorial moment densities. It is shown that in the second method, closest to traditional transport theory (the equation for the density of the first factorial moment coincides with the usual kinetic Boltzmann equation), multiparticle adjoint functions do not coincide with multiparticle importance, a connection has been established between these functions. Multiparticle densities and readings of weakly non-additive detectors (dimensions 2 and 3) are expressed in terms of single-particle functions using the diagrammatic technique. A perturbation theory for non-additive detectors is constructed, and a general scheme for solving the inverse problem is formulated. The problem of fluctuations of linear stochastic importance is considered in detail, equations are obtained for the distribution function and moments of this quantity (variance, covariance) of the detector readings, which makes it possible to carry out a physical analysis of the main sources of fluctuations in cascades (fluctuations in the ranges of particles of different types, fluctuations in the distribution of secondary particles arising in various interaction processes) and highlight the energy regions that make the main contribution to the fluctuations. A generalization of the backward (adjoint) equations is the most convenient way to describe branching processes in the case of an arbitrary additive detector, whose reading is formed during the entire time of the development of the cascade, and is not determined only by the state of the cascade at the time of measurement.

Two most developed areas of application of branching process model in physics undoubtedly stand out: physics of nuclear reactors and physics of cosmic rays. This is not only because of difference in tasks, but also of a significant difference in the processes occurring in these energy areas, separated by many orders of magnitude. As it indicates the title of the book, the second application will be at the center of attention in its second part.

An important position in the cosmic ray physics is placed by the theory of electron–photon cascades (EPC). The main processes that determine development of EPC are bremsstrahlung, production of electron–positron pairs, Compton scattering of photons, and scattering of electrons (elastic and inelastic) on atoms of the medium are considered in Chap. 5. Using both, classical and quantum languages (some sort of quasi-classical approach), author reveals electron and photon peculiarities intertwining them into tangled objects called *electron–photon* (or *electromagnetic*) *cascades* (*showers*). Thus, a correspondence arises between this process and the mathematical model of the Markov branching process, which underlies the mathematical apparatus of the cascade theory.

Knowledge of the probabilities of elementary events (cross sections) helps us to unravel the cascade, decompose it into conditionally independent subcascades, and thereby prepare the equations for solving. Chapter 6 introduces the reader to the basic concepts of cascade theory, including one-dimensional cascade equations, energy degradation equations (Belenky), the method of moments (Ivanenko), accounting for particle scattering in the small-angle approximation (Landau-Rumer), classical results of the Nishimura–Kamata theory. In contrast to the previous chapter dealing with ensemble-averaged cascade characteristics. Chapter 7 demonstrates an original numerical SBCE (Spline Backward Cascade Equation) method in applications to calculations of various EPC averaged characteristics. Chapter 8 relates to EPC—fluctuations from cascade to cascade, whereas Chap. 9 contains description of the transition effect—the behaviour of EPC when crossing a border with a different chemical composition. Chapter 9 describes the EPC development in heterogeneous (layered) structures such as scintillation and Cherenkov spectrometers, ionization calorimeters, X-ray emulsion, and so on. The attention is focused on the transition of EPC across layer boundaries, investigated by means of the Uchaikin–Ryzhov generalized perturbation theory.

Chapter 10 gives a general idea of the Monte Carlo method specifics, bearing in mind its further use in subsequent chapters of the book.

Complicated computational problems arise in astrophysical experiments with ultrahigh-energy cosmic rays, where energy spectrum, nuclear composition of primary cosmic rays, as well as the characteristics of their interactions with the nuclei of the atmosphere are the desired CR characteristics. These particles, born in distant cosmic sources and eventually reaching the Earth's atmosphere, generate in the latter bunches of particles of different types correlated in time and space, rushing to the Earth's surface and even plunging into underground depths. Such phenomena consisting mainly from EPCs are called *extensive air showers* (EAS). Sometimes, many thousands (less often—millions, very rarely billions) of particles arrive at ground level during short period of time. Most of them arrive within some hundred meters from the axis of motion of the original particle (the shower axis), but some particles can be found even kilometers away. Along the axis, most particles can be found in a kind of disk only a few meters thick and moving almost at the speed of light.

The most convenient for measurements are charged particles of the electron–photon cascades that are part of the EAS. The author witnessed and participated in a lively discussion in the late 1970s (ICRC Kyoto) regarding disagreements between several scientific groups regarding the spatial distribution of EPC electrons in the atmosphere. As for the calculations of the nuclear active component of the EAS, in view of the absence of a complete theory of strong interactions, their calculations are carried out within the framework of various models of multiple processes, which to a certain extent are phenomenological, simplified constructions, often built on the extrapolation of accelerator data to the region of ultrahigh energies, while obeying natural probabilistic restrictions. Some of these models are listed in Chap. 11, where the author's model is presented as well.

The general scheme for solving the above problems is presented today which appears as follows [28, 29]. Modern accelerators provide detailed data on the interaction of nucleons with energies up to 10^{15} eV in the laboratory coordinate system.

For the first time, the problem of calculating fluctuations in an EAS with a fixed primary energy originated in connection with the interpretation of the observed characteristics of showers at a steeply falling energy spectrum of primary radiation [30]. According to standard approach to the fluctuations problem, it is necessary to know at least a two-particle inclusive function of a multiple production process. Certainly, in contrast to EPC, where energies of two final particles are uniquely determined by one of them, two-particle function in multiple processes is not unambiguously linked with one-particle function. In this situation, a rough estimate of fluctuations can be obtained by highlighting the main factors affecting their magnitude ("sources" of fluctuations).

The initial assumption that the main effect is given by fluctuations of the point of the first interactions, "since in subsequent interactions the number of particles involved is so large, that fluctuations of individual interactions more or less compensate each other" [31], had not been justified. The reason for this was the leading effect, due to which one particle makes a large contribution to the measured quantity, the fluctuations of which cannot be neglected. Authors of works [31, 32] calculated EAS fluctuations taking into account events with only leading particle. Calculation [33–35] showed that, along with the fluctuations, fluctuations of the mileage, you need to take into account fluctuations of the coefficient inelasticity factors in collisions which are the leading particle.

Some authors [36], however, noted that an important role in the development of EAS should be played by fluctuations of the fraction of energy transferred to the electromagnetic component (partial coefficient of inelasticity K_γ). Experimental studies of fluctuations K_γ, at accelerators and in cosmic rays, have shown that their value weakly depends on the energy of the incident nucleon and the mass number target nucleus and is close to unity, which is approximately in is twice as large as the fluctuations of the total coefficient of inelasticity [37, 38].

Thus, if in the process of simulation, K_γ occurs equal to $K/3$ or becomes close to this value, and as a result of multiplicity charged particles simulations, the fluctuations of the shower characteristic under study can be significantly underestimated. In this situation, the derived and third chapter equations for the variance (covariance) of the cascade turned out to be especially useful. They made it possible to study separately the contributions of runs and interactions into the resulting variance. To calculate the contribution of multiple processes, a method has been developed based on the use of eventual (collective) variable distributions of partial inelasticity coefficients and partial multiplicities), which are easily extendable to the ultrahigh-energy region.

Probability of EAS registration by using some of its component (nucleon, electron, muon) depends on the width of its lateral distribution: the wider it, the higher this probability. In this relation, the championship belongs to the Cherenkov radiation of ESP electrons, available for observation on clear moonless nights. Methods of the observed characteristics are described in Chap. 11, the main attention of which is focused on the ALTAI code, elaborated by my colleagues Plyasheshnikov and Konopelko from Altai State University (Russia).

Chapter 12 is devoted to CR transport in the interstellar medium (ISM), or more precisely speaking, to physical properties of this medium. Talking about this process, they usually call it the CR-diffusion in ISM, but this term is not very appropriate in this context, since it provokes the idea of something resembling the Brownian motion (BM) along continuous but nowhere differentiable trajectories. This process is described by a diffusion equation—a parabolic differential equation with the first-order derivative with respect to time, and the second one (represented by the Laplace operator) with respect to coordinates. In fact, the trajectories of this process look different, differing from Brownian intermittency: each of them consists of random periods of smooth (ballistic) motion, intermitted by regions similar to Brownian. The physical reason to leave the Brownian model and pass to other lies in the highly inhomogeneous structure of interstellar magnetic fields, having a key effect on the charged particles motion.

As Dorman writes in his book [38, p. 57], CR in space is essentially confined within magnetic traps of one or another scale and fails to propagate freely (excluding for CR gamma-quanta and neutrinos which cannot be trapped by magnetic fields). Enormous CR traps are very diverse in their properties and the charged particle behavior in such traps is essentially energy dependent. The trap in the Earth's environment formed by the near-dipolar magnetic field exhibits a high degree of stability and considerable lifetime of particles in the trap. At the same time, the traps in the vicinities of chromospheric flares or in solar corpuscular streams of magnetized plasma are much more transparent for particles and the mode of particle ejection from such traps resembles the diffusion in irregular magnetic fields. Traps of various kinds are also formed in the vicinities of normal stars, in particular, in the solar system and supernova shells. On the other hand, the Galaxy (the galactic disk and halo) also forms a peculiar trap of dimensions of many thousands of parsecs which can safely (with a $\sim 10^7$ years life time) retain the particles of moderate and high energies and is very transparent for the super-high-energy particles. It is quite possible that the clusters of galaxies form even more enormous traps of the super-high-energy particles. Possessing self-similar property, such media are often referred to as the *fractal media*.

Chapter 13 takes the reader beyond the boundaries of the Galaxy itself, where the main processes that form the EFC are inverse Compton scattering and synchrotron. In the concluding chapter, Chap. 14, such facultative topics as cascade ages, scaling, universality, and similarity are discussed.

Three appendixes end the book, containing last article with participation with Ivanenko participation and a few images from his archive.

References

1. Bell G (1965) On the stochastic theory of neutron transport. Nucl Sci Engin 21:390–401
2. Balescu R (1975) Equilibrium and nonequilibrium statistical mechanics. NASA STI/Recon Tech Rep A 76:32809
3. Liboff RL (1979) Introduction to the Theory of Kinetic Equations. John Willey and Sons Inc., New York
4. Kolchuzhkin AM, Uchaikin VV (1975) On statistical fluctuations of detector readings in a radiation field. Izv Vyssh Uchebn Zaved Fisika 10:90–96 (in Russian)
5. Kolchuzhkin AM, Uchaikin VV (1978) Introduction to the theory of particle passage through the matter. Atomizdat, Moscow (in Russian)
6. Uchaikin VV (1978) Fluctuations in cascade processes of particles passing through matter. Izv Vyssh Uchebn Zaved Fisika 12:30–34 (1979); 2:7–12; 8:14–21 (in Russian)
7. Bharucha Reid AT (1960) Elements of the theory of Markov processes and their applications. Mc Graw-Hill Company Inc, New York
8. Ramakrishnan A (1962) Elementary particles and cosmic rays. Pergamon Press, Oxford
9. Weinberg A, Wigner E (1958) The physical theory of neutron chain reactors. The University of Chicago Press

10. Kadomtsev BB (1957) On the influence function in the theory of radiant energy transfer. Dokl. Acad Sci USSR 3:541–543 (in Russian)
11. Ivanov LD (1970) The Green's function method in the theory of neutron transport. Fizika yadernykh reaktorov Moscow Atomizdat 2:144–152 (in Russian)
12. Agu M (1977) On the concept of importance in markoffian processes. Prog Theor Phys 58(2):526–536
13. Lewins J (1978) Linear stochastic neutron transport theory. Proc Roy Soc A 362:537–558
14. Janossy L (1950) Note on the fluctuation problem of cascades. Proc Phys Soc A 63:241–249
15. Moyal JE (1955) Theory of ionization fluctuations. Phil Mag-Ser 46:203–280
16. Gedalin EV (1969) Cascade theory of showers. Nuclear Interact High Energ. Tbilisi 64–73
17. Messel G (1965) Nuclear cascade process. In: Physics of Cosmic Rays, vol 2. Moscow, p 69
18. Gedalin EV (1966) Cascade theory of showers. Nuclear Interact High Energ. Tbilisi 2:5–12 (in Russian)
19. Gerasimova NM (1964) Fluctuations in the number of particles in electron-photon lithium. Proc Phys Inst Acad Sci USSR 26:192–206 (in Russian)
20. Janossy L (1955) Statistical problems of an electron multiplier. Zh Eksp Teor Fiz 1(3):520–532
21. Srinivasan SK et al (1961) Fluctuation problem in electromagnetic cascades. Zeitschrift fur Physik 161:346–352; 1964; 177:164–173
22. Pal L (1962) Statistical theory of neutron chain reactions III. Acta Php Hung 14:369–380
23. Bell G (1965) On the stochastic theory of neutron transport. Nucl Sci Eng 21:390–401
24. Williams MR (1977) A fluctuation problem in particle transport theory. Physica 86A:535–548; 88A:561–573
25. Uchaikin VV (1977) On the representation of the expectation of symmetric functionals in the theory of particle transport. Izv Vyssh Uchebn Zaved Fizica 7:131–136 (in Russian)
26. Uchaikin VV (1977) General equation for the generating functional and factorial moments in the cascade theory. In: Proceedings of 15-th IGRC, vol 7. Plovdiv, pp 484–489
27. Baiburina SG, Borisov AS, Guseva ZM et al (1984) Investigation of nuclear interactions in the energy range $10^{14} - 10^{17}$ eV by the method of X-ray emulsion chambers in cosmic rays (experiment "Pamir"). Bull Phys Inst Acad Sci USSR 154:3–141 (in Russian)
28. Gaisser TK (1982) Cosmic rays and particle physics. Proc 17 Rencontre Moriond. Les Arcs - Savoie Gif - sur - Yvette 2:465–469
29. Zatsepin GT, Sarycheva LI (1954) About high-rise during extensive air showers. Dokl AN the USSR. 99:951–954 (in Russian)
30. Hayakawa S (1969) Cosmic ray physics. Nuclear and astrophysical aspects. In: Interscience monographs and texts in physics and astronomy. Wiley-Interscience, New York
31. Dedenko LG (1961) Calculation of some characteristics Extensive air showers with allowance for fluctuations. JETP 13(2):439–445
32. Vernov SN, Khristiansen GV, Abrosimov AG et al (1965) Investigation of fluctuations in the development of extensive air showers with a fixed total number of charged particles and with fixed total number of muons. Izv Acad Sci USSR Ser Phys 29(9):1676–1681 (in English)
33. Atrashkevich VB, Fomin YA, Khristiansen GB (1965) Calculation of fluctuations in the development of extensive air showers using the Monte Carlo method. Izv Acad Sci USSR Ser Phys 29(9):1696–1701 (in English)
34. Vernov SN, Khrenov BA, Christiansen GB (1967) The structure of the central region of the muon shower at a depth of 40 m in e. Izv Acad Sci USSR Ser Phys 31(9):1523 (in Russian)
35. Karakula S, Kempa Y, Vdovchik E (1969) Theoretical studies of fluctuations in development EAS. Izv Acad Sci USSR Ser Phys 32(9):1495–1497 (in English)
36. Nam RA, Nikolsky SI, Pavlyuchenko VP et al (1977) Verification of scale invariance in the energy range 1–40 TeV using ionization calorimeter, p 23. Moscow (Preprint/Phys Inst Acad Sci USSR; No. 104) (in English)
37. Grigorov NL, Shestoperov VY (1980) Main characteristics of the generation of γ-quanta by nucleons with an average energy of 5 TeV and multiparticle correlations in the region of fragmentation. Izv Acad Sci USSR Ser Phys 44(3):512–513 (in Russian)
38. Dorman L (2006) Cosmic ray interactions, propagation, and acceleration in space plasmas. Springer Science & Business Media

For Further Reading

39. Degond P, Pareschi L, Russo G (eds) (2004) Modeling and computational methods for kinetic equations. Springer Science & Business Media
40. Jenkins TM, Nelson WR, Rindi A (eds) (2012) Monte Carlo transport of electrons and photons vol. 38. Springer Science & Business Media
41. Gaisser TK, Engel R, Resconi E (2016) Cosmic rays and particle physics. Cambridge University Press

Chapter 2
Basic Concepts of Cascade Theory

2.1 The General Scheme of the Cascade Process

In accordance with the general principles of the particle transport theory, their motion in between collisions will be described in the classical manner [1]. The state of a single particle at a fixed time t is defined by a set of thus particle variables, which include the type of particles a (including its mass, charge, etc.), its spatial coordinates $\mathbf{r}$, and momentum $\mathbf{p}$. Taking these variables of all particles, we may consider the complex of all these sets as a state of the CP itself at the measurement time t. According to Markovian property of CP, the CP state at some of preceding time θ can be considered as an initial condition of the process.

To visualize a CP, let us consider a small time interval $(t, t + dt)$. Casting a glance at a CP at time t, we will see some number μ of particles with the parameters $a_1, \mathbf{r}_1, \mathbf{p}_1, \ldots, a_\mu, \mathbf{r}_\mu, \mathbf{p}_\mu$ (the numbering order is arbitrary). If the ith particle does not subject to collision during the interval, then it will be at the point $(a_i, \mathbf{r}_i + \dot{\mathbf{r}}_i dt, \mathbf{p}_i + \dot{\mathbf{p}}_i dt)$ of phase space at $t + dt$. If a collision occurs during this interval, then a number ν of secondary particles with the parameters $a'_1, \mathbf{p}'_1; \ldots; a'_\nu, \mathbf{p}'_\nu$ appear in place of a single particle with parameters $a_i, \mathbf{p}_i + \dot{\mathbf{p}}_i dt$.

In what follows, we will write x for the set $(a, \mathbf{r}, \mathbf{p})$ and X for the set of all possible values of x (the phase space), to make the formulas and equations less cumbersome. At the same time, one must bear in mind that one of these variables (a) entering the set takes discrete values. The continuous part of this set contains two vectors, one of which $(r(t))$ is continuous (and even time-differentiable) between collisions, another $(\mathbf{p})$ breaks at collision points. Note, also, that in what follows x is not necessary to understand exactly $(a, \mathbf{r}, \mathbf{p})$. Firstly, a complete dynamic description of the particle can be obtained by using other sets of equivalent variables. In particular, instead of the momentum $\mathbf{p}$ we can use the unit vector $\boldsymbol{\Omega} = \mathbf{p}/p$, specifying the direction of the particle motion, and its energy $E = E_a(p)$, or its velocity $\mathbf{v}_a = \boldsymbol{\Omega} v_a(p)$. Secondly, in many cases, reduced sets of variables are used, simplifying the calculation process due to decreasing the number of variables.

© Higher Education Press Limited Company 2024
V. V. Uchaikin, *Electron-Photon Cascades*, Nonlinear Physical Science,
https://doi.org/10.1007/978-981-99-7524-2_2

In addition, we recall that for Markovian time is not necessary to be a real physical time. So, if all cascade particles move with the same velocity along, say, x-axis, then this coordinate x can play a role of Markovian time (in a homogeneous medium, naturally).

A CP state at any moment t can be described in terms of an integer-valued *random measure*, generated by a collection of isolated points randomly distributed over phase space X. Recall that the random measure $\widehat{\mu}$ in a measurable space $(X, \mathscr{A})$ is a collection of random variables $\widehat{\mu}(A)$, defined for all $A \in \mathscr{A}$ and satisfying (almost certainly) conditions of non-negativity, and countable additivity [2].

We introduce the density $\mu(x)$ of the random measure $\widehat{\mu}(x)$:

$$\mu(x) = \widehat{\mu}(dx)/dx, \tag{2.1}$$

so that

$$\int_A \mu(x)dx = \int_X 1(x, A)\mu(x)dx = \widehat{\mu}(A), \tag{2.2}$$

where

$$1(x, A) = \begin{cases} 1, & x \in A; \\ 0, & x \notin A \end{cases}$$

is the indicator function for domain A. Unlike the measure (2.2) being a function of a domain A, the density measure is a (generalized) function of a point x in the phase space.

The functional of a random measure density

$$q = Q(\mu(\cdot)) \tag{2.3}$$

is the generalization of the concept of a function of a random variable. Applying to CP state, it can be interpreted as a reading of the instant response detector, immersed into cascade (the *instantaneous detector*, for brevity), whose reading depends on spatial arrangement of all cascade particles at a given instant t. Later, we will consider another (*cumulative*) type of detectors, the readings of which are formed during passage of the cascade particles through the sensitive volume of the detector. This process requires a certain finite time, during which the resulting signal accumulation is going on.

Coming back to the instant detector model, we expand (2.3) into a functional series

$$q = \sum_{n=0}^{\infty} \frac{1}{n!} \int dx_1 \dots \int dx_n \varphi_n(x_1, \dots, x_n)\mu(x_1) \dots \mu(x_n), \tag{2.4}$$

where

$$\varphi_n(x_1, \dots, x_n) = \left. \frac{\delta^n Q(\cdot)}{\delta\mu(x_1) \dots \delta\mu(x_n)} \right|_{\mu=0} \tag{2.5}$$

is a set of functions characterizing this functional. Equations (2.4)–(2.5) express the detector reading through the CP random state characteristics $\mu(x)$ and the detector properties, described by collection of the *response functions* $\varphi_n(x_1, \ldots, x_n)$. Observe, that if all $\varphi_n = 0$ excepting φ_1,

$$q = \int dx\varphi(x)\mu(x) \tag{2.6}$$

and we deal with a linear functional.

In the framework of the model under consideration, two particles cannot occupy the same point in phase space (more strictly, the probability of this event is zero), so the formula

$$\widehat{\mu}(A) = \sum_{i=1}^{m} 1(x_i, A) \tag{2.7}$$

makes a one-to-one mapping between the integer-valued measure and a random point distribution [3]. Note that here and below

$$m = \widehat{\mu}(X) \tag{2.8}$$

stands for a total number of points in the phase space X.

In view of this correspondence, any functional of the random integer-valued measure can be represented as a functional of the corresponding point distribution. Introducing the δ-function in X-space by definition

$$\int_A \delta(x, x')dx' = 1(x, A), \tag{2.9}$$

we obtain from (2.1) and (2.8):

$$\mu(x) = \sum_{i=1}^{m} \delta(x, x_i). \tag{2.10}$$

The next substitution of Eq. (2.10) into Eq. (2.4) yields the desired presentation:

$$q = \sum_{n=0}^{\infty} \frac{1}{n!} \sum_{i,\ldots,i_n=1}^{M} \varphi_n(x_{i_1}, \ldots, x_{i_n}). \tag{2.11}$$

To simplify the expressions like (2.11), we will use the abbreviations encountered in the statistical mechanics [4]

$$\varphi_n(1, \ldots, n) \equiv \varphi_n(x_1, \ldots, x_n), \tag{2.12}$$

etc. This way of abbreviation (replacing variables by their indices) will be extended to multiple integrals as well. Equations (2.4) and (2.11) become then

$$q = \sum_{n=0}^{\infty} \frac{1}{n!} \int d1 \ldots \int dn \varphi_n(1, \ldots, n)\mu(1) \ldots \mu(n) \tag{2.13}$$

and

$$q = \sum_{n=1}^{m} \frac{1}{n!} \sum_{i_1,\ldots,i_m=1}^{m} \varphi_n(i_1, \ldots, i_m) \equiv h_m(1, \ldots, m). \tag{2.14}$$

Representation of an integer-valued random measure by means of a point distribution allows us to specify an arbitrary functional by using a set of measurable functions $h_n(1, 2, \ldots, n)$ determining its value for any possible realization of the random arrangement $(1, \ldots, m)$:

$$q = Q(\mu(\cdot)) = h_m(1, \ldots, m). \tag{2.15}$$

From a physical point of view, the function $h_n(1, \ldots, n)$ represents the response of the measuring device (detector) to cascade, consisting of n particles at points $1, 2, \ldots, n$ at the measurement moment. In case of a linear measurement

$$h_n(1, 2, \ldots, n) = h_1(1) + h_1(2) + \cdots + h_1(n).$$

Because the points can be enumerated in various ways, the functions h_n are assumed to be invariant under permutation of their arguments:

$$h_n(1, 2, \ldots, n) = h_n(2, 1, \ldots, n) = \ldots = h_n(n, \ldots, 2, 1).$$

Note that the same is true of the functions φ_n introduced above. Such functions are called the *symmetrical functions*.

2.2 Mathematical Expectation of a Measurand

The typical problem in the cascade modeling of high-energy (cosmic ray or acceleration physics) processes is the measurement of some their characteristics for the subsequent comparison with the correspondent experimental data. Because of stochastic character of the process, the comparison requires knowing not only an average (the mathematical expectation) value of the characteristics under investigation

$$\bar{q} = \mathsf{M}q = \mathsf{M}Q(\mu(\cdot)),$$

but their statistical fluctuations as well. The simplest numerical characteristic of random variable q fluctuations is the *variance*

$$\sigma_q^2 = \overline{q^2} - \bar{q}^2,$$

although the exhaustive one is its probability distribution

$$F_q(y) = \mathsf{P}(q < y) = \mathsf{M}1(Q(\mu(\cdot)) < y),$$

or its Fourier transform

$$\tilde{F}_q(k) = \mathsf{M}e^{iqk},$$

knowing in mathematical statistics as the *characteristic function*.

Note that in all these cases we are talking about the mathematical expectation (m.e.) of some functional from the random point distribution. There are three representations of this value. One of them is based on Eq. (2.4). By applying operator M to both sides, we get

$$\bar{q} = \sum_{n=0}^{\infty} \frac{1}{n!} \int d1 \ldots \int dn \varphi_n(1, \ldots, n) m_n(1, \ldots, n), \tag{2.16}$$

where

$$m_n(1, \ldots, n) d1 \ldots dn = \mathsf{M}\widehat{\mu}(d1) \ldots \widehat{\mu}(dn) \tag{2.17}$$

is the mean value of product of random numbers of particles in $d1, \ldots, dn$, and the integral

$$\int_X d1 \ldots \int_X dn m_n(1, \ldots, n) = \mathsf{M}\widehat{\mu}^n(X), \tag{2.18}$$

so that the function $m_n(1, \ldots, n)$ can be recognized as the n-th moment multiparticle density of the random measure $\widehat{\mu}$. In a case of integer-valued measure, the density can be written through

$$m_n(1, \ldots, n) = \mathsf{M} \sum_{i_1, \ldots, i_n}^{M} \delta(1, i_1') \ldots \delta(n, i_n'). \tag{2.19}$$

Note that in accordance with our notation, this means

$$m_n(x_1, \ldots, x_n) = \mathsf{M} \sum_{i_1, \ldots, i_n} \delta(x_1, x_{i_1}') \ldots \delta(x_n, x_{i_n}'),$$

where x_i' are positions of random points (whereas x_i are fixed arguments of the deterministic multiparticle density m_n). During summation, each index runs through all integer values from 1 to $M \equiv \hat{\mu}(X)$, regardless of the values of other indices.

The second representation for the measurand expectation can be obtained by using Eq. (2.15) rewritten in the form

$$q = h_n(1, \ldots, M) = \sum_{n=0}^{\infty} \frac{1}{n!} \sum_{\langle i_1 \ldots i_n \rangle}^{M} h_n(i_1, \ldots, i_n)\delta_{nM}.$$

The subscript $\langle i_1 \ldots i_n \rangle$ under the sum sign says that in contrast to above summation, this is performed over all sets of integer indices, excluding those containing two or more coincident indices. Applying operator M to both sides of this expression yields

$$\bar{q} = \sum_{n=0}^{\infty} \frac{1}{n!} \int d1 \ldots \int dn h_n(1, \ldots, n) p_n(1, \ldots, n), \tag{2.20}$$

where

$$p_n(1, \ldots, n) = \mathsf{M} \sum_{\langle i_1 \ldots i_n \rangle}^{M} \delta(1, i_1') \ldots \delta(n, i_n')\delta_{nM}. \tag{2.21}$$

The subscript $i_1 \ldots i_n$ under the summation sign means that the summation is over only sets $(i_1, \ldots, i_n)$, not containing pairwise matching elements (i.e., the same index cannot occur two or more times in this set).

Finally, the third representation of the m.e. can be obtained by rewriting the functional (2.15) in the Balescu form (Eq. (3.1.7) from [4, p. 59]):

$$q = h_M(1, \ldots, M) = b_0 + \sum_{i=1}^{M} b_1(i) + \frac{1}{2!} \sum_{i \neq j}^{M} b_2(i, j) + \ldots$$

$$+ b_M(1, \ldots, M) = \sum_{n=0}^{\infty} \frac{1}{n!} \sum_{\langle i_1 \ldots i_n \rangle}^{M} b_n(i_1, \ldots, i_n)\varepsilon_{nM}, \tag{2.22}$$

where

$$\varepsilon_{nM} = \begin{cases} 1, & n \leq M, \\ 0, & n > M. \end{cases} \tag{2.23}$$

Symmetric functions b_n can be determined by sequential inserting $M = 0, 1$, and so on into equality

$$b_n(1, \ldots, n) = \sum_{k=0}^{n} \frac{(-1)^{n-k}}{k!} \sum_{\langle i_1 \ldots i_k \rangle}^{n} h_k(i_1, \ldots, i_k). \tag{2.24}$$

We meet special cases of this expression, when raising a linear functional to the square or the higher degrees:

$$\left[\sum_{i=1}^{M}\varphi(i)\right]^2 = \sum_{i=1}^{M}\varphi^2(i) + \sum_{i\neq j}^{M}\varphi(i)\varphi(j), \tag{2.25}$$

$$\left[\sum_{i=1}^{M}\varphi(i)\right]^3 = \sum_{i=1}^{M}\varphi^3(i) + 3\sum_{i\neq j}^{M}\varphi^2(i)\varphi(j) + \sum_{ijk}^{M\ \prime}\varphi(i)\varphi(j)\varphi(k), \tag{2.26}$$

etc. Comparing (2.25) (2.26) with the general formula (2.22), we see that in the case of the second degree of this functional $b_0 = 0$; $b_1(1) = \varphi^2(1)$; $b_2(1,2) = 2\varphi(1)\varphi(2)$; $b_n = 0, n > 2$, and in the case of the third degree

$$\begin{gathered} b_0 = 0, b_1(1) = \varphi^3(1), \\ b_2(1,2) = (3/2)[\varphi^2(1)\varphi(2) + \varphi(1)\varphi^2(2)], \\ b_3(1,2,3) = 3!\varphi(1)\varphi(2)\varphi(3), \\ b_n = 0,\ n > 3. \end{gathered} \tag{2.27}$$

The functionals, for which the expansion (2.22) terminated at the N-th term ($b_N \neq 0, b_n = 0, n > N$), are called *$N$-dimensional symmetric functionals*.

Acting on Eq. (2.22) by operator M, we obtain

$$\bar{q} = \sum_{n=0}^{\infty}\frac{1}{n!}\int d1\ldots\int dn b_n(1,\ldots,n) f_n(1,\ldots,n), \tag{2.28}$$

where

$$f_n(1,\ldots,n) = \mathsf{M}\sum_{\langle i_1\ldots i_n\rangle}^{M}\delta(1,i_1')\ldots\delta(n,i_n')\varepsilon_{nM}. \tag{2.29}$$

2.3 Multiparticle Densities

Let us now explain the meaning of the above functions p_n and f_n. Multiplying Eq. (2.21) by $d1\ldots dn$ and using Eq. (2.9) yield

$$p_n(1,\ldots,n)d1\ldots dn = \mathsf{M}\omega_n(d1,\ldots,dn), \tag{2.30}$$

where

$$\omega_n(d1, \dots, dn) = \sum_{\langle i_1 \dots i_n \rangle}^{M} 1(i'_1, d1) \dots 1(i'_n, dn)\delta_{nM} \tag{2.31}$$

is a random variable taking only two values, 0 and 1. At fixed set of elements $d1, \dots, dn$, its value is determined by the distribution of random points $i'_1, \dots, i'_\mu$. If at least one of these elements turns out to be empty, then $\omega_n(d1, \dots, dn) = 0$. The value of ω_n is nonzero only if every element contains one or more particles.

If $M = n$ and $\omega_n \neq 0$ in case that elements $d1, \dots, dn$ do not intersect, then each of these elements contains exactly one particle. To find the value of ω_n in this case, we note that the summation in (2.31) runs over all possible orders of numbering n points. Only one sequence from all these permutations will be such that the particle number 1 will be in the cell $d1$, the particle 2 in the cell $d2$, ..., the particle n in the cell dn, so that only one term in the sum (2.31) becomes nonzero: it will be equal to 1.

Integrating both sides of (2.30) over all variables yields

$$\int d1 \dots \int dn p_n(1, \dots, n) = n! P_n, \tag{2.32}$$

where

$$P_n = \mathsf{M}\delta_{n\mu} = \mathsf{P}\{M = n\}$$

is the probability that the total number of particles is n. Since $\sum_{n=0}^{\infty} P_n = 1$ in (2.32), the functions p_n are normalized as

$$p_0 + \sum_{n=1}^{\infty} \frac{1}{n!} \int d1 \dots \int dn p_n(1, ..., n) = 1. \tag{2.33}$$

Strictly speaking, $p_n(1, \dots, n)$ is not a probability density, because the matching elements of the random variable (2.31) take values different from 0 and 1 (for example, if all di coincide, and all points are in this element, then $\omega_n = n!\delta_{n\mu}$). This explains the emergence of the factor of $n!$ on the right-hand side of (2.32). It would not appear if the particles were numbered and we consider the probability that the particle with the number 1 would be in the $d1$, the particle 2 in the $d2$, etc. If more and the number of particles was fixed, then $p_n(1, \dots, n)$ represented an ordinary probability density function in n-dimensional space, and the integral over all variables would be equal to 1.

In case of unlabeled particles, the factor $(n!)^{-1}$ is a consequence of the particles indistinguishability. Nevertheless, in statistical mechanics, the function $p_n(1, \dots, n)$, satisfying (for a fixed total number of particles) the condition

$$\int d1 \dots \int dn p_n(1, \dots, n) = n!$$

is called the probability density [5]. We will use this term even when the number of particles is random and normalization (2.33) takes place.

Let us now discuss the expressions

$$f_n(1, \dots, n)d1 \dots dn = \mathsf{M}\Omega_n(d1, \dots, dn), \tag{2.34}$$

$$\Omega_n(d1, \dots, dn) = \sum_{\langle i_1 \dots i_n \rangle}^{M} 1(i'_1, d1) \dots 1(i'_n, dn)\varepsilon_{nM}. \tag{2.35}$$

In the case of disjoint elements of the coincidence of the two indices (e.g., i'_1 and i'_2) reduces the corresponding terms of sum (2.35) to zero, because a point cannot belong to two non-intersecting elements $d1$ and $d2$, and one of the indicators $1(i'_1, d1)$, $1(i'_2, d2)$ vanishes. Restriction on summation in (2.35) can be removed in this case, and the random function

$$\Omega_n(d1, \dots, dn) = \sum_{i_1=1}^{M} 1(i'_1, d1) \dots \sum_{i_n=1}^{M} 1(i'_n, dn)\varepsilon_{n\mu}$$

becomes a product of random numbers of particles in $d1, \dots, dn$:

$$\Omega_n(d1, \dots, dn) = \widehat{\mu}(d1) \dots \widehat{\mu}(dn).$$

Therefore, the function f_n with distinct arguments gives an average value of this product and coincides with the nth moment density at corresponding points (2.17) although the integral of (2.34) over all variables differs from (2.18).

Let us integrate (2.35) over the region $\underbrace{A \times \dots \times A}_{n}$:

$$\int_A \dots \int_A \Omega(d1, \dots, dn) = \sum_{\langle i_1 \dots i_n \rangle}^{M} 1(i'_1, A) \dots 1(i'_n, A)\varepsilon_{nM}.$$

The summation in the right side of this expression is running now over indices of only those particles that are in the domain A (the rest terms of the sum are equal to 0). Taking into account that all indicators included in this sum are equal to unity, and the number of terms

$$\widehat{\mu}^{[n]}(A) = \widehat{\mu}(A)[\widehat{\mu}(A) - 1] \dots [\widehat{\mu}(A) - n + 1],$$

we get

$$\Omega_n(A, \dots, A) = \int_A \dots \int_A \Omega(d1, \dots, dn) = \mathsf{M}\widehat{\mu}^{[n]}(A), \tag{2.36}$$

so

$$\int_A d1 \ldots \int_A dn f_n(1, \ldots, n) \mathsf{M}\widehat{\mu}^{[n]}(A), \tag{2.37}$$

i.e., the integral of f_n over A^n is the n-th factorial moment of a random number $\hat{\mu}(A)$ (recall, $\hat{\mu}(X) = M$) . Consequently, the functions f_n may be called the factorial moments density (FMD). In the statistical mechanics, they are called n-particle function distribution. If the volumes $d1, \ldots, dn$ are so small that the appearance of two or more particles in each of them can be neglected, then $f_n(1, \ldots, n)d1 \ldots dn$ can be interpreted as a probability to find each of these volumes filled with only one particle regardless of the total number of particles (as opposed to $p_n(1, \ldots, n)d1 \ldots dn$, when the total number of particles is assumed to be equal to n). Observe that the same sense is prescribed (under the above conditions) to $m_n(1, \ldots, n)d1 \ldots dn$. However, if two or more arguments coincide, the functions f_n differ from the corresponding functions m_n by absence of singularities.

Thus, an arbitrary functional of an integer-valued random measure can be represented in three equivalent forms (2.16), (2.20), and (2.28). We call them, respectively, m-, p-, and f-representations.

In conclusion, we rewrite this formulas in a more convenient form using their symmetry property. Note that the operator

$$\mathsf{S}_n(1, \ldots, n)_{\ldots} \equiv \frac{1}{n!} \sum_{\langle i_1 \ldots i_n \rangle}^{n} \ldots \tag{2.38}$$

in Eq. (2.22) and others is the symmetrizing operator with respect to variables $1, \ldots, n$. Symmetric functions $f_n(1, ..., n)$ are eigenfunctions of this operator:

$$\mathsf{S}_n(1, \ldots, n) f_n(1, \ldots, n) = f_n(1, \ldots, n).$$

Acting on an arbitrary function $f'_n(1, \ldots, n)$ of n variables, this operator transforms it into a symmetric one:

$$\mathsf{S}_n(1, \ldots, n) f'_n(1, \ldots, n) = f_n(1, \ldots, n). \tag{2.39}$$

It may be that the dimension n of the symmetrized operator does not coincide with the number of variables k of the function function to which it applies. If $n < k$, then the summation is carried out only on those variables that are specified in the arguments of S_n. If $n > k$, it is possible to carry out a partial summation, in which we get

$$\mathsf{S}_n f'_k(1, \ldots, k) = \frac{(n-k)!}{n!} \sum_{\langle i_1 \ldots i_k \rangle}^{n} f'_k(i_1, \ldots, i_k). \tag{2.40}$$

In particular,

$$\mathsf{S}_n 1 = 1,$$

$$\mathsf{S}_n f_1'(1) = \frac{1}{n}\sum_{i=1}^{n} f_1'(i),$$

$$\mathsf{S}_n f_2'(1,2) = \frac{1}{n(n-1)}\sum_{i\neq j}^{n}{}' f_2'(i,j),$$

etc. Note also that $n \geq k$

$$\mathsf{S}_n(1,\ldots,n)\mathsf{S}_k(1,\ldots,k) = \mathsf{S}_n(1,\ldots,n).$$

With these notation, Eqs. (2.21), (2.22), (2.29), and (2.24) can be presented as follows:

$$p_n(1,\ldots,n) = \mathsf{S}_n\mathsf{M}[M!\delta(1,1')\ldots\delta(n,n')\delta_{nM}],$$

$$h_M(1,\ldots,M) = \mathsf{S}_M\sum_{n=0}^{\infty}\binom{\mu}{n} b_n(1,\ldots,n)\varepsilon_{nM},$$

$$f_n(1,\ldots,n) = \mathsf{S}_n\mathsf{M}[M^{[n]}\delta(1,1')\ldots\delta(n,n')\delta_{nM}]$$

and

$$b_n(1,\ldots,n) = \mathsf{S}_n\sum_{k=0}^{\infty}(-1)^{n-k}\binom{n}{k} h_k(1,\ldots,k)\varepsilon_{n\mu},$$

respectively.

For the sake of brevity, the symbol of the symmetrizing operator (2.39) will be omitted in front of the functions which are symmetric by definition

$$f_n(1,\ldots,n) = f_n'(1,\ldots,n) \quad \text{instead of} \quad f_n(1,\ldots,n) = \mathsf{S}_n f_n'(1,\ldots,n). \tag{2.41}$$

With this convention, the above formulas take the form

$$p_n(1,\ldots,n) = \mathsf{M}[M!\delta(1,1')\ldots\delta(n,n')\delta_{nM}],$$

$$m_n(1,\ldots,n) = \mathsf{M}[M!\delta(1,1')\ldots\delta(n,n')\delta_{nM}],$$

$$h_M(1,\ldots,M) = \sum_{n=0}^{\infty}\binom{M}{n} b_n(1,\ldots,n)\varepsilon_{nM},$$

$$b_n(1,\ldots,n) = \sum_{k=0}^{\infty}(-1)^{n-k}\binom{n}{k} h_k(1,\ldots,k)\varepsilon_{nM},$$

$$f_n(1,\ldots,n) = \mathsf{M}[M^{[n]}\delta(1,1')\ldots\delta(n,n')\delta_{nM}].$$

2.4 Fluctuations of a Functional from a Random Measure

Because of stochastic nature of the process, any result of measurements can be considered as a random variable, and its comparison with a non-random theoretical value makes sense only if statistical fluctuations of the measurand are known. Comprehensive information about fluctuations of a functional gives us the density distribution probability

$$\psi(q) = \mathsf{M}\delta[q - Q(\mu(\cdot))]$$

or a total set of its moments (if they exist):

$$\overline{q^n} = \mathsf{M}[Q(\mu(\cdot))]^n.$$

Thus, the formal statement of the fluctuations problem does not take us beyond the expectation computing, but the desired functional becomes nonlinear.

The computational complexity is determined by a specific form of the functional under consideration. For a linear functional

$$q = \int \varphi(x)\mu(x)dx$$

$$\psi(q) = \mathsf{M}\delta\left[q - \int \varphi(x)\mu(x)dx\right],$$

$$\overline{q^n} = \mathsf{M}\left[\int \varphi(x)\mu(x)dx\right]^n.$$

In accordance with the above relations, moments of the linear functional are expressed through multiparticle functions m_n and f_n as

$$\bar{q} = \int d1\varphi(1)m_1(1) = \int d1\varphi(1)f_1(1), \tag{2.42}$$

$$\overline{q^2} = \int d1 \int d2\varphi(1)\varphi(2)m_2(1,2) = \int d1\varphi^2(1)f_1(1)$$

$$+ \int d1 \int d2\varphi(1)\varphi(2)f_2(1,2);$$

$$\overline{q^3} = \int d1 \int d2 \int d3\varphi(1)\varphi(2)\varphi(3)m_3(1,2,3) = \int d1\varphi^3(1)f_1(1)$$

$$+ 3\int d1 \int d2\varphi^2(1)\varphi(2)f_2(1,2) + \int d1 \int d2 \int d3\varphi(1)\varphi(2)\varphi(3)f_3(1,2,3), \tag{2.43}$$

etc.

A suitable approach to the fluctuation problem is based on the cumulant analysis. In contrast to the ordinary moments, the cumulants have a clearly expressed independent statistical meaning and can be given to a certain extent independently of each other, as some "normal coordinates" of statistical description. With the help of cumulants, one can easily estimate a degree of non-Gaussianity of random variables and processes and approximate the distribution function [6]. The four first cumulants are the mean

$$\bar{q} = \mathsf{M}q,$$

the variance

$$\sigma_q^2 \equiv \mathsf{M}(q - \bar{q})^2 = \overline{q^2} - \overline{q}^2, \tag{2.44}$$

the skewness

$$\alpha_q \equiv \overline{(q - \bar{q})^3}/\sigma^3 = (\overline{q^3} - 3\overline{q^2}\bar{q} + 2\bar{q}^3)/\sigma_q^3,$$

and the excess:

$$\beta_q \equiv \overline{(q - \bar{q})^4}/\sigma_q^4 = (\overline{q^4} - 3\left(\overline{q^2}\right)^2 - 4\bar{q}\overline{q^3} + 12\overline{q^2}\bar{q}^2 - 6\bar{q}^4)/\sigma_q^4. \tag{2.45}$$

Substituting (2.42), (2.43) into (2.44), (2.45), we obtain

$$\sigma_q^2 = \int d1 \int d2\varphi(1)\varphi(2)k_2(1,2),$$

$$\alpha_q = \sigma^{-3} \int d1 \int d2 \int d3\varphi(1)\varphi(2)\varphi(3)k_3(1,2,3)$$

etc., where

$$k_2(1,2) = m_2(1,2) - m_1(1)m_1(2), \tag{2.46}$$

and

$$k_3(1,2,3) = m_3(1,2,3) - 3m_2(1,2)m_1(3) + 2m_1(1)m_1(2)m_1(3) \tag{2.47}$$

are cumulant densities, and their integrals are equal cumulants of the total number of the particles in the phase space:

$$\int d1 \int d2k_2(1,2) = \sigma_M^2,$$

$$\int d1 \int d2 \int d3k_3(1,2,3) = \alpha_M.$$

Recall that according to (2.41), (2.47) means

$$k_3(1,2,3) = m_3(1,2,3) - m_2(1,2,3)m_1(3)$$
$$- m_2(1,3)m_1(2) - m_2(2,3)m_1(1) + 2m_1(1)m_1(1)m_1(3).$$

With the help of $k_2(1,2)$, the covariance of two linear functionals

$$q = \sum_{i=1}^{\mu} \varphi(i)$$

and

$$r = \sum_{i=1}^{\mu} g(i)$$

can be written as

$$\sigma_{qr} \equiv \overline{qr} - \bar{q}\,\bar{r} = \int d1 \int d2\varphi(1)g(2)k_2(1,2). \tag{2.48}$$

2.5 Characteristic and Generating Functionals

The most important tools for study of random measures are generating functionals. Here, we focus on two of them: on the *characteristic functional* (CF) defined as

$$C(s(\cdot)) = \mathsf{M}\exp\left\{i\int s(x)\mu(x)dx\right\} \tag{2.49}$$

whenever $\mathsf{Im}s \geq 0$ and the integral in the exponent exists, and on the *generating functional* (GF)

$$F(u(\cdot)) = \mathsf{M}\exp\left\{\int \mu(x)\ln u(x)dx\right\} \tag{2.50}$$

for integer-valued random measures which always exists for $|u| \leq 1$. These functionals are linked to each other via relations

$$C(s(\cdot)) = F(e^{is(\cdot)}), \tag{2.51}$$

and

$$F(u(\cdot)) = C(-i\ln u(\cdot)). \tag{2.52}$$

They possess the following properties:

$$C(0) = 1, \quad |C(s(\cdot))| \leqslant 1, \quad C(-s(\cdot)) = C^*(s(\cdot)), \tag{2.53}$$

and

$$F(1) = 1, \quad |F(u(\cdot))| \leqslant 1, \quad |u(x)| \leqslant 1. \tag{2.54}$$

An important property of the functionals under consideration is their multiplicativity with respect to independent measures. This means, that if $\mu = \sum_k \mu_k$ where μ_k are mutually independent measures and $C(s(\cdot))$, $F(u(\cdot))$, $C_k(s(\cdot))$, and $F_k(u(\cdot))$ are corresponding CFs and GFs, then

$$C(s(\cdot)) = \prod_k C_k(s(\cdot)), \tag{2.55}$$

and

$$F(u(\cdot)) = \prod_k F_k(u(\cdot)). \tag{2.56}$$

Really,

$$\exp\left\{i \int s(x) \sum_k \mu_k(x) dx\right\} = \prod_k \exp\left\{i \int s(x)\mu_k(x) dx\right\}.$$

Acting on both sides of this equality by operator M and taking into account the independence of μ_k, we arrive at Eq. (2.55). Equation (2.56) can be proved in a similar way. This property of the functionals is very important for study of branching processes because it allows to reflect the main property (the mutual independence of branches) in a compact form [2, 7, 8].

Just as the characteristic and generating functions give to determine the initial moments and factorial moments of a random variable, the CF and the GF can be used to compute the corresponding measure densities, but the usual derivatives should be replaced by the functional ones

$$\mathscr{D}_s^{(n)}(1, \ldots, n) Q(s(\cdot)) = \frac{\delta^n Q(s(\cdot))}{\delta s(1) \ldots \delta s(n)} \equiv Q^{(n)}(1, \ldots, n; s(\cdot)). \tag{2.57}$$

Let us expand the exponential function under the expectation sign in Eq. (2.49) in the Taylor series and change the order of expectation and integration operators. As a result, we get

$$C(s(\cdot)) = 1 + \sum_{k=1}^{\infty} \frac{1}{k!} \int d1 \ldots \int dk C^{(k)}(1, \ldots, k; 0) s(1) \ldots s(k). \tag{2.58}$$

Taking the n-th functional derivative with respect to $s(x)$ from both sides of Eq. (2.58) yields

$$C^{(n)}(1, \ldots, n; s(\cdot)) = \delta_{n0}$$

$$+\sum_{k=1}^{\infty} \frac{i^k}{k!} \int d1' \ldots \int dk' C^{(k)}(1', \ldots, k'; 0) \mathscr{D}_s^{(n)}(1, \ldots, n)[s(1') \ldots s(k')]. \tag{2.59}$$

In calculating the derivative of the product, we use the formula (2.57), putting it $Q_i(s(\cdot)) = s(i')$. As

$$\mathscr{D}_0^{(n)}(1, \ldots, n)s(i') = \delta_{n0}s(i') + \delta_{n1}\delta(1, i'), \tag{2.60}$$

then at $s = 0$ in (2.57) will be the only one term corresponding to $n_1 = n_2 = \ldots = n_k = 1$:

$$\mathscr{D}_0^{(n)}(1, \ldots, n) \prod_{i=1}^{k} s(i') = n!\delta(1, 1') \ldots \delta(k, k')\delta_{nk}. \tag{2.61}$$

Putting in (2.59) $s = 0$ and using (2.61), we obtain

$$\mathscr{D}_0^{(n)}(1, \ldots, n)C(s(\cdot)) = C^{(n)}(1, \ldots, n; 0) = i^n m_n(1, \ldots, n). \tag{2.62}$$

As a result

$$C(s(\cdot)) = 1 + \sum_{k=1}^{\infty} \frac{i^k}{k!} \int d1 \ldots \int dk m_k(1, \ldots, k)s(1) \ldots s(k).$$

It is easy to see that formula (2.62) is a generalization of the well-known relation between the moments and characteristic function random variable.

To describe the probabilistic characteristics of random integer measures, the GF (2.50) is used. In the point distribution representation,

$$F(u(\cdot)) = \mathsf{M} \exp\left\{\sum_{i=1}^{\mu} \ln u(i)\right\} = \mathsf{M} \prod_{i=1}^{\mu} u(i). \tag{2.63}$$

Taking into account, that the product with $\mu = 0$ is considered to be equal to 1, and using for calculations of the Eq. (2.20), we obtain

$$F(u(\cdot)) = p_0 + \sum_{n=1}^{\infty} \frac{1}{n!} \int d1 \ldots \int dn p_n(1, \ldots, n)u(1) \ldots u(n). \tag{2.64}$$

Thus, the properties (2.53), (2.54) can be supplemented by $F(0) = p_0$.

The p-representation is suitable when the population consists predominantly of small number of particles, and large number events are unlikely. Otherwise, we have to deal with integrals of increasing multiplicity. There exists another representation,

f-representation, when the number of terms in the series is determined by the degree of nonlinearity of the measurand: the weaker nonlinearity effect is, the less number of terms can be used even if the total population is large.

In order to derive this representation formula, we start with the known expansion of product formula

$$\prod_{k=1}^{n}(1+a_k) = \sum_{k=0}^{n} E_k,$$

where

$$\begin{aligned}
E_0 &= 1,\\
E_1 &= a_1 + a_2 + \ldots + a_n,\\
E_2 &= a_1a_2 + a_1a_3 + \ldots + a_{n-1}a_n,\\
E_3 &= a_1a_2a_3 + a_1a_2a_4 + \ldots + a_{n-2}a_{n-1}a_n,\\
&\ldots\ldots\\
E_n &= a_1a_2\ldots a_n.
\end{aligned}$$

The common term of this expansion can be represented as

$$E_k = \frac{1}{k!}\sum_{i_1\ldots i_k}^{n}{}' a_{i_1}\ldots a_{i_k},$$

so

$$\prod_{k=1}^{n}(1+a_k) = \sum_{k=0}^{n}\frac{1}{1!}\sum_{i_1\ldots i_k}^{n}{}' a_{i_1}\ldots a_{i_k}. \tag{2.65}$$

Introducing the notation $\bar{u}(x) = 1 - u(x)$ and using (2.65), we get

$$F(u(\cdot)) = \mathsf{M}\prod_{i=1}^{\mu}[1-\bar{u}(i)] = \mathsf{M}\sum_{k=0}^{\infty}\frac{1}{k!}\sum_{i_1\ldots i_k}^{\mu}{}'(-1)^k\bar{u}(i_1)\ldots\bar{u}(i_k)\varepsilon_{k\mu}.$$

Comparing the expression under the sign of M with the expansion (2.22), we see that for a given functional

$$g(1,\ldots,k) = (-1)^k\bar{u}(1)\ldots\bar{u}(k),$$

so that Eq. (2.28) leads to the following representation of the GF:

$$F(u(\cdot)) = 1 + \sum_{k=1}^{\infty}\frac{(-1)^k}{k!}\int d1\ldots\int dk f_k(1,\ldots,k)\bar{u}(1)\ldots\bar{u}(k). \tag{2.66}$$

The functional differentiation of Eqs. (2.64) and (2.66) yields

$$\mathscr{D}_u^{(n)}(1,\ldots,n)F(u(\cdot)) = \delta_{n0}p_0 + \sum_{h=n}^{\infty}\frac{1}{k!}\int d1'\ldots\int dk'\; p_k(1',\ldots,k')$$

$$\times \sum_{N_k=n}\binom{n}{n_1\ldots n_k}\prod_{i=1}^{k}\left[\delta_{n_i0}u(i') + \delta_{n_i1}\delta(N_{i-1}+1,i')\right], \tag{2.67}$$

$$\mathscr{D}_u^{(n)}(1,\ldots,n)F(u(\cdot)) = \delta_{n0} + \sum_{h=n}^{\infty}\frac{(-1)^k}{k!}\int d1'\ldots\int dk'\; f_k(1',\ldots,k')$$

$$\times \sum_{N_h=n}\binom{n}{n_1\ldots n_k}\prod_{i=1}^{k}\left[\delta_{n_i0}\overline{u}(i') - \delta_{n_i1}\delta(N_{i-1}+1,i')\right]. \tag{2.68}$$

Setting $u = 0$ in Eq. (2.67) and $u = 1$ in Eq. (2.68), we arrive at the relations

$$p_n(1,\ldots,n) = F^{(n)}(1,\ldots,n;0), \tag{2.69}$$

$$f_n(1,\ldots,n) = F^{(n)}(1,\ldots,n;1) \tag{2.70}$$

for all p_n and f_n, which are similar to Eq. (2.62) for the moment densities m_n.

2.6 Interrelations Between Multiparticle Densities

In order to establish a relationship between the functions p_n and f_n,, we generalize Eq. (2.65). Let us multiply both sides by $b_1\ldots b_n$ and write c_k for $a_k b_k$:

$$\prod_{k=1}^{n}(b_k + c_k) = \sum_{k=0}^{n}\frac{1}{k!}\sum_{i_1\ldots i_k}^{n} a_{i_1}\ldots a_{i_k}b_1\ldots b_n.$$

After some algebra, we obtain

$$\prod_{k=1}^{n}(b_k + c_k) = \sum_{k=0}^{n}\frac{1}{k!(n-k)!}\sum_{i_1\ldots i_n}^{n}{}' c_{i_1}\ldots c_{i_k}b_{i_{k+1}}\ldots b_{i_n},$$

or, assuming the symmetrization procedure,

$$\prod_{k=1}^{n}(b_k + c_k) = \sum_{k=0}^{n}\binom{n}{k}c_1\ldots c_k b_{k+1}\ldots b_n. \tag{2.71}$$

In accordance with this

$$\prod_{i=1}^{k}[\delta_{n_i0}u(i') + \delta_{n_i1}\delta(N_{i-1}+1, i')]$$

$$= \sum_{i=0}^{k}\binom{k}{i}[\delta_{n_1,1}\delta(1,1')\ldots\delta_{n_i1}\delta(N_{i-1}+1,i')\delta_{n_{i+1}0}u((i+1)')\ldots\delta_{n_k0}u(k')]. \tag{2.72}$$

Substituting Eq. (2.72) into the right side of Eq. (2.67) by setting $u = 1$ and using Eq. (2.70) yields

$$f_n(1,\ldots,n) = \delta_{n0}P_0 + \sum_{k=n}^{\infty}\frac{1}{k!}\int d1'\ldots\int dk' p_h(1',\ldots,k')$$

$$\times\; n!\sum_{i=0}^{k}\binom{n}{k}\delta(1,1')\ldots\delta(i,i')\delta_{ni} = \delta_{n0}p_0$$

$$+ \sum_{h=n}^{\infty}\frac{1}{(k-n)!}\int d(n+1)\ldots\int dk p_k(1,\ldots,k). \tag{2.73}$$

Similarly, from Eq. (2.68) with $u = 0$ we find

$$p_n(1,\ldots,n) = f_n(1,\ldots,n) + \sum_{h=n+1}^{\infty}\frac{(-1)^{k-n}}{(k-n)!}\int d(n+1)\ldots\int dk f_k(1,\ldots,k). \tag{2.74}$$

The relations (2.73), (2.74), establishing the link between the probability densities and the factorial moment densities, can be derived directly from the definitions (2.21) and (2.29). Using the identities $\varepsilon_{k\mu} \equiv \sum_{n=k}^{\infty}\delta_{n\mu}$ and $1 \equiv \int d(n+1)'\ldots\int dn'\delta(k+1,(k+1)')\ldots\delta(n,n')$, we rewrite (2.29) as

$$f_n(1,\ldots,n) = \sum_{k=n}^{\infty}\frac{1}{(k-n)!}\int d(n+1)'\ldots\int dk'\mathsf{M}[k!\delta(1,1')\ldots\delta(k,k')\delta_{k\mu}].$$

Comparing the right-hand side of this equation with the definition (2.21) shows that f_n are expressed through p_n via Eq. (2.73). To find the inverse relation (2.74), one needs to use an obvious consequence of Eq. (2.23)

$$\delta_{k\mu} = \varepsilon_{k\mu} - \varepsilon_{k+1,\mu}$$

and the formula

$$\varepsilon_{k+1,\mu} = -\sum_{n=1}^{\infty}(-1)^n \binom{\mu-k}{n} \varepsilon_{k+n,\mu}$$

based on the binomial expansion.

Interrelation between f_n and m_n are found by using Eqs. (2.51) and (2.52), connecting the CF and GF. Differentiating Eq. (2.52) with respect to $u(x)$, using expansion (2.58), Eq. (2.57), and setting $u(x) = 1$, we get the general expression of the multiparticle factorial moment densities f_n through the multiparticle ordinary moment densities m_n.

$$f_n(1,\ldots,n) = \sum_{k=n}^{\infty} \frac{(-1)^{k+n}}{k!} \sum_{N_k=n} \binom{n}{n-1\ldots n_k} m_k(1, n_1+1, \ldots, N_{k-1}+1)$$

$$\times\ [\delta(1,2)\ldots\delta(1,n_1)]\ldots[\delta(N_{k-1}+1, N_{k-1}+2)\ldots\delta(N_{k-1},n)]. \tag{2.75}$$

Functional differentiation of Eq. (2.51) with respect to $s(x)$ at $s = 0$ yields the general inverse relation:

$$m_n(1,\ldots,n) = \sum_{k=1}^{n} \frac{1}{k!} \sum_{N_k=n} \binom{n}{n\ldots n_k}$$

$$\times\ f_k(1,\ldots,k)[\delta(1,k+1)\ldots\delta(1,k+n_1-1)]\ldots[\delta(k,n-n_k+2)\ldots\delta(k,n)].$$

For the first three densities, we have

$$f_1(1) = m_1(1),$$

$$f_2(1,2) = m_2(1,2) - m_1(1)\delta(1,2),$$

$$f_3(1,2,3) = m_3(1,2,3) - 3m_2(1,2)\delta(3,2) + 2m_1(1)\delta(2,1)\delta(3,1)$$

and

$$m_2(1,2) = f_2(1,2) + f_1(1)\delta(2,1), \tag{2.76}$$

$$m_3(1,2,3) = f_3(1,2,3) + 3f_2(1,2)\delta(3,2) + f_1(1)\delta(2,1)\delta(3,1).$$

References

1. Wigner EP (1992) Mathematical problems of nuclear reactor theory. In: Weinberg AM (eds) Nuclear energy. The collected works of Eugene Paul Wigner, vol A/5. Springer, Berlin, Heidelberg

2. Sevast'yanov BA, Chistyakov VP (1971) Multidimensional renewal equations and moments of branching processes. Theory Probab Appl 16(2):199–214
3. Harris TE (1963) The theory of branching processes. Springer, Berlin
4. Balescu R (1975) Equilibrium and non-equilibrium statistical mechanics, vol 2. Wiley-Interscience, New York
5. Akhiezer AI, Peletminsky SV (1977) Methods of statistical physics. Nauka, Moscow (in Russian)
6. Malakhov AN (1978) Cumulant analysis of random non-Gaussian processes and their transformations. Sov, Radio, Moscow (in Russian)
7. Jirina M (1967) General branching processes with continuous time parameter. In: Proceedings fifth Berkeley symposium on mathematical statistics and probability, vol 2, pp 389–399
8. Mullikin TW (1963) Limiting distributions for critical multitype branching processes with discrete time. Trans Am Math Soc 103(3):469–494

For Further Reading

9. Daley DJ, Vere-Jones D (2003) An introduction to the theory of point processes, volume I: elementary theory and methods. Springer, New York, pp 211–275
10. Kroese DP, Botev ZI (2014) Spatial process simulation. Stochastic geometry, spatial statistics and random fields: models and algorithms. Springer International Publishing, Cham, pp 369–404
11. Snyder DL, Miller MI (2012) Random point processes in time and space. Springer Science & Business Media

Chapter 3
Forward (Basic) Equation for CP

3.1 The Generating Functional

Now we consider an instantaneous detector whose response is uniquely determined by the location of all particles in the phase space at the measurement moment T. Mathematical expectation of this response is computed with the use of multiparticle densities p_n or f_n for the process at the time T:

$$\mathsf{M}q(T) = \sum_{n=1}^{\infty} \frac{1}{n!} \int d1 \ldots \int dn h_n(1, \ldots, n) p_n(T, 1, \ldots, n)$$

$$= \sum_{n=1}^{\infty} \frac{1}{n!} \int d1 \ldots \int dn g_n(1, \ldots, n) f_n(T, 1, \ldots, n), \tag{3.1}$$

where

$$g_n(1, \ldots, n) = \sum_{k=1}^{n} \binom{n}{k} (-1)^{n-k} h_k(1, \ldots, k). \tag{3.2}$$

These functions can be obtained from the FAP GF

$$F(T, u(\cdot)) = \mathsf{M} \exp\left\{ \int \mu(T, x) \ln u(x) dx \right\} \tag{3.3}$$

via its functional differentiation (see (2.69) and (2.70)).

Let $\mu(T, x)$ be the random density of particles at the measurement moment T and $\mu(t, i \to T, x)$ be the part of this density formed by descendants of the particle which was located at point x_i at time $t < T$. We assume that external sources do not produce particles during the interval t, T. Under this condition,

© Higher Education Press Limited Company 2024

V. V. Uchaikin, *Electron-Photon Cascades*, Nonlinear Physical Science,
https://doi.org/10.1007/978-981-99-7524-2_3

$$\mu(T, x) = \sum_{i=1}^{M(t)} \mu(t, i \to T, x). \tag{3.4}$$

Substituting Eq. (3.4) into Eq. (3.3) yields

$$F(T, u(\cdot)) = \mathsf{M} \prod_{i=1}^{M(t)} \exp\left\{ \int \mu(t, i \to T, x) \ln u(x) dx \right\}. \tag{3.5}$$

Observe that random variables

$$\exp\left\{ \int \mu(t, i \to T, x) \ln u(x) dx \right\}, \quad i = 1, \ldots, M$$

are mutually independent. Therefore, in accordance with the theorem on a total mathematical expectation,

$$\mathsf{M} = \mathsf{M}\mathsf{M}_{1,\ldots,\mu},$$

the right-hand side of Eq. (3.5) can be transformed as follows:

$$\mathsf{M}\mathsf{M}_{1,\ldots,\mu} \prod_{i=1}^{M(t)} \exp\left\{ \int \mu(t, i \to T, x) \ln u(x) dx \right\}$$

$$= \mathsf{M} \prod_{i=1}^{M(t)} \mathsf{M}_i \exp\left\{ \int \mu(t, i \to T, x) \ln u(x) dx \right\}. \tag{3.6}$$

Each multiplier in this product is a GF of the subprocess (cascade) generated by the single particle arising at point x_i at time t:

$$\mathsf{M}_i \exp\left\{ \int \mu(t, i \to T, x) \ln u(x) dx \right\} = F(t, i \to T, u(\cdot)). \tag{3.7}$$

Substituting this expression into Eq. (3.6) and taking into account that

$$\prod_{i=1}^{M(t)} F(t, i \to T, u(\cdot)) = \exp\left\{ \sum_{i=1}^{M(t)} \ln F(t, i \to T, u(\cdot)) \right\}$$

$$= \exp\left\{ \int \mu(t, x) \ln F(t, x \to T, u(\cdot)) dx \right\},$$

we obtain instead of (3.5)

$$F(T, u(\cdot)) = \mathsf{M} \exp\left\{ \int \mu(t, x) \ln F(t, x \to T, u(\cdot)) dx \right\}.$$

In its structure, the right side of this expression is similar to (3.3), but T is replaced by t, and $u(x)$ is replaced by $F(t, i \to T, u(\cdot))$. Consequently, the following equality takes place:

$$F(T, u(\cdot)) = F(t, F(t, \cdot \to T, u(\cdot))). \tag{3.8}$$

The functional equation (3.8) links the GFs of the process at two different moments t and T, expressing a defining property of the branching process in its most compact form [1]. GF $F(t, x \to T, u(\cdot))$ relates to a cascade, that is to a process formed by one primary particle located at time t at point x, while the GF $F(t, u(\cdot))$ describes the process with an arbitrary the initial state $\mu(\theta, x)$. In particular, it may also be a single-particle determinated state. In this case, Eq. (3.8) takes the form

$$F(\theta, x \to T, u(\cdot)) = F(\theta, x \to t, F(t, \cdot \to T, u(\cdot))), \quad \theta < t < T. \tag{3.9}$$

Taking the functional derivative of both sides of this expression with respect to $u(x)$ and setting $u(x) = 1$, we obtain

$$f_1(\theta, x \to T, x') = \int f_1(\theta, x \to t, y) f_1(t, y \to T, x') dy \tag{3.10}$$

where

$$f_1(\theta, x \to T, x') = \mathscr{D}_{u=1} F(\theta, x \to T, u(\cdot))$$

is the average density of particles at x' at time T belonging to the cascade formed by the primary particle, appeared at point x at time θ (Green's function). It is easy to see that Eq. (3.10) is a generalization of the Kolmogorov–Chapman equation to the case of a branching process; in the absence of branching, $f_1(\theta, x \to T, x')$ becomes the $x \to x'$ transition probability density.

3.2 Equations for GF

In order to establish the relationship between the GF and the probabilistic characteristics of elementary fragmentation processes, the intermediate time t has to be taken close to the θ or to the T.

Suppose that at time t the FAP consists of a single particle, located at the point x. During the interval $(t, t + \Delta t)$, the particle can be fragmented with probability $\sigma \delta t$ producing a random distribution of secondary particles $(1, \ldots, \nu)$, the density of which is written as

$$\nu(x \to y) = \sum_{i=1}^{\nu} \delta(y, y_i).$$

However, there exists a non-zeroth probability, $1 - \sigma \Delta t$, that the particle has moved from x to $x + \Delta x$ along a differentiable curve $x = x(t)$, defined by the motion equation. So, the increment of the phase coordinates can be written as

$$\Delta x \approx \dot{x}(t,x)\Delta t. \tag{3.11}$$

Recall that a vector x includes, in general, a discrete variable a, denoting the type of a particle, which can vary only in collisions (fragmentations), but formula (3.11) describes the change of coordinates in the absence of collisions. In accordance with this we will assume $\dot{a} = 0$.

Thus, for the random density of particles at time $t + \Delta t$, produced by a single particle arisen at time t at x, we obtain

$$\mu(t, x \to t+\Delta t, y) = \begin{cases} \delta(y, x+\Delta x) & \text{with probability} \quad 1-\sigma\Delta t, \\ \nu(x \to y) & \text{with probability} \quad \sigma\Delta t. \end{cases} \tag{3.12}$$

With the help of Eq. (3.12) and the definition of GF (3.7), the infinitesimal functional can be written as

$$F(t, x \to t+\Delta t, u(\cdot)) = \mathsf{M}_x \exp\left\{\int \mu(t, x \to t+\Delta t, y)\ln u(y)dy\right\}$$

$$\approx (1-\sigma\Delta t)u(x+\Delta x) + \Delta t W(x \to u(\cdot)), \tag{3.13}$$

where

$$W(x \to u(\cdot)) = \sigma\mathsf{M}_x \exp\left\{\int \nu(x \to y)\ln u(y)dy\right\} \tag{3.14}$$

is the GF of fragments producing by a particle at the phase point x with Cartesian coordinates x_α per unite time. With this definition, we have instead of (1.54)

$$W(x \to 1) = \sigma(x), \qquad |W(x \to u(\cdot))| \le \sigma(x). \tag{3.15}$$

Note also that in the general case, probabilistic characteristics of the fragmentation distribution $\nu(x \to y)$ may depend on time, although the argument t in the functional W and its derivatives is omitted for the sake of brevity.

Using Eq. (3.11), we rewrite Eq. (3.13) as

$$F(t, x \to t+\Delta t, u(\cdot)) \approx u(x) + \dot{x}\partial_x u(x)\Delta t + [W(x \to u(\cdot)) - \sigma(x)u(x)]\Delta t, \tag{3.16}$$

where

$$x\partial_x u(x) = \sum_\alpha \dot{x}^\alpha \frac{\partial u}{\partial x^\alpha}.$$

Now, we come back to the functional relation (3.9) and replace T by $t + \Delta t$:

$$F(\theta, x \to t+\Delta t, u(\cdot)) = F(\theta, x \to t, F(t, \cdot \to t+\Delta t, u(\cdot))). \tag{3.17}$$

Applying Eq. (3.16) to the inner GF in the right-hand side of Eq. (3.17), we obtain (3.16)

$$F(\theta, x \to t + \Delta t, u(\cdot)) = F(\theta, x \to t, u(\cdot) + \Delta u(\cdot)), \tag{3.18}$$

where

$$\Delta u(x) = [\dot{x}\partial_x u(x) + W(x \to u(\cdot)) - \sigma(x)u(x)]\,\Delta t.$$

When $\Delta t \to 0$, the functional relation Eq. (3.18) is converted into the following integro-differential equation for F:

$$\frac{\partial F}{\partial t} = \int \frac{\delta F(\theta, x \to t, u(\cdot))}{\delta u(y)} \left[\dot{y}\partial_y u + W(y \to u(\cdot)) - \sigma(y)u(y)\right] dy. \tag{3.19}$$

The right-hand side of this equation, namely, the first term on the right can be transformed as follows: observing that ∂_y is the gradient operator in a multidimensional space, we get

$$\int \frac{\delta F}{\delta u(y)} \dot{y} \frac{\partial u}{\partial y} dy = \int \nabla_y \left(\frac{\delta F}{\delta u(y)} u(y)\dot{y} \right) dy - \int u(y)\nabla_y \left(\dot{y} \frac{\delta F}{\delta u(y)} \right) dy.$$

The first term in this expression being a volume integral of the divergence of some vector can be transformed into a surface integral taken over the surface bounding this volume. From the GF definition (3.7), it follows that

$$\frac{\delta F}{\delta u(y)} = \mathsf{M}_x \exp\left\{ \int \mu(\theta, x \to T, x') \ln u(x')dx' \right\} \mu(\theta, x \to T, y)/u(y),$$

therefore, at the points y which are free from the particles, $\delta F/\delta u(y) = 0$ and the surface integral vanishes. In most physical problems, such a surface can easily be established. Thus, if the source emits particles with energies not exceeding a certain maximum value, the projection of such a surface on the momentum subspace becomes a sphere of radius corresponding to a given energy. The coordinate space occupied by a fragmenting system is usually limited to a natural or artificial (say, in laboratory conditions) boundary.

As a result, Eq. (3.19) becomes

$$\frac{\partial F}{\partial t} + \int u(y) \frac{\partial}{\partial y} \left(\dot{y} \frac{\delta F}{\delta u(y)} \right) dy = \int \frac{\delta F}{\delta u(y)} [W(y \to u(\cdot)) - \sigma(y)u(y)]dy. \tag{3.20}$$

Equation (3.16) shows that the GF (3.7) satisfies the following initial condition:

$$F(\theta, x \to \theta + 0, u(\cdot)) = u(x). \tag{3.21}$$

Moreover, the causality principle says that the particle appeared at the time θ, cannot create particles in the previous times, so

$$\mu(\theta, x \to \theta - 0, y) = 0$$

and

$$F(\theta, x \to \theta - 0, u(\cdot)) = 1.$$

Equation (2.22) [2] can be called the *direct* or *basic* equation by analogy with the corresponding Kolmogorov equation. Let us stress that the differentiation is with respect to the observation time $t = T$ whereas the time θ of the birth of the primary particle remains fixed (the same is valid for the phase variable).

Evidently, if we started with Eq. (3.8) instead of (3.9), we would obtain for $F(\theta, u(\cdot))$ the same equation but with arbitrary initial condition $F(\theta + 0, u(\cdot)) = F_0(u(\cdot))$. For this reason, we will not specify further the initial parameters θ and x.

3.3 One-Particle Direct Equations

The equation for the GF is an effective tool for theoretical analysis of branching processes [1, 3, 4] and is a convenient starting point for the derivation of the multiparticle equations for p_n, f_n and other moment densities.

Let us put $u = 0$ in Eq. (3.20). Taking into account that $v_0(y) = W(y \to 0)$ means the probability of the y-particle disappearance per unit time, we obtain for the probability of the process degeneration the following relation:

$$\partial_t p_0 = \int p_1(y) v_0(y) dy. \tag{3.22}$$

Here $p_1(y) \equiv p_1(y, t)$ is the one-particle probability density[1]. The sense of Eq. (3.22) is clear: $p_1(y)dy$ is the probability that currently the process consists of a one particle, which is located in dy, and $v_0(y)$ is the probability of its disappearance per unit time. Summing the product of these probabilities for all elements of the phase space, we find the increment of the probability of degeneration per unit time.

Putting $u = 1$ in Eq. (3.20), we obtain an identity.

Now, we take the first-order functional derivative from both sides of Eq. (3.20):

$$\partial_t F^{(1)}(x; u(\cdot)) + \partial_x \left[\dot{x} F^{(1)}(x; u(\cdot))\right] + \int u(y) \partial_y \left[\dot{y} F^{(2)}(x, y; u(\cdot))\right] dy$$

$$= \int F^{(2)}(x, y; u(\cdot)) \left[W(y \to u(\cdot)) - \sigma(y) u(y)\right] dy$$

$$+ \int F^{(1)}(y; u(\cdot)) W^{(1)}(y \to x; u(\cdot)) dy - \sigma(x) F^{(1)}(x, u(\cdot)). \tag{3.23}$$

[1] Hereinafter, the time-variable t in argument is dropped.

Setting here $u = 0$, we obtain the equation for one-particle probability density $p_1(x)$:

$$\partial_t p_1 + \partial_x(\dot{x} p_1) + \sigma p_1 = \int p_1(y) v_1(y \to x) dy + \int p_2(x, y) v_0(y) dy. \quad (3.24)$$

Here, $v_1(y \to x)dx = W^{(1)}(y \to x; 0)dx$ is the probability that the y-particle disappears per unit time generating only one secondary particle in the element dx of the phase space. The first term in Eq. (3.24) describes the rate of change of the probability density of finding the particle in the unite vicinity of the point x per unit time, the second—a part of this change, due to the difference between the probabilities of arrival and departure particles from this vicinity, the following describes the decay of the probability due to the possible interaction of the particles in this domain, result of which it is either absorbed or jumps to another region of the phase space. The first term on the right of Eq. (3.24) corresponds to the appearance in the x-vicinity of the particle produced by fragmentation of a parent particle, and the last term describes the transition from a two-particle state into the single-particle one. If the initial state contains only one particle, and fragmentation is absent, the multiplication is impossible, $p_2(x, y) = 0$ and Eq. (3.24) coincides with the direct Kolmogorov equation for a wandering particle.

Now, we set $u(x) = 1$ in Eq. (3.23). First of all, let us pay attention to the terms containing the FMD (factorial moment density) of the second order. The first of them $\int \partial_y[\dot{y} f_2(x, y)]dy$ is convertible into a surface integral and under the conditions discussed in the previous section vanishes. The integrand of the second such term is also equal to zero ($W(y \to a) = \sigma(y)$). Thus, the equation for FMD of the first order is closed and requires for their solution no more information:

$$\partial_t f_1 + \partial_x(\dot{x} f_1) + \sigma f_1 = \int f_1(y) w_1(y \to x) dy. \quad (3.25)$$

Here $w(y \to x)dx = W^{(1)}(y \to x; 1)dx$ stands for the mean number of particles which are born in the element dx per unit by parent particle fragmented at the point y. The all said about the terms of Eq. (3.24) remains valid for the corresponding terms of Eq. (3.25), but $p_1(x)$ should be replaced by $f_1(x)$. Equation (3.25) for the number density has the same form as the direct Kolmogorov equation for the probability density, differing from it only by the kernel and the solution normalization.

We conclude this subsection by rewriting Eq. (3.25) in the "standard" phase variables $(m, \mathbf{r}, \mathbf{p})$,

$$f_1(x) = f(m, \mathbf{r}, \mathbf{p}),$$

where m is the mass of a fragment and $\mathbf{p} = m\dot{\mathbf{r}}$:

$$\partial_t + \partial_{\mathbf{p}}(\dot{\mathbf{p}} f) + \partial_{\mathbf{r}}(\dot{\mathbf{r}} f) + \sigma f = \int dm' \int d\mathbf{p}' f(t, m', \mathbf{r}, \mathbf{p}') w(m' \to m; \mathbf{p}' \to \mathbf{p}). \quad (3.26)$$

Here is taken into account that the secondary particles start their histories at the point of fragmentation ($\mathbf{r} = \mathbf{r}'$)with the produced in the primary interaction point particles $w_1(a', \mathbf{r}', \mathbf{p}' \to a, \mathbf{r}, \mathbf{p}) = w_{a' \to a}(\mathbf{p}' - \mathbf{p})\delta(\mathbf{r}' - \mathbf{r})$. Obviously, (3.26) is a conventional system of kinetic equations of a linear transport theory for the mean concentration of particles at point $\mathbf{r}$ with parameters m, $\mathbf{p}$ in a region free from sources [5]. Thus, the equations of conventional transport theory are only the first term of the expansion of Eq. (3.20) with respect of factorial moments and in accordance with (3.1) can be used to calculate reading expectation of only additive detectors, described by linear functionals.

3.4 Multiparticle Direct Equations

Differentiating Eq. (3.23) with respect to $u(x)$ again, we obtain the equation for two-particle densities:

$$\begin{aligned}\partial_t p_2 &= -\partial_{x_1}(\dot{x}_1 p_2) - \partial_{x_2}(\dot{x}_2 p_2) - \sigma(x_1) p_2(x_1, x_2) - \sigma(x_2) p_2(x_1, x_2) \\ &+ \int dx_1' p_2(x_1', x_2) v_1(x_1' \to x_1) + \int dx_2' p_2(x_1, x_2') v_1(x_2' \to x_2) \\ &+ \int dx_1' p_1(x_1') v_2(x_1' \to x_1, x_2) + \int dx_3 p_3(x_1, x_2, x_3) v_0(x_3), \qquad (3.27)\end{aligned}$$

and

$$\begin{aligned}\partial_t f_2 &= -\partial_{x_1}(\dot{x}_1 f_2) - \partial_{x_2}(\dot{x}_2 f_2) - \sigma(x_1 1) f_2(x_1, x_2) - \sigma(x_2) f_2(x_1, x_2) \\ &+ \int dx_1' f_2(x_1', x_2) w_1(x_1' \to x_1) + \int dx_2' f_2(x_1, x_2') w_1(x_2' \to x_2) \\ &+ \int dx_1' f_1(x_1') w_2(x_1' \to x_1, x_2). \qquad (3.28)\end{aligned}$$

Here $v_n(x \to x_1, \ldots, x_n) = W^{(n)}(x \to x_1, \ldots, x_n; 0)$ and $w_n(x \to x_1, \ldots, x_n) = W^{(n)}(x \to x_1, \ldots, x_n; 1)$ are multiparticle PD and FMD of fragments produced by a single parent particle at point x per unit time. In the high-energy nuclear physics, these functions are called, respectively, *exclusive* and *inclusive distributions*. The relationship between them is established by (2.73) (2.74) [6].

Each term of Eq. (3.28) can be given by a special physical interpretation. As noted in Sect. 3.1,

$$f_2(x_1, x_2) dx_1 dx_2 = \mathsf{M}\Omega_2(dx_1, dx_2),$$

where

$$\Omega_2(dx_1, dx_2) = \begin{cases} \hat{\mu}(dx_1)\hat{\mu}(dx_2), & dx_1 \neq dx_2; \\ \hat{\mu}(dx)[\hat{\mu}(dx) - 1], & dx_1 = dx_2 = dx \end{cases}$$

is the number of pairs of particles, one of which is in dx_1, and the other is in dx_2. Hence, $f_2(x_1, x_2)$ denotes the average number of such pairs in the corresponding unit volumes. The left side of Eq. (3.28) gives the increment of the number per unit time. It is due to collisionless moving through the observable volumes in phase space (the first and second terms on the right), result of interactions of particles with a substance (third and fourth members), the appearance in one of the amount of new particles (fifth and sixth terms) and, finally, such interactions that result in secondary particles arise in each of the volumes (the last term). Note that Eq. (3.28) coincides (in the relevant notation) with the equation derived in [7]. A special case of Eq. (3.28) is the equations for "the product of probability densities of the second degree", discussed in [8, 9].

The corresponding terms of Eq. (2.29) have the same meaning only instead of an average number of pair of particles one should talk about the probability of finding in the phase space exactly two particles, one of which is in dx_1, and the other in dx_2. Let us look at the last term in this equation, which has no counterpart in Eq. (3.28). It describes the increment of the probability of the two-particle states due to the transition process from a three-particle state as a result of the death of one of the particles. It is easy to see that the equation for p_n in contrast to the equations for f_n in addition to the functions of the lower orders of $(p_1, \ldots, p_{n-1})$ will include the term p_{n+1}. The higher order terms will be absent, since the particles "die alone", whereas they can be born by pairs, triples, etc., which explains the presence in Eq. (3.27) of PDs with all lower orders.

Introducing differential operators

$$\partial_t = \frac{\partial}{\partial t}, \quad \partial_i f_n = \frac{\partial}{\partial x_i}(\dot{x}_i f_n), \quad d_n = \partial_t + \sum_{i=1}^{n} \partial_i, \tag{3.29}$$

we will write Eqs. (3.27) and (3.28) in shorthand form:

$$d_2 p_2 = -2\sigma(1) p_2(1, 2) + 2 \int d1' p_2(1', 2) v_1(1' \to 1)$$

$$+ \int d1' p_1(1') v_2(1' \to 1, 2) + \int d3 p_3(1, 2, 3) v_0(3),$$

$$d_2 f_2 = -2\sigma(1) f_2(1, 2) + 2 \int d1' f_2(1', 2) w_1(1' \to 1)$$

$$+ \int d1' f_1(1') w_2(1' \to 1, 2).$$

Performing the necessary transformations leads to the following equations for n-particle densities [10, 11]:

$$[d_n + n\sigma(1)]p_n(1, \dots, n) = \sum_{h=1}^{n} \binom{n}{k-1} \int dk' p_k(1, \dots, k')$$

$$\times\ v_{n-k+1}(k' \to k, \dots, n) + \int d(n+1)p_{n+1}(1, \dots, n+1)v_0(n+1), \quad (3.30)$$

$$[d_n + n\sigma(1)]f_n(1, \dots, n) = \sum_{h=1}^{n} \binom{n}{k-1} \int dk' f_k(1, \dots, k')w_{n-k+1}(k' \to k, \dots, n). \quad (3.31)$$

Equations (3.30) and (3.31) must be supplemented by initial conditions that define the relevant characteristics of the process at the initial time. An important case for us is the cascade produced by a single primary particle with fixed coordinates x. The functions corresponding to this case will be denoted by $p_n(x \to 1, \dots, n)$ and $f_n(x \to 1, \dots, n)$. They satisfy the obvious initial conditions:

$$p_n(x \to 1, \dots, n) = f_n(x \to 1, \dots, n) = \delta_{n1}\delta(x, 1), \qquad t = \theta. \quad (3.32)$$

3.5 Green's Function

As seen from comparison of Eqs. (3.30) and (3.31), FMDs are more convenient for the analysis of FAP than PDs. This follows from the physical sense of these functions: a first-order FMD is the average density of particles in phase space, i.e., the characteristic which is quite sufficient for many purposes. In order to solve the problem of fluctuations of an additive detector reading, one has to calculate only one additional function: two-particle FMD, whereas the PD-functions should be used in a large amount. In this regard, we restrict ourselves to further consideration of only FMDs.

Following the methods of nonequilibrium statistical mechanics, we consider the set of FMDs of different orders as a vector-function

$$f(t) = \begin{pmatrix} f_1 \\ f_2 \\ \vdots \\ f_n \\ \vdots \end{pmatrix},$$

and the corresponding equation written in the operator form:

$$\partial_t f = \mathscr{L} f, \quad f|_{t=\theta} = f(\theta). \tag{3.33}$$

Here, $\mathscr{L}$ is a matrix operator with elements $\mathscr{L}_{nk}$:

$$\mathscr{L}_{nk} f_k = -\sum_{i=1}^{n} [\partial_i + \sigma(i)] f_k \delta_{nk}$$

$$+ \binom{n}{k-1} \int dk' f_k(1, \ldots, k') w_{n-k+1}(k' \to k, \ldots, n) \varepsilon_{kn}. \tag{3.34}$$

Using the corresponding Green operator $\mathscr{G}(t, \theta)$ (propagator),

$$\partial_t \mathscr{G}(t, \theta) = \mathscr{L}(t) \mathscr{G}(t, \theta), \qquad \mathscr{G}(\theta, \theta) = I,$$

we write the formal solution of Eq. (3.33) as

$$f(t) = \mathscr{G}(t, \theta) f(\theta) = \exp\left\{ \int_0^t \mathscr{L}(t') dt' \right\} f(\theta). \tag{3.35}$$

According to (3.34), the kinetic operator is the sum of two terms,

$$\mathscr{L} = \mathscr{L}^0 + \mathscr{L}', \tag{3.36}$$

the second of which is a diagonal operator:

$$\mathscr{L} q^0_{nk} = \left(\sum_i \mathscr{L}_i^0 \right)_{nk} \delta_{nk}, \tag{3.37}$$

$$(\mathscr{L}_i^0)_{nn} f_n = -[\partial_i + \sigma(i)] f_n + \int di' f_n(1, \ldots, i', \ldots, n) w_1(i' \to i), \tag{3.38}$$

and the second is a triangular matrix with zero diagonal:

$$\mathscr{L}'_{nk} f_k = \begin{cases} \binom{n}{k-1} \int dk' f_k(1, \ldots, k') w_{n-k+1}(k' \to k, \ldots, n), & k < n; \\ 0, & k \geqslant n. \end{cases} \tag{3.39}$$

Let us first consider the equation containing only the diagonal operator (in the statistical dynamics, this equation describes evolution of an ideal system without interaction of its constituent particles [12]):

$$\partial_t f^0 = \mathscr{L}^0 f^0, \qquad f^0(\theta) = f(\theta). \tag{3.40}$$

As follows from (3.37) (3.38), the operator $\mathscr{L}^0$ has the form of a sum of operators, each of which acts on one of the arguments of f. Representing the solution of Eq. (3.40) in form (3.35), we obtain

$$f^0(t) = \mathscr{G}^0(t, \theta) f(\theta), \tag{3.41}$$

where $\mathscr{G}^0_{nk}(t, \theta) = \prod_{i=1}^{n} (\mathscr{G}^0_i)_{nk}\delta_{nk}$ is the product of one-particle propagators $\mathscr{G}^0_i = \exp\left\{\int_\theta^t \mathscr{L}^0_i dt'\right\}$. Using the integral representation

$$\mathscr{G}^0_i(t, \theta) f_1(\theta) = \int di' f_1(\theta, i') g(\theta, i' \to t, i), \tag{3.42}$$

we write the solution (3.41) as follows:

$$f^0_n(t, 1, \ldots, n) = \int d1' \ldots \int dn' f^0_n(\theta, 1', \ldots, n') \prod_{i=1}^{n} g(\theta, i' \to t, i). \tag{3.43}$$

Now, we substitute Eq. (3.36) into the original equation (3.33) and move the diagonal component to the left:

$$\partial_i f - \mathscr{L}^0 f = \mathscr{L}' f. \tag{3.44}$$

This first-order differential equation is transformed into the integral one:

$$f(t) = f^0(t) + \int_0^t dt' \mathscr{G}^0(t, t') \mathscr{L}' f(t'). \tag{3.45}$$

The equivalence of (3.44) and (3.45) can be easily verified by differentiating the last equation with respect to t:

$$\partial_t f = \partial_t f^0 + \int_0^t dt' \partial_t \mathscr{G}^0(t, t') \mathscr{L}' f(t') + \mathscr{L}' f(t)$$

$$= \mathscr{L}^0 \left[f^0(t) + \int_0^t dt' \mathscr{G}^0(t, t') \mathscr{L}' f(t') \right] + \mathscr{L}' f(t) = \mathscr{L}^0 f(t) + \mathscr{L}' f(t).$$

Given the structure of operators $\mathcal{G}^0$ and $\mathcal{L}'$ in Eq. (3.45), we obtain the following formula for FMD of the n-th order:

$$f_n(t) = f_n^0(t) + \int_\theta^t dt' \mathcal{G}_{nn}^0(t, t') \sum_{k=1}^{n-1} \mathcal{L}'_{nk} f_k(t'). \tag{3.46}$$

This gives an algorithm for a consecutive calculation of functions $f_n(t)$, starting from $f_1(t) = f_1^0(t)$. Using Eqs. (3.39) and (3.41)–(3.43), we write the recurrent relation (3.46) in an expanded form:

$$f_n(t, 1, \ldots, n) = \int d1' \ldots \int dn' f_n(\theta, 1', \ldots, n') \prod_{i=1}^n g(\theta, i' \to t, i)$$
$$+ \int_\theta^t dt' \int d1' \ldots \int dn' \sum_{k=1}^{n-1} \left[\binom{n}{k-1} \int dk'' f_k(t', 1', \ldots, k'') \right.$$
$$\left. \times\, w_{n-k+1}(k'' \to k', \ldots, n') \right] \prod_{i=1}^n g(t', i' \to t, i). \tag{3.47}$$

In particular,

$$f_1(t, 1) = \int d1' f_1(\theta, 1') g(\theta, 1' \to t, 1), \tag{3.48}$$

$$f_2(t, 1, 2) = \int d1' \int d2' \Big[f_2(\theta, 1', 2') g(\theta, 1' \to t, 1) g(\theta, 2' \to t, 2)$$
$$+ \int_\theta^t dt' \int d3 f_1(t', 3) w_2(3 \to 1', 2') g(t', 1' \to t, 1) g(t', 2' \to t, 2) \Big], \tag{3.49}$$

$$f_3(t, 1, 2, 3) = \int d1' \int d2' \int d3' \Big\{ f_3(\theta, 1', 2', 3') g(\theta, 1' \to t, 1)$$
$$\times\, g(\theta, 2' \to t, 2) g(\theta, 3' \to t, 3) + \int_0^t dt' \int d4 [3 f_2(t', 1', 4) w_2(4 \to 2', 3')$$
$$+ f_1(t', 4) w_3(4 \to 1', 2', 3')] g(t', 1' \to t, 1) g(t', 2' \to t, 2) g(t', 3' \to t, 3) \Big\}, \tag{3.50}$$

etc. Note that due to permutation symmetry, the contents of square brackets in Eq. (3.47) can be replaced by sum of such terms over all possible partitions of the set of variables $1', \ldots, n'$ into two groups with $k-1$ and $n-k+1$ variables in each, serving as arguments of the first and second integrands, respectively. $\binom{n}{k-1}\left[\int dk'' f_k(t', a', \ldots, k'')w_{n-k+1}(k'' \to k', \ldots, n')\right]$ can be replaced by the sum of the square brackets over all possible partitions of the variables $1', \ldots, n'$ into two groups of $k-1$ and $n-k+1$ variables in each, being the arguments, respectively, of first and second factors in the integrand.

3.6 The Diagrammatic Representation

Many of the above formulas will become more obvious if to use the diagram technique. Following Balescu, we compile a "dictionary" of the diagrammatic language (Table 3.1).

Substituting Eq. (3.48) into Eqs. (3.49) and (3.49) in Eq. (3.50), we can express f_2 and f_3 through the initial conditions, FDMs of elementary processes and one-particle Green's function. It is not difficult in principle to write the corresponding expressions for higher moments. The formulas become cumbersome and more convenient way is to represent these relations graphically. We shall represent $g(\theta), i' \to t, i$ by a line segment the left end of which corresponds to the point i', and the right - i, $W_n(i \to 1, \ldots, n)$ by vertex with one incoming line i and n outgoing ones $(1, \ldots, n)$, $f_n(\theta, 1, \ldots, n)$ by a similar vertex but without outgoing lines, $f_n(t, 1, \ldots, n)$ by a rectangle with n outgoing lines (Fig. 2.1). With the help of these diagrams, the expressions (3.49) and (3.50) are represented in Figs. 3.2 and 3.3. The number of diagrams of each type is determined by the above number of partitions. Substituting decomposition shown in Fig. 3.2, the right-hand side, shown in Fig. 3.3, we get a visual representation of the formula (3.50), expressing FMD through the third-order Green's function (Fig. 3.4). Using the given rules, we can easily obtain expressions for FMD's higher orders.

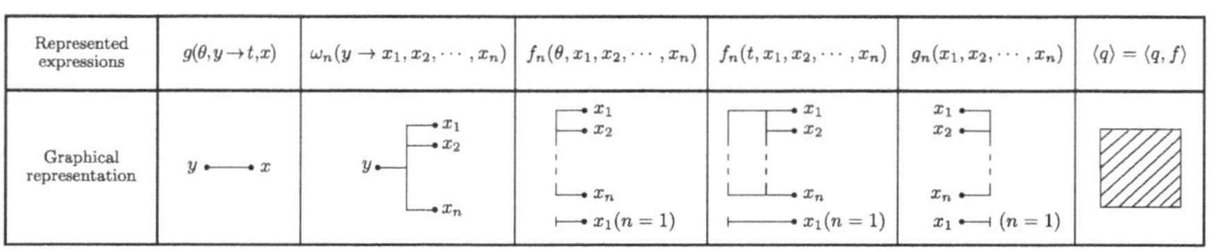

Fig. 3.1 Table of correspondence for diagram elements

Fig. 3.2 Graphical representation of (3.49)

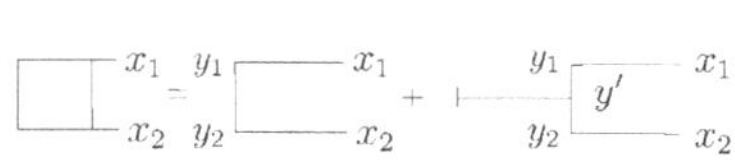

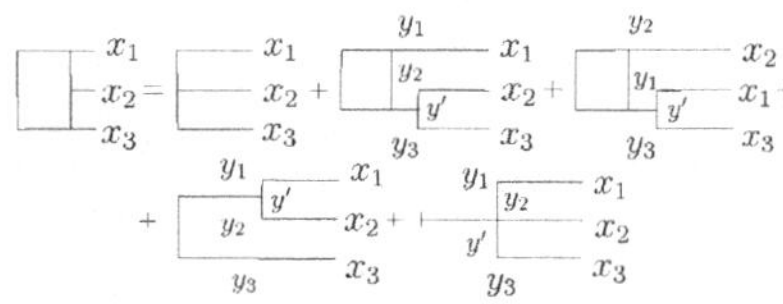

Fig. 3.3 Graphical representation of (3.50)

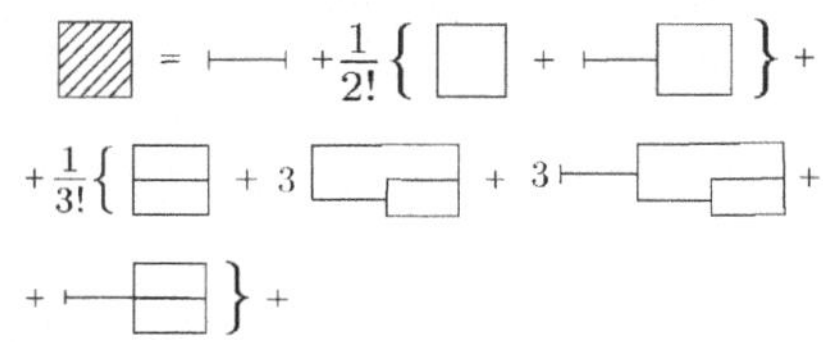

Fig. 3.4 Graphical representation of a measurand expectation

If the factor $g_n(1, \ldots, n)$ is represented by a vertex with n incoming lines, but no outgoing ones, and the functional expectation by a rectangle (see Fig. 3.1), we can give a graphical representation for the measurand expectation itself. Figure 3.4 shows a few first terms of this expansion.

3.7 Dynamics of Correlation Forms

The simplest type of random point distributions is such that consists of independent random points forming the *Poisson ensemble* [1]:

$$\mathsf{P}\{\hat{\mu}(A) = n\} = \frac{[m(A)]^n}{n!} e^{-m(A)}, \qquad m(A) = \mathsf{M}\hat{\mu}(A) < \infty. \tag{3.51}$$

To establish the Poisson ensemble FMDs, we divide the phase space X into disjoint domains $A_1, \ldots, A_r$ and choose $u(x)$ in the form

$$u(x) = u_k, \quad x \in A_k. \tag{3.52}$$

Substituting (3.52) into Eq. (2.50) and noting that, due to Eq. (3.52)

$$\ln u(x) = \sum_{k=1}^{r} \ln u_k 1(x, A_k)$$

and

$$\int 1(x, A_k)\mu(x)dx = \hat{\mu}(A_k),$$

we get

$$F(u(\cdot)) = \mathsf{M}\exp\left\{\sum_{k=1}^{r}\hat{\mu}(A_k)\ln u_k\right\} = \mathsf{M}\prod_{k=1}^{r}u_k^{\hat{\mu}(A_k)}.$$

By properties of Poisson's measure the numbers of particles in the disjoint domains are mutually independent, so,

$$F(u(\cdot)) = \prod_{k=1}^{r}\mathsf{M}u_k^{\hat{\mu}(A_k)}. \tag{3.53}$$

Using Eq. (3.51), we easily obtain the generating function of the distribution the number of particles in A:

$$\mathsf{M}u^{\hat{\mu}(A_k)} = \sum_{n=0}^{\infty}\frac{[um(A)]^n}{n!}e^{-m(A)} = e^{(u-1)m(A)}. \tag{3.54}$$

Substituting Eq. (3.54) into Eq. (3.53) yields

$$F(u(\cdot)) = \exp\left\{\sum_{k=1}^{r}(u_k - 1)m(A_k)\right\}.$$

Since this expression is valid for any function of the form (3.52), one can show that

$$F(u(\cdot)) = \exp\left\{\int[u(x) - 1]f_1(x)dx\right\}, \tag{3.55}$$

where $f_1(x)$, as before, is the average particle density. Applying the n-fold functional differentiation operator to Eq. (3.55), one can see that in case of Poissonian measure the FMD of n-th order is equal to the product of the average densities in corresponding points:

$$f_n(1,\ldots,n) = \prod_{i=1}^{n}f_1(i). \tag{3.56}$$

Such a state of the system is called the *uncorrelated state*. A quantitative measure of correlations is given by the differences:

$$f_n(1,\ldots,n) - \prod_{i=1}^{n}f_1(i) = \varphi_n'(1,\ldots,n).$$

The function $\varphi_2'(1,2) \equiv \varphi_2(1,2)$ reflects the simplest type of correlations—the pair correlation. For $n = 3$ two types of correlations are possible: when the two particles are correlated but the third is not correlated with them, and when all three particles are correlated. The corresponding terms in $\varphi'_3(1,2,3)$ are given by $f_1(i)\varphi_2(j,k)$ and $\varphi_3(1,2,3)$. To analyze the possible types in the general case, one has to consider all ways of partitioning the set $\{1,\ldots,n\}$ into disjoint subsets, containing at least one particle. By interpreting each subset or group (cluster) as a set of mutually correlated particles, which are statistically independent of the particles of other subsets, we obtain the following representation for factorial moment densities, called the *cluster expansion*:

$$f_2(1,2) = f_1(1)f_1(2) + \varphi(1,2), \tag{3.57}$$

$$f_3(1,2,3) = f_1(1)f_1(2)f_1(3) + f_1(1)\varphi_2(2,3) + $$
$$+ f_1(2)\varphi_2(1,3) + f_1(3)\varphi_2(1,2) + \varphi_3(1,2,3) \tag{3.58}$$

and so on. R. Balescu called $\varphi_n(1,\ldots,n)$ the *irreducible n-particle functions*, and each term in the expansions (3.57), (3.58) has the correlation form. Following [12], we will apply the notation

$$f_1(1) = \pi_1(1), \quad f_1(1)f_1(2) = \pi_2(1|2), \quad \varphi_2(1,2) = \pi_2(1,2),$$

$$f_1(1)f_1(2)f_1(3) = \pi_3(1|2|3), \quad f_1(1)\varphi_2(2,3) = \pi_3(1|2,3),$$

$$\varphi_3(1,2,3) = \pi_3(1,2,3),$$

etc. Observe that

$$\int_A\int_A \varphi_2 d1d2 = \mathsf{D}\mu(A) - \mathsf{M}\mu(A)$$

is a measure of difference between measures μ and Poisson one. The correlation forms have special symmetry properties, arising from the general symmetry of multiparticle densities: any of them is invariant with respect to permutation of indices within each group, and under permutations of the groups themselves. In this regard, we use the standard representation [12], in which all the subsets of the partition from the left to right in order increase in size, while the indices in each subset from smaller to larger one. The expansions for the first three densities are

$$f_1(1) = \pi_1(1), \qquad f_2(1,2) = \pi_1(1|2) + \pi_2(1,2),$$

$$f_3(1,2,3) = \pi_3(1|2|3) + \pi_3(1|2,3) + \pi_3(2|1,3) + \pi_3(3|1,2) + \pi_3(1,2,3).$$

Now, we derive equations for the correlation forms. Naturally, the equation for $\pi_1(1)$ coincides with (3.25):

$$\partial_t\pi_1(1) - \mathscr{L}_{11}(1)\pi_1(1) = 0. \tag{3.59}$$

In parentheses after an operator, the index of the variable on which it acts is shown. Similarly for $\pi_1(2)$ we have

$$\partial_t \pi_1(2) - \mathscr{L}_{11}(2)\pi_1(2) = 0. \tag{3.60}$$

Multiplying Eq. (3.59) by $\pi_1(2)$, and Eq. (3.60) by $\pi_1(1)$, summing them and taking into account that

$$\mathscr{L}_{11}(1) + \mathscr{L}_{11}(2) = \mathscr{L}_{22}(1, 2),$$

we get

$$\partial_t \pi_2(1|2) - \mathscr{L}_{22}(1, 2)\pi_2(1|2) = 0. \tag{3.61}$$

Subtracting (3.61) from (3.28), we find an equation for the other form of the correlation—the two-particle irreducible correlation functions:

$$\partial_t \pi_2(1, 2) - \mathscr{L}_{22}(1, 2)\pi_2(1, 2) = \mathscr{L}_{21}(1)\pi_1(1). \tag{3.62}$$

To obtain the equation for $\pi_3(1|2|3) = \pi_2(1|2)\pi_1(3)$, we write the equation for $\pi_1(3)$, multiply it by $\pi_2(1|2)$, and (3.61) by $\pi_1(3)$, and add the resulting expression:

$$\partial_t \pi_3(1|2|3) - \mathscr{L}_{33}(1, 2, 3)\pi_3(1|2|3) = 0.$$

Similarly to (3.62), we obtain

$$\partial_t \pi_3(1|2, 3) - \mathscr{L}_{33}(1, 2, 3)\pi_3(1|2, 3) = \mathscr{L}_{21}(1)\pi_1(1|2).$$

Finally, for the three-particle function

$$\partial_t \pi_2(1, 2, 3) - \mathscr{L}_{33}(1, 2, 3)\pi_3(1, 2, 3) = \mathscr{L}_{31}(1)\pi_1(1) + \mathscr{L}_{32}(1, 2)\pi_2(1, 3).$$

The process of deriving the equations for the correlation of forms can be continued.

Thus, we have passed from description in terms of FMDs to its representation through the correlation forms. The total set of correlation forms completely characterizes the FAP. Various forms, even if they correspond to the same number of particles, subject to different equations of motion. This difference is determined by the non-diagonal terms of the kinetic operator. In their absence, the equations for various forms are independent and the specific type of correlations is determined by initial conditions. The presence of non-diagonal terms leads to the creation of new correlation forms. Description of the process evolution of the system as a birth, transformation, and destruction of the correlation forms is called *dynamic correlations*.

The above equations produce some patterns in the correlation dynamics. One can see, for example, that the form $\pi_3(1|2|3)$ is generated only by similar forms of smaller dimensions $\pi_2(1|2)$ and $\pi_1(1)$, and the form $\pi_3(1, 2, 3)$, respectively, by $\pi_2(1, 2)$ and $\pi_1(1)$, etc.

From the practical point of view, the forms with the low dimensions are more convenient for calculations. Therefore, the cluster expansion described here can be useful in those problems where the initial approximation (3.56) gives a result more or less close to the exact one. In other words, fluctuations in the number of particles in any region of phase space should be close to Poissonian (the *weak-nonlinear approximation*).

References

1. Sevastyanov BA (1971) Branching processes. Nauka, Moscow (in Russian)
2. Uchaikin VV (1979) The theory of cascade process nonlinear measurements and the statement of backward problems. In: Proceedings of the 16-th ICRC Kyoto, vol 7, pp 24–29
3. Jirina M (1967) General branching processes with continuous time parameter. Math Stat Prob 2:389–399
4. Mullikin TW (1963) Limiting distributions for critical multitype branching processes with discrete time. Trans Am Math Soc 103(3):469–494
5. Kolchuzhkin AM, Uchaikin VV (1978) Introduction to the theory of particle passage through the matter. Atomizdat, Moscow (in Russian)
6. Collins P (1977) An introduction to Regge theory and high energy physics. Cambridge University Press
7. Congdon M, Albrecht R (1970) An adjoint-weighting in neutron fluctuation analysis with emphasis on an effective detector efficiency. Nucl Sci Engng 39:207–214
8. Barucha-Reid AT (1960) Elements of the theory of markov processes and their applications. MC Graw-Hill Book Company Inc, New York, Toronto, London
9. Ramakrishnan A (1962) Elementary particles and cosmic rays. Pergamon Press, Oxford, London, New York, Paris
10. Uchaikin VV (1984) Theory of nonlinear measurements in high-energy cascade processes. Inelastic interactions of hadrons and nuclei at high energies, Alma-Ata, pp 172–179 (in Russian)
11. Uchaikin VV (1979) The theory of cascade process nonlinear measurements and the statement of backward problems. In: Proceedings of the 16-th ICRC, Kyoto, vol 7, pp 24–29
12. Balescu R (1978) Equilibrium and nonequilibrium statistical mechanics. Wiley-Blackwell, New York, London, Sidney, Toronto

For Further Reading

13. Lewins J (1965) Importance: The adjoint function. The physical basis of variational and perturbation theory in tranport and diffusion problems. Pergamon Press, Oxford, London
14. Weinberg AM, Wigner EP (1958) The physical theory of neutron chain reactors. University of Chicago Press
15. Bell GI, Glasstone S (1970) Nuclear reactor theory (No TID-25606). US atomic energy commission, Washington, DC (USA)

Chapter 4
Backward (Adjoint) Equation

4.1 Processes with a Single-Particle Initial State

Markov branching nature of fragmentation-advection processes provides opportunity to restrict our attention to processes with initial states consisting only of one particle. We call these processes cascades. According to Eq. (3.8), the GFs at times θ and $T > \theta$ of a FAP are linked with each other through the cascade GF $F(\theta, x \to T, u(\cdot))$ via the Chapman–Kolmogorov relation

$$F(t, u(\cdot)) = F(\theta, F(\theta, \cdot \to T, u(\cdot))). \tag{4.1}$$

Equation (4.1) generates the corresponding relations for multiparticle densities.

Using representations (2.64) and (2.66) in the right-hand side of Eq. (4.1), we obtain

$$F(T, u(\cdot)) = p_0(\theta) \sum_{k=1}^{\infty} \frac{1}{k!} \int d1 \ldots \int dk p_k(\theta, 1, \ldots, k) \prod_{i=1}^{k} F(\theta, i \to T, u(\cdot)),$$

$$F(T, u(\cdot)) = 1 + \sum_{k=1}^{\infty} \frac{(-1)^k}{k!} \int d1 \ldots \int dk f_k(\theta, 1, \ldots, k) \prod_{i=1}^{k} \overline{F}(\theta, i \to T, u(\cdot)).$$

Differentiating (using Eq. (2.58)) both of these expressions with respect to $u(x)$ and setting $u = 0$ in the first one and $u = 1$ in the second, we arrive at the following relations for multiparticle probability densities (PD):

© Higher Education Press Limited Company 2024
V. V. Uchaikin, *Electron-Photon Cascades*, Nonlinear Physical Science,
https://doi.org/10.1007/978-981-99-7524-2_4

$$p_n(T, 1, \dots, n) = \sum_{k=1}^{\infty} \frac{1}{k!} \int d1' \dots \int dk' p_k(\theta, 1', \dots, k')$$

$$\times \sum_{N_k=n} \binom{n}{n_1 \dots n_k} \prod_{i=1}^{k} p_{n_i}(\theta, i' \to T, N_{i-1} + 1', \dots, N_i), \tag{4.2}$$

$$f_n(T, 1, \dots, n) = \sum_{k=1}^{n} \frac{1}{k!} \int d1' \dots \int dk' f_k(\theta, 1', \dots, k')$$

$$\times \sum_{\substack{N_k=n \\ (n_i \neq 0)}} \binom{n}{n_1 \dots n_k} \prod_{i=1}^{k} f_{n_i}(\theta, i' \to T, N_{i-1} + 1, \dots, N_i). \tag{4.3}$$

We have taken into account that $\overline{F}(\theta, i \to T, 1) = 0$ so the last sum consists only of terms with all n_i being nonzero.

Thus, to calculate the multiparticle FAP functions at the measurement moment T, one has to know the set of FAP function at the initial time θ and their cascade counterparts. Note that Eq. (4.3) for the FMD differs from Eq. (4.2) for PD in that to compute $f_n(T, 1, \dots, n)$ one has to have sets of functions $f_k(\theta, 1, \dots, k)$ and $f_k(\theta, x \to T, 1, \dots, k)$ of order k at most n, whereas the expression for $p_n(T, 1, \dots, n)$ contains the functions $p_k(\theta, 1, \dots, k)$ and $p_k(\theta, i \to T, 1, \dots, k)$ for all k 1 to ∞.

4.2 Backward Equations for GFs and Multiparticle Densities

Return to the functional relation (3.9). Replacing the θ by $t - \Delta t$ and using the infinitesimal form (3.16) for the functional on the right side, we arrive at the backward equation for GF

$$[d^+ + \sigma(x)]F(t, x \to T, u(\cdot)) = W(x \to F(t, \cdot \to T, u(\cdot))), \tag{4.4}$$

with

$$d^+ \equiv -\partial_t - \dot{x}\partial x,$$

and initial condition (3.21):

$$F(T - 0, x \to T, u(\cdot)) = u(x).$$

In contrast to Eq. (3.20), here the initial variables are active and the observed ones remain passive.

Note that replacing the GF by CF does not change Eq. (4.4). Indeed, assuming $u(x) = e^{is(x)}$ and using (2.51), we obtain

$$[d^+ + \sigma(x)]C(t, x \to T, s(\cdot)) = W(x \to C(t, \cdot \to T, s(\cdot))). \tag{4.5}$$

Let us write here a similar equation for the functional

$$K(t, x \to T, s(\cdot)) = \ln C(t, x \to T, s(\cdot))$$

generating cumulant (semi-invariant) moments

$$d^+ K(t, x \to T, s(\cdot)) = Q(x \to K(t, \cdot \to T, s(\cdot))) - \sigma(x). \tag{4.6}$$

Here,

$$Q(x \to s(\cdot)) = e^{-s(x)} W(x \to e^{s(\cdot)}). \tag{4.7}$$

Equations (4.4) and (4.5) are obtained in the stochastic theory neutron transport with reproduction in relation to nuclear reactors [1, 2]. Analogous equations appeared in the theory of high-energy particle cascades are known as the G-equations Janossy [3].

Putting in (4.4) $u = 0$, we obtain an equation for the extinction probability (to the measurement time T)[1]

$$[d^+ + \sigma(x)]p_0(x) = W(x \to p_0(\cdot)).$$

Expanding the right side of the equation in a power series with respect to p_0 yields

$$[d^+ + \sigma(x)]p_0(x) = v_0(x) + \sum_{n=1}^{\infty} \frac{1}{n!} \int d1 \ldots \int dn v_n(x \to 1, \ldots, n) p_0(1) \ldots p_0(n)$$

with initial condition $p_0 = 0, t = T$.

The first term on the right-hand side gives the probability of absorption of the primary particles per unit time, and the remainder is the cascade extinction probability provided that the interaction of a primary particle produces one or more secondary particles. At time T, the cascade will be empty (degenerated) if each cascade produced by secondary particles $1, \ldots, n$ will be empty. The cascades are mutually independent so the probability of this event is determined by the product $p_0(1) \ldots p_0(n)$ under the integral. The physical meaning of the initial condition is obvious: the probability of absorption during zero time interval is zero. If $u = 1$, Eq. (4.4) becomes identity $f_0 \equiv 1$.

---

[1] Again, for brevity, we omit the time arguments.

Let us take the functional derivative of both sides of the Eq. (4.4), using the known rule for differentiating a functional with respect to a functional:

$$[d^+ + \sigma(x)]F^{(1)}(x \to 1; u(\cdot)) = \int d1' W^{(1)}(x \to 1'; F(\cdot \to u(\cdot)))F^{(1)}(1' \to 1; u(\cdot)), \tag{4.8}$$

$$F^{(1)}(x \to 1; u(\cdot)) = \delta(x, x_1), \qquad t = T.$$

In case $u = 0$, we obtain the equation for $p_1(x \to 1)$:

$$[d^+ + \sigma(x)]p_1(x \to 1) = \int d1' W^{(1)}(x \to 1'; p_0(\cdot))p_1(1' \to 1).$$

Expanding $W^{(1)}$ in power series with respect to $p_0(x)$ we represent this equation in the form

$$[d^+ + \sigma(x)]p_1(x \to 1) = \int d1' v_1(x \to 1')p_1(1' \to 1)$$

$$+ \sum_{n=2}^{\infty} \frac{1}{(n-1)!} \int d1' \ldots \int dn' v_n(x \to 1', \ldots, n')p_1(1' \to 1)p_0(2') \ldots p_0(n').$$

It is easy to see that the right-hand side is equal to the rate of such an interaction, which produces eventually exactly one particle being in unit volume near point 1 at measurement moment T. Indeed, the first term gives the probability of birth in this interaction of one particle, generating a cascade in which at the time of observation will be only one particle at the point 1. The term with index n in the next sum is equal to the probability that n particles are born in this collision, but only one of these cascades will survive to T and will contain exactly one particle in the unit vicinity of the point 1.

Putting in (4.8) $u = 1$, we obtain the backward equation for the particle density $f_1(x \to 1)$:

$$[d^+ + \sigma(x)]f_1(x \to 1) = \int d1' w_1(x \to 1')f_1(1' \to 1). \tag{4.9}$$

The initial conditions for both functions coincide:

$$p_1(x \to 1) = f_1(x \to 1) = \delta(x, x_1), \qquad t = T. \tag{4.10}$$

Equation (4.9) with initial conditions (4.10) coincides with adjoint equation for Green's function of the transport theory [4].

We now turn to the derivation of equations for two-particle densities. Differentiating (4.8), we obtain

$$[d^+ + \sigma(x)]F^{(2)}(x \to 1, 2; u(\cdot)) = \int d1' \int d2' W^{(2)}(x \to 1', 2';$$

$$F(\cdot \to u(\cdot)))F^{(1)}(1' \to 1; u(\cdot))F^{(1)}(2' \to 2; u(\cdot))$$

$$+ \int d1' W^{(1)}(x \to 1'; F(\cdot \to u(\cdot)))F^{(2)}(1' \to 1, 2; u(\cdot));$$

$$F^{(2)}(x \to 1, 2; u(\cdot)) = 0, \qquad t = T. \tag{4.11}$$

This yields the following equation for the second-order PD:

$$[d^+ + \sigma(x)]p_2(x \to 1, 2) = \sum_{n=2}^{\infty} \frac{1}{n!} \int d1' \ldots \int dn' v_n(x \to 1', \ldots, n')$$

$$\times [n(n-1)p_1(1' \to 1)p_1(2' \to 2)p_0(3') \ldots p_0(n')$$

$$+ np_2(1' \to 1, 2)p_0(2') \ldots p_0(n')].$$

The first term in square brackets describes such events in which two of the particles born in the first interaction give rise to two cascades, each of which will give one particle at measurement time (so that the pair of particles is in the points 1, 2 phase space), and cascades of other particles to this time are empty. The second term refers to the events in which all secondary cascades are empty except one which to the moment T consists of two particles being at points 1 and 2.

If $u = 1$ (4.11), we obtain the equation for the second-order FMD:

$$[d^+ + \sigma(x)]f_2(x \to 1, 2) = \int d1' \int d2' w_2(x \to 1', 2')f_1(1' \to 1)$$

$$\times f_1(2' \to 2) + \int d1' w_1(x \to 1')f_2(1' \to 1, 2).$$

The interpretation of individual members of this equation is similar to that of PDs. Since the initial time stage consists of a single particle, these equations must be supplemented by the zero initial conditions.

Multiple differentiating Eq. (4.4) with respect to $u(x)$ times and operating with the right-hand side as was done in deriving Eqs. (4.2) and (4.3), we arrive at the general equations for the densities of arbitrary orders:

$$[d^{+}+\sigma(x)]p_n(x\to 1,\dots,n)=\sum_{k=1}^{\infty}\frac{1}{k!}\int d1'\dots\int dk'v_k(x\to 1',$$

$$\dots,k')\sum_{N_k=n}\binom{n}{n_1\dots n_k}\prod_{i=1}^{k}p_{n_i}(i'\to N_{i-1}+1,\dots,N_i),\qquad(4.12)$$

$$[d^{+}+\sigma(x)]f_n(x\to 1,\dots,n)=\sum_{k=1}^{n}\frac{1}{k!}\int d1'\dots\int dk'w_k(x\to 1',\dots,k')$$

$$\times\sum_{\substack{N_k=n\\(n_i\neq 0)}}\binom{n}{n_1\dots n_k}\prod_{i=1}^{k}f_{n_i}(i'\to N_{i-1}+1,\dots,N_i).\qquad(4.13)$$

Observe that the backward equations differ from their forward counterparts by presence of only one active phase variable x instead of the set $\{x_1,\dots,x_n\}$ and do not contain the densities of $(n+1)$-th and higher orders, although the equations for p_n have as before a more complicated structure than those of the f_n.

4.3 Multiparticle Importance Functions

The term "importance function" (synonym: "adjoint function", "influence function") appeared in the middle of the former century in connection with development of the nuclear reactor theory. Its embedding into nuclear reactors calculations are linked with such names as K.Fucks, G.Soodak, H.Hurwitz, A.Weinberg, E.Wigner, D.Blokhintsev, B.B.Kadomtsev, who interpreted this function as a solution of the equation adjoint (in the Lagrange sense) to the basic reactor equation[2]. The physical interpretation of the solution as a mean number of neutrons in the FAP, generated by a single primary neutron with phase coordinates indicated in the argument, was given by L.N.Usachev, A.Weinberg, E.Wigner. A further generalization of the importance concept has been performed in works of G.I.Marchuk, V.V.Orlov, and others and found applications in many physical problems.

In [5], this formalism was extended to statistical description of branching processes. The multiparticle stochastic importance has been determined as a total contribution into a detector reading of a family of random cascades generated by primary particles with phase coordinates $1,2,\dots,n$. Averaging this function over all possible realizations with the fixed initial coordinates, we omit the adjective "stochastic" and talk about a multiparticle importance $f_n^*(1,\dots,n)$.

[2] The terms "forward equations" and "backward equations" came from the theory of random processes.

Evidently, in the case of an additive detector

$$f_n^*(1,\dots,n) = \sum_{i-1}^{n} f_1^*(i).$$

In the case of non-additive detector, this relation does not hold. Let us consider the simplest case of non-additive detector, when its reading

$$q = h_\mu(1,\dots,\mu) \tag{4.14}$$

is described by a two-dimensional functional, that is,

$$h_1(1) = g_1(1),$$

$$h_2(1,2) = g_1(1) = g_1(2) + g_2(1,2),$$

$$\dots$$

$$h_n(1,\dots,n) = \sum_{i=1}^{n} g_1(i) + \frac{1}{2}\sum_{ij}^{n}{}' g_2(i,j).$$

The one-particle importance $f_1^*(x)$ being an expected value of functional (4.14) of the FAP state measured at time T, provided at time t the cascade consisted of a single particle at point x, is expressed through corresponding FMDs by means of Eq. (3.1) as

$$f_1^*(x) = f_{11}^*(x) + f_{12}^*(x),$$

$$f_{11}^*(x) = \int d1 f_1(x \to 1) g_1(1),$$

$$f_{12}^*(x) = \frac{1}{2}\int d1 \int d2 f_2(x \to 1,2) g_2(1,2).$$

The first term refers to the additive part of the functional (4.14). The equation for it can be obtained by multiplying both sides of (4.9) $g_1(1)$ and integrating with respect to 1:

$$[d^+ + \sigma(x)] f_{11}^*(x) = \int d1 w_1(x \to 1) f_{11}^*(1), \qquad f_{11}^*(x)|_{t=T} = g_1(x). \tag{4.15}$$

Expectation of the non-additive part of the functional (the second term) satisfies the equation that follows from (4.13):

$$[d^+ + \sigma(x)]f_{11}^*(x) = \int d1 w_1(x \to 1) f_{12}^*(1)$$

$$+\frac{1}{2}\int d1 \int d2 w_2(x \to 1,2) g_2'(1,2), \qquad f_{12}^*(x)|_{t=T} = 0. \tag{4.16}$$

Here

$$g'_2(1,2) = \int d1' \int d2' f_1(1 \to 1') f_2(2 \to 2') g_2(1',2').$$

Summing Eqs. (4.15) and (4.16) term by term, we arrive at the following equation for the one-particle importance:

$$[d^+ + \sigma(x)]f_1^*(x) = \int d1 w_1(x \to 1) f_1^*(1)$$

$$+\frac{1}{2}\int d1 \int d2 w_2(x \to 1,2) g_2'(1,2), \qquad f_1^*(x)|_{t=T} = g_1(x). \tag{4.17}$$

When $g_2 = 0$, the functional (4.14) becomes additive, and Eq. (4.17) turns into equation for one-particle importance relatively to the additive detector reading, which coincides with (4.15). The same result is obtained in the case of absence of fragmentation (that is when $w_2 = 0$). The reason for this is quite clear: if the fragmentation is absent, the cascade cannot consist of more than one particle and the detector reading is determined by only one-particle response g_1 of the device.

Two-particle importance f_2^* is the expected reading of a detector in case the cascade is formed by two initial particles beginning their motion from the points 1 and 2. In accordance with the general formula (3.1), for a two-dimensional functional, we have

$$f_2^*(1,2) = f_{22}^*(1,2) + f_{22}^*(1,2), \tag{4.18}$$

$$f_{21}^*(1,2) = \int d1' f_1(1,2 \to 1') g_1(1'), \tag{4.19}$$

$$f_{22}^*(1,2) = \frac{1}{2}\int d1' \int d2' f_2(2,2 \to 1',2') g_2(1',2'). \tag{4.20}$$

To express the FAP FMDs through the cascade ones, we turn again to the GF. According to Eq. (2.56), GF is the product of the GFs related to the descendants of each of the primary particles:

$$F_2(1,2 \to u(\cdot)) = F_1(1 \to u(\cdot)) F_1(2 \to u(\cdot)).$$

Differentiation with respect to u at $u = 1$ yields

$$f_1(1, 2 \to 1') = f_1(1 \to 1') + f_1(2 \to 1'),$$

$$f_2(1, 2 \to 1', 2') = f_2(1 \to 1', 2') + f_2(2 \to 1', 2')$$

$$+ \; f_1(1 \to 1') f_2(2 \to 2') + f_1(2 \to 1') f_1(1 \to 2').$$

Substituting these expressions into (4.18)–(4.20) and using the above notation, we find

$$f_2^*(1, 2) = f_1^*(1) + f_1^*(2) + g_2'(1, 2). \tag{4.21}$$

In accordance with its definition, the function g_2' satisfies equation following from Eq. (4.9):

$$[d_2^+ + 2\sigma(1)] g_2'(1, 2) = 2 \int d1' w_1(1 \to 1') g_2'(1', 2) \tag{4.22}$$

with initial condition

$$g_2'(1, 2) = g_2(1, 2), \qquad t = T.$$

Here we have introduced the differential operator

$$d_n^+ = -\partial_t - \sum_{i=1}^{n} \dot{x}_i \frac{\partial}{\partial x_i}.$$

Similarly, we obtain the equation for f_2^*:

$$[d_2^+ + 2\sigma(1)] f_2^*(1, 2) = 2 \int d1' w_1(1 \to 1') f_2^*(1', 2)$$

$$+ \int d1' \int d2' w_2(1 \to 1', 2') g_2'(1', 2') + 2\langle \nu - 1 \rangle \sigma(1) f_1^*(2),$$

$$f_2^*(1, 2)|_{t=T} = g_1(1) + g_1(2) + g_2(1, 2).$$

Here $\langle \nu \rangle = \int d1' w_1(1 \to 1') / \sigma(1)$ is an average multiplicity of fragmentation products (that is, the number of particles produced in the collision).

4.4 Adjoint Equations and Adjoint Functions

Adjoint equations play an important role in the transport theory. They cannot only simplify to solve many specific problems but also formulate a number of general statements in framework of the transport theory (variational principle, continuity theorem, reciprocity theorem, etc. [6, 7]). In addition, the concept of the adjoint function lies in the basis of the perturbation theory, theory of measurements, formulations of inverse problems [8], the fluctuation theory, and other sections of the modern transport theory.

Taking Eq. (3.33) as a basic system for cascade description and following a common method, we derive the corresponding system of coupled equations. With this purpose, we introduce the scalar product in the space of vector-valued functions,

$$\langle \varphi, f \rangle = \sum_{n=1}^{\infty} \frac{1}{n!} \int d1 \ldots \int dn \varphi_n(1, \ldots, n) f_n(1, \ldots, n) \tag{4.23}$$

and define the operator $\mathscr{L}^+$, adjoint to $\mathscr{L}$, via condition

$$\langle \varphi, \mathscr{L} f \rangle = \langle \mathscr{L}^+ \varphi, f \rangle. \tag{4.24}$$

According to (3.1) and (4.23), expected detector readings can be rewritten as

$$\mathsf{M}q = \langle g, f \rangle_T, \tag{4.25}$$

where the subscript t indicates the argument values in time-dependent multipliers (in this case, in f).

Along with (3.33), let us consider the adjoint equation

$$-\partial_t \varphi = \mathscr{L}^+ \varphi \tag{4.26}$$

with boundary condition given at the measuring time T:

$$\varphi|_{t=T} = \varphi_T. \tag{4.27}$$

Multiplying Eq. (3.33) by φ, and Eq. (4.26) by f, subtracting the second result from the first one and then using the property (4.24), we get

$$\frac{d}{dt} \langle \varphi, f \rangle = 0, \qquad \theta \leqslant t \leqslant T.$$

Integrating the resulting expression with respect to t from θ to T, we obtain

$$\langle \varphi, f \rangle_t = \langle \varphi, f \rangle_\theta. \tag{4.28}$$

If the condition (4.27) is chosen in such a way that $\varphi_T = g$, then the left-hand side of Eq. (4.28), according to (4.25) will be an expectation of detector readings. Denoting the corresponding solution of the adjoint equation in terms of f^+:

$$-\partial_t f^+ = \mathscr{L}^+ f^+, \qquad f^+|_{t=T} = g, \tag{4.29}$$

write the formula following from Eq. (4.28) as

$$\mathsf{M}q = \langle f^+, f\rangle_\theta. \tag{4.30}$$

Thus, as in the case of additive detectors [4], the expectation of an arbitrary detector reading can be represented in two forms: in the form of Eq. (4.25), which expresses the desired value by multiparticle density at the time of measurement and characteristics detector, and in the form of Eq. (4.30), which includes multiparticle distributions at the initial time (the characteristics of source) and multiparticle adjoint functions.

Based on the definition (4.24), we find an explicit form of the adjoint operator. In accordance with Eq. (3.34), we represent the kinetic operator $\mathscr{L}$ as a sum of two operators

$$\mathscr{L} = \mathscr{L}^{(1)} + \mathscr{L}^{(2)}, \tag{4.31}$$

defined by the formulas

$$\mathscr{L}^{(1)}_{nk} f_k = -n[\partial_1 + \sigma(1)] f_k \delta_{nk}$$

and

$$\mathscr{L}^{(2)}_{nk} f_k = \binom{n}{k-1} \int dk' f_k(1, \ldots, k') w_{n-k+1}(k' \to k, \ldots, n)\varepsilon_{kn}.$$

The operator being adjoint to some of operators (4.31) is the sum of the adjoint operators,

$$\mathscr{L}^+ = \mathscr{L}^{(1)+} + \mathscr{L}^{(2)+}. \tag{4.32}$$

Let us apply the definition (4.24) to the first term,

$$\langle \varphi, \mathscr{L}^{(1)} f\rangle = \langle \mathscr{L}^{(1)+}\varphi, f\rangle \tag{4.33}$$

and expand the left-hand side of this equation:

$$\langle \varphi, \mathscr{L}^{(1)} f\rangle = -\sum_{n=1}^{\infty} \frac{1}{n!} \int d1 \ldots \int dn \varphi_n(1, \ldots, n) n[\partial_1 + \sigma(1)] f_n(1, \ldots, n)$$

$$= -\sum_{n=1}^{\infty} \frac{1}{n!} \int d1 \ldots \int dn f_n(1, \ldots, n) n[\partial_1^+ + \sigma(1)]\varphi_n(1, \ldots, n), \tag{4.34}$$

$$\partial_1^+ = -\dot{x}_1 \frac{\partial}{\partial x_1}.$$

Comparing the result with the expanded right-hand side of Eq. (4.34)

$$\langle \mathcal{L}^{(1)+}\varphi, f\rangle = \sum_{n=1}^{\infty}\frac{1}{n!}\int d1\ldots\int dnf_n(1,\ldots,n)\sum_{k=1}^{\infty}\mathcal{L}_{nk}^{(1)+}\varphi_k(1,\ldots,k),$$

we see that

$$\mathcal{L}_{nk}^{(1)}\varphi_k(1,\ldots,k) = k[\partial_1^+ + \sigma(1)]\varphi_k(1,\ldots,k)\delta_{nk}. \tag{4.35}$$

Similarly, expanding the inner product, which contains the integral part of the kinetic operator, we obtain

$$\langle\varphi, \mathcal{L}^{(2)} f\rangle$$

$$= \sum_{n=1}^{\infty}\frac{1}{n!}\int d1\ldots\int dn\varphi_n(1,\ldots,n)\sum_{k=1}^{n}\binom{n}{k-1}\int dk' f_k(1,\ldots,k-1,k').$$

Changing the order of summation and redesignating the indexes, we obtain

$$\mathcal{L}_{nk}^{(2)+}\varphi_k = \frac{n}{(k-n+1)!}\int d(n+1)\ldots\int d(k+1)$$

$$\times\, w_{k-n+1}(1\to n+1,\ldots,k+1)\varphi_k(2,\ldots,k+1)\varepsilon_{nk}. \tag{4.36}$$

Now, we substitute the operators defined by Eqs. (4.35) and (4.36) into the sum (4.32) and use the operator obtained in (4.29), arrive at the following system of coupled equations for multiparticle adjoint functions [9, 10]:

$$[d_n^+ + n\sigma(1)]f_n^+(1,\ldots,n) = \sum_{k=n}^{\infty}\frac{n}{(k-n+1)!}\int d(n+1)\ldots\int d(k+1)$$

$$\times\; w_{k-n+1}(1\to n+1,\ldots,k+1)f_k^+(2,\ldots,k+1) \tag{4.37}$$

with initial conditions

$$f_n^+(1,\ldots,n)|_{t=T} = g_n(1,\ldots,n). \tag{4.38}$$

According to Eq. (4.30), the expected detector reading is expressed in terms of this system solution as

$$\mathsf{M}q = \sum_{n=1}^{\infty}\frac{1}{n!}\int d1\ldots\int dnf_n^+(1,\ldots,n)f_n(1,\ldots,n)|_{t=\theta}.$$

Comparing this relation with the formula that expresses the given value through multiparticle importances

$$\mathsf{M}q = \sum_{n=1}^{\infty} \frac{1}{n!} \int d1 \ldots \int dn f_n^*(1, \ldots, n) p_n(1, \ldots, n)|_{t=\theta},$$

we see that f_n^+ is the decomposition (2.22) of the functions f_n^* (see the concluding formulas in Sect. 1.3):

$$f_n^*(1, \ldots, n) = \sum_{k=1}^{\infty} \binom{n}{k}, \tag{4.39}$$

$$f_n^+(1, \ldots, n) = \sum_{k=1}^{n} (-1)^{n-k} \binom{n}{k} f_k^*(1, \ldots, k). \tag{4.40}$$

Thus, in case of a non-additive detector, the multiparticle importances do not coincide with multiparticle adjoint functions but are linked with them via relations (4.39)–(4.40). This is due to the fact that, namely, the FMDs are chosen as the basic functions. Taking instead of them the PDs, we obtain the adjoint functions being at same time the multiparticle importances:

$$[d_n^+ + n\sigma(1)] p_n^+(1, \ldots, n) = \sum_{k=n}^{\infty} \frac{n}{(k-n+1)!} \int d(n+1) \ldots \int d(k+1)$$

$$\times\ \upsilon_{k-n+1}(1 \to n+1, \ldots, k+1) p_k^+(2, \ldots, k+1) + n\upsilon_0(n) p_{n-1}^+(1, \ldots, n-1) \varepsilon_{2n}.$$

However, the latter representation does not include the most important characteristics of the process: the average density particles f_1 in the phase space. Instead, we see the multiparticle probability densities, the equations for which are more cumbersome. Meanwhile, the first choice leads to the system of equations, and the first of them contains f_1 and for this reason is the basic equation of the transport theory. It greatly simplifies both the general theoretical analysis and the formulation and solution of concrete problems.

A characteristic feature of the adjoint equation system is that the equation for n-th function contains in its right-side functions of all higher orders $k > n$. Using finite-dimensional functionals restrict this system. For N-dimensional functional

$$g_k(1, \ldots, k) = 0, \qquad k > N. \tag{4.41}$$

Respectively, the solutions of Eqs. (4.37) satisfying the zero initial conditions (4.38) and (4.41) vanish: $f_k^+(1, \ldots, k) = 0$, $k > N$. In this case, the solving of the system may be started with the equation for f_N^+, then proceed to the equation for f_{N-1}^+, and so on. In particular, for $N = 2$ the adjoin equations coincide with Eqs.(3.17), (3.22), and $f_1^+(1) = f_1^*(1)$, $f_2^+(1,2) = g_2^{'}(1,2)$.

4.5 The Perturbation Theory

The dual representation of a non-additive detector response is a convenient tool for constructing perturbation theory [8, 10].

Let $\mathscr{L}$ be an operator of the main (unperturbed) problem, and $\mathscr{L}' = \mathscr{L} + \Delta\mathscr{L}$ be its perturbation counterpart with the same initial conditions problems (i.e., the source and detector are the same, only the medium properties are different). Write down the direct equation for the main problem and the adjoint one for perturbed case,

$$\partial_t f = \mathscr{L} f, \qquad f|_{t=\theta} = f_0,$$

$$-\partial_t f^{'+} = \mathscr{L}^{'+} f^{'+}, \qquad f^{'+}|_{t=T} = f_T^+.$$

And after inner multiplying first of them by f'^+ and second one by f, subtract one from another:

$$\langle f'^+, \partial_t f\rangle + \langle \partial_t f^+, f\rangle = \langle f'^+, \mathscr{L} f\rangle - \langle \mathscr{L}'^+ f'^+, f\rangle. \tag{4.42}$$

After evident transformations, expression (4.42) takes the form

$$\partial_t \langle f^{'+}, f\rangle = -\langle f^{'+}, \Delta\mathscr{L} f\rangle. \tag{4.43}$$

Time-integrating Eq. (4.43) in limits from θ to T yields

$$\langle f^{'+}, f\rangle_T - \langle f^{'+}, f\rangle_\theta = -\int_\theta^T \langle f^{'+}, \Delta\mathscr{L} f\rangle dt. \tag{4.44}$$

Taking into account, that $\langle f^{'+}, f\rangle_\theta = \bar{q}'$ is the m.e. of detectors response in the perturbed medium, and $\langle f^{'+}, f\rangle_T = \bar{q}$ is the same in the main medium, from (4.44) we find

$$\Delta\overline{q} = \overline{q}' - \overline{q} = -\int_\theta^T \langle f^{'+}, \Delta\mathscr{L} f\rangle dt. \tag{4.45}$$

Using the basic (forward) equation for the main problem and the adjoint (backward) one for the perturbed problem, arrive at other representations of the same quantity

$$\Delta\overline{q} = \int_{\theta}^{T} \langle f^+, \Delta\mathscr{L} f' \rangle dt. \tag{4.46}$$

By using the adjoint operation definition, one can obtain two more formulas:

$$\Delta\overline{q} = -\int_{\theta}^{T} \langle \Delta\mathscr{L}^+ f'^+, f \rangle dt, \tag{4.47}$$

and

$$\Delta\overline{q} = -\int_{\theta}^{T} \langle \Delta\mathscr{L}^+ f^+, f' \rangle dt. \tag{4.48}$$

Each of the four formulas (4.45)–(4.48) is an exact expression for change in m.e. of measured cascade quantity caused by the medium replacement $\mathscr{L} \to \mathscr{L} + \Delta\mathscr{L}$. Under small perturbations of the operator $\delta\mathscr{L}$ corresponding changes $\delta f, \delta f^+$ of f vector functions and f^+ are also small and the terms of the second order of smallness in (4.45)–(4.48) can be neglected. As a result, all four formulas take the same form known as a *small perturbation formula*

$$\delta\overline{q} = -\int_{\theta}^{T} \langle f^+, \delta\mathscr{L} f \rangle dt = -\int_{\theta}^{T} \langle \delta\mathscr{L}^+ f^+, f \rangle dt, \tag{4.49}$$

expressing change in m.e. of the measured functional through perturbation of the operator and solution of unperturbed equations.

Using the matrix representation of the kinetic operator, we write down the formula of the theory of small perturbations in an explicit form. Properties of the medium in which the cascade develops are described by the cross-sectional interactions σ, w_n, reflecting the influence of atomic and nuclear fields, as well as the coefficients $\dot{x} = \dot{x}(t, x)$, containing information about external fields. Replacing one environment with another means the change of coefficients: $\dot{x} \to \dot{x} + \delta\dot{x}$, $\sigma \to \sigma + \delta\sigma$, $w_n \to w_n + \delta w_n$. According to this (see Eq. (3.31))

$$(\delta\mathscr{L} f)_n + n[\delta\partial_1 + \delta\sigma(1)] f_n$$
$$= \sum_{k=0}^{n-1} \binom{n}{k} \int d(k+1)' f_{k+1}(1, \ldots, (k+1)') \delta w_{n-k}((k+1)' \to k+1, \ldots, n), \tag{4.50}$$

where $\delta\partial_1 = \partial_{x_1} \delta\dot{x}_1$. Inserting (4.50) into (4.49) yields:

$$\delta\overline{q} = -\sum_{n=1}^{\infty} \frac{1}{(n-1)!} \times \int d1 \ldots \int dn f_n^+(1,\ldots,n)[\delta\partial_1 + \delta\sigma(1)] f_n(1,\ldots,n)$$

$$+ \sum_{n=1}^{\infty} \frac{1}{n!} \sum_{k=0}^{n-1} \binom{n}{k} \int d1 \ldots \int dn f_n^+(1,\ldots,n)$$

$$\times \int d(k+1)' f_{k+1}(1,\ldots,(k+1)') \delta w_{n-k}((k+1)' \to k+1,\ldots,n).$$

Passing to the adjoint perturbation operator, we rewrite the m.e. of perturbation in another form:

$$\delta\overline{q} = -\sum_{n=1}^{\infty} \frac{1}{(n-1)!} \int d1 \ldots \int dn f_n(1,\ldots,n)[\delta\partial_1^+ + \delta\sigma(1)] f_n^+(1,\ldots,n) +$$

$$+ \sum_{n=1}^{\infty} \frac{1}{(n-1)!} \int d1 \ldots \int dn \sum_{k=n}^{\infty} \frac{1}{(k-n+1)!} \int d(n+1) \ldots \int d(k+1) \times$$

$$\times\ \delta w_{k-n+1}(1 \to n+1,\ldots,k+1) f_k^+(2,\ldots,k+1).$$

The theory of small perturbations can be used for various purposes: to calculate real perturbations of the detector readings, caused by a change in the properties of the environment as a result of any physical processes, to compare the readings of two detectors, placed in environments that differ little from each other, to evaluate the influence of the inhomogeneity of the medium on the detector reading, to take into account the influence of the error of the given sections on the results of calculations, etc. We will focus on one problem: the formulation of the inverse problem.

In the simplest case, this problem looks like this. Let the kinetic operator $\mathscr{L}$ be given as a function of some parameter set, $\mathscr{L} = \mathscr{L}(\alpha_1,\ldots,\alpha_k)$. These may include, for example, parameters describing the dependence of cross sections and multiplicities on energy, etc. We have at our disposal only approximate estimates of the parameters. It is necessary to refine these values, knowing the detector system readings. Let us also assume that these readings are practically not subject to fluctuations (or we have the possibility repeat the experiment, reducing the statistical error results to the desired level). Denote by $q_1,\ldots,q_r$ detector readings, and by $f^{+(1)},\ldots,f^{+(r)}$ corresponding adjoint vector functions computed with known parameter estimates $\alpha_1,\ldots,\alpha_k$. Measurement results $q_1',\ldots,q_r'$ differ from the calculated values due to the fact that the actual (but unknown) parameter values are different from those used in the calculations.

Let us change the parameters values, $\alpha_1 \to \alpha_1 + \delta\alpha_1,\ldots,\alpha_k \to \alpha_k + \delta\alpha_k$, taking increments $\delta\alpha_1,\ldots,\delta\alpha_k$ so small, that the perturbation operator may be represented in the form

$$\delta \mathscr{L} = \mathscr{L}(\alpha_1 + \delta\alpha_1, \ldots, \alpha_k + \delta\alpha_k) - \mathscr{L}(\alpha_1, \ldots, \alpha_k) = \sum_{i=1}^{k} \frac{\partial \mathscr{L}}{\partial \alpha_i} \delta\alpha_i,$$

admitting the use of the small perturbation formulas:

$$\delta \overline{q}_m = \int_{\theta}^{T} \left\langle f^{+(m)}, \sum_{i=1}^{k} \frac{\partial \mathscr{L}}{\partial \alpha_i} \delta\alpha_i f \right\rangle dt = \sum_{i=1}^{k} \Lambda_{mi} \delta\alpha_i, \quad m = 1, \ldots, r. \tag{4.51}$$

Here

$$\Lambda_{mi} = \int_{\theta}^{T} \left\langle f^{+(m)}, \frac{\partial \mathscr{L}}{\partial \alpha_i} f \right\rangle dt$$

are coefficients including only the initial information about parameters $\alpha_1, \ldots, \alpha_k$. For $r = k$, when the number detectors coincide with the number of sought parameters, the linear system (4.51) allows one to uniquely determine the corrections $\delta\alpha_i$ to initial parameter values. For $r > k$ one can use least squares method or similar methods.

If necessary, this procedure can be repeated by calculating with updated $\alpha_1 + \delta\alpha_1, \ldots, \alpha_k + \delta\alpha_k$ readings detectors and related functions and again using the system Eq. (4.51).

Consider a special case: perturbation of the indication additive detector, provided that the cascade is formed by one particle with deterministic initial coordinates:

$$f_1(\theta, x) = \delta(x, x_0).$$

In this case

$$\overline{q}_m \equiv f_1^{+(m)}(\theta, x_0)$$

and

$$\Lambda_{mi} = \int_{\theta}^{T} dt \int dx f_1^{+(m)} \frac{\partial \mathscr{L}_{11}}{\partial \alpha_i} f_1 = \int_{\theta}^{T} dt \int dx \left[\frac{\partial \mathscr{L}_{11}^{+}}{\partial \alpha_i} f_1^{+(m)} \right] f_1.$$

Taking into account that

$$f_1(t, x) = \mathscr{G}_1^0 f_1(\theta, x),$$

we get

$$\Lambda_{mi} = \frac{\partial f_1^{+(m)}}{\partial \alpha_i} = \int dx \delta(x, x_0) \int_{\theta}^{T} dt \mathscr{G}_1^{0+} \frac{\partial \mathscr{L}_{11}^{+}}{\partial \alpha_i} f_1^{+(m)}.$$

Using, further, Eqs. (3.44), (3.45), we arrive to the following equation for the matrix Λ_{mi} :

$$-\partial_t \Lambda_{mi} - \mathscr{L}_{11}^{+} \Lambda_{mi} = \frac{\partial \mathscr{L}_{11}^{+}}{\partial \alpha_i} f_1^{+(m)}. \tag{4.52}$$

Observe that it may be derived by means of plain differentiation of the adjoint equation (4.29) with respect to parameter α_i.

4.6 Fluctuations and Correlations in Linear Functionals

As can be seen from a comparison of Eqs.(4.4) and (4.5), the backward equations for moment densities m_n must have the same form as those for factorial moment densities f_n. Specifically, for the second-order MD, we have

$$[d^+ + \sigma(x)]m_2(x \to 1, 2) = \int d1' \int d2' w_2(x \to 1', 2') m_1(1' \to 1)$$

$$\times\ m_1(2' \to 2) + \int d1' w_1(x \to 1') m_2(1' \to 1, 2). \tag{4.53}$$

The latter is not difficult to get directly from Eqs. (4.9) and (4.13) by using the relation (2.76).

The backward equations for cumulant densities are obtained by functional differentiating Eq. (4.6):

$$[d^+ + \sigma(x)]k_1(x \to 1) = \int d1' w_1(x \to 1') k_1(1' \to 1), \tag{4.54}$$

$$[d^+ + \sigma(x)]k_2(x \to 1, 2) - \int d1' w_1(x \to 1') k_2(1 \to 1, 2)$$

$$= \sigma(x) k_1(x \to 1) k_1(x \to 2) - 2k_1(x \to 1) \int d2' w_1(x \to 2') k_1(2' \to 2)$$

$$+ \int d1' w_1(x \to 1') k_1(1' \to 1) k_1(1' \to 2)$$

$$+ \int d1' \int d2' w_2(x \to 1', 2') k_1(1' \to 1) k_1(2' \to 2). \tag{4.55}$$

Multiplying Eq. (4.54) by $\varphi(1)$, Eqs. (4.53) and (4.55) by $\varphi(a)\varphi(2)$, and integrating results with respect to the variables 1 and 2, we arrive at the backward equations for moments $\bar{q}$, $\overline{q^2}$ and the variance $\sigma_q^2(x) \equiv D(x)$ of the functional $q = \sum_{i=1}^{\mu} \varphi(i)$:

$$[d^{+} + \sigma(x)]\bar{q}(x) = \int d1 w_1(x \to 1)\bar{q}(1), \tag{4.56}$$

$$[d^{+} + \sigma(x)]\overline{q^2}(x) = \int d1 w_1(x \to 1)\overline{q^2}(1) + \int d1 \int d2 w_2(x \to 1,2)\bar{q}(1)\bar{q}(2), \tag{4.57}$$

$$[d^{+} + \sigma(x)]D(x) = \int d1 w_1(x \to 1)D(1) + \sigma(x)\overline{q^2}(x)$$

$$- 2\bar{q}(x)\int d1 w_1(x \to 1)\bar{q}(1) + \int d1 w_1(x \to 1)\overline{q^2}(1)$$

$$+ \int d1 \int d2 w_2(x \to 1,2)\bar{q}(1)\bar{q}(2). \tag{4.58}$$

If we multiply Eqs. (4.53) and (4.55) by $\varphi(a)g(2)$ and integrate, we obtain the equation for the mixed moment $\overline{qr}$ and the covariance $\sigma_{qr}(x)$ of functionals q $r = \sum_{i=1}^{\nu} g(i)$:

$$[d^{+} + \sigma(x)]\overline{qr}(x) + \int d1 w_1(x \to 1)\overline{qr}(1) = \int d1 \int d2 w_2(x \to 1,2)\bar{q}(1)\bar{r}(2), \tag{4.59}$$

$$[d^{+} + \sigma(x)]C(x) = \int d1 w_1(x \to 1)C(1) + \sigma(x)\bar{q}(x)\bar{r}(x)$$

$$- \bar{q}(x)\int d1 w_1(x \to 1)\bar{r}(1) + \bar{r}(x)\int d1 w_1(x \to 1)\bar{q}(1)$$

$$+ \int d1 w_1(x \to 1)\bar{q}(1)\bar{r}(1) + \int d1 \int d2 w_2(x \to 1,2)\bar{q}(1)\bar{r}(2). \tag{4.60}$$

It is convenient to transform the dispersion and covariance equations as follows. We denote by q' and r' value functionals of q and r, provided that the first interaction occurred at x. Then $q' = \sum_{i=1}^{\nu} q_i$ and $r' = \sum_{i=1}^{\nu} r_i$ where the q_i and r_i are contributions in the functionals q and r of i-th secondary particle produced in this interaction. It is easy to understand the following relations:

$$\overline{q'}(x) = \int d1 W_1(x \to 1)\bar{q}(1),$$

$$\overline{r'}(x) = \int d1 W_1(x \to 1)\bar{r}(1),$$

$$C'(x) = \mathrm{cov}(q', r') = \int d1 W_1(x \to 1)C(1) + \int d1 W_1(x \to 1)\bar{q}(1)\bar{r}(1)$$

$$+ \int d1 \int d2 w_2(x \to 1, 2)\bar{q}(1)\bar{r}(2) - \overline{q'}(x)\overline{r'}(x),$$

where W_1 and W_2 are corresponding inclusive distributions of particles arising in the collision: $w_1(x \to 1) = \sigma(x)W_1(x \to 1)$, $w_2(x \to 1, 2) = \sigma(x)W_2(x \to 1, 2)$. Returning to (4.60) and noting that by Eq. (4.56)

$$\overline{q'}(x) = \frac{1}{\sigma(x)}[d^+ + \sigma(x)]\bar{q}(x), \tag{4.61}$$

$$\overline{r'}(x) = \frac{1}{\sigma(x)}[d^+ + \sigma(x)]\bar{r}(x), \tag{4.62}$$

we obtain the relation between the covariances of the functional from cascades with a fixed point of birth of a primary particle and the covariance with a fixed point interaction of its first fragmentation:

$$[d^+ + \sigma(x)]C(x) = \sigma(x)C'(x) + \frac{1}{\sigma(x)}[d^+\bar{q}(x)d^+\bar{r}(x)]. \tag{4.63}$$

For variances, it takes the form

$$[d^+ + \sigma(x)]D(x) = \sigma(x)D'(x) + \frac{1}{\sigma(x)}[d^+\bar{q}(x)]^2. \tag{4.64}$$

Given the initial condition $C(x) = D(x) = 0$, $t = T$, this means that a variance (covariance) can be written as the sum of two terms satisfying the equations

$$[d^+ + \sigma(x)]D^{\mathrm{I}}(x) = \frac{1}{\sigma(x)}[d^+\bar{q}(x)]^2$$

and

$$[d^+ + \sigma(x)]D^{\mathrm{II}}(x) = \sigma(x)D'(x).$$

The first one describes a contribution to the variance of the measured value of fluctuations of the primary particle path to the point of fragmentation, the second gives the contribution of fluctuations in the rest part of the cascade [11].

Formulas (4.61)–(4.64) make it easy to find $\bar{q}'$, $\bar{r}'$, C', D' by the known $\bar{q}$, $\bar{r}$, C, D. In order to use them for calculation of the C and D, we consider functionals different from q' and r' by that random contributions q_i, r_i are replaced by their mean values, but the distribution of secondary particles is still random. Obviously, the average values of the corresponding functionals coincide,

$$\bar{q}'' = \bar{q}',$$

$$\bar{r}'' = \bar{r}',$$

but covariances are different:

$$C'(x) = \mathsf{M}\sum_{ij} q_i r_i - \overline{q'}\,\overline{r'} = \mathsf{M}\sum_{i}[q_i r_i - \bar{q}(i)\bar{r}(i)] + \mathsf{M}\sum_{ij}\bar{q}(i)\bar{r}(j) - \overline{q''}\,\overline{r''}$$

$$= \int d1 W_1(x \to 1)C(1) + C''(x). \tag{4.65}$$

Substituting (4.65) into (4.63), we obtain

$$[d^+ + \sigma(x)]C(x) - \int d1 w_1(x \to 1)C(1) = \frac{1}{\sigma(x)}[d^+\bar{q}(x)d^+\bar{r}(x)] + \sigma(x)C''(x). \tag{4.66}$$

The equation for the variance has the similar form:

$$[d^+ + \sigma(x)]D(x) - \int d1 w_1(x \to 1)D(1) = \frac{1}{\sigma(x)}[d^+\bar{q}(x)]^2 + \sigma(x)D''(x). \tag{4.67}$$

Observe that the functions C'' and D'' in the right-hand sides of these equations contain only information about average characteristics of the cascade and about fluctuations in one fragmentation event:

$$C''(x) = \mathrm{cov}(q'', r'') = \int d1 W_1(x \to 1)\bar{q}(1)\bar{r}(1)$$

$$+ \int d1 \int d2 W_2(x \to 1, 2)\bar{q}(1)\bar{r}(2) - \int d1 W_1(x \to 1)\bar{q}(1) \int d2 W_1(x \to 2)\bar{r}(2). \tag{4.68}$$

Equation (4.66) allows us to separate fluctuations in two other components C_1 and C_2:

$$[d^+ + \sigma(x)]C_i(x) - \int d1 w(x \to 1)C_i(1) = B_i(x), \tag{4.69}$$

where

$$B_1(x) = \frac{1}{\sigma(x)}[d^+\bar{q}(x)d^+\bar{r}(x)],$$

$$B_2(x) = \sigma(x)C''(x).$$

It is not hard to see that the first of them reflects the influence of fluctuations in path lengths of all particles, and the second does the effect of fluctuations in fragmentation. If there are several possible processes of fragmentation, $B_2(x) =$

$\sum_k \sigma_k(x)C''^{(k)}(x)$, $C_2(x) = \sum_k C_2^{(k)}(x)$, where $C_2^{(k)}(x)$ is the contribution of k-th elementary process in the studied correlations:

$$[d^+ + \sigma(x)]C_2^{(k)}(x) - \int d1 w_1(x \to 1)C_2^{(k)}(1) = \sigma_k(x)C''^{(k)}(x).$$

4.7 Equations for Moments and Distribution of a Linear Functional

The approach based on backward equations allows us to consider not only moments, but also the distribution of the linear functional of a cascade process:

$$\psi(q|x) = \mathsf{M}\delta\left(q - \sum_{i=1}^{\mu}\varphi(i)\right).$$

To obtain an equation for this characteristic, we insert $s(x) = s_1\varphi(x)$ in Eq. (2.49). As a result, the characteristic functional becomes the characteristic function of the random variable $q = \sum_{i=1}^{\mu}\varphi(i)$. Designating it by $\chi(t, x \to T, s_1) \equiv C(t, x \to T, s_1\varphi(\cdot))$ and substituting in Eq. (4.5), we obtain the corresponding equation for the characteristic function. On using representation (2.64) for the functional W, we rewrite this equation in the form

$$[d^+ + \sigma(x)]\chi(t, x \to T, s_1) = v_0(x) + \sum_{h=1}^{\infty}\frac{1}{k!}\int d_1$$

$$\dots \int dk\, v_k(x \to 1, \dots, n)\chi(t, 1 \to T, s_1)\dots\chi(t, n \to T, s_1),$$

whence follows equations for the moments of the linear functional:

$$\overline{q^n}(x) = \left(\frac{1}{i}\frac{\partial}{\partial s_1}\right)\chi(t, x \to T, s_1)|_{s_1=0}.$$

According to (2.64)–(2.66), one can rewrite the right-hand side of the equation in the form

$$v_0(x) + \sum_{k=1}^{\infty}\frac{1}{k!}\int d_1 \dots \int dk\, v_k(x \to 1, \dots, k)\prod_{i=1}^{k}\chi(t, i \to T, s_1)$$

$$= \sigma(x) + \sum_{k=1}^{\infty}\frac{(-1)^k}{k!}\int d_1 \dots \int dk\, w_k(x \to 1, \dots, k)\prod_{i=1}^{k}\bar{\chi}(t, i \to T, s_1).$$

Applying the differentiation formula of Binet in the symmetrized form

$$\left(\frac{\partial}{\partial s_1}\right)^n \prod_{i=1}^{k} \bar{\chi}(t, i \to T, s_1) = \sum_{\substack{N_k=n \\ (n_i \neq 0)}} \binom{n}{n_1 \dots n_k} \left(\frac{\partial}{\partial s_1}\right)^{n_i} \bar{\chi}(t, i \to T, s_1),$$

we get (for $n > 0$)

$$[d^+ + \sigma(x)]\overline{q^n}(x) - \int d1 w_1(x \to 1)\overline{q^n}(1) = d^{(n)}(x),$$

$$d^{(n)}(x) = \sum_{k=2}^{n} \frac{1}{k!} \int d_1 \dots \int dk\, w_k(x \to 1, \dots, k) \sum_{\substack{N_k=n \\ (n_i \neq 0)}} \binom{n}{n_1 \dots n_k} \prod_{i=1}^{k} \overline{q^{n_i}}(i). \tag{4.70}$$

It is easy to see that for $n = 2$, this equation coincides with Eq. (4.57).

Returning to the equation for the characteristic function, let us perform the inverse transformation leading to the distribution function $\psi(q|t, x)$:

$$[d^+ + \sigma(x)]\psi(q|t, x) = v_0(x)\delta(q) + \sum_{k=1}^{\infty} \frac{1}{k!} \int d1$$

$$\dots \int dk v_k(x \to 1, \dots, k)\psi(q|t, 1) * \dots * \psi(q|t, k). \tag{4.71}$$

The nonlinear integro-differential Eq. (4.71) together with the initial condition $\psi(q|T, x) = \delta(q - \varphi(x))$ defines a distribution function (more strictly, the probability density function) of a linear functional of a cascade produced when the primary particle begins to move from point x at time t.

If we take $s(x) = s_1\varphi(x) + s_2 g(x) + \dots$ and repeat the above calculations, we obtain the equation for joint distribution function of linear functionals $q = \sum_{i=1}^{\mu} \varphi(i)$, $r = \sum_{i=1}^{\mu} g(i)$. It can be written in the form (4.71), if q is understood as a vector $\mathbf{q} = (q, r \dots)$, and convolution

$$\psi(q|t, 1) * \dots * \psi(q|t, k)$$

as a convolution of all components of the vector:

$$\psi(\mathbf{q}|1) * \psi(\mathbf{q}|2) = \int d\mathbf{q}' \psi(\mathbf{q} - \mathbf{q}'|1)\psi(\mathbf{q}'|2)$$

$$= \int dq' \int dr' \dots \psi(q - q', r - r', \dots, |1)\psi(q', r', \dots, |2).$$

Multiplying the resulting equation by $qrdqdr...$ and integrating, we arrive at Eq. (4.59) for the mixed moment.

In an important special case, when the function $s(x)$ is a simple function, that is,

$$s(x) = \sum_{n=1}^{m} s_n 1(x, A_n),$$

where A_n form the system of disjoint regions ("cells") of the phase space X so that $\bigcup_{n=1}^{m} = X$. Each of m number of particles $N_n = \tilde{\mu}(A_n)$ in the relevant domain ("cell") can be considered as a linear functional. In this presentation of occupation numbers, Eq. (4.71) takes the form

$$[d^+ + \sigma(x)]\psi(N|x) = v_0(x)\delta_{N_0} + \sum_{k=1}^{\infty} \frac{1}{k!} \int d_1$$

$$\ldots \int dk v_k(x \to 1, \ldots, k)\psi(N|1) * \ldots * \psi(N|k), \tag{4.72}$$

where

$$\mathbf{N} = (N_1, \ldots, N_m),\ \delta_{N0} = \delta_{N_1 0} \ldots \delta_{N_m 0},\ \psi(N|1) * \psi(N|2)$$

$$= \sum_{N'} (\mathbf{N} - \mathbf{N}'|1)\psi(N'|2) = \sum_{N_1' \ldots N_m'} \psi(N_1 - N_1', \ldots, N_m - N_m'|1)\, \psi(N_1', \ldots, N_m'|2).$$

A similar equation was used in [3] for deriving the main equation equations of G-theory in case of one cell ($m = 1$).

4.8 Fluctuations in an Arbitrary Additive Functional

Above, we considered a class of detectors whose response is determined by the state of the measured cascade at the measurement moments and called these functionals linear. Here, we shall show that the backward equation can be generalized to a more wide class of functionals, called *additive*, describing detectors whose responses are formed by all particle fragmentation events and paths during the whole measurement period. We define an additive functional $q_{t,x}$ of the cascade, formed by one primary particle produced at time $t (\theta \le t \le T)$ at point x, via the relation

$$q_{t,x} = \Delta Q + \sum_{i=1}^{\nu} q_{t+\Delta t, i}, \tag{4.73}$$

where ΔQ is the part of the detector reading produced by the cascade during interval $(t, t+\Delta t)$. When Δt is small, one of the two events is only possible here: free (collisionless) motion with probability $1-\sigma(x)\Delta t$ and fragmentation collision $(x \to x_1, \ldots, x_\nu)$ with probability $\sigma(x)\Delta t$. In the first case, the contribution of the element Δt into $q_{t,x}$ is equal to $a(t,x)\Delta t$, thus

$$q_{t,x}\big|_{\mathrm{I}} = a(t,x)\Delta t + q_{t+\Delta t, x+\dot{x}\Delta t} + o(\Delta t) \quad \text{with probability } 1-\sigma\Delta t.$$

In the second case, the fragmentation process contributes $b_\nu(x' \to 1, \ldots, \nu)$ and the remaining part of the cascade consisting of ν daughter's cascades, so

$$q_{t,x}\big|_{\mathrm{II}} \equiv q'_{t,x} = b_\nu(x' \to 1, \ldots, \nu) + \sum_{i=1}^{\nu} q_{t,x_i} + O(\Delta t) \quad \text{with probability } \sigma\Delta t. \tag{4.74}$$

Raising now $q_{t,x}$ in n-th degree and averaging it with the use of the total expectation theorem, we obtain the relation

$$\overline{q^n}(t,x) = (1-\sigma\Delta t)\left[\overline{q^n}(t+\Delta t, x+\dot{x}\Delta t) + na\Delta t\overline{q^{n-1}}(t,x)\right] + \sigma\Delta t\overline{q'^n}(t,x) + o(\Delta t)$$

which leads to

$$\left[d^+ + \sigma(x)\right]\overline{q^n}(t,x) = \sigma(x)\overline{q'^n}(t,x) + na\overline{q^{n-1}}(t,x) \tag{4.75}$$

as $\Delta t \to 0$. The latter equation links the moments of an additive functional calculated under two different conditions: at a fixed point and time of birth of the primary particle $(\overline{q^n}(t,x))$ and at a fixed point and time of the first fragmentation $(\overline{q'^n}(t,x))$. The difference between these characteristics is due to the influence of the first path.

A corresponding equation can be derived for the distribution functions as well:

$$[d^+ + \sigma(x)]\psi(q|t,x) = \sigma(x)\psi'(q|t,x) - a\partial_q\psi. \tag{4.76}$$

From Eq. (4.74), one can obtain the inverse relation expressing ψ' through ψ:

$$\sigma(x)\psi'(q|t,x) = v_0(x)\delta(q-b_0(x)) + \sum_{k=1}^{\infty}\frac{1}{k!}\int d1$$

$$\ldots\int dk v_k(x \to 1, \ldots, k)\psi(q|1) * \ldots * \psi(q|k) * \delta(q-b_k). \tag{4.77}$$

Substituting (4.77) into (4.76), we obtain the nonlinear integro-differential equation for the function distribution of an additive functional of a cascade process:

$$\left[d^{+} + \sigma(x) + a\partial_q\right]\psi(q|t,x) = v_0(x)\delta(q - b_0(x))$$

$$+\sum_{k=1}^{\infty}\frac{1}{k!}\int d1\ldots\int dk v_k(x \to 1,\ldots,k)\psi(q|1) * \ldots * \psi(q|k) * \delta(q - b_k). \tag{4.78}$$

The corresponding equations for the moments have the form

$$[d^{+} + \sigma(x)]\overline{q^n}(x) - \int d1 w_1(x \to 1)\overline{q^n}(1) = d^{(n)}(x). \tag{4.79}$$

The right-hand sides of the equations include information on the properties of the detector and on the lower order moments. In particular, the quantity

$$d^{(1)}(x) = a(x) + v_0(x)b_0(x) + \sum_{k=1}^{\infty}\frac{1}{k!}\int d_1\ldots\int dk v_k(x \to 1,\ldots,k)b_k(x \to 1,\ldots,k)$$

$$= a(x) + \sigma(x)\overline{b(x)} \tag{4.80}$$

is equal to the average value of the contribution per unit time of the particle passing through x. The first term corresponds to the track part of the contribution, the second to the collision one. The right-hand side of the equation for the second moment is more complicated:

$$d^{(2)}(x) = 2a(x)\bar{q}(x) + \sigma(x)\overline{b^2}(x) + 2\int d_1 c(x \to 1)\bar{q}(1)$$

$$+\int d1\int d2 w_2(x \to 1,2)\bar{q}(1)\bar{q}(2), \tag{4.81}$$

where

$$c(x \to x') = v_1(x \to x')b_1(x \to x')$$

$$+\sum_{n=1}^{\infty}\frac{1}{n!}\int d_1\ldots\int dn v_{n+1}(x \to 1,\ldots,n,x')b_{n+1}(x \to 1,\ldots,n,x').$$

In accordance with Sect. 4.7, replacing $q \to (q_1, q_2, \ldots)$, $a \to (a_1, a_2, \ldots)$, $b \to (b_1, b_2, \ldots)$ in Eq. (4.78) allows us to describe the joint probability distribution of various functionals from the same cascade. In particular, the equation for the mixed moment $\overline{q_1 q_2}$ has a form being similar to Eq. (4.79) with the right-hand side

$$d^{+}(x) = a_1(x)\bar{q}_2(x) + a_2(x)\bar{q}_1(x) + \sigma(x)\overline{b_1 b_2} + \int d1c_1(x \to 1)\overline{q_2}(1)$$

$$+ \int d1c_2(x \to 1)\overline{q_1}(1) + \int d1 \int d2w_2(x \to 1, 2)\overline{q_1}(1)\overline{q_2}(2). \tag{4.82}$$

Performing operations similar to those carried out for Eq. (4.66), we obtain an equation for the covariance readings of the detectors:

$$[d^{+} + \sigma(x)]C(x) - \int d1w_1(x \to 1)C(1) = \frac{1}{\sigma(x)}(a_1 - d^{+}\overline{q_1})$$

$$\times (a_2 - d^{+}\overline{q_2}) + \sigma(x)C''(x),$$

$$C''(x) = \mathrm{cov}(q_1'', q_2''),$$

$$q_k'' = b_{\nu k}(x \to 1, \ldots, \nu) + \sum_{i=1}^{\nu} \overline{q_k}(i).$$

Remark in conclusion, that the solutions of equations of the form (4.79) (i.e., Eqs. (4.53)–(4.54), (4.56)–(4.60), and others) can be written using the adjoint one-particle Green's functions:

$$\overline{q^n}(t, x) = \int_t^T dt' \int dx' d^{(n)}(t', x')g^{+}(t', x' \to t, x)$$

$$+ \int dx' \overline{q^n}(T, x')g^{+}(T, x' \to t, x). \tag{4.83}$$

By the reciprocity theorem, the adjoint and basic Green's functions are linked by the relation

$$g^{+}(t', x' \to t, x) = g(t, x \to t', x'),$$

so Eq. (4.83) can be represented as

$$\overline{q^n}(t, x) = \int_t^T dt' \int dx' g(t, x \to t', x')d^{(n)}(t', x')$$

$$+ \int dx' g(t, x \to T, x')\overline{q^n}(T, x'). \tag{4.84}$$

The physical meaning of this relation is easily understood when taking into account that $d^{(n)}(t, x)dt$ is an average contribution to this quantity of one particle during dt, and $\overline{q^n}(T, x)$ means its contribution at the time of measurement.

References

1. Pal L (1962) Statistical theory of neutron chain reactions. II. Acta Physica Hungaricae 14(4):369–380
2. Bell G (1965) On the stochastic theory of neutron transport. Nucl Sci Engng 21:390–401
3. Uchaikin VV (1978) Fluctuations in cascade processes of particle passage through the matter. Izv Vyssh Uchebn Zaved, Fizika 12:30–34, 2:7–12, 8:14–21 (in Russian)
4. Kolchuzhkin AM, Uchaikin VV (1978) Introduction to the theory of particle passage through the matter. Atomizdat, Moscow (in Russian)
5. Uchaikin VV, Ryzhov VV (1988) Stochastic theory of high-energy particle transport. Nauka, Novosibirsk (in Russian)
6. Marchuk GI, Orlov VV (1964) On the theory of adjoint functions. Neutron physics, Atomizdat, Moscow, pp 30–45 (in Russian)
7. Lewis J (1965) Importance. The Adjoint Function. Pergamon Press, Oxford, London
8. Marchuk GI (1977) Methods of computational mathematics. Nauka, Moscow (in Russian)
9. Uchaikin VV (1984) Theory of nonlinear measurements in high-energy cascade processes. Inelastic interactions of hadrons and nuclei at high energies, Alma-Ata, pp 172–179 (in Russian)
10. Uchaikin VV (1979) The theory of cascade process nonlinear measurements and the statement of backward problems. In: Proceedings of 16-th ICRC Kyoto, vol 7, pp 24–29
11. Vetoshkin VV, Lagutin AA, Uchaikin VV (1980) On the influence of the first run on fluctuations in an electromagnetic cascade, Izv. Academy of Sciences of the USSR. Ser Phys, vol 44, pp 575–578 (in Russian)

For Further Reading

12. Marchuk GI, Shutyaev V, Bocharov G (2005) Adjoint equations and analysis of complex systems: application to virus infection modelling. J Comput Appl Math 184(1):177–204
13. Cao Y, Li S, Petzold L, Serban R (2003) Adjoint sensitivity analysis for differential-algebraic equations: the adjoint DAE system and its numerical solution. SIAM J Sci Comput 24(3):1076–1089

Chapter 5
Electrons and Photons

5.1 Photon Birth (Classical Sketch)

The passage of a charged particle passage through the matter is accompanied by emission of electromagnetic radiation. According to classical electrodynamics, the charge $Z_1 e$, moving through a medium with permittivity ε according to the law $\mathbf{r} = \mathbf{r}(t)$, radiates

$$dI = \frac{Z_1^2 e^2}{4\pi^2 c\sqrt{\varepsilon}} \left| \int\limits_{-\infty}^{\infty} \frac{[\boldsymbol{\Omega}[(c/\sqrt{\varepsilon})\boldsymbol{\Omega} - \mathbf{v}, \dot{\mathbf{v}}]]}{(c/\sqrt{\varepsilon} - \mathbf{v}\boldsymbol{\Omega})^2} e^{i\omega(t-\boldsymbol{\Omega}\mathbf{r}\sqrt{\varepsilon}/c)} dt \right|^2 d\omega d\boldsymbol{\Omega} \tag{5.1}$$

energy in the frequency range $d\omega$ into the solid angle $d\boldsymbol{\Omega}$ [1]. The vector part of the integrand is a total differential:

$$\frac{[\boldsymbol{\Omega}[(c/\sqrt{\varepsilon})\boldsymbol{\Omega} - \mathbf{v}, \dot{\mathbf{v}}]]}{(c/\sqrt{\varepsilon} - \mathbf{v}\boldsymbol{\Omega})^2} = \frac{d}{dt} \frac{[\boldsymbol{\Omega}[\boldsymbol{\Omega}, \mathbf{v}]]}{(c/\sqrt{\varepsilon} - \mathbf{v}\boldsymbol{\Omega})^2}. \tag{5.2}$$

Actually, the right-hand side of Eq. (5.2) reduces to the form

$$\frac{(c/\sqrt{\varepsilon})[\boldsymbol{\Omega}[\boldsymbol{\Omega}, \dot{\mathbf{v}}]] - [\boldsymbol{\Omega}[\boldsymbol{\Omega}, \dot{\mathbf{v}}]](\mathbf{v}, \boldsymbol{\Omega}) + [\boldsymbol{\Omega}[\boldsymbol{\Omega}, \mathbf{v}]](\dot{\mathbf{v}}, \boldsymbol{\Omega})}{(c/\sqrt{\varepsilon} - \mathbf{v}\boldsymbol{\Omega})^2}. \tag{5.3}$$

Expanding the double vector products, we find that the numerator in (5.3) is equal to

$$((c/\sqrt{\varepsilon})\boldsymbol{\Omega} - \mathbf{v})(\boldsymbol{\Omega}, \dot{\mathbf{v}}) - \dot{\mathbf{v}}(\boldsymbol{\Omega}, (c/\sqrt{\varepsilon})\boldsymbol{\Omega} - \mathbf{v}) = [\boldsymbol{\Omega}[(c/\sqrt{\varepsilon})\boldsymbol{\Omega} - \mathbf{v}, \dot{\mathbf{v}}]]. \tag{5.4}$$

Equality (5.4) proves identity (5.2). Integrating (5.1) by parts and taking into account that

$$de^{i\omega(t-\boldsymbol{\Omega}\mathbf{r}\sqrt{\varepsilon}/c)} = i\omega\sqrt{\varepsilon}/c(c/\sqrt{\varepsilon} - \mathbf{v}\boldsymbol{\Omega})e^{i\omega(t-\boldsymbol{\Omega}\mathbf{r}\sqrt{\varepsilon}/c)}dt,$$

© Higher Education Press Limited Company 2024
V. V. Uchaikin, *Electron-Photon Cascades*, Nonlinear Physical Science,
https://doi.org/10.1007/978-981-99-7524-2_5

we have

$$dI = \frac{Z_1^2 e^2 \omega^2 \sqrt{\varepsilon}}{4\pi^2 c^3} \left| \int_{-\infty}^{\infty} [\boldsymbol{\Omega}[\boldsymbol{\Omega}, \mathbf{v}]] e^{i\omega(t - \boldsymbol{\Omega}\mathbf{r}\sqrt{\varepsilon}/c)} dt \right| d\omega d\boldsymbol{\Omega}. \tag{5.5}$$

By using Eq. (5.5), we calculate the charged particle energy loss as a result of its interaction with the continuous matter environment as a continuum, excluding direct collisions with individual atoms of the medium. With this aim, we consider such a small segment Δz of its path, that the velocity $\mathbf{v}$ of the particle is almost constant on it, but at the same time satisfies the condition

$$\Delta z \gg v/\omega. \tag{5.6}$$

Replacing the upper limit in (5.5) by $\Delta z/2v$, the lower by $-\Delta z/2v$ and, denoting via Θ the angle between $\mathbf{v}$ and $\boldsymbol{\Omega}$, after time integration, we get

$$\frac{dI}{d\omega d\boldsymbol{\Omega}} = \frac{Z_1^2 e^2 v\omega\sqrt{\varepsilon}\sin^2\Theta}{4\pi c^3}$$

$$\times \left\{ \frac{2\Delta z}{\pi(1 - v\sqrt{\varepsilon}\cos\Theta/c)^2 \omega\Delta z/v} \frac{\sin^2[\omega\Delta z(1 - v\sqrt{\varepsilon}\cos\Theta/c)/2v]}{1} \right\}. \tag{5.7}$$

Provided (5.6), contents of curly braces in (5.7) can be replaced by $\delta(1 - v\sqrt{\varepsilon}\cos\Theta/c)$, therefore

$$dI = \frac{Z_1^2 e^2 \omega \sin^2\Theta}{2\pi c^3} \delta(\cos\Theta - cv\sqrt{\varepsilon})\Delta z d\omega d\boldsymbol{\Omega}. \tag{5.8}$$

Thus, if

$$v > c/\sqrt{\varepsilon}, \tag{5.9}$$

i.e., the charged particle velocity exceeds the phase velocity of electromagnetic waves in this medium, then it radiates the EM-energy along a conical surface with a central angle Θ, such that

$$\cos\Theta = c/v\sqrt{\varepsilon}. \tag{5.10}$$

This radiation (directly observed in transparent media) is called the *Vavilov–Cherenkov radiation* . If the condition (5.9) fails, the delta-function in (5.8) equals zero for all angles Θ and the radiation is absent. Integrating (5.8) over angles and taking into account (5.10), we get

$$dI = \frac{Z_1^2 e^2 \omega}{c^2} \left[1 - \frac{c^2}{v^2 \varepsilon(\omega)} \right] d\omega\Delta z. \tag{5.11}$$

Integration (5.11) over all frequencies, satisfying the condition (5.9) leads to the *Frank–Tamm formula* for the total energy, emitted by the charge on its path Δz:

$$\Delta I = \Delta z \left(\frac{Z_1 e}{c}\right)^2 \int\limits_{\varepsilon(\omega)>(c/v)^2} \left[1 - \frac{c^2}{v^2 \varepsilon(\omega)}\right] \omega d\omega. \tag{5.12}$$

The properties of Cherenkov radiation are widely used in nuclear physics and cosmic ray physics to measure energy of fast charged particles (Cherenkov spectrometers).

Remaining within the classical radiation theory, we consider now the photon emission process accompanying the collision of a charged particle with one of atoms of the medium. Neglecting influence of all other atoms, put in (5.1) $\varepsilon = 1$. In case of non-relativistic particle energy, this formula becomes

$$dI = \left(\frac{Z_1 e}{2\pi}\right)^2 \frac{1}{c} \left| \int\limits_{-\infty}^{\infty} [\boldsymbol{\Omega}[\boldsymbol{\Omega}, \dot{\mathbf{v}}]] e^{i\omega(t - \boldsymbol{\Omega}\mathbf{r})} dt \right|^2 d\omega d\boldsymbol{\Omega}.$$

Let us take the origin of coordinates at the center of the atom and, neglecting the term $r/c \sim vt/c$ in the exponential function (as is usually done in the dipole approximation [1]), we represent the frequency–angular distribution of the radiation intensity in the form

$$dI \approx \left(\frac{Z_1 e}{2\pi c}\right)^2 \frac{1}{c} \left| \int\limits_{-\infty}^{\infty} [\boldsymbol{\Omega}[\boldsymbol{\Omega}, \dot{\mathbf{v}}]] e^{i\omega t} dt \right|^2 d\omega d\boldsymbol{\Omega}. \tag{5.13}$$

Actually, the main fraction of the energy is radiated when the charge flies near the nucleus: in this section of its trajectory, the field of the nucleus is maximum; remote parts of its trajectory (before and after the closest approach) make a small contribution, which in the dipole approximation may be neglected, and the time integration may be reduced to integration over a finite time interval τ, called the *interaction time*. The time can be estimated through the impact parameter ρ and the particle velocity v as $\tau \sim \rho/v$ (in the relativistic case $\tau \sim (\rho/v)\sqrt{1-(v/c)^2}$).

At low frequencies, $\omega\tau \ll 1$, the exponent in (5.13) changes slowly during the acceleration time τ, so one can take

$$\int\limits_{-\infty}^{\infty} \dot{\mathbf{v}}(t) e^{i\omega t} dt \approx \int\limits_{-\infty}^{\infty} \dot{\mathbf{v}}(t) dt = v(\infty) - v(-\infty) = \Delta v.$$

Thus, the energy radiated (5.13) is estimated as

$$dI \approx \left(\frac{Z_1 e}{2\pi c}\right)^2 \frac{|\Delta v|^2}{c} \sin^2 \Theta d\omega d\boldsymbol{\Omega}, \quad \omega\rho/v \ll 1 \tag{5.14}$$

where the angle Θ is measured from the direction of the vector $\Delta\mathbf{v}$. At higher frequencies ($\omega\tau \gg 1$), the integrand quickly oscillates, and radiation is small. The integral of (5.14) over the total solid angle can be approximated as

$$dI(\omega) \approx \begin{cases} \frac{2Z_1^2 e^2}{3\pi c^3} |\Delta\mathbf{v}|^2 d\omega, & \rho < v/\omega; \\ 0, & \rho > v/\omega. \end{cases} \tag{5.15}$$

In perturbation theory approximation, the total change of incident particle velocity with impact parameter ρ is

$$\Delta v = \frac{2Z_1 Z e^2}{m_1 v \rho}.$$

Inserting this quantity into Eq. (5.15) yields the spectral density $dI(\omega)/d\omega$ in the form:

$$\frac{dI(\omega)}{d\omega} = \frac{8}{3\pi} Z_1^4 Z^2 \left(\frac{e^2}{m_1 c^2}\right)^2 \frac{e^2 c}{v^2} \frac{1}{\rho^2}, \quad \rho < v/\omega.$$

In order to compute the differential with respect to ω cross section, one should multiply it by $2\pi\rho d\rho$, integrate over the impact parameter, and divide by $\hbar\omega$ (for passage from radiation energy to photon numbers):

$$\frac{d\sigma}{d\omega} = 2\pi \int_{\rho_{min}}^{\rho_{max}} \frac{dI(\omega)}{d\omega} \frac{\rho d\rho}{\hbar\omega}.$$

The upper limit $\rho_{max} = v/\omega$, and the lower one can be stated from quantum mechanical uncertainty principle:

$$\rho_{min} \simeq \frac{\hbar}{m_1 v}.$$

As a result, we obtain

$$\frac{d\sigma}{d\omega} = \frac{16}{3} \frac{Z^2 e^2}{\hbar c} \left(\frac{Z_1^2 e^2}{m_1 c^2}\right)^2 \left(\frac{c}{v}\right)^2 \frac{1}{\omega} \ln\left(\frac{m_1 v^2}{\hbar\omega}\right).$$

5.2 Electromagnetic Field Operators

The quantum theory of radiation, built by Dirac, considers the electromagnetic (EM) field (in the area, free from sources) as a set of independent harmonic oscillators, and the processes of absorption or emission of photons by an atom as transitions in a single atom+field system. Because of smallness of the fine structure constant

$\alpha = 1/137$ compared to the unit, calculations of these transitions probability can be performed with the use of perturbation methods.

Let us denote by $\xi = x\sqrt{\omega/\hbar}$ the dimensionless coordinate of one of the oscillators with frequency ω and unit mass. Quantization of the EM-field yields the following expression for the vector potential operator, relating to the given oscillator:

$$\hat{\mathbf{A}} = c\sqrt{2\pi\imath/\omega}\eta\{\hat{a}^{i\mathbf{qr}} + \hat{a}^{+}e^{-i\mathbf{qr}}\}. \tag{5.16}$$

Here η stands for the polarization vector perpendicular to the wave vector $\mathbf{q}$, and the normalization volume not affecting the final result is taken equal to 1. Operators $\hat{a}$ and $\hat{a}^{+}$ are of the form:

$$\hat{a} = \frac{1}{\sqrt{(2)}}\left(\xi + \frac{\partial}{\partial\xi}\right); \qquad \hat{a}^{+} = \frac{1}{\sqrt{(2)}}\left(\xi - \frac{\partial}{\partial\xi}\right). \tag{5.17}$$

Interaction of non-relativistic with a given oscillator is described by operator

$$\hat{W} = \frac{e}{mc}\hat{\mathbf{A}}\hat{\mathbf{p}} + \frac{e^2}{2mc^2}\hat{\mathbf{A}}^2, \tag{5.18}$$

whose form is determined by the classical expression for interaction energy

$$W = \frac{e}{mc}\mathbf{A}\mathbf{p} + \frac{e^2}{2mc^2}\mathbf{A}^2.$$

Note that due to the orthogonality of vectors $\boldsymbol{\eta}$ and $\mathbf{q}$ operators $\hat{\mathbf{A}}$ and $\hat{\mathbf{p}} = -\hbar\nabla$ commute, so the order in which they appear in (5.18) is not significant. Let n denote a state of atomic electron and N stand for an oscillator state. As a result of interaction, the system under consideration can go from the state characterized by parameters n, N, into another state with parameters n', N'. For the matrix element of such transition, the following notation is introduced:

$$\langle n', N'|W|n, N\rangle = \iint \psi^{*}_{n',N'}(\mathbf{r}, \xi)\hat{W}\psi_{n,N}(\mathbf{r}, \xi)d\mathbf{r}d\xi. \tag{5.19}$$

Using here (5.18), we get

$$\langle n, N'|W|n, N\rangle = \langle n', N'\left|\frac{e}{mc}\hat{\mathbf{A}}\hat{\mathbf{p}}\right|n, N\rangle + \langle n', N'\left|\frac{e^2}{2mc^2}\hat{\mathbf{A}}^2\right|n, N\rangle. \tag{5.20}$$

The last term contains the square of the operator $\hat{\mathbf{A}}$, which determines the measure of the interaction of an electron with the EM-field, and may be omitted if the first term of the sum is not equal to 0.

Unperturbed wave functions in (5.19) are calculated in the absence of interaction between the electron and the field and therefore they are equal to the products of the electron wave function ψ_n and the oscillator wave function ψ_N, expressed through Hermite polynomials $H_N(\xi)$

$$\psi_{n,N}(\mathbf{r},\xi) = \psi_n(\mathbf{r})\varphi_N(\xi); \tag{5.21}$$

$$\varphi_N(\xi) = [2^N N!\sqrt{\pi}]^{-1/2} e^{-\xi^2/2} H_N(\xi). \tag{5.22}$$

Using (5.16) and (5.21), we write the first term of the matrix element (5.20) as

$$\langle n', N' \left| \frac{e}{mc^2} \hat{\mathbf{A}}\hat{\mathbf{p}} \right| n, N\rangle = \frac{e}{m}\sqrt{2\pi\hbar\omega}\,\times$$

$$\eta\left\{\int \psi_{n'}^* e^{i\mathbf{qr}}\hat{\mathbf{p}}\psi_n d\mathbf{r}\int \psi_{N'}^*\hat{a}\varphi_N d\xi + \int \psi_{n'}^* e^{-i\mathbf{qr}}\hat{\mathbf{p}}\psi_n d\mathbf{r}\int \psi_{N'}^*\hat{a}^+\varphi_N d\xi\right\}. \tag{5.23}$$

Taking into account the Hermite polynomials property

$$H_N'(\xi) = 2\xi H_N(\xi) - H_{N+1}(\xi) \tag{5.24}$$

and

$$H_{N+2}(\xi) - 2\xi H_{N+1} - 2(N+1)H_N(\xi) = 0, \tag{5.25}$$

we find that

$$\hat{a}\varphi_N = \sqrt{N}\varphi_{N-1}; \tag{5.26}$$

$$\hat{a}^+\varphi_N = \sqrt{N+1}\varphi_{N+1}. \tag{5.27}$$

Applying operator $\hat{a}$ (5.17) to the wave function $\varphi_N(\xi)$ (5.22) yields

$$\hat{a}\varphi_N = [2^{N+1}N!\sqrt{\pi}]^{-1/2} e^{-\xi^2/2} H_N'(\xi). \tag{5.28}$$

Finding $2\xi H_N(\xi)$ from (5.25) and inserting the result into the right-hand side of Eq. (5.24), we see that

$$H_N'(\xi) = 2NH_{N-1}(\xi). \tag{5.29}$$

Substituting (5.29) into (5.28) and taking into account (5.22), we are convinced of the fairness of equality (5.26). Formula (5.27) is proved by applying operator $\hat{a}^+$ to the wave function φ_N and using the property (5.24). Thus, operator $\hat{a}$ causes the harmonic oscillator to make a transition from a state with the energy level E_N into the state with the energy level $E_{N-1} = E_N - \hbar\omega$, whereas another operator relates to transition $E_N \to E_{N+1} = E_N + \hbar\omega$. In the first transition, the harmonic oscillator loses energy $\hbar\omega$, in the second one it acquires such an energy. Quantum electrodynamics associates these processes with annihilation and creation of an elementary particle *photon*, and the operators $\hat{a}$ and $\hat{a}^+$ are called the *annihilation* and *creation* operators, respectively.

Taking into account (5.26)–(5.27) and eigenfunctions orthogonality, we rewrite (5.23) as

$$\langle n', N' \left| \frac{e}{mc^2} \hat{\mathbf{A}}\hat{\mathbf{p}} \right| n, N \rangle = \frac{e}{m}\sqrt{2\pi\hbar/\omega} \times$$

$$\eta \left\{ \sqrt{N}\delta_{N',N-1} \int \psi_{n'}^{*} e^{i\mathbf{qr}} \hat{\mathbf{p}} \psi_n d\mathbf{r} + \sqrt{N+1}\delta_{N',N+1} \int \psi_{n'}^{*} e^{-i\mathbf{qr}} \hat{\mathbf{p}} \psi_n d\mathbf{r} \right\}. \quad (5.30)$$

The first term in this expression describes the photon disappearance process as a result, for example, of photoelectric absorption by the atom, containing the electron.

$$\langle n', N-1 \left| \frac{e}{mc} \hat{\mathbf{A}}\hat{\mathbf{p}} \right| n, N \rangle = \frac{e}{m}\sqrt{2\pi\hbar/\omega} \int \psi_{n'}^{*} e^{i\mathbf{qr}} \boldsymbol{\eta} \hat{\mathbf{p}} \psi_n d\mathbf{r}. \quad (5.31)$$

The second term describes an inverse process: the photon due to transition of the electron from n-th n'-th level:

$$\langle n', N+1 \left| \frac{e}{mc} \hat{\mathbf{A}}\hat{\mathbf{p}} \right| n, N \rangle = \frac{e}{m}\sqrt{2\pi\hbar(N+1)/\omega} \int \psi_{n'}^{*} e^{-i\mathbf{qr}} \boldsymbol{\eta} \hat{\mathbf{p}} \psi_n d\mathbf{r}. \quad (5.32)$$

An example of such a process is the *bremsstrahlung*.

If the wavelength of the radiation is large enough, then arguments of the exponential functions in (5.31)–(5.32) remain small in the integration domain, and the integral function itself can be replaced by unity:

$$\int \psi_{n'}^{*} e^{\pm i\mathbf{qr}} \boldsymbol{\eta} \hat{\mathbf{p}} \psi_n d\mathbf{r} \approx \int \psi_{n'}^{*} \boldsymbol{\eta} \hat{\mathbf{p}} \psi_n d\mathbf{r} = (\boldsymbol{\eta}\mathbf{p})_{n'n}.$$

But the momentum matrix element is linked with the coordinate matrix element as follows:

$$\mathbf{p}_{n,n'} = m\dot{\mathbf{r}}_{n',n} = im\omega\mathbf{r}_{n',n}; \qquad \omega = \frac{E_n - E_{n'}}{\hbar}. \quad (5.33)$$

Considering also that $e\mathbf{r}$ is the dipole moment $\mathbf{d}$ of the system, we obtain a *dipole approximation*, in which matrix element (5.32) becomes

$$\langle n', N+1 \left| \frac{e}{mc} \hat{\mathbf{A}}\hat{\mathbf{p}} \right| n, N \rangle = -i\sqrt{2\pi\hbar\omega(N+1)}(\boldsymbol{\eta}d)_{n'n}. \quad (5.34)$$

Unlike the first term in (5.20), describing single-photon processes, the second term corresponds to two-photon processes. Analyzing two-photon process, we have to consider two field oscillators with parameters $\boldsymbol{\eta}_1, \mathbf{q}_1, \omega_1$ and $\boldsymbol{\eta}_2, \mathbf{q}_2, \omega_2$, respectively:

$$\hat{\mathbf{A}} = c\sqrt{2\pi\hbar/\omega_1}\boldsymbol{\eta}_1\{\hat{a}_1 e^{i\mathbf{q}_1\mathbf{r}} + \hat{a}_1^{+} e^{-i\mathbf{q}_1\mathbf{r}}\} + c\sqrt{2\pi\hbar/\omega_2}\boldsymbol{\eta}_2\{\hat{a}_2 e^{i\mathbf{q}_2\mathbf{r}} + \hat{a}_2^{+} e^{-i\mathbf{q}_2\mathbf{r}}\}. \quad (5.35)$$

Substituting this expression into the second term in (5.20), we obtain the sum of four terms, containing following combinations of annihilation-creation operators

$$\hat{a}_2^+\hat{a}_2^+,\ \hat{a}_2^+\hat{a}_2,\ \hat{a}_2\hat{a}_2^+,\ \hat{a}_2\hat{a}_2,\ldots,$$

pertaining to different versions of two-photon processes. In particular, the scattering of a photon by an electron can be treated as annihilation of the primary photon ($\boldsymbol{\eta}_1$, $\mathbf{q}_1$, ω_1) and creation of a second photon ($\boldsymbol{\eta}_2$, $\mathbf{q}_2$, ω_2). This process is represented by combination $\hat{a}_2^+\hat{a}_2$, numbers $N_1 = 1, N_2 = 0$ and the matrix element [1, 2]:

$$\langle n', N_1' - 1, N_2 + 1\,|W|\,n, N_1, N_2\rangle = \frac{e^2}{mc^2}\int \frac{2\pi\hbar c}{\sqrt{q_1 q_2}}\boldsymbol{\eta}_1\boldsymbol{\eta}_2 e^{i(\mathbf{q}_1-\mathbf{q}_2)\mathbf{r}}\psi_{n'}^*\psi_n d\mathbf{r}. \tag{5.36}$$

5.3 Bremsstrahlung (Quantum Sketch)

Quantum theory treats the radiation process, called here *bremsstrahlung*, as a transition of a primary electron to a new state accompanied by the emission of a photon. An analysis of the conservation laws shows that for free electron, such a process is impossible. Indeed, in the coordinate system where before radiation the electron has been at rest, the energy and momentum conservation laws claim

$$mc^2 = c\sqrt{m^2c^2 + p'^2} + \hbar\omega; \tag{5.37}$$

$$0 = \mathbf{p}' + \mathbf{p}_\gamma. \tag{5.38}$$

Inserting $\mathbf{p}_\gamma = \hbar\omega/c$ into (5.37)–(5.38) yields

$$mc^2 = c\sqrt{m^2c^2 + (\hbar\omega c)^2} + \hbar\omega.$$

Moving the summand $\hbar\omega$ to the left and squaring both sides, we get

$$2mc^2\hbar\omega = 0.$$

Thus, the system (5.37)–(5.38) does not have positive solutions for $\hbar\omega$, so to execute conservation laws in this process another body should participate. Usually it is the atomic nucleus, in the field of which a charge particle emits the bremsstrahlung photon.

Including the nucleus into conservation laws, one can show that if the mass of an incident particle is negligible compared to the mass of the nucleus, then the energy conservation law (5.37) remains in effect, and violated only the relation (5.38), including the momentum transferred to the nucleus is violated.

Let us calculate the bremsstrahlung of a non-relativistic charged particle (say, electron) in the field of the nucleus, using the dipole approximation of the perturbation theory. Setting in the matrix element (5.34) $N = 0$ (there are no photons in the initial state), we have

$$\langle n', 1 \left| \frac{e}{mc} \hat{\mathbf{A}}\hat{\mathbf{p}} \right| n, 0\rangle = -i\sqrt{2\pi\hbar}\eta \mathbf{d}_{n'n}, \tag{5.39}$$

where $\mathbf{d}_{n'n}$ is the matrix element of the dipole moment of the system incident particle + nucleus. Assuming the nucleus to be infinitely heavy compared to an electron, we write the dipole moment as

$$\mathbf{d} = -e\mathbf{r}. \tag{5.40}$$

The double time-differentiation of (5.40) and describing the particle motion in the Coulomb field of the nucleus in a classical way lead to the formula

$$\ddot{\mathbf{d}} = -e\ddot{\mathbf{r}} = \frac{1}{m}\frac{Ze^3}{r^2}\frac{\mathbf{r}}{r}. \tag{5.41}$$

On the other hand, according to (5.33)

$$\ddot{\mathbf{d}}_{n'n} = -\omega^2 \mathbf{d}_{n'n}; \quad \omega = \frac{E_n - E_{n'}}{\hbar} = \frac{p^2 - p'^2}{2m\hbar}, \tag{5.42}$$

where $\mathbf{p} = \hbar\mathbf{k}$ and $\mathbf{p}' = \hbar\mathbf{k}'$ are the incident particle momentums before and after photon emission correspondingly. The wave functions of these states are of the form

$$\psi_n = e^{i\mathbf{kr}}; \qquad \psi_{n'} = e^{i\mathbf{k}'\mathbf{r}}. \tag{5.43}$$

According to (5.41)–(5.43), we get

$$\mathbf{d}_{n'n} = -\ddot{\mathbf{d}}_{n'n}\omega^2 = -\frac{Ze}{m\omega^2}\int e^{i\mathbf{qr}}(\mathbf{r}/r^3)d\mathbf{r}, \quad \mathbf{q} = \mathbf{k} - \mathbf{k}'. \tag{5.44}$$

Taking into account that

$$\int e^{i\mathbf{qr}}(\mathbf{r}/r^3)d\mathbf{r} = \frac{4\pi\mathbf{q}}{iq^2}, \tag{5.45}$$

inserting this expression into (5.44), and using the result in (5.39), we will get

$$\langle n', 1 \left| \frac{e}{mc} \hat{\mathbf{A}}\hat{\mathbf{p}} \right| n, 0\rangle = 4\pi\sqrt{2\pi\hbar\omega}\frac{Ze}{mq^2\omega^2}\mathbf{q}\boldsymbol{\eta}. \tag{5.46}$$

To calculate the bremsstrahlung cross section, one should use the *Fermi golden rule*, which requires to find the final states density for the electron+photon system. If the momentum $\mathbf{p}$ and the frequency E of each particle were independent, the number of states of the system, in which the electron energy belongs to dE', its momentum direction belongs to $d\boldsymbol{\Omega}'$, the photon frequency belongs to $d\omega$, and its momentum is directed into elementary solid angle $d\boldsymbol{\Omega}$ would be equal to the product of the quantity

$$\rho_e dE' = \frac{\sqrt{2mE'}md\boldsymbol{\Omega}'}{(2\pi\hbar)^3}dE' = \frac{mp'd\boldsymbol{\Omega}'}{(2\pi\hbar)^3}dE'$$

and

$$\rho_\gamma d\omega = \frac{k^2dkd\boldsymbol{\Omega}}{(2\pi c)^3} = d\omega.$$

Recall that in the latter equality k means a wave number of the photon, participating in this process. Due to the energy conservation law, which in the non-relativistic case has the form

$$\frac{p^2}{2m} = \frac{p'^2}{2m} + \hbar\omega,$$

fixing the interval $(\omega, \omega + d\omega)$, we uniquely set the position of the interval $(E', E' + dE')$, and that $E' = E - \hbar\omega$.

Therefore, in this case, we have

$$\rho = \frac{mp'\omega^2 d\omega}{(2\pi)^6(c\hbar)^3}d\boldsymbol{\Omega}'d\boldsymbol{\Omega}. \tag{5.47}$$

Taking into account that the wave function ψ_n (5.43) corresponds to the electron flux density $v = p/m$, we find the differential in ω, $\boldsymbol{\Omega}$, $\boldsymbol{\Omega}'$ cross section of a photon emission by an electron with polarization $\boldsymbol{\eta}$:

$$d\sigma(\omega, \boldsymbol{\Omega}, \boldsymbol{\Omega}', \boldsymbol{\eta}) = \frac{2\pi}{\hbar}\frac{1}{v}\left|\langle n', 1 \left| \frac{e}{mc}\hat{\mathbf{A}}\hat{\mathbf{p}} \right| n, 0\rangle^2 \rho\right|$$

$$= \frac{(Ze^3)^2}{\pi^2(\hbar c)^3 q^4}|\boldsymbol{\eta}\mathbf{q}|^2 \frac{p'}{p}\frac{d\omega}{\omega}d\boldsymbol{\Omega}d\boldsymbol{\Omega}'. \tag{5.48}$$

To find the bremsstrahlung photons energy spectrum regardless of their polarization and particle momentum directions after radiation, one should sum expression (5.48) over two possible values of η ($\boldsymbol{\eta}_1$, $\boldsymbol{\eta}_2$ are orthogonal to each other and to the photon wave vector) and integrate over all angular variables. For this aim, we choose the coordinate system (5.1), as shown in Fig. 5.1. The Oz-axis is directed along momentum of the incident particle, so that

$$k_x = k_y = 0; \qquad k_z = k.$$

Let us direct the Oy-axis perpendicular to the photon momentum:

$$\Omega_x = \sin\vartheta; \quad \Omega_y = 0; \quad \Omega_z = \cos\vartheta.$$

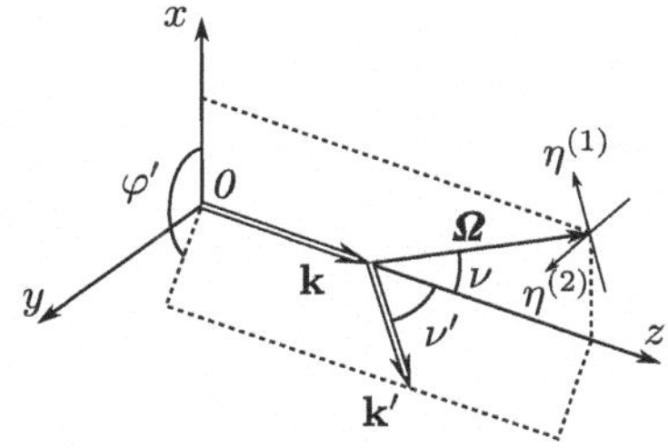

Fig. 5.1 Calculation of the bremsstrahlung photon spectrum

As to the polarization vector, we choose one of its component in plane zOx, in which the vector $\boldsymbol{\Omega}$ lies

$$\eta_x^{(1)} = -\cos\vartheta; \quad \eta_y^{(1)} = 0; \quad \eta_z^{(1)} = -\sin\vartheta,$$

then the other will be parallel to the Oy-axis:

$$\eta_x^{(2)} = 0; \qquad \eta_y^{(2)} = 1; \qquad \eta_z^{(2)} = 0.$$

The electron wave vector after radiation $\mathbf{k}'$ in this system has the following projections:

$$k_x' = k' \sin\vartheta' \cos\varphi'; \quad k_y' = k' \sin\vartheta' \sin\varphi'; \quad k_z' = k' \cos\vartheta'.$$

Thus, the matter is reduced to the calculation of the radiation intensity

$$I = \int d\boldsymbol{\Omega} \int d\boldsymbol{\Omega}' q^{-4} \sum_{i=1}^{2} |\mathbf{q}\boldsymbol{\eta}^{(i)}|^2,$$

where

$$\begin{aligned}
\mathbf{q}\boldsymbol{\eta}^{(1)} &= k' \sin\vartheta' \cos\varphi' \cos\vartheta + (k - k' \cos\vartheta') \sin\vartheta; \\
\mathbf{q}\boldsymbol{\eta}^{(2)} &= k' \sin\vartheta' \sin\varphi'; \\
q^4 &= (\mathbf{k} - \mathbf{k}')^4 = (k^2 - 2kk' \cos\vartheta' + k'^2)^2.
\end{aligned}$$

It is convenient to start the calculations with integrating over φ', during which members containing $\sin\varphi'$ or $\cos\varphi'$ in the first power, vanish, and the integrals of the second powers of these functions will give π:

$$I = \int d\boldsymbol{\Omega} \int d\cos\vartheta' q^{-4} \pi [k'^2 \sin^2\vartheta' \cos^2\vartheta + 2(k - k' \cos\vartheta')^2 \sin^2\vartheta + k'^2 \sin^2\vartheta'].$$

Then, changing the order of operations, we obtain

$$I = \frac{16\pi^2}{3} \int q^{-4}[k^2 - 2kk' \cos\vartheta' + k'^2]d\cos\vartheta' = \frac{16\pi^2}{3} \int_{-1}^{1} \frac{d\cos\vartheta'}{k^2 - 2kk'\cos\vartheta' + k'^2}.$$

The last integral is calculated elementarily:

$$\int_{-1}^{1} \frac{d\cos\vartheta'}{k^2 - 2kk'\cos\vartheta' + k'^2} = \frac{1}{kk'} \log\left(\frac{k+k'}{k-k'} = \frac{\hbar^2}{pp'} \log\frac{p+p'}{p-p'}\right).$$

As a result of these calculations, the differential in frequency cross section (i.e., the frequency spectrum of emitted photons) is obtained [3]:

$$d\sigma = \frac{16}{3} Z^2 \alpha r_0^2 \left(\frac{c}{v}\right)^2 \log\frac{p+p'}{p-p'} \frac{d\omega}{\omega}, \tag{5.49}$$

where $\alpha = e^2/\hbar c \approx 1/137$ is the *fine structure constant*. Considering the relationship between energy and momentum non-relativistic particle and the conservation law (5.46), one can rewrite the resulting formula as follows:

$$d\sigma = \frac{16}{3} Z^2 \alpha r_0^2 \left(\frac{c}{v}\right)^2 \log\left[\frac{(\sqrt{E} + \sqrt{E + \hbar\omega})^2}{\hbar\omega}\right] \frac{d\omega}{\omega}. \tag{5.50}$$

Note that for $\hbar\omega = E$ (the electron radiates all its kinetic energy), the cross section vanishes, and for $\hbar\omega =\rightarrow 0$ grows indefinitely (*infrared catastrophe*). The differential cross section for bremsstrahlung in general case (of a relativistic particle) was obtained by Bethe and Heitler by the methods of quantum electrodynamics [4]. Relevant formulas as well as analysis of various corrections are given in the review [5]. In the ultrarelativistic case, when conditions are met:

$$E \gg mc^2; \qquad E' = E - \hbar\omega \gg mc^2,$$

photons are emitted mainly inside the cone with an angle $\sim mc^2/E$ around the incident particle direction.

Bremsstrahlung can also be emitted in the field of the nucleus, but also in the field of atomic electrons. However, the cross section of this process is proportional to the number of electrons Z, while the cross section on the nucleus is proportional to Z^2. Therefore, the contribution of bremsstrahlung in electrons can become noticeable only in substances with small Z. It can be taken into account by replacing in sections factor Z^2 on $Z(Z + l)$.

5.4 Compton Scattering

In classical electrodynamics, the process of scattering of the harmonic electromagnetic waves limited by a system of unbound point charges (we will talk about electrons) is interpreted very simply: the wave controls the oscillatory motion of the electron with its own frequency, and with the same frequency it re-radiates the wave. The angular distribution of intensity is described by the *Thomson formula*, in terms of the differential cross section (per atom with atomic number Z) which has the form

$$\frac{d\sigma}{d\Omega} = \frac{r_0^2}{2}(1 + \cos^2 \vartheta)Z,$$

where r_0, as noticed earlier, is the electron classical radius. All electrons emit waves with the same frequency, equal to the frequency of the incident radiation:

$$\omega = \omega_0.$$

This description is valid if the wavelength is not too small ($\lambda \geq 10^{-8}$ cm). At shorter wavelengths, corpuscular properties of radiation begin to play a significant role.

The scattering cross section for this case is obtained in the framework of quantum electrodynamics. Quantum theory considers scattering electromagnetic radiation on free electrons as the result of elastic collisions of photons with electrons. The energy and momentum conservation laws show that with angle scattering ϑ increasing, the scattered photon energy $\hbar\omega$ decreases and, accordingly, the wavelength shift increases:

$$\lambda' - \lambda = \frac{2\pi\hbar}{mc^2}(1 - \cos\vartheta).$$

Such angle dependence of shift was called the *Compton effect* , and the process itself the *Compton scattering*.

In the framework of non-relativistic quantum theory, the primary photon with wave vector $\mathbf{q}_1$ and polarization $\boldsymbol{\eta}_1$ (denoted below by $\mathbf{q}$, $\boldsymbol{\eta}$) and a secondary photon is created with the wave vector $\mathbf{q}_2$ and polarization $\boldsymbol{\eta}_2$ (denoted below by $\mathbf{q}'$, $\boldsymbol{\eta}'$). In this case, the electron passes from the state with wave function

$$\psi_n = e^{i\kappa\mathbf{r}} \tag{5.51}$$

into a state with a wave function

$$\psi_{n'} = e^{i\kappa'\mathbf{r}}. \tag{5.52}$$

Substituting these functions into expression (5.36) for the matrix element yields

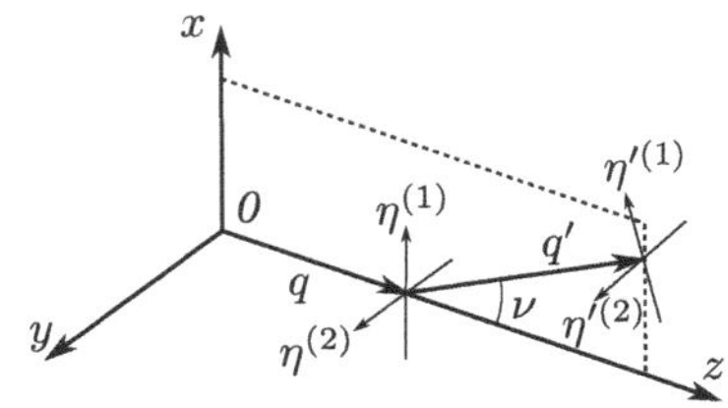

Fig. 5.2 Calculation of Compton scattering cross section

$$\langle\ldots\rangle = \frac{e^2}{mc^2}\int \frac{2\pi\hbar c}{\sqrt{qq'}}\boldsymbol{\eta}\boldsymbol{\eta}' e^{i(\mathbf{k}+\mathbf{q}-\mathbf{k}'-\mathbf{q}')\mathbf{r}} d\mathbf{r}. \tag{5.53}$$

According to the momentum conservation law

$$\hbar\mathbf{k} + \hbar\mathbf{q} = \hbar\mathbf{k}'\hbar\mathbf{q}',$$

therefore

$$\langle\ldots\rangle = \frac{2\pi\hbar e^2}{mc\sqrt{qq'}}\boldsymbol{\eta}\boldsymbol{\eta}', \tag{5.54}$$

since the integration in (5.53) is carried out over the cube of unit volume. With this normalization (one photon per unit volume), the golden rule becomes

$$d\sigma = \frac{2\pi}{\hbar c}|\langle\ldots\rangle|^2\rho(E), \tag{5.55}$$

where the final state density is given by

$$\rho(E) = \frac{q'^2 d\boldsymbol{\Omega}'}{(2\pi)^3}\frac{dq'}{dE}, \tag{5.56}$$

where

$$E = \hbar c\left[q' + \frac{1}{2x}(\mathbf{q}' - \mathbf{k} - \mathbf{q})^2\right] x = \frac{mc}{\hbar} \tag{5.57}$$

means the total energy of the photon and electron in the end state of the event.

To consider the polarization, let us draw through vectors $\mathbf{q}$, $\mathbf{q}'$ the plane in which we place the vectors $\boldsymbol{\eta}^{(1)}$ and $\boldsymbol{\eta}^{(2)}$ polarizations (Fig. 5.2); vectors $\boldsymbol{\eta}^{(2)}$ and $\boldsymbol{\eta}^{(2)'}$ will be perpendicular to it. The photon scattering angle (i.e., the angle between the vectors $\mathbf{q}$ and $\mathbf{q}'$) is denoted via ϑ. The same angle will be between the vectors $\boldsymbol{\eta}^{(1)}$ and $\boldsymbol{\eta}^{(1)'}$ perpendicular to the corresponding vectors $\mathbf{q}$ and $\mathbf{q}'$, so

$$\boldsymbol{\eta}^{(1)}\boldsymbol{\eta}'^{(1)} = \cos\vartheta; \qquad \boldsymbol{\eta}^{(1)}\boldsymbol{\eta}'^{(2)} = 0;$$

$$\boldsymbol{\eta}^{(2)}\boldsymbol{\eta}'^{(1)} = 0; \qquad \boldsymbol{\eta}^{(2)}\boldsymbol{\eta}'^{(2)} = 1.$$

As can be seen from these formulas, only those transitions are possible in this process, in which both vectors $\boldsymbol{\eta}$ and $\boldsymbol{\eta}'$ lie in the plane built on vectors $\mathbf{q}$ and $\mathbf{q}'$, or both are perpendicular to it. In the first case,

$$d\sigma^{(1)} = \frac{(2\pi)^3 e^4 \hbar}{m^2 c^3 q q'} \rho(E) \cos^2 \vartheta,$$

in the second

$$d\sigma^{(2)} = \frac{(2\pi)^3 e^4 \hbar}{m^2 c^3 q q'} \rho(E).$$

If the radiation incident on the electron is not polarized, it is necessary to average the cross section over the initial polarization:

$$d\sigma = \frac{d\sigma^{(1)} + d\sigma^{(2)}}{2} = \frac{(2\pi)^3 e^4 \hbar}{m^2 c^3 q q'} \rho(E) \, \frac{1 + \cos^2 \vartheta}{2} \, \rho(E). \tag{5.58}$$

If the electron was initially at rest, $k = 0$, then we get from formulas (5.57)

$$\frac{dE}{dq'} = \frac{d}{dq'} \hbar c \left[q' + \frac{1}{2x} (q'^2 + q^2 - 2qq' \cos \vartheta) \right] = \hbar c \left[1 + \frac{1}{2x} (q' - q \cos \vartheta) \right]. \tag{5.59}$$

Inserting Eq. (5.59) into (5.56) and coming back to Eq. (5.58), we find

$$d\sigma = \frac{r_0^2}{2} (1 + \cos^2 \vartheta)(q'/q) \left[1 + \frac{1}{2x} (q' - q \cos \vartheta) \right]^{-1} d\boldsymbol{\Omega}'. \tag{5.60}$$

The quantity

$$q' = q \cos \vartheta - x + \sqrt{2xq(1 - \cos \vartheta) - q^2 \sin^2 \vartheta} \tag{5.61}$$

can easily be found from the energy conservation law and Eq. (5.57), which lead to the equality

$$E = \hbar c \left[q' + \frac{1}{2x} (q'^2 + q^2 - 2qq' \cos \vartheta) \right] = \hbar c q.$$

In non-relativistic approximation

$$p'^2 / 2m = \hbar c (q - q') \ll mc^2,$$

i.e.,

$$q - q' \ll \frac{mc}{\hbar} = \kappa,$$

therefore, expressions (5.60)–(5.61) can be simplified by using the series expansion with respect to $\varepsilon = q/\kappa = \hbar\omega/mc^2$:

$$q'\kappa = \varepsilon - \varepsilon^2(1 - \cos\vartheta) + \dots$$

and

$$d\sigma = \frac{r_0^2}{2}(1 + \cos^2\vartheta)\,[1 - 2\varepsilon(1 - \cos\vartheta)]d\boldsymbol{\Omega}'. \tag{5.62}$$

Integrating over the total solid angle yields

$$\sigma = \frac{8\pi r_0^2}{3}\,[1 - 2\varepsilon]. \tag{5.63}$$

Expressions (5.62)–(5.63) differ from their classical counterparts by a factor in square brackets. These corrections increase the concentration of the angular distribution in the direction of the primary photon momentum, reduce the transverse dimensions of the distribution, and decrease the total cross section of the Compton scattering.

The exact formulas for this scattering process, being valid for any primary energy, are derived by the methods of the relativistic quantum electrodynamics (the *Klein–Nishina–Tamm formula*) [6]:

$$\frac{d\sigma}{d\boldsymbol{\Omega}} = \frac{r_0^2}{2}\left(\frac{\varepsilon'}{\varepsilon}\right)\left(\frac{\varepsilon}{\varepsilon'} + \frac{\varepsilon'}{\varepsilon}\sin^2\vartheta\right), \tag{5.64}$$

where $\varepsilon = \hbar\omega/mc^2$; $\varepsilon' = \hbar\omega'/mc^2$. Expressing ε' through the scattering angle ϑ via the relation

$$\varepsilon' = \frac{\varepsilon}{1 + \varepsilon(1 - \cos\vartheta)},$$

representing the Compton effect through the scattered photon energy, we obtain energy distribution of scattered photons:

$$\frac{d\sigma}{d\varepsilon'} = \pi r_0^2 \frac{1}{\varepsilon'^2}\left[\frac{\varepsilon}{\varepsilon'} + \frac{\varepsilon'}{\varepsilon} + \left(\frac{1}{\varepsilon'} - \frac{1}{\varepsilon}\right)^2 - 2\left(\frac{1}{\varepsilon'} - \frac{1}{\varepsilon}\right)\right]. \tag{5.65}$$

After some manipulations and passing to a special *cascade unite* of length, one can reduce Eq. (5.62) to the differential cross section of Compton scattering with energy change $E \to E'$ per c.u.

$$w(E \to E')dE' = \frac{g}{E}\frac{dE'}{E'}\left[1 + \left(\frac{E'}{E}\right)^2\right], \tag{5.66}$$

where g is a material constant (= 1.53 MeV for C, 0.45 MeV for Fe, 0.175 MeV for Pb, 1.32 MeV for air).

In the nonrelativistic region, the cross section

$$\sigma = \frac{8\pi}{3} r_0^2 (1 - 2\varepsilon + \ldots)$$

coincides with (5.63). When $\varepsilon \gg mc^2$

$$\sigma = \frac{\pi r_0^2}{\varepsilon} (\log 2\varepsilon + 1/2),$$

i.e., in the high-energy region, it decreases almost inversely proportional to the photon energy.

The scattering cross section per atom is usually obtained, multiplying the expression (5.64) by the number Z of electrons in the atom. It should be noted, however, that during derivation of these formulas the electron-nucleus connection was neglected, so they are applicable if the energy transferred in the scattering is much more than the binding energy. For this, energy of incident photon and its scattering angle should be large enough.

An ingenious approach to the bremsstrahlung interpretation was suggested by Landau in his article [7]. Instead of considering the coordinate system in which the nucleus is at rest and the electron flies at near-light speed, Landau took the coordinate system in which the electron is approximately at rest, so that the uncertainty of its velocity Δv is small compared to c and is related to the uncertainty of the impact parameter $\Delta\rho$ (the same in both coordinate systems):

$$\Delta\rho\Delta v \sim \hbar/m.$$

The interaction time τ, in the rest system being $\sim \rho/c$, becomes shorter:

$$\tau' \sim \rho/c\varepsilon, \tag{5.67}$$

where

$$\varepsilon = \frac{E}{mc^2} = \frac{1}{\sqrt{1 - (v/c)^2}}.$$

The frequency, relating to this time,

$$\omega = c\varepsilon/\rho. \tag{5.68}$$

In a moving system, the electron is under the influence of a plane wave with frequency (5.68), intensity

$$E' = \frac{Ze}{\rho^2}\varepsilon$$

and action duration (5.67). Thus, the emission of electromagnetic radiation observed in the rest system can be interpreted as the scattering of this wave by the electron. It gives us the reason for the KNT formula application. Under the conditions $\hbar\omega \gg mc^2$

(when $\rho \ll \frac{\hbar}{mc}\varepsilon$) the total KNT cross section becomes

$$\sigma = \left(\frac{e^2}{mc^2}\right)^2 \frac{\ln(\hbar\omega/mc^2)}{(\hbar\omega/mc^2)},$$

and for total number of scattered photons, we get

$$\Delta N \sim \frac{Z^2 e^6}{\hbar^2 m^2 c^4 \varepsilon\rho} \ln\frac{\hbar\varepsilon}{mc\rho}.$$

The mean energy radiated

$$\Delta E \sim \frac{Z^2 e^6}{\hbar^2 c^2 \rho} \ln(\hbar\varepsilon/mc\rho).$$

Along with the direct Compton process considered above, we meet in the modern astrophysics the inverse Compton process: the scattering of electrons in a photon field. It was considered as a mechanism for the loss of energy of cosmic ray electrons and was investigated in this context by many authors in 1950–1960 years.

A new role for this process as a source of high-energy photons was suggested by [8–11].

Most of the calculations of photon spectra have been based on a rather simple approximation. It has been noted that the average energy transferred to a photon in a Compton collision is proportional to the initial energy of the photon and the square of the electron energy. The latter dependence is reminiscent of the synchrotron process and for this reason the radiated photon spectra for a single electron energy is approximated by a delta function spike at the average radiated energy. This spectrum was then folded into the distribution of electron energies to produce the resultant photon spectrum.

5.5 Photo Absorption and Pair Production

In addition to the scattering considered above, the interaction of photons with matter can lead both to their traceless disappearance and to new particles creation. It is easy to show that the absorption of a photon (as well as its emission) by a free electron is impossible, because it conflicts with the conservation laws. To verify this statement, let's take its scheme $\gamma + e \rightarrow e$ and write both conservation laws assuming that before collision the electron is at rest: for the energy

$$\hbar\omega + mc^2 = \frac{mc^2}{\sqrt{1-(v/c)^2}},$$

and for the momentum

$$q = p'.$$

Observe, left-hand sides of both these equations relate to the system before collision, the right sides—after it, so

$$q = \hbar\omega/c, \qquad p = mv/\sqrt{1-(v/c)^2}.$$

From the energy conservation law, we find

$$\frac{\hbar\omega}{mc^2} = \frac{1}{\sqrt{1-(v/c)^2}} - 1,$$

and from the momentum conservation law we see

$$\frac{\hbar\omega}{mc^2} = \frac{v/c}{\sqrt{1-(v/c)^2}}.$$

The left sides of these equalities are the same. Equating the right sides, we find the possible values of $\beta = v/c$:

$$\frac{1}{\sqrt{1-\beta^2}} - 1 = \frac{\beta}{\sqrt{1-\beta^2}},$$

or

$$1 - \frac{1}{\sqrt{1-\beta^2}} = \beta, \quad 1-\beta^2 = (1-\beta)^2.$$

This equation has two solutions: $\beta = 0$, which corresponds to the trivial $\hbar\omega = 0$; and $\beta = 1$, impossible for particles with non-zero rest mass.

If a photon interacts with an electron, which is bound to the atom, then the process of absorption is possible. At the same time, the energy photon goes to detach the electron from the atom and to give him kinetic energy:

$$\hbar\omega = I + T_e', \tag{5.69}$$

where I is the ionization potential. The weaker the connection electron with an atom, the less likely photoabsorption, so the most probable is tearing out electrons from the lowest shells. With growth energy of the photon, the cross section of the photoelectric effect should decrease, since the influence of the bonding forces decreases in this case. Similarly, as Z increases, the cross section must increase. Quantum calculation confirms the correctness of these qualitative reasoning.

Let us consider of the hydrogen atom [12]. Initially, the atomic electron is in state with wave function

$$\psi_n = \frac{1}{\sqrt{\pi a^3}} e^{-r/a}, \tag{5.70}$$

where $a = a_0 Z$, $a_0 = \hbar^2/me^2$ is the Bohr radius. As a result of the photoelectric effect, the electron goes into a state with continuous spectrum and leaves the atom. The photon disappears and the electron leaves the atom. If the energy of the emitted electron is large compared to the ionization potential I:

$$\frac{mv^2}{2} \gg I = \frac{me^4 Z^2}{2\hbar^2}$$

or

$$\xi = \frac{Ze^2}{\hbar v} \ll 1,$$

this interaction can be ignored and chosen wave function of the final state of the electron has the form

$$\psi_{n'} = e^{i\mathbf{k}'\mathbf{r}}. \tag{5.71}$$

Substituting (5.70) and (5.71) into the matrix element (5.31) and using the Hermitianity of the momentum operator

$$\langle \ldots \rangle = -(e/m)\sqrt{2\hbar/\omega a^3} \int e^{-r/a} \boldsymbol{\eta} \hat{\mathbf{p}} e^{i\mathbf{K}\mathbf{r}} d\mathbf{r}, \tag{5.72}$$

where $\mathbf{K} = \mathbf{q} - \mathbf{k}'$. Obviously

$$\boldsymbol{\eta} \hat{\mathbf{p}} e^{i\mathbf{K}\mathbf{r}} = \hbar \boldsymbol{\eta} \mathbf{K} e^{i\mathbf{K}\mathbf{r}} = -\boldsymbol{\eta} \mathbf{p} e^{i\mathbf{K}\mathbf{r}}, \quad \mathbf{p} = \hbar \mathbf{K},$$

since the polarization vector $\boldsymbol{\eta}$ is perpendicular wave vector $\mathbf{q}$. With this in mind, (5.72) takes the form

$$\langle \ldots \rangle = (e/m)\sqrt{2\hbar/\omega a^3} (\boldsymbol{\eta}\mathbf{p}) \int e^{-(r/a - i\mathbf{K}\mathbf{r})} d\mathbf{r}. \tag{5.73}$$

To calculate the integral, we choose a spherical coordinate system with the Oz-axis directed along $\mathbf{K}$. Integrating first over the angles and then over the radius, we get

$$2\pi \iint e^{-(1/a - iK\cos\Theta)r} r^2 dr d\cos\Theta = \frac{4\pi a^3}{[1 + a^2 K^2]^2}.$$

Thus, instead of (5.73) we have

$$\langle \ldots \rangle = 8\pi (e/m)\sqrt{2\pi a^3/\omega} (\boldsymbol{\eta}\mathbf{p}) [1 + a^2 K^2]^{-2}.$$

Considering that the final state of the system is completely determined by the momentum p of a free electron, write the density of states

$$\rho(E) = \frac{mpd\Omega}{(2\pi\hbar)^3}.$$

Taking into account that in the chosen normalization the incident photon flux density is c, we arrive at the following expression for differential cross section of the photoelectric effect:

$$d\sigma = \frac{2\pi}{\hbar c}\frac{128\pi^2(e/m)^2\hbar a^3/\omega}{(1+a^2K^2)^4}(\boldsymbol{\eta}\mathbf{p})^2\frac{mpd\boldsymbol{\Omega}}{(2\pi\hbar)^3}. \tag{5.74}$$

Let's analyze the result obtained by entering spherical coordinate system with the Oz-axis along the direction primary photon momentum $\mathbf{q}$ and the Ox-axis—along polarization vector $\boldsymbol{\eta}$. In this case

$$\hbar^2K^2 = |\hbar\mathbf{q} - \mathbf{p}|^2 = \hbar^2q^2 - 2\hbar qp\cos\vartheta + p^2$$

and

$$\boldsymbol{\eta}\mathbf{p} = p\sin\vartheta\cos\varphi.$$

$$1 + a^2K^2 = \frac{a^2}{\hbar^2}\left[\frac{\hbar^2}{a^2} + \hbar^2K^2\right]$$

$$= \frac{a^2}{\hbar^2}\left[2m\left(\frac{\hbar^2}{2ma^2} + \frac{p^2}{2m}\right) + \hbar^2q^2 - 2\hbar qp\cos\vartheta\right]. \tag{5.75}$$

Because

$$\frac{\hbar^2}{2ma^2} = \frac{me^4Z^2}{2\hbar^2} = I,$$

then, according to the law of conservation of energy (5.69), the content parentheses in (5.75) equals $\hbar\omega$. So

$$1 + a^2K^2 = \frac{2\hbar^2\omega}{me^4Z}\left(1 - \frac{p}{mc}\cos\vartheta + \frac{\hbar\omega}{2mc^2}\right).$$

In the pump-perlative case under consideration, the last term can be neglected in comparison with one, and replace p with mv:

$$1 + a^2K^2 = \frac{2\hbar^3\omega}{me^4Z^2}(1 - v\cos\vartheta/c). \tag{5.76}$$

Since, on the other hand, the kinetic energy ejected electron is much larger than the potential ionization, we replace $p^2/2m$ by $p^2/2m + I = \hbar\omega$ everywhere. As a result of these simplifications, we get

$$\frac{d\sigma}{d\boldsymbol{\Omega}} = 4\sqrt{2}r_0^2 a^4 Z^5 \left(\frac{mc^2}{\hbar\omega}\right)^{7/2} \frac{\sin^2\vartheta\cos^2\varphi}{\left(1 - \frac{v}{c}\cos\vartheta\right)^4}, \tag{5.77}$$

where $r_0 = e^2/mc^2$ is the classical electron radius; $a = e^2/\hbar c$ is the fine structure constant. From this formula, it follows that in the direction of the wave vector of the incident photon no electrons are emitted, maximum emitted electrons fall on the photon polarization direction ($\theta = \pi/2, \varphi = 0$). With an increase in the photon energy this maximum shifts forward. The total cross section should be integrated over directions and multiplied by 2 (the number of electrons in this shell).

Observe that the term $\frac{v}{c}\cos\vartheta$ in the denominator gives after integration, the correction $\sim (v/c)^2$, which can be neglected:

$$\sigma = \frac{32\sqrt{2}}{3}\pi r_0^2 a^4 Z^5 \left(\frac{mc^2}{\hbar\omega}\right)^{7/2}. \tag{5.78}$$

It can be seen that as Z increases, the cross section of the photoelectric effect increases $\sim Z^5$ and decreases with increasing photon energy $\hbar\omega \sim (\hbar\omega)^{-7/2}$, so the photoelectric effect is most significant at low photon energies in heavy substances. The contribution to the photoelectric effect of the remaining shells of the atom usually is taken into account by means of additional factor equal to 5/4.

In conclusion, we note that the reaction of the photoelectric effect has a threshold character: according to (5.69), this process becomes possible only when $\hbar\omega > I$. During the transition $\hbar\omega$ in terms of $I_K, I_L, I_M, \ldots$photoelectric effect cross section increases abruptly, then immediately continues to decrease.

The similar (threshold) nature is revealed by another event also ending with the photon disappearance. If the photon energy $\hbar\omega$ exceeds $2mc^2 \approx 1.02$ MeV, the energy and charge conservation laws allow the process of converting a photon into an *electron–positron pair*:

$$\gamma \to e^+ + e^-.$$

$$E = \pm\sqrt{m^2c^4 + c^2p^2}.$$

But for its implementation, it is also necessary to fulfill the law of conservation of momentum. A detailed analysis of this event is possible only within the framework of quantum electrodynamics.

Recall that the Dirac equation , which describes the motion of relativistic electrons, has eigenfunctions that correspond not only to positive, but also to negative energy values:

$$E = \pm\sqrt{(mc^2)^2 + (cp)^2}.$$

To explain why an ordinary electron (with positive energy) does not go into a state with negative energy, it was hypothesized that all negative states in vacuum are filled with electrons. In accordance with the Pauli principle, the transition to these states is disabled.

From this point of view, the mechanism of pair formation is similar to the mechanism of the photoelectric effect: a photon translates electron from a state with negative energy to state of positive energy. Arising from this unfilled negative state ("hole") is interpreted as a positron.

In the ultra-relativistic case, the eigenvalue functions describing negative states (positrons) differ little from the eigenfunctions, corresponding to positive states, since the influence of the Coulomb field of the nucleus is small here. Neglecting this difference, we can consider the formation of electron–positron pairs as a process inverse bremsstrahlung: an electron passes from one state with positive energy into another, absorbing at this photon. According to the principle of detailed balance, the cross sections of the forward and reverse processes can differ only due to the density of the final states. In the case of bremsstrahlung in the final state is an electron and a photon, and in the case of the formation of pairs—electron and positron. Densities of positron and photon states at high energies are approximately equal, so the formation cross section pair but coincides in order of magnitude with the cross section bremsstrahlung integrated over energies quanta of the order of electron energy [13].

As in the case of bremsstrahlung, the direction impulses of secondary particles (electron and positron) in the case of a high-energy photon $\hbar\omega \gg mc^2$ lie in narrow cone near the primary direction with an angle solution $\sim mc^2/\hbar\omega$. Denote by E_- the kinetic electron energy, and through $E_+ = \hbar\omega - E_- - 2mc^2$—the kinetic energy of the positron. In the ultrarelativistic case, the differential in E_- section of the formation steam looks like

$$d\sigma = 4ar_0^2 Z^2 \frac{dE_-}{\hbar\omega} G\left(\hbar\omega, \frac{E_- + mc^2}{\hbar\omega}\right), \tag{5.79}$$

where $G(u, v)$ is a slowly varying function arguments $u = \hbar\omega, v = \frac{E_- + mc^2}{\hbar\omega}$ depending on the screening parameter. In the absence of screening ($\xi \gg 1$)

$$G(u, v) = \left[v^2 + (1 - v)^2 + \frac{2}{3}v(1 - v)\right]\left[\ln \frac{2u}{mc^2}v(1 - v) - \frac{1}{2}\right],$$

in case of total screening ($\xi \approx 0$)

$$G(u, v) = \left[v^2 + (1 - v)^2 + \frac{2}{3}v(1 - v)\right] \log\left(191Z^{-1/3} - \frac{1}{9}v(1 - v)\right).$$

The function G is symmetrical with respect to permutation $v, (1-v)$, therefore $d\sigma/dE_-dE_+$ is symmetrical relatively electron and positron energies.

Differential cross-sectional integration (5.79) along E_- up to $\hbar\omega - 2mc^2$ gives the total cross section of the formation steam. In the limiting cases of low- and high-energy photons (corresponding to the absence of screening and full screening), we get [13]

$$\sigma = 4ar_0^2Z^2\left[\frac{7}{9}\ln\frac{2\hbar\omega}{mc^2} - \frac{109}{54}\right], \qquad mc^2 \ll \hbar\omega \ll 137mc^2Z^{-1/3}; \tag{5.80}$$

$$\sigma = 4ar_0^2Z^2\left[\frac{7}{9}\ln(191Z^{-1/3}) - \frac{1}{54}\right], \qquad \hbar\omega \gg 137mc^2Z^{-1/3}. \tag{5.81}$$

From (5.81), it follows that the total pair production cross section at a high photon energy is constant for given substances. Bremsstrahlung (5.3) and the production of electron–positron pairs play an important role in the process of passage of electrons or photons of high energies through matter: they give rise to electron–photon cascades (EPC).

5.6 Continuous Energy Loss Model

Unlike photons, which can cross millions of atoms without interacting with them, and whose trajectories are broken lines consisting of straight free paths, charged particles (both electrons and heavier particles—muons, protons, ions, and nuclei) make frequent collisions with atoms of the medium, accompanied by very small energy losses, mainly for ionization and excitation of atoms. The deviations from the initial directions arising in these processes, as well as displacements relative to the axis specified by the primary momentum of the particle, are also small. These features provide the conditions for the applicability of the central limit theorem, which reduces the computational problem to the determination of two parameters—the mean value and the variance (or the correlation coefficient when studying joint distributions).

Let's start with consideration of the energy losses. A particle with an electric charge Z_1e and a kinetic energy E much greater than the binding energy of an electron with a nucleus transmits to it the momentum $p = 2Z_1e^2/\rho v$ and the corresponding amount of kinetic energy transmitted

$$Q(\rho) = \frac{p^2}{2m} = \frac{2Z_1^2e^4}{mv^2\rho^2}, \tag{5.82}$$

where m is the electron mass and ρ is the impact parameter.

The transferred energy is inversely proportional to the square of the impact parameter and tends to infinity as $\rho \to 0$. However, when for small ρ, the formula (5.82) is not valid, since for its derivation, it was assumed that the trajectory of the incident particle is rectilinear, and the particle-target in the process collision does not move. The minimum value of the impact parameter, at which the formula (5.82) is still valid, can be found by equating Q to the maximum possible value transmitted energy. If the incident particle is not an electron, then $m_1 \gg m_2$ and

$$Q_{\max} \approx 2m_2c^2 \frac{(v/c)^2}{1-(v/c)^2},$$

therefore

$$\rho_{\min} = \frac{Ze^2}{mv^2}\sqrt{1-(v/c)^2}.$$

The expression (5.82) also needs to be corrected for large ρ. In this case, the interaction time

$$\tau = \frac{\rho}{v}\sqrt{1-(v/c)^2}$$

is so large, that the electron can no longer be considered as an immobile particle and the action of an incident particle will lead to adiabatic change in the atomic electron motion without resulting energy transfer. Let ω be the average frequency of rotation of electrons in an atom. Equating it to $1/\tau = \frac{v}{\rho}[1-(v/c)^2]^{-1/2}$, we obtain the estimate of the maximum value of the impact parameter.

$$\rho_{\max} = v/\omega\sqrt{1-(v/c)^2}. \tag{5.83}$$

At $\rho < \rho_{\min}$, $Q \to Q_{\max}$, and if $\rho > \rho_{\max}$, the energy losses quickly decrease to zero. The mean value of the energy loss per unite path is determined by expression

$$\beta = -\frac{dE}{dz} = n_e \int Q \frac{d\sigma}{dQ} dQ, \tag{5.84}$$

where $n_e = Zn$ is the atomic electron number per matter unit. Passing to integration over ρ and taking into account that

$$\frac{d\sigma}{dQ} = \frac{d\sigma}{d\rho} / \left|\frac{d\sigma}{d\rho}\right| = 2\pi\rho / \left|\frac{dQ}{d\rho}\right|,$$

we have

$$\beta = 2\pi Zn \int_{\rho_{\min}}^{\rho_{\max}} Q(\rho)\rho d\rho.$$

Continuing by inserting here (5.82) and integrating, we arrive at the *Bohr formula*

$$\beta = 4\pi n Z \frac{Z_1^2 e^4}{mv^2} \ln B, \tag{5.85}$$

where

$$B = \frac{\rho_{\max}}{\rho_{\min}} = \frac{mv^3}{\omega Z_1 e^2} \cdot \frac{1}{1-(v/c)^2}; \tag{5.86}$$

m and e are electron mass and charge. It describes with a good degree of accuracy the loss of alpha particles and heavy nuclei. For electrons, mesons and protons, it overestimates the energy losses because of quantum mechanical effects, in particular, the uncertainty principle, which changes a value $\rho_{\min}$ in the formula (5.86).

The quantum theory considers particles as, in some sense, "smeared" in a domain with dimensions of order of De Broglie wavelength $\lambda = 2\pi\hbar/p$, so one can expect that the minimum value of the impact parameter will be determined by the relation $\rho_{\min} \sim \frac{\hbar}{\rho}$. For a given speed of relative motion of two particles, the light particle momentum will be smaller and the uncertainty in their relative positions will be determined mainly by this particle. When colliding a heavy particle with an atom, the light particle is the atomic electron. Substituting instead of p its relativistic counterpart

$$p = mv/\sqrt{1-(v/c)^2},$$

we get

$$\rho_{\min} = \frac{\hbar}{mv}\sqrt{1-(v/c)^2},$$

therefore

$$B = mv^2\hbar\omega\sqrt{1-(v/c)^2}. \tag{5.87}$$

Observe that parameters $\rho_{\min}$ and $\rho_{\max}$ are in (5.85) under the logarithm sign, so some inaccuracies in their definition will have little effect on the final result. Interrelation (5.85) with quantum mechanical the value of B given by Eq. (5.87) is a good approximation to the *Bethe–Bloch* formula obtained within the framework of quantum mechanics and taking into account relativistic effects:

$$-\frac{dE}{dx} = \frac{4\pi Z_1^2 e^4}{mv^2} nZ \left\{ \ln\frac{2mv^2}{I} - \ln\left(1-\frac{v^2}{c^2}\right) - \frac{v^2}{c^2} \right\}, \tag{5.88}$$

with I being an averaged atomic ionization potential. Equations (5.85) and (5.88) show that ionization losses first decrease with increasing energy and then increase due to the relativistic term $\ln(1-(v/c)^2)$. Physically, this increase is due to the fact that in relativistic energy region, the maximum energy increases, which particle can transfer in a collision. In addition, as v increases, the maximum is the value of the impact parameter, on which energy is transferred, which in turn with Lorentz contraction of the electric field moving particle, due to which it becomes possible energy transfer at large ρ. Ultimately ionization losses with increasing energy will tend to saturation, and the maximum loss will depend on the density of the medium (*density effect*).

If the incident particle is an electron, an essential role in energy losses will be played by influence of spin and exchange forces. The Bethe–Bloch formula in this case takes the form

$$-\left(\frac{dE}{dx}\right)_{\text{ion}} = \frac{2\pi n r_0^2 mc^2 Z}{(v/c)^2}\left\{\log\left[\frac{\varepsilon^2(\varepsilon+2)}{2(I/mc^2)^2}\right] + F(\varepsilon) - \delta\right\}; \qquad (5.89)$$

$$F(\varepsilon) = 1 - (v/c)^2 + \left[\frac{\varepsilon^2}{8} - (2\varepsilon+1)\ln 2\right]/(\varepsilon+1)^2,$$

where ε is the kinetic energy (in units mc^2), δ is the density effect correction, and

$$r_0 = \frac{e^2}{mc^2}$$

is a fundamental constant called the *electron classical radius*.

In the case of light particles (electrons) passing through matter, an important competition to the ionization loss mechanism is its alternative analog, the bremsstrahlung already mentioned above, whose contribution increases with increasing particle energy, decreasing its mass, and increasing Z of matter. The energy radiative loss per unit path is

$$\beta_{\text{rad}} = n\int_0^{\omega_{\max}} \hbar\omega\frac{d\sigma}{d\omega}d\omega,$$

where $\omega_{\max} = E/\hbar$. Referring to section with bremsstrahlung of charged particles, we get

$$\beta_{\text{rad}} = \frac{16}{3}n\left(\frac{Z^2e^2}{\hbar c}\right)\left(\frac{Z_1^2e^2}{m_1c^2}\right)^2\frac{c^2}{v^2}\int_0^{E/\hbar}\hbar\omega\ln\left[\frac{(\sqrt{E}+\sqrt{E-\hbar\omega})^2}{\hbar\omega}\right]d\omega.$$

Observe that, we omitted the coefficient 1/2 under the sign of the logarithm, since it is not included in the more precise expression (5.50). Making the change of variables $x = \hbar\omega/E$ and calculating integral

$$\int_0^{E/\hbar}\hbar\ln\left[\frac{(\sqrt{E}+\sqrt{E-\hbar\omega})^2}{\hbar\omega}\right]d\omega = 2E\int_0^1\ln\left[\frac{1+\sqrt{1-x}}{\sqrt{x}}\right]dx = 2E,$$

we get the next formula for radiative losses:

$$\beta_{\text{rad}} = \frac{16}{3} n \left(\frac{Z^2 e^2}{\hbar c} \right) \left(\frac{Z_1^4 e^4}{m_1 c^2} \right).$$

The ratio of radiation losses to ionization ones (5.85) equals

$$\beta_{\text{rad}} / \beta_{\text{ion}} = \frac{4}{3n} Z_1^2 Z \alpha \frac{m}{m_1} \left(\frac{v}{c} \right)^2 \frac{1}{\ln B}.$$

Note that in this approximation β_{rad} does not depend on energy, and the ratio $\beta_{\text{rad}} / \beta_{\text{ion}}$ increases with increasing velocity of the primary particle, and for particles with a mass m_1 it is m_1 / m times less than that for electrons.

In the framework of the quantum relativistic theory (see (5.3)), the energy loss can only be found by numerical integration of the relevant cross section [13]. These calculations show that radiation losses are growing with energy is almost linear, while ionization first decreases and then increases in a logarithmic manner. Therefore, at low energies, ionization processes predominate losses, and at large—radiation. An energy, at which these losses become equal, is called the *critical energy*. For various substances, it can be calculated using the formula

$$E_{\text{cr}} = \frac{800}{Z + 1.2} \text{ MeV}.$$

Linear growth of radiation losses with energy means that in this region of electron energies, where radiation losses are predominant, the electron energy decreases exponentially law. Therefore, the distance at which the energy decreases by a factor of e is a convenient unit of length. It is called the *radiation unit*.

Significant excess of the critical energy means that we are entering the region of composite processes, including elementary bremsstrahlung events, intermittent with *electron–positron pairs production*. Following one after another, such events form a more complicated phenomena called the *electron–photon cascades* (EPC). To complete the EPC description, we must add one more process to the cascade scheme—the *Compton scattering*, in which the disappearance of a photon is compensated by the formation of a fast electron and a new photon (in other words, the incident photon scattering on electron sets is in motion). We consider this event sequence in the next section, ending this one with some conclusion about stochastic nature of the ionization losses of electron between radiation events.

5.7 Electron Multiple Scattering

5.7.1 Energy Losses and Straggling (After Bohr)

Considering the high-energy photon motion on macroscopic scales, we may interpret its trajectory as a broken line consisting of rectilinear segments of random length connecting the points of photon scattering on the atoms of the medium. The entire modern theory of gamma radiation transfer, also called the theory of multiple scattering, is based on this assumption. The cross sections for the interaction of photons with atoms are small, so the lengths of their free paths are many orders of magnitude greater than the interatomic distances. The situation is different in the case of electrons or other charged particles: their free paths are commensurate with these distances, but each scattering only slightly changes the direction of particle motion. On a macroscopic scale (for example, in a cloud chamber or photographic emulsion), we will see a continuous uneven curve, the tangents to which at close points will deviate by small angles, which also indicates low energy losses at these points. One can see here an analogy with the motion of a solid particle in a liquid, when its collisions with atoms become indistinguishable, and one can assume the hydrodynamic approach, called here the slowing-down approximation.

Let us denote by $Q(l)$ the energy loss of a particle that has traveled the path l in the substance, and by ΔQ the energy loss at Δl, so that

$$Q(l + \Delta l) = Q(l) + \Delta Q. \tag{5.90}$$

Due to the random nature of the interaction of particles with atoms of matter, the quantities included in (5.90) are random. Averaging this relation

$$\overline{Q}(l + \Delta l) = \overline{Q}(l) + \overline{\Delta Q} \tag{5.91}$$

leads to equation for average losses

$$\frac{d\overline{Q}(l)}{dl} = \beta,$$

where

$$\beta = \frac{\overline{\Delta Q}}{\Delta l} = n \int Q d\sigma(Q) \tag{5.92}$$

stands for the average energy loss per unit path length. The value of β depends on the energy of the particle at the point l, which is approximately equal to $E_0 - \overline{Q}$. In case of small change of β in the interval under consideration,

$$\overline{Q}(l) = \int_0^l \beta(E(l'))dl' = \beta(E_0)l. \tag{5.93}$$

The variance of the random variable $Q(z)$ due to fluctuations in elementary collisions is determined by the difference

$$\sigma_Q^2(l) = \overline{Q^2}(l) - \overline{Q}^2(l). \tag{5.94}$$

Squaring both parts of equality (5.90) and averaging, we get

$$\overline{Q^2}(l + \Delta l) = \overline{Q^2}(l) + 2\overline{Q(l)\Delta Q} + \overline{\Delta Q^2}. \tag{5.95}$$

If the scattering cross section, characterizing random variable ΔQ, changes little in $(E_0 - Q, E_0)$, then the quantities $Q(l)$ and ΔQ can be considered statistically independent and put

$$\overline{Q(l)\Delta Q} = \overline{Q(l)}\ \overline{\Delta Q}. \tag{5.96}$$

Substituting (5.96) into (5.95) and subtracting the squared equation (5.91), according to (5.94), we get

$$\sigma_Q^2(l + \Delta l) = \sigma_Q^2(l) + \overline{\Delta Q^2} - \overline{\Delta Q}^2 \cong \gamma \Delta l - (\beta \Delta l)^2.$$

Here

$$\gamma = n \int Q^2 d\sigma(Q)$$

is a mean square of energy losses per unit length of path. Neglecting the change of γ in $(E_0, E_0 - \overline{Q})$ leads to

$$\sigma_Q^2 \simeq \gamma l. \tag{5.97}$$

For a heavy ($m \gg m_e$) particle $\gamma = 4\pi n Z_1^1 Z^2 e^4$ (Bohr), a more accurate formula is derived by Bethe–Livingston (see [14]).

In accordance with the basic laws of probability theory, the distribution of a random variable Q, consisting of a large number of statistically independent, in this case, small term ΔQ, having a finite variance, tends to normal (Gaussian distribution), the average value of which is given by formula (5.93), and the variance is given by formula (5.97). Fluctuations in energy loss are closely related to fluctuations in the path of a particle (so-called *straggling*). Consider a random trajectory in range-energy coordinates, assuming that $z = z(E)$ is a differentiable function, so that the length of path Δx, passed by the particle while its energy belongs to interval ΔE, can be determined via relation

$$\Delta x = -\frac{dl}{dE}\Delta E = -\frac{\Delta E}{dE/dl}. \tag{5.98}$$

Referring to the evident fact, that $E(l) = E_0 - Q(l)$, one can interpret Eq. (5.98) rewritten in the form

$$\Delta l = \frac{\Delta E}{\Delta Q/\Delta l}, \tag{5.99}$$

as a link between two random variables Δl and ΔQ. Introducing

$$\delta Q = \Delta Q - \overline{\Delta Q} \tag{5.100}$$

and letting the relative fluctuations $\delta Q/\overline{\Delta Q}$ be small, represent (5.99) in the form

$$\Delta l = \frac{\Delta E}{\beta}\left(1 - \frac{\delta Q}{\overline{\Delta Q}}\right). \tag{5.101}$$

Taking into account Eq. (5.100), according to which $\overline{\delta Q} = 0$, we arrive at the formula

$$\overline{\Delta l} = \frac{\Delta E}{\beta}.$$

The second moment of random variable (5.101) is of the form

$$\overline{\Delta l^2} = \frac{(\Delta E)^2}{\beta^2}\left(1 + \frac{\overline{(\delta Q)^2}}{\overline{\Delta Q}^2}\right),$$

so that the variance of the particle path in ΔE

$$\sigma^2_{\Delta l}(\Delta E) = \overline{(\Delta l)^2} - \overline{\Delta l}^2 = \left(\frac{\Delta E}{\beta}\right)^2 \frac{\overline{\delta Q^2}}{\overline{\Delta Q}^2}.$$

One can see from (5.93), (5.97), and (5.100) that

$$\overline{\delta Q^2} = \sigma_{\Delta Q} = \gamma \Delta l, \quad \overline{\Delta Q}^2 = \beta^3 \Delta l^2,$$

therefore

$$\sigma^2_{\Delta l} = \frac{\gamma}{\beta^3}\Delta E. \tag{5.102}$$

Assuming the traces in non-overlapping energy intervals to be statistically independent, we find the total path R variance as the sum of variances (5.102) over all energy intervals from 0 to E_0:

$$\sigma^2_R = \int\limits_0^{E_0} \frac{\gamma(E)}{\beta^3(E)} dE. \tag{5.103}$$

5.7.2 Fluctuations in Energy Losses (After Landau)

In reality, the elementary charged particle motion is not continuous, but consists of statistically interconnected local events—collisions with atoms of this medium. This process is successfully described by the integro-differential kinetic equation of the Boltzmann type. We show here the Landau solving this equation for fast particles, i.e., particles having a sufficiently high energy to be able to apply the usual theory of ionization. The layer of matter is considered not too thick, so that the total average energy loss is small compared to the initial energy E_0.

The kinetic equation for the probability density $f(l, Q)$ of the random loss energy Q on the path l has the form

$$\partial_l f(l, Q) = \mu \int_0^\infty [f(l, Q - Q') - f(l, Q)]w(Q')dQ', \tag{5.104}$$

with initial condition

$$f(0, Q) = \delta(Q). \tag{5.105}$$

The equation assumes that the sequence of collision points of a particle with atoms of the medium forms a Poisson process with the rate μ, and the energy losses Q' in individual collisions are independent and obey the same probability density w. The upper limit is ∞, because $f(l, Q) = 0$ as $Q < 0$ and $w(Q') = 0$ as $Q' > E_0$. Applying the Laplace transformation

$$f(l, Q) \mapsto \tilde{f}(l, p) = \int_0^\infty f(l, Q)e^{-pQ}dQ \tag{5.106}$$

yields the equation

$$\partial_l \tilde{f}(l, p) = -A(p)\tilde{f}(l, p)$$

with

$$A(p) = \mu[1 - \tilde{w}(p)] = \mu \int_0^\infty [1 - \exp(-pQ)]w(Q)dQ \tag{5.107}$$

and the initial condition $\tilde{f}(0, p) = 1$. Its solution reads

$$\tilde{f}(l, p) = \exp(-lA(p))$$

and produces the original

$$f(l, Q) = \frac{1}{2\pi i} \int_{c-i\infty}^{c+i\infty} \exp\{pQ - lA(p)\}dp. \tag{5.108}$$

The probability density $w(Q)$ quickly decreases with growing Q, and applying the Taylor expansion yields

$$\tilde{w}(p) = 1 - p\beta + (1/2)p^2\gamma + \ldots \tag{5.109}$$

Keeping here only two terms reduces (5.108) to

$$f(l, Q) = \delta(Q - \beta l), \tag{5.110}$$

where $\beta = \mu \int_0^\infty Qw(Q)dQ$ is the mean energy loss per unit length of path. This represents the continuous slowing-down model, relating to the *large number law* in the probability theory: *the arithmetic mean of independent random terms tends to their mathematical expectation* (*if the latter exists*).

Keeping the third term in the above expansion yields the *diffusion* (in energy) *approximation*

$$\begin{aligned} f(l, Q) &= \frac{1}{2\pi} \int_{-\infty}^{\infty} \exp\left\{-\gamma l/2)p^2 + i(Q - \beta l)p\right\} dp \\ &= \frac{1}{\sqrt{2\pi\gamma l}} \exp\{-(Q - \beta l)^2/2\gamma l\}. \end{aligned} \tag{5.111}$$

It is easy to recognize here the second limit law of the probability theory, namely, the *central limit theorem*, which supplements the large number law by indicating the Gaussian nature of the limit sum distribution and introducing the additional limitation: the *existence of variance*.

However, it should be noted that in a wide range of energy lost in one event, its distribution is written in the form that does not satisfy the conditions of the above limiting laws:

$$w(Q) = \frac{a}{Q^2}, \tag{5.112}$$

so, the improper integrals

$$\overline{Q} = \int_0^\infty w(Q)QdQ, \quad \overline{Q^2} = \int_0^\infty w(Q)Q^2dQ,$$

contained in coefficients β and γ divergent, and distributions (5.110)–(5.111) strictly speaking are incompatible with (5.112).

There are two ways to overcome this difficulty. The first is to refine the limits of integration based on physical considerations, the second involves a generalized limit theorem that allows the use of Eq. (5.112) as the main term of the asymptotic expansion for large Q. The first one was used by Landau [15].

Following this way, we come back to Eq. (5.107) and choose such a value Q_1 exceeding the binding energy Q', that for $Q > Q_1$ one can use the Rutherford formula, whereas for $Q < Q_1$ the ansatz $\exp(-pQ) \approx 1 - pQ$ is admissible. After replacing the zero limit by Q', Eq. (5.107) becomes

$$A(p) = p \int_{Q'}^{Q_1} \mu w(Q) Q dQ + \mu a \int_{Q_1}^{\infty} \frac{1 - e^{-pQ}}{Q^2} dQ, \tag{5.113}$$

with $a = 2\pi n_0 Z e^4 / m v^2$. The first integral

$$J_1 = \int_{Q'}^{Q_1} \mu w(Q) Q dQ$$

represents the specific losses due to collisions with small energy transfers ($Q < Q_1$):

$$J_1 = \beta_{(Q<Q_1)} = a \ln(Q_1/Q'),$$

where $\ln Q' = \ln[(I^2(1 - v^2/c^2)/(2mv^2)] + (v/c)^2$ and I denotes the ionization potential of an atom determined in a known way (its value is usually taken equal to $I = Z \times 13.5$ eV). After passage to the new integration variable $x = pQ$ and double integration by parts, the second integral

$$J_2 = \int_{Q_1}^{\infty} \frac{1 - e^{-pQ}}{Q^2} dQ$$

becomes

$$J_2 = p \left\{ \frac{1 - e^{-pQ_1}}{pQ_1} - \ln(pQ_1) e^{-pQ_1} + \int_{pQ_1}^{\infty} e^{-x} \ln x dx \right\}.$$

Referring to the smallness of Q_1, we expand the expression in curly brackets into a series and keep the terms that do not vanish at $Q_1 \to 0$. This reduces J_2 to asymptotic form

$$J_2 \sim p[1 - \ln(pQ_1) - C], \quad Q_1 \to 0,$$

where $C = -\int_0^\infty e^{-x} \ln x dx$ is the Euler constant. Thus,

$$A(p) \sim ap[1 - C - \ln(pQ')],$$

and

$$f(l, Q) \cong \frac{1}{2\pi i} \int_{-i\infty}^{i\infty} e^{p\{Q - al[1 - C - \ln(pQ')]\}} dp.$$

Opening the brackets and regrouping the terms, we rewrite the exponential function argument $\varphi(p)$ as

$$\Phi(l, p) = p[Q - al(1 - C) + al \ln Q'] + pal \ln p. \tag{5.114}$$

By means of using identity $\ln p = \ln(pal) - \ln(al)$, it is reduced to the form

$$\Phi(l, p) = u \ln u + \lambda$$

with $u = pal$, $\lambda = \{Q - al[\ln(al/Q') + 1 - C]\}/\xi$, and $\xi = al$. Consequently, the distribution of electron energy losses in a thin layer $f(l, Q)$ is expressed through the universal function

$$\phi(\lambda) = \frac{1}{2\pi i} \int_{-i\infty}^{i\infty} e^{u \ln u + u\lambda} du \tag{5.115}$$

via the relation

$$f(l, Q) = \frac{1}{\xi} \phi(\lambda),$$

which will be referred to as the *Landau distribution*.

5.7.3 The Link with the Generalized Limit Theorem

To recover the relation of this result to the generalized limit theorem (the *stable law theory*, see, for detail, [16]), let us make the change of integration variable $u = -it$, $\lambda = x$. Using identities

$$\ln(-it) = \ln|t| + i\arg(-it) = -i(\pi/2)\mathrm{sign}t + 2k\pi i,$$

we obtain instead of Eq. (5.108):

$$f(x) = \frac{1}{2\pi} \int_{-\infty}^{\infty} \exp\{-ixt - |t|\theta - it \ln|t|\} dt,$$

where θ is a constant. One should recall that the characteristic function of a stable law with parameters $\alpha \in (0, 2]$, $\beta \in [-1, 1]$ has the form

$$\varphi_{\alpha,\beta}(t) = \exp\left\{-|t|^{\alpha}\left[1 - i\beta\frac{t}{|t|}\tan\frac{\pi\alpha}{2}\right]\right\}, \quad \alpha \neq 1, \tag{5.116}$$

$$\varphi_{1,\beta} = \exp\left\{-|t|\theta - i\beta t \ln|t|\right\}, \quad \theta > 0. \tag{5.117}$$

Designating corresponding cumulative distribution and density functions by $F_{\alpha,\beta}(x)$ we conclude that the Landau distribution is a stable law with index $\alpha = 1$ and parameter $\beta = 1$:

$$f_L(x) = f_{(1,1)}(x). \tag{5.118}$$

The result (5.118) becomes more obvious, if to bear in mind that Eq. (5.104), together with the condition (5.105), describes the distribution of a sum $S_N = \sum_{i=1} Q_i$ of a random number N of independent random numbers Q_i such that N obeys the Poisson distribution with expectation $\bar{N} \equiv \mathsf{M}N = \mu l$, and Q_i possess the same distributions, satisfying the condition:

$$\mathsf{P}(Q_i \geq x) = ax^{-1} + o(x^{-1}), \quad x \to \infty, \tag{5.119}$$

$$\mathsf{P}(Q_i < x) = 0, \quad x < 0. \tag{5.120}$$

This is only a part of the generalized limit theorem, relating to positive summands. A more general case assumes following conditions:

$$\mathsf{P}(Q_i \geq x) = ax^{-\alpha} + o(x^{-\alpha}), \quad x \to \infty, \tag{5.121}$$

$$\mathsf{P}(Q_i < x) = b|x|^{-\alpha} + o(|x|^{-\alpha}), \quad x \to -\infty, \tag{5.122}$$

where $\alpha \in (0, 2]$. Under these conditions, the following theorem holds.

Generalized Limit Theorem. *Let $\xi_1, \xi_2, \ldots, \xi_n, \ldots$ be an infinite sequence of independent random variables with the same distribution, obeying the above conditions, and*

$$S_n = \begin{cases} \sum_{i=1}^{n} \xi_i, & 0 < \alpha \leq 1; \\ \sum_{i=1}^{n} (\xi_i - \bar{\xi}), & 1 < \alpha < 2. \end{cases}$$

Then, for $\alpha \neq 1$

$$\lim_{n\to\infty} \mathsf{P}\left(\frac{S_n}{[nc(\alpha)(a+b)]^{1/\alpha}} < x\right) \Rightarrow F_{\alpha,\beta}(x), \tag{5.123}$$

where $\beta = (a-b)/(a+b)$, $c(\alpha) = \Gamma(1-\alpha)\sin[(1-\alpha)\pi/2]$, and for $\alpha = 1$ there exists a sequence $A_n = (a-b)\ln n + o(\ln n)$, such that

$$\lim_{n\to\infty} \mathsf{P}\left(\frac{S_n}{n(a+b)} - A_n < x\right) \Rightarrow F_{1,\beta}(x). \tag{5.124}$$

In order to extend this result to a sum of a random number of terms, the following lemma was proved in [17].

Lemma. *Let $\tau,\ \tau_2,\ \ldots,\ \tau_n, \ldots$ be a sequence of random variables, independent of each other and independent of ξ_i, having a Poisson distribution with the mathematical expectation $\overline{\tau_n} = n$. Then, the theorem presented above remains valid, if n is replaced by τ_n. To prove this statement it is sufficient to show that for $n \to \infty$*

$$\frac{S_n - S_{\tau_n}}{[nc(\alpha)(a+b)]^{1/\alpha}} \xrightarrow{p} 0 \ \ for \ \ \alpha \neq 1, \tag{5.125}$$

and

$$\frac{S_n - S_{\tau_n}}{n(a+b)} \xrightarrow{p} 0 \ \ for \ \ \alpha = 1. \tag{5.126}$$

We will limit ourselves to proving Eq. (5.126) (the preceding one can be proved in a similar way).

Let us introduce designations $\Delta_n = [n - \sqrt{n}\ln n, n + \sqrt{n}\ln n]$ for the interval indicated and

$$\chi_n(x) = \begin{cases} 1, \ x \in \Delta_n; \\ 0, \ \text{otherwise} \end{cases}$$

for its indicator function. Considering τ_n as a sum of n independent Poisson random quantities having mathematical expectation equal to 1, and using the central limit theorem, we find that $\xi_n(\tau_n) \xrightarrow{p} 1$ for $n \to \infty$. Therefore, Eq. (5.125) will be proved, if we prove that for $n \to \infty$

$$\frac{S_n - S_{\tau_n}}{n(a+b)} \chi_n(\tau_n) \xrightarrow{p} 0. \tag{5.127}$$

For $\tau \in \Delta_n$

$$|S_n - S_{\tau_n}| \leq \tilde{S}_n, \tag{5.128}$$

where

$$\tilde{S}_n = \sum_{i \in \Delta_n} |\xi_i|$$

is the sum of independent random variables $\tilde{\xi}_i = |\xi_i|$ whose cumulative distribution function $\tilde{F}(x)$ satisfies the conditions

$$1 - \tilde{F}(x) = (a+b)x^{-1}, \quad x \to \infty,$$

$$\tilde{F}(x) = 0, \quad x < 0.$$

Table 5.1 Levy stable law with index $\alpha = 1$

x	$F_{1,1}$	$f_{1,1}$	x	$F_{1,1}$	$f_{1,1}$	x	$F_{1,1}$	$f_{1,1}$
−3.0	0.0001	0.0007	2.5	0.6237	0.0882	8.0	0.8535	0.0181
−2.5	0.0020	0.0096	3.0	0.6642	0.0742	8.5	0.8620	0.0166
−2.0	0.0141	0.0440	3.5	0.6984	0.0627	9.0	0.8697	0.0146
−1.5	0.0498	0.1005	4.0	0.7272	0.0533	9.5	0.8767	0.0132
−1.0	0.1136	0.1514	4.5	0.7519	0.0455	10.0	0.8829	0.0120
−0.5	0.1969	0.1773	5.0	0.7730	0.0392	10.5	0.8887	0.0109
0.0	0.2868	0.1788	5.5	0'7913	0.0339	11	0.8939	0.0103
0.5	0.3733	0.1652	6.0	0.8071	0.0295	12	0.9030	0.0084
1.0	0.4510	0.1452	6.5	0.8209	0.0259	13	0.9108	0.0372
1.5	0.5183	0.1242	7.0	0.8331	0.0229	14	0.9175	0.0062
2.0	0.5755	0.1049	7.5	0.8439	0.0203	15	0.9233	0.0054

Let $N \equiv N(\Delta_n)$ be the number of terms in $\tilde{S}_n$ which is equal to a number of integer numbers in Δ_n. According to the above version of the central limit theorem, we have

$$\lim_{n\to\infty} \mathsf{P}\left(\frac{\tilde{S}_n}{(a+b)N} - A_N < x\right) \Rightarrow F_{1,\beta}(x),$$

where $A_N = (a+b)[(1/2)\ln n + o(\ln n)], \quad n \to \infty$. Thus, for $n \to \infty$

$$\frac{\tilde{S}_n}{n(a+b)} = \frac{N\tilde{S}_n}{n(a+b)N} - \frac{NA_N}{n} + \frac{NA_N}{n} = \frac{N}{n}\left[\frac{\tilde{S}_n}{(a+b)N} - A_N\right] + \frac{NA_N}{n} \xrightarrow{P} 0.$$

Together with Eq. (5.128), this proves the assertion (5.126), and consequently the lemma itself.

Thus, as applied to the ionization losses problem considered, we obtain

$$\mathsf{P}\left(\frac{Q}{a\mu l} - a\ln(\mu l) < x\right) \Rightarrow F_{1,1}(x), \quad \mu l \to \infty.$$

The numerical values of the cumulative distribution function $F_{1,1}(x)$ and the probability density $f_{1,1}$ calculated in [18] are presented in Table 5.1.

The asymmetric form of the one-sided Levy distribution and its infinite (including even the first) moments significantly distinguished it from the expected Gaussian distribution. Uchaikin and Topchii [19] represent an attempt to bridge the gap between the Landau spectrum and the Gaussian limit. In order to go beyond the diffusion approximation, Vavilov included a next (third-order) term in the expansion (5.109), dropped all subsequent terms, but unlike Landau, did not ignore the finiteness of the upper limit. Although Vavilov's ansatz, involving the Airy function, recovered a trend toward the Gaussian shape with increasing thickness, significant skewness was retained at the largest thickness. The further development of the theory includes complications of kinetic equations, consideration of other elementary processes, and application of numerical methods. Some of these aspects will be discussed later in connection with cascade phenomena.

5.7.4 The Angular Distribution (After Fermi)

Now we consider the angular distribution of a charged particle that has traveled the path l as a result of a large number of small deviations in the course of consequent collisions with uniformly distributed atoms of medium. The particle energy is supposed to be high enough and the layer is sufficiently thin in order to neglect the difference between z-coordinate along the initial direction of the particle and the path length l. The motion direction of the particle traveled the layer thickness of z characterized by the polar angle $\vartheta(z)$ (with respect to z-axis), axial distribution is assumed to be uniform in $(0, \pi)$.

Reducing the well-known formula of spherical trigonometry

$$\cos\vartheta(z+\Delta z) = \cos\vartheta(z)\cos\Theta + \sin\vartheta(z)\sin\Theta\cos\chi$$

to small-angle approximation yields

$$\vartheta^2(z+\Delta z) = \vartheta^2(z) + \Theta^2 + 2\vartheta(z)\Theta\cos\chi\,.$$

Averaging this equation and taking into account, that χ is independent of scattering angles Θ and χ, we get

$$\overline{\vartheta^2}(z+\Delta z) = \overline{\vartheta^2}(z) + \langle\Theta^2\rangle,$$

whence

$$\frac{d\overline{\vartheta^2}(z)}{dz} = \langle\Theta^2\rangle = \sigma_{\text{scatt}}\langle\theta^2\rangle \tag{5.129}$$

with $\langle\Theta^2\rangle$ being the mean-square scattering angle per unit of length $\langle\theta^2\rangle$ per one scattering and σ is the total cross section of this process. For the Rutherford scattering averaging integral over θ diverges and needs in limiting the integration domain to some finite interval. Making this from physical reasons yields for relativistic particle with momentum p_0 and velocity v_0

$$\langle\Theta^2\rangle = \frac{1}{X_s}\left(\frac{E_s}{pv}\right)^2 \tag{5.130}$$

where X_s is close to the radiation length X_0 and E_s (denoted often by K in the cascade theory) to the critical energy $K = 21 \cdot Z_1$ MeV. Thus, from (5.129) follows

$$\overline{\vartheta^2}(z) = \left(\frac{E_s}{p_0 v_0} K\right)^2 t,$$

where $t = z/X_s$ is a layer thickness in radiation units and $\kappa = K/(p_0 v_0)$ is dimensionless factor. If the energy loss in the layer cannot be neglected (i.e., it is not thin enough), one should use the integral form

$$\overline{\vartheta^2}(z) = \frac{E_s^2}{X_s} J, \quad J \equiv \int_0^z \frac{dz'}{p^2 v^2}.$$

Taking for integrating variable the momentum p, we rewrite the integral in the form

$$J = -\int_{p_0}^{p} \frac{dp}{|dp/dz|(pv)^2}.$$

For sufficiently large energy one may take $v \simeq c$ and $dp/dz \approx$ const, then the integral becomes easy for computing, and we get

$$J = \frac{p_0 - p}{c^2 |dp/dz| p p_0}.$$

Because $|dp/dz| = (p_0 - p)/z$, the mean square of scattering angle takes the form (see [20])

$$\overline{\vartheta^2}(z) = \frac{E_s^2}{X_s} \frac{z}{c^2 p(z) p_0}.$$

Similar results can also be obtained for the mean square of the transverse displacement of a particle from the primary direction, which characterizes the distribution of particles along the radius.

Because θ consists of large number of small independent $2d$ vectors with identical finite variances, the central limit theorem is applicable, and we arrive at the Fermi small-angle distribution

$$f(\theta, t) = \frac{1}{2\pi t \kappa^2} \exp(-\theta^2/2\kappa^2 t).$$

Observe that $f(\theta, t) = f(\theta_x, \theta_y, t) = f(\theta_x, t) f(\theta_y, t)$, i.e., the random projections are independent and obey identical distributions. One more comment should be done: the CLT-produced distribution decreases exponentially at large scattering angles; however, the first orders of scattering can play a significant role there due to the power-law form of the Rutherford cross section (see Moliere theory, where their contributions are computed separately, but in the framework of small-angle approximation as well). A detailed study of the small-angle approximation was undertaken by L. Janossy in his remarkable book [21]. In particular, he derived the joint angular-lateral distribution density: $f(\boldsymbol{\theta}, \boldsymbol{\rho})$.

5.8 Electron–Photon Cascades

Quantum electrodynamics considers the propagation of an electromagnetic wave in this medium as a stream of photons (quants of electromagnetic field) interacting with electrons and atomic nuclei. As a result, the interacting particles do not only exchange by momenta and energies, but also can disappear and appear (at a sufficiently high energy). Positively charged electrons (*positrons*) may also appear among the latter. Because of the charge conservation law, positively charged electrons (positrons) may also appear among the latter. By virtue of the law of conservation of electric charge, they appear in composition of the electron–positron pair. A sequence of such transformations, which may start with any of these particles (photon, electron, or positron) is called the *electron–photon cascade* (*EPC*, for short).

An inherent feature of cascade processes accompanying the passage of high-energy particles through matter is small-angle deviations from the primary direction. So, in electron–photon cascades, the multiplication of particles occurs mainly due to bremsstrahlung processes and formation of electron–positron pairs. However, the angular deflections of particles in these processes are negligible at high energies ($\sim mc^2/E$). The main expansion mechanism EPC in transverse directions is the multiple Coulomb scattering of charged particles, but its RMS angle decreases with energy growing as E^{-1}. Significantly smaller contribution is given by Compton scattering. In the case of a primary hadron (proton or nucleus), the main mechanism responsible for particle number increasing in the cascade becomes the multiple hadron generation in hadron-hadron collisions. It is characterized by limited transverse momentum of secondary particles [22, 23].

Small deviations of the cascade particles from the primary direction (shower axis) make it possible to calculate the longitudinal development of the cascade regardless of lateral one. Neglecting the lateral development of the process reduces its consideration to the *one-dimensional cascade theory*, when a particle spatial position is characterized by one coordinate, counted along the cascade axis. Such a coordinate, continuously increasing at particle motion, can play the role of Markov time. For this reason, there is no need for a special passage from non-stationary equations as above to stationary ones (not containing the physical time), it suffices to assume that t means not a physical time, but the longitudinal coordinate, while the phase variable includes only the particle type and its energy: $x = (a, E)$.

The corresponding equations follow from general ones by means of replacements

$$x \to (a, E), \quad \dot{x} \to \dot{E} = -\beta, \quad \int dx \to \sum_a \int dE,$$

$$f_1(t, x) \to f_a(t, E), \; f_2(t, x_1, x_2) \to f_{a_1 a_2}(t, E_1, E_2)$$

and so on.

The theory of fluctuations of cascade processes can be developed in three ways. The **first way** is connected with the further development of forward equations approach by increasing the number of arguments, that is, involving in the process of describing many-particle distribution functions along the lines of the standard approach in statistical physics. Thus, the forward kinetic equations for differential energy spectrum $f_a(t, E)$ of a-particles and factorial moment density $f_{a,b}(t, E_a, E_b)$ of a, b-particles are followed from Eqs. (3.25), (3.28) via adequate reducing the phase space $x \to a, t, E$ of the process:

$$[\partial_t + \sigma_a(E)]f_a(t, E) - \partial_E[\beta_a(E)f_a(t, E)] = \sum_b \int_E^{E_0} dE' f_b(t, E')w_{b\to a}(E' \to E), \tag{5.131}$$

and

$$[\partial_t + \sigma_a(E_1) + \sigma_b(E_2)]f_{ab}(t, E_1, E_2)$$

$$- \partial_{E_1}[\beta_a(E_1)f_{ab}(t, E_1, E_2)] - \partial_{E_2}[\beta_b(E_2)f_{ab}(t, E_1, E_2)]$$

$$+ \sum_c \int_{E_1}^{E_0 - E_2} dE_1' f_{cb}(t, E_1', E_2)w_{c\to a}(E_1' \to E_1)$$

$$+ \sum_c \int_{E_2}^{E_0 - E_1} dE_2' f_{ac}(t, E_1, E_2')w_{c\to b}(E_2' \to E_2)$$

$$+ \sum_c \int_{E_1 + E_2}^{E_0} dE_1' f_c(t, E_1')w_{c\to ab}(E_1' \to E_1, E_2). \tag{5.132}$$

This process must end with the calculation of sufficient numbers of functionals like

$$\int dE_1 \ldots \int dE_n \varphi_n(E_1, \ldots, E_n) f_n(E_1, \ldots, E_n)$$

with predetermined (and, generally speaking, non-additive) detector responses to each of the possible sets of particle energies falling at the registration plane.

Equations (5.131), (5.132) are written for case when the energy spectrum is bounded from above by a finite limit E_0. Then, according to the law of energy conservation, $f_a(t, E) = 0$ for $E > E_0$ and $f_{ab}(t, E_1, E_2) = 0$ for $E_1 + E_2 > E_0$, as one can see in integral limits. Initial (more exactly, boundary) conditions are determined by the source spectrum at $t = 0$. If the cascade is generated by a single particle (a, E_0),

$$f_b(\theta, E) = \delta_{ab}\delta(E - E_0), \tag{5.133}$$

$$f_{ab}(\theta, E_1, E_2) = 0. \tag{5.134}$$

Obviously, involving higher order multiparticle densities into statistical description of EPC cascade may be continued. As a result, we will arrive at the system of integro-differential equations with increasing number of arguments $E_1,\ E_2,\ E_3 \ldots.$

The **second way** to build a statistical theory of cascade processes, choosing for this book, is based on the modernization of backward (adjoint) equation approach. According to Eq. (4.79), the n-th (statistical) moment of detector with the threshold energy E^* responses

$$[-\partial_t + \beta_a \partial_E + \sigma_a]\overline{q_a^n}(t, E)$$

$$-\sum_b \int_{E^*}^{E} dE' w_{a\to b}(E' \to E')\overline{q_b^n}(t, E') = d_a^{(n)}(t, E); \tag{5.135}$$

$$\overline{q_a^n}(t, E) = 0, \quad E < E^*.$$

Their right-hand sides and boundary conditions are determined by detector properties. For a track detector, responsive to d-particles,

$$d_a^{(1)}(t, E) = a_d(t, E)1(E - E^*)\delta_{ad}, \tag{5.136}$$

$$d_a^{(2)}(t, E) = 2a_a(t, E)\bar{q}_a(z, E)$$

$$+\sum_{bc} \int_{E^*}^{E_1+E_2<E} dE_1 \int_{E^*} dE_2 w_{a\to bc}(E \to E_1, E_2)\overline{q_b}(E_1)\overline{q_c}(E_2), \tag{5.137}$$

$$\overline{q_a}(T, E) = \overline{q_a^2}(T, E) = 0. \tag{5.138}$$

If the detector measures the number of particles at the depth $t = T$, then one should put $a_a = 0$ in Eqs. (5.136), (5.137) and change boundary condition (5.138) to:

$$\overline{q_a}(t_d, E) = \overline{q_a^2}(T, E) = 1(E - E^*)\delta_{ad}.$$

Observe that this second way also leads us to an infinite system of equations for the statistical moments and the rather old problem of reconstructing the distribution function from its moments. Fortunately, modern transport theory has developed a number of efficient algorithms for solving this ill-posed problem. In our presentation, the detector is formalized using a single function, but this is not a necessary condition for applying this method: one can take a vector representation of the sensitivity corresponding to a set of detectors that simultaneously measures several parameters and investigates the correlation properties of various characteristics. We will see this method in action discussing measurements of EAS components.

Finally, let us dwell on the **third way** of developing the theory, the main feature of which is the emphasis on the global nature of the information extracted from a specific implementation of the cascade, including the direction of the shower axis, the position of the maximum, the width of the distribution, the shape of the cascade curves, and the display of the signal from the shower in Cherenkov telescopes. If we talk only about the longitudinal characteristics of the cascade, then there is a clear tendency to present its realization in terms of random functions, and in this respect is indicative, in which the concept of a family of longitudinal moments is extended from averaged cascade curves to individual random realizations, which makes it possible to clarify the formulation of the question of mutual correlations of the parameters of individual cascades.

The functions, saving information on individual random cascades of the whole ensemble, play the role of a kind of collective variables and can become the basis of a non-local version of the statistical theory of cascade processes as a new perspective direction of the cascade theory development. Examples of this idea implication can be found in Chaps. 8 and 10.

5.9 Cascade Equations in Mellin's Transforms

The integral transformation method is the most popular analytical approach underlying the cascade theory. Originating from the first works on the theory of electron–photon cascade (EPC), we begin our review with so-called *A-approximation*, taking into account only two processes, being the most significant for EPC development: bremsstrahlung and pair production. All other processes are neglected. The most important factor for successful application of the Mellin transform in this theory is that cross sections of the electromagnetic processes in the high-energy region are described in terms of homogeneous functions:

$$w_{a\to b}(E \to E')dE' = V_{a\to b}(x)dx, \quad x = E'/E; \tag{5.139}$$

$$w_{a\to bc}(E \to E_1, E_2)dE_1dE_2 = V_{a\to bc}(x_1, x_2)dx_1dx_2,$$

$$x_1 = E_1/E, \quad x_2 = E_2/E. \tag{5.140}$$

Due to the two-particle nature of elementary processes involved in the EPC's development, the differential sections (5.139) and (5.140) are expressed through distributions of the energy share transferred to the secondary electron in bremsstrahlung ($V_e(x)$) and pair formation ($V_\gamma(x)$) events. Rewritten per unit radiation length (for air, the radiation length is about 37g/cm^2), the distributions are given by the Bethe–Heitler formulas derived in the *complete screening approximation*:

$$V_e(x) = 1 - x + \frac{x\delta}{1-x}, \tag{5.141}$$

$$V_\gamma(x) = 1 - x\delta + x^2\delta, \tag{5.142}$$

where δ is the only parameter dependent (very weakly) on the atomic number Z of the matter. Since at high energies the cross sections of electrons and positrons interactions with matter differ little from each other, and positron annihilation is significant only at low energies, both kinds of particles are usually (but not always) considered as belonging to the same kind called *electron*. Accordingly, the index a takes two values e and γ, and the inclusive cross sections (5.139), (5.140) are written as

$$W_{e\to e}(x) = V_e(x),\ \ W_{e\to\gamma}(x) = V_e(1-x),\ \ W_{\gamma\to e}(x) = 2V_\gamma(x),$$

$$W_{e\to e\gamma}(x_1, x_2) = V_e(x_1)\delta(x_2 - 1 + x_1),$$

$$W_{\gamma\to ee}(x_1, x_2) = 2V_\gamma(x_1)\delta(x_2 - 1 + x_1).$$

In a homogeneous medium, solutions of equations (5.131), (5.132) depend only on the difference in coordinates $t - \theta$, which we will further denote by the same letter t (i.e., assuming that $\theta = 0$). It is easy to show that in the case of one primary particle (boundary conditions (5.133), (5.134)) solutions of the equations depend on the ratio of primary and threshold energies

$$x = E/E_0, \quad x_1 = E_1/E_0, \quad x_2 = E_2/E_0. \tag{5.143}$$

Introducing new designations for dependent variables as functions of new independent ones t and x,

$$\Phi_a(t, x) = E_0 f_a(t, E_0, x),$$

$$G_{a\to b}(t, x) = E_0 g_{a\to b}(t, E_0 \to E_0 x),$$

$$\Phi_{ab}(t, x_1, x_2) = E_0^2 f_{ab}(t, E_0 x_1, E_0 x_2),$$

from Eqs. (5.131)–(5.132) in A-approximation we obtain

$$[\partial_t + \sigma_b]G_{a\to b}(t,x) = \sum_c \int_x^1 dx' G_{a\to c}(t,x')W_{c\to b}(x/x')/x',$$

and

$$[\partial_t + \sigma_a + \sigma_b]\Phi_{ab}(t,x_1,x_2) = \sum_c \int_{x_1}^{1-x_2} dx_1' \Phi_{cb}(t,x_1',x_2)W_{c\to a}(x_1/x_1')/x_1'$$

$$+ \sum_c \int_{x_2}^{1-x_1} dx_2' \Phi_{ac}(t,x_1,x_2')W_{c\to b}(x_2/x_2')/x_2'$$

$$+ \sum_c \int_{x_1+x_2}^{1} dx' G_{a\to c}(t,x')W_{c\to ab}(x_1/x',x_2/x')/x'^2.$$

According to (3.49), the solution of the last equation (taking into account the boundary condition (5.134)) has the form

$$\Phi_{ab}(t,x_1,x_2) = \sum_{cde} \int_0^t dt' \int\int\int dx' dx_1' dx_2' G_{a\to c}(t',x')W_{c\to de}(x_1'/x',x_2'/x')$$
$$\times G_{d\to a}(t-t',x_1/x_1')G_{e\to b}(t-t',x_2/x_2')1(1-x_1-x_2)$$
$$\times 1(x'-x_1'-x_2')1(x_1'-x_1)1(x_2'-x_2)[x_1'x_2'x_3']^{-1}. \quad (5.144)$$

The set of unit functions under the integral shows the integrating region restrictions following from the energy conservation law.

Using [21], we apply the Mellin transform, corresponding to the one-sided Laplace transform, with respect to x

$$\bar{G}_{a\to b}(t,s) = \int_0^1 x^s G_{a\to b}(t,x)dx,$$

$$\bar{\Phi}_{ab}(t,s_1,s_2) = \int_0^1\int_0^1 x_1^{s_1}x_2^{s_2}\Phi_{ab}(t,x_1,x_2)dx_1dx_2$$

to the above expressions. By changing the order of integration and taking into account that $W_{c\to ab}(x_1,x_2) = 0$, $x_1 + x_2 > 1$, we arrive at the following equations for the transforms:

$$[\partial_t + \sigma_b]\bar{G}_{a\to b}(t,s) - \sum_c \bar{G}_{a\to c}(t,s)\bar{W}_{c\to b}(s) = 0, \tag{5.145}$$

$$[\partial_t + \sigma_a + \sigma_b]\bar{\Phi}_{ab}(t,s_1,s_2) - \sum_c [\bar{\Phi}_{cb}(t,s_1,s_2)\bar{W}_{c\to a}(s_1)$$

$$+ \bar{\Phi}_{ac}(t,s_1,s_2)\bar{W}_{c\to b}(s_2) + \bar{G}_{a\to c}(t,s_1+s_2)\bar{W}_{c\to ab}(s_1,s_2)] = 0,$$

with boundary conditions $\bar{G}_{a\to b}(0,s) = \delta_{ab}$, $\bar{\Phi}_{ab}(0,s_1,s_2) = 0$. The Mellin transformation of the expression (5.144) leads to the following relation for the transforms:

$$\bar{\Phi}_{ab}(t,s_1,s_2) = \sum_{cde}\int_0^t dt'\bar{G}_{a\to c}(t',s_1+s_2)\bar{W}_{c\to de}(s_1,s_2)$$

$$\times\ \bar{G}_{e\to a}(t-t',s_1)\bar{G}_{e\to b}(t-t',s_2). \tag{5.146}$$

When dealing with backward equation, the inverse independent variable $x = E^*/E$ is more convenient. Observe, its values belong to the same region as (5.143). Since the right sides and solutions of Eq. (5.135) are not energy distributions, passing to a new variable is carried out by a simple substitution of argument:

$$F_a^{(n)}(t,x) = d_a^{(n)}(t,E^*/x), \quad Q_a^{(n)}(t,x) = \overline{q_a^n}(t,E^*/x).$$

Taking as a spatial variable the distance between the birth point of the primary particle and the boundary position $T - t$ and replacing this variable by the same letter t, we obtain

$$[\partial_t + \sigma_a]Q_a^{(n)}(t,x) = \sum_b \int_x^1 dx' W_{a\to b}(x')Q_b^{(n)}(t,x/x') + F_a^{(n)}(t,x). \tag{5.147}$$

In case of track detectors, the right parts for the first two moments are

$$F_a^{(1)}(t,x) = a_a(t,x),$$

and

$$F_a^{(2)}(t,x) = 2a_a(t,x)Q_a^{(1)}(t,x) + \sum_{bc} \int_x^{x_1+x_2<1} dx_1 \int_x dx_2 W_{a\to bc}(x_1,x_2) \times$$

$$Q_b^{(1)}(t,x/x_1)Q_c^{(1)}(t,x/x_2). \tag{5.148}$$

With (4.84), we can express the second moment in terms of Green's function

$$Q_a^{(2)}(t,x) = \sum_{bcd} \int_0^t dt' \int\int\int dx' dx_1 dx_2 G_{a\to b}(t-t', x/x') W_{b\to cd}(x'/x_1, x'/x_2)$$

$$\times\ Q_c^{(1)}(t', x_1) Q_d^{(1)}(t', x_2) x[x_1 x_2]^{-2} 1(x'-x) 1\left(\frac{1}{x} - \frac{1}{x_1} - \frac{1}{x_2}\right)$$

$$\times\ 1(x_1 - x_1') 1(x_2 - x_2') + 2a_a Q_a, \qquad (5.149)$$

where $1(x)$ denotes the unit step (Heaviside) function.

It is convenient to transform inverse equations with parameter s shifted by 1,

$$\bar{Q}_a(t,s) = \int_0^1 x^{s-1} Q_a(t,x) dx,$$

$$Q_a(t,x) = \frac{1}{2\pi i} \int x^{-s} \bar{Q}_a(t,s) ds,$$

leaving the former definition for transforms of sections. Then for (5.147) and (5.149), respectively, we get

$$[\partial_t + \sigma_a] \bar{Q}_a^{(n)}(t,s) = \sum_b \bar{W}_{a\to b}(s) \bar{Q}_b^{(n)}(t,s) = \bar{F}_a^{(n)}(t,s),$$

$$\bar{Q}_a^{(n)}(t,s) = \sum_b \int_0^t \bar{G}_{a\to b}(t-t', s) \bar{F}_b^{(n)}(t', s) dt'. \qquad (5.150)$$

In order to write down the transform of the right side (5.148), let us bring it to the form

$$F_a^{(2)}(t,x) = \frac{1}{(2\pi i)^2} \int ds_1 \int ds_2 [2\bar{a}_a(t, s_1)$$

$$\times\ \bar{Q}_a^{(1)}(t, s_2) x^{-s_1-s_2} + \sum_{bc} \bar{Q}_b(t, s_1) \bar{Q}_c(t, s_2) x^{-s_1-s_2} \widehat{W}_{a\to bc}(s_1, s_2; x)], \qquad (5.151)$$

where

$$\widehat{W}_{a\to bc}(s_1, s_2; x) = \int_x^{x_1+x_2<1} dx_1 \int_x dx_2 W_{a\to bc}(x_1, x_2) x_1^{s_1} x_2^{s_2}. \qquad (5.152)$$

Observe that when $x \to 0$ this expression becomes the Mellin transform of the two-particle cross section

$$\widehat{W}_{a\to bc}(s_1, s_2; x) \to \int_0^1 dx_1 \int_x^1 dx_2 W_{a\to bc}(x_1, x_2) x_1^{s_1} x_2^{s_2} = \bar{W}_{a\to bc}(s_1, s_2).$$

Transforming now (5.151) with taking into account (5.152) yields

$$\bar{F}_a^{(2)}(t, s) = \frac{1}{(2\pi i)^2} \int ds_1 \int ds_2 [2\bar{a}_a(t, s_1)\bar{Q}_a^{(1)}(t, s_2)$$

$$+ \sum_{bc} \bar{Q}_b^{(1)}(t, s_1)\bar{Q}_c^{(1)}(t, s_2)\bar{W}_{a\to bc}(s_1, s_2; s)](s - s_1 - s_2)^{-1}, \quad (5.153)$$

$$\widehat{W}_{a\to bc}(s_1, s_2; s) = \int_x^{x_1+x_2<1} dx_1 \int_x dx_2 W_{a\to bc}(x_1, x_2)[\min\{x_1, x_2\}]^{s-s_1-s_2} x_1^{s_1} x_2^{s_2}.$$

References

1. Jackson J (1999) Classical electrodynamics. Wiley, NJ, USA
2. Flügge S (2012) Practical quantum mechanics. Springer Science & Business Media
3. Berestetsky VB, Lifshitz EM, Pitaevsky LP (1968) Relativistic quantum theory, part 1. Nauka, Moscow (in Russian)
4. Bethe HA, Heitler W (1934) On the stopping of fast particles and on the creation of positive electrons. Proc Roy Soc A146:83–112
5. Koch HW, Motz JW (1959) Bremsstrahlung cross-section formulas and related Data. Rev Mod Phys 31(4):920–955
6. Tamm I, Belenky S (1939) On the soft component of cosmic rays at see level. J Phys USSR 1:177–198
7. Landau LD (1935) Loss of energy of fast electrons by radiation. Zh Eksp Teor Fiz 5:255
8. Felten JE, Morrison P (1963) Recoil photons from scattering of starlight by relativistic electrons. Phys Rev Lett 453:453
9. Shklovsky IS (1964) The inverse compton effect as a possible cause of the X-ray radiation of solar flares. Nature 202:275–276
10. Ginzburg VL, Syrovatskii SI (1964) Cosmic magnetobremsstrahlung. J Expt Theoret Phys (USSR), 46:1865
11. Ginzburg VL, Syrovatskii SI (1964) The origin of Cosmic Rays. The Pergamon Press Inc, Oxford
12. Jones FC (1967) The calculated spectrum of inverse Compton scattering photons. Goddard Space Flight Center, Greenbelt, MD
13. McConnell J (1958) Quantum particle dynamics. North-Holland Publishing Company, Amsterdam
14. Migdal AB, Krainov VP (1966) Approximate methods of quantum mechanics. Nauka, Moscow (in Russian)
15. Berger MJ, Seltzer SM (1964) Tables of energy losses and ranges of electrons and positrons, NASA SP-3012

16. Goldberger ML, Watson KM (1964) Collision theory. Wiley
17. Sternheimer RM (1960) Range straggling of charged particles in Be, C, Al, Cu, Pb, and Air. Phys Rev 117:485–488
18. Uchaikin VV, Zolotarev VM (1999) Chance and stability. Stable Distributions and their Applications, The Netherlands, Utrecht, VSP
19. Uchaikin VV, Topchii VA (1978) Stable law with index $\alpha = 1$ in the problem of fluctuations of ionization losses of charged particles. Izv Vyssh Uchebn Zaved Fizika 4:60–64 (in Russian)
20. Janossy L (1965) Theory and practice of the evaluation of measurements. Oxford University Press, Budapest
21. Fano U, Spencer LV, Berger MJ (1959) Penetration and diffusion of X rays. Handbuch der Physik, XXXVIII/2. Goettingen-Heidelberg
22. Vavilov PV (1957) Ionization losses of high-energy heavy particles. Zh Eksp Teor Fiz 5:749–751
23. Nikitin YP, Rozental IL (1976) Theory of multiple processes. Atomizdat, Moscow (in Russian)
24. Ivanenko IP (1972) Electromagnetic cascade processes. Moscow State University, Moscow (in Russian)
25. Landau L (1944) On the energy loss of fast particles by ionization. J Phys USSR 8:201–205
26. Srinivasan SK et al (1961) Fluctuation problem in electromagnetic cascades. Zeitschrift fur Physik 161:346–352; 1964, 177:164–173
27. Rossi B (1952) High energy particles. Prentice Hall, New York

For Further Reading

28. Ginzburg VL, Syrovatskii SI (2013) The Origin of Cosmic Rays. Elsevier
29. Berezinsky VS, Bulanov SV, Ginzburg VL et al (1990) Astrophysics of Cosmic Rays. Physmatgiz, North-Holland, Amsterdam, Moscow. English translation
30. Gaisser TK, Engel R, Resconi E (2016) Cosmic Rays and particle physics. Cambridge University Press
31. Schlickeiser R (2013) Cosmic Ray Astrophysics. Springer Science & Business Media
32. Case CT, Battle EL (1968) Molière's theory of multiple scattering. Phys Rev 169(1):201

Chapter 6
Analytical Theory

6.1 Basic Equation of EPC Theory

Most of analytical calculations of in the EPC theory are related to the number of particles at a given depth performed in the framework of the straightforward model.

Going to deal with the electron-photon cascade theory, we slightly change designations of basic cascade functions, characterizing differential (with respect to observed energy E) spectra particles in cascades initiated by single-particle sources, and make replacements

$$G_{a\to e}(t, E_0 \to E) \mapsto P(t, E|E_0), \quad G_{a\to\gamma}(t, E_0 \to E) \mapsto \Gamma(t, E|E_0).$$

Information about the primary particles of the cascade will be stored in the initial conditions of the equations for these functions:

$$P(0, E) = \delta(E - E_0), \quad \Gamma(0, E) = 0 \tag{6.1}$$

which means that the cascade is initiated by primary electron with energy E_0, whereas conditions

$$P(0, E) = 0, \quad \Gamma(0, E) = \delta(E - E_0)$$

say, that the primary particle is a photon.

As a rule, we will omit the common symbol E_0 for primary particle energy, being a constant parameter (as long as we remain within the forward equations approach). For the same purpose of shortening the notation of formulas, we will denote the transforms by the same letters as the originals, distinguishing them only by the symbols of the transform arguments: s for the Mellin transform

$$\int_0^\infty E^s P(t, E)dE = P(t, s), \quad \int_0^\infty E^s \Gamma(t, E)dE = \Gamma(t, s)$$

© Higher Education Press Limited Company 2024

V. V. Uchaikin, *Electron-Photon Cascades*, Nonlinear Physical Science,
https://doi.org/10.1007/978-981-99-7524-2_6

and so is for joint transforms:

$$\int_0^\infty dEE^s \int_0^\infty e^{-\lambda t} P(t,E)dt = P(\lambda,s), \quad \int_0^\infty dEE^s \int_0^\infty e^{-\lambda t} \Gamma(t,E)dt = \Gamma(\lambda,s).$$

Using standard functions of EPC theory, arising after inserting Bethe–Heitler's cross sections and Mellin's transformation,

$$A(s) = \sigma_e - \bar{W}_{e\to e}(s) = \delta \int_0^1 \frac{1-x^{s+1}}{1-x} dx - \frac{1}{(s+1)(s+2)} + \frac{1}{2} - \delta, \tag{6.2}$$

$$B(s) = \bar{W}_{\gamma\to e}(s) = 2\left[\frac{1}{s+1} - \frac{\delta}{(s+2)(s+3)}\right], \tag{6.3}$$

$$C(s) = \bar{W}_{e\to\gamma}(s) = \frac{1}{s+2} + \frac{\delta}{s(s+1)}, \tag{6.4}$$

$$\sigma_0 = \sigma_\gamma = 1 - \frac{\delta}{6} = 0.773, \tag{6.5}$$

we obtain the following system for Mellin's transforms:

$$[\partial_t + A(s)]P(t,s) - B(s)\Gamma(t,s) = 0, \tag{6.6}$$

$$[\partial_t + \sigma_0]\Gamma(t,s) - C(s)P(t,s) = 0. \tag{6.7}$$

In what follows, it may be convenient to consider the coefficient functions $A(s)$, $B(s)$, and $C(s)$ to be the Mellin images of integral operators $\mathscr{A}$, $\mathscr{B}$, $\mathscr{C}$ and write the original counterpart of the system in the form

$$[\partial_t + \mathscr{A}]P(t,E) - \mathscr{B}\Gamma(t,E) = 0, \tag{6.8}$$

$$[\partial_t + \sigma_0]\Gamma(t,E) - \mathscr{C}P(t,E) = 0. \tag{6.9}$$

The algebraic system (6.6)–(6.7) has the solution

$$P(t,s) = \sum_{k=1}^{2} a_k(s)e^{\lambda_k(s)t}, \quad \Gamma(t,s) = \sum_{k=1}^{2} b_k(s)e^{\lambda_k(s)t}, \tag{6.10}$$

where

$$\lambda_{1,2} = \frac{1}{2}\left[-A - \sigma_0 \pm \sqrt{(A-\sigma_0)^2 + 4BC}\right], \tag{6.11}$$

and coefficients a_k, b_k being linked among themselves via relations

$$b_1(s) = \frac{C(s)}{\sigma_0 + \lambda_1(s)} a_1(s), \quad b_2(s) = \frac{C(s)}{\sigma_0 + \lambda_2(s)} a_2(s) \tag{6.12}$$

are determined through the boundary conditions, i.e., Mellin's transforms of electron and photon spectra at $t = 0$:

$$a_1(s) + a_2(s) = P_0(s) \equiv \int_0^\infty E^s P(0, E) dE,$$

$$\frac{C(s)}{\sigma_0 + \lambda_1(s)} a_1(s) + \frac{C(s)}{\sigma_0 + \lambda_2(s)} a_2(s) = \Gamma_0(s) \equiv \int_0^\infty E^s \Gamma(0, E) dE.$$

If, for example, the cascade is generated by a single electron with energy E_0, Eq. (6.1), then

$$a_1(s) = \frac{\sigma_0 + \lambda_1(s)}{\lambda_1(s) - \lambda_2(s)} E_0^s, \quad a_2(s) = \frac{\sigma_0 + \lambda_2(s)}{\lambda_2(s) - \lambda_1(s)} E_0^s. \tag{6.13}$$

Let us come back to Eqs. (6.6)–(6.7) and perform the Laplace transformation with respect to t:

$$\lambda P(\lambda, s) - E_0^s + A(s) P(\lambda, s) = B(s) \Gamma(\lambda, s), \tag{6.14}$$

$$(\sigma_0 + \lambda) \Gamma(\lambda, s) = C(s) P(\lambda, s). \tag{6.15}$$

Eliminating $\Gamma(\lambda, s)$ reduces the system to equation

$$\psi(\lambda, s) P(\lambda, s) = E_0^s$$

with function

$$\psi(\lambda, s) = \lambda + A(s) - \frac{B(s) C(s)}{\sigma_0 + \lambda}, \tag{6.16}$$

which may be represented in the form

$$\psi(\lambda, s) = \frac{[\lambda - \lambda_1(s)][\lambda - \lambda_2(s)]}{\sigma_0 + \lambda}.$$

Using it in solving the equation yields the Mellin–Laplace transform

$$P(\lambda, s) = \frac{\sigma_0 + \lambda}{[\lambda - \lambda_1(s)][\lambda - \lambda_2(s)]} E_0^s.$$

Recall that integrating $P(t, E)$ over energy is tantamount to dividing its Mellin transform $P(t, E)$ by parameter s,

$$N(\lambda, s) = \frac{P(\lambda, s)}{s},$$

the inversion formula (with using the Jordan lemma under closing the integration contour around poles) yields

$$N(t, s) = \frac{1}{2\pi i} \frac{E_0^s}{s} \int_{c-i\infty}^{c+i\infty} \frac{[\sigma_0 + \lambda]e^{\lambda t}}{[\lambda - \lambda_1(s)][\lambda - \lambda_1(s)]} d\lambda$$

$$= \frac{E_0^s}{s} \left[\frac{\sigma_0 + \lambda_1(s)}{\lambda_1(s) - \lambda_2(s)} e^{\lambda_1(s)t} + \frac{\sigma_0 + \lambda_2(s)}{\lambda_2(s) - \lambda_1(s)} e^{\lambda_2(s)t} \right].$$

In order to return to the energy variable, one should perform the inverse Mellin transformation of this function. As one can see from the above formulas and Table 1, the contribution of the second term decreases rapidly with increasing t beyond $t \sim 1$ (because of exponential factor $e^{\lambda_2(s)t}$ with negative $\lambda_2(s)$). Below, we restrict ourselves by consideration of only this term,

$$\frac{\sigma_0 + \lambda_1(s)}{\lambda_1(s) - \lambda_2(s)} e^{\lambda_1(s)t},$$

rewritten as

$$H_1(s)e^{\lambda_1(s)t}$$

with

$$H_1(s) = \frac{\sigma_0 + \lambda_1(s)}{\lambda_1(s) - \lambda_2(s)}. \tag{6.17}$$

The most significant for the inversion of the Laplace–Mellin transform is its behavior in the region $(0 < s < 2) \times (-0.6 < \lambda < 2)$. In this region, $\psi(\lambda, s)$ has two essential peculiarities: a simple pole at $s = 0$ and zero at the curve $\lambda = \lambda_1(s)$ (or $s = s_1(\lambda)$). Here, the function $\psi(\lambda, s)$ may be successfully approximated by expression

$$\psi_0(\lambda, s) = f(\lambda) \frac{s - s_1(\lambda)}{s}, \quad f(\lambda) = \left(s, \frac{\partial \psi}{\partial s} \right)_{s=s_1(\lambda)}. \tag{6.18}$$

It is not difficult to verify that

$$f(\lambda) = -\frac{1}{\sigma_0 + \lambda} \left\{ s \frac{d\lambda_1}{ds} [\lambda - \lambda_1(s)] \right\}_{s=s_1(\lambda)}.$$

This approximation was suggested in [1].

Inverting the Mellin–Laplace image

$$N(\lambda, s) = \frac{E_0^s}{f(\lambda)[s - s_1(\lambda)]},$$

we obtain

$$N^{\mathrm{A}}(t, E) = \frac{1}{2\pi i} \int\limits_{c-i\infty}^{c+i\infty} e^{\lambda_1(s)t + ys} \frac{H_1(s)}{s} ds, \tag{6.19}$$

where $y = \ln(E_0/E)$. Symbol A says, that the expression is obtained in the framework of A-approximation, i.e., without taking into account ionization losses of electrons. The complex integral (6.36) which represents a most important characteristic of the cascade process depicting the number of particles with energies higher E at a depth t is quite accessible for calculation by numerical methods, but a preliminary analysis with approximate methods is always useful and can give an idea leading to fairly accurate quantitative results.

We estimate this integral by using the saddle-point method. Rewriting the integrand in the form $e^{\varphi}(s)$ with

$$\varphi(s) = \lambda_1(s)t + ys + \ln \frac{H_1(s)}{s}, \tag{6.20}$$

one can see that in both cases, $s \to 0$ and $s \to \infty$ the function goes to infinity, so it should have a minimum at some real positive value $\bar{s}$. Consequently, changing along the imaginary axis z this function must pass through a maximum, the sharpness of which depends on the values of the parameters y and t. The position of the saddle point s on the real axis x is determined by equation

$$\varphi'(\bar{s}) = 0.$$

Let's draw an integration contour through this point, then the exponent will take in it a maximum value. Expanding it into a series in powers of the difference $iz = s - \bar{s}$, we get

$$\varphi = \varphi(\bar{s}) - \frac{z^2}{2}\varphi''(\bar{s}) - \cdots .$$

Inserting these derivatives into (6.20) and retaining in result terms of order $1/t$, after some tedious but primitive calculations we arrive at expression

$$N(t) = \frac{H_1(\bar{s})}{\bar{s}} \frac{\exp(y\bar{s} + \lambda_1'(\bar{s})t)}{\sqrt{2\pi\lambda_1''(\bar{s})t}} \left[1 + \frac{\alpha(\bar{s})}{t}\right], \tag{6.21}$$

where

$$\alpha(s) = -\frac{1}{2\lambda_1''(\bar{s})}\left(\ln\frac{H_1(\bar{s})}{\bar{s}}\right)'' + \frac{1}{8}\frac{\lambda_1^{iv}(\bar{s})}{[\lambda_1''(\bar{s})]^2} - \frac{5}{24}\frac{[\lambda_1'''(\bar{s})]^2}{[\lambda_1''(\bar{s})]^3}, \tag{6.22}$$

and parameter $\bar{s}$ is determined from equation

$$t = -\frac{1}{\lambda_1(\bar{s})}\left[y + \left(\ln\frac{H_1(\bar{s})}{\bar{s}}\right)'\right], \quad y = \ln\frac{E_0}{E}. \tag{6.23}$$

6.2 Cascade Curves

The model of the electron–photon cascade considered above takes into account only two processes responsible for the multiplication of particles: the production of electron–positron pairs and bremsstrahlung, the cross sections of which are represented by simplified Bethe–Heitler formulas. The ionization losses of charged particles derived in the total screening approximation are neglected, deviations of particle motion directions from the shower axis (in which the Rutherford scattering of electrons should play the main role) are also not taken into account: the cascade is represented as a one-dimensional process developing along a straight line determined by the initial position of the particle and the direction of its momentum. Such a model, called approximation-A of the cascade theory, was considered by us above. For the indicated reasons, it has limited application, mainly in the study of the high-energy part and in heavy substances (in this case Compton scattering begins to play a significant role).

The theoretical construction of a model of a cascade process is convenient to start with the assumption that all particles move in the same direction, but as a result of the exchange of energy, many particles of lower energies appear, which are scattered more intensively in the medium. In many cases, the experimental conditions themselves (for example, in accelerator experiments with bubble chambers or in cosmic rays with X-ray emulsion devices and Cherenkov telescopes) make it possible to separate some effects from others and use particular solutions of the cascade theory, or their combinations (as we will show in chapter on the transition effects). Here we return to the model of one-dimensional motion of shower particles in one direction, adding to only ionization energy losses by electrons, assuming these losses to be a continuous process, given by a constant value β, which characterizes the average loss of energy by a single electron on one t-unit.

For the physical interpretation of further modifications of the theory, it is convenient to modify the system (6.8)–(6.9) by introducing operators $\mathscr{L}_1$ and $\mathscr{L}_2$, acting on pairs of functions,

$$\mathscr{L}_1[P, \Gamma] = -\mathscr{A}P + \mathscr{B}\Gamma,$$

$$\mathscr{L}_2[P, \Gamma] = \mathscr{C}P - \sigma_0\Gamma,$$

so that Eqs. (6.8)–(6.9) take the form

$$\partial_t P = \mathscr{L}_1[P, \Gamma], \tag{6.24}$$

and

$$\partial_t \Gamma = \mathscr{L}_2[P, \Gamma], \tag{6.25}$$

respectively. What remains to be done is to take into account the production of particles by independent sources. If this is a single particle, and we count the longitudinal coordinate of the cascade from the place of its birth, then it suffices to set the corresponding initial condition at $t = 0$, which has the meaning of the energy spectrum of the primary particle. In the case of a source continuously distributed over depth, which describes the production of electrons and photons by other particles that do not belong to the cascade, it is necessary to put the corresponding distributions on the right-hand sides of these equations

$$\partial_t P = \mathscr{L}_1[P, \Gamma] + S_e, \tag{6.26}$$

and

$$\partial_t \Gamma = \mathscr{L}_2[P, \Gamma] + S_\gamma. \tag{6.27}$$

In order to take into account the electron ionization losses, we consider elementary energy interval $(E, E + dE)$ around E. Due to this losses, a certain number of charged particles $P(E + dE)\beta dt$ on the way dt enter the energy interval, while other particles in the amount $P(E)\beta dt$ leave it. The change in its number is determined by the expression

$$(\partial_t P(t, E))_{\text{ion}} = \beta \partial_E P(t, E),$$

and with this contribution Eq. 6.24 becomes

$$\partial_t P = \mathscr{L}_1[P, \Gamma] + \beta \partial_E P(t, E). \tag{6.28}$$

Performing the Mellin transformation of this term by integration by parts

$$\int_0^\infty E^s \beta \partial_E P dE = \beta \left[E^s P(E) \Big|_0^\infty - s \int_0^\infty E^{s-1} P(E) dE \right] = -\beta s P(s-1)$$

(here $\lim E^s P(E) = 0$ is assumed in both cases $E \to 0$ and $E \to \infty$), add it to the right-hand side of Eq. (6.7). Suppose also, that the only primary particle, induced the cascade is an electron with energy E_0, i.e., $P_0(E) = \delta(E - E_0)$, and performing the Laplace transformation of system (6.14)–(6.15) with respect to t, we obtain instead of it:

$$\lambda P(\lambda, s) - E_0^s + A(s)P(\lambda, s) = B(s)\Gamma(\lambda, s) - \beta s P(\lambda, s-1), \tag{6.29}$$

$$(\sigma_0 + \lambda)\Gamma(\lambda, s) = C(s)P(\lambda, s). \tag{6.30}$$

Elimination from these equations $\Gamma(\lambda, s)$ leads us

$$P(\lambda, s)\psi(\lambda, s) + \beta s P(\lambda, s-1) = E_0^s.$$

Setting $\beta = 0$ reduces Eq. 6.29 to a simple algebraic relation. In case of non-zero β, it becomes the equation in finite differences, whose solution,

$$P(\lambda, s) = \frac{E_0^{s+1}}{\beta} \sum_{n=0}^{\infty} \frac{(-E_0/\beta)^n \psi(\lambda, s+1) \ldots \psi(\lambda, s+n)}{(s+1)(s+2)\ldots(s+n)}. \tag{6.31}$$

This a strict formal solution of the problem, but is not very convenient for practical analysis. Tamm and Belenky proposed to replace it by ansatz of the type (6.18), which was proved to be successful in the A-approximation, i.e., in case of high energies, when ionization losses do not play significant role. Because (6.31) is considered as a fairly accurate result, except for the energy dependence of ionization losses and scattering effects, it is not surprising that the use of TB-ansatz in this problem also shows good results, consistent with both numerical calculations and experimental data.

So, admitting

$$\psi(\lambda, s) = f(\lambda)\frac{s - s_1(\lambda)}{s} \tag{6.32}$$

and substituting it into Eq. (6.31), we arrive at the following expression being the basis of standard result in framework of B-approximation of the cascade theory in terms of Mellin–Laplace images:

$$P(\lambda, s) = \frac{E_0^{s+1}}{\beta} \sum_{n=0}^{\infty} \frac{[-f(\lambda)E_0/\beta]^n [\Gamma(s+1)]^2 \Gamma(s+n+1-s_1)}{\Gamma(s+n+1)\Gamma(s+m+2)\Gamma(s+1-s_1)}. \tag{6.33}$$

Observe that when $s \to 0$ transform $P(t, s)$ becomes $\int\limits_0^{\infty} E^0 P(E, t)dE = N(t)$, i.e., the total cascade curve, counting particles of all energies. At the same time, according to inverse Laplace transform

$$N(t) = \frac{1}{2\pi i} \int\limits_{d-i\infty}^{d+\infty} \lim_{s\to 0} P(\lambda, s) e^{\lambda t} d\lambda,$$

recall that

$$N(E) = \frac{1}{2\pi i} \int\limits_{d-i\infty}^{d+i\infty} \frac{P(0, s)}{s} E^{-s} ds. \tag{6.34}$$

6.2.1 Electron Cascade Curve (From Primary Electron)

Some manipulations with the use (6.32)–(6.33) led Tamm and Belenky to the B-approximation of the total number of electrons at depth t cascade curve from primary electron (the exact formula):

$$N^{\mathrm{B}}(t, 0) = \frac{1}{2\pi i} \int\limits_{c-i\infty}^{c+i\infty} e^{\lambda_1(s)t+ys} \frac{H_1(s)D(s)}{s} ds, \tag{6.35}$$

differing from its A-counterpart (6.35) by an additional integrand factor

$$D(s) = \left[-\frac{s\lambda_1'(s)}{H_1(s)} \right]^s \frac{1}{\Gamma(s+1)}. \tag{6.36}$$

Inserting this multiplier will obviously affect the transformations associated with the saddle-point method, leading to easily predictable B-analogs of the previously obtained formulas (6.21) and (6.23):

$$N^{\mathrm{B}}(t) = \frac{H_1(\bar{s})D(\bar{s})}{\bar{s}} \frac{\exp(y\bar{s} + \lambda_1(\bar{s})t)}{\sqrt{2\pi\lambda_1''(\bar{s})t}} \left[1 + \frac{\alpha(s)}{t} \right], \tag{6.37}$$

$$t = -\frac{1}{\lambda_1(\bar{s})} \left[y + \left(\ln \frac{H_1(\bar{s})D(\bar{s})}{\bar{s}} \right)' \right], \quad y = \ln \frac{E_0}{\beta}. \tag{6.38}$$

Note that some authors prefer to use instead of β critical energy $\varepsilon_{\mathrm{cr}}$, at which radiation losses are equal to ionization ones.

At $\bar{s} = 1$

$$N^{\mathrm{B}}(t) = -\frac{0.3}{\ln(E_0/\beta) - 1.16}, \quad t = 1.01(y - 1.16).$$

6.2.2 Electron Cascade Curve (From Primary Photon)

For a primary photon of high energy, the following exact formula for the electron cascade curve is valid:

$$N^{\mathrm{B}}(t, 0) = \frac{1}{2\pi i} \int_{c-i\infty}^{c+i\infty} H_1(s) D_1(s) e^{\lambda_1(s)+ys} ds/s, \tag{6.39}$$

where $y = \ln \frac{E_0}{\beta}$, and (in contrast to (6.36))

$$D_1(s) = \frac{2\sigma_0 [f(\lambda_1(s))]^s}{(\lambda_1(s) + \sigma_0)\Gamma(s-2)}, \quad H_1(s) = \frac{\sigma_0 + \lambda_1(s)}{\lambda_1(s) - \lambda_2(s)}.$$

The relevant saddle-point approximation is

$$N^{\mathrm{B}}(t, 0) = \frac{H_1(\bar{s}) D_1(\bar{s})}{\bar{s}} \frac{\exp(y\bar{s} + \lambda_1(\bar{s})t)}{\sqrt{2\pi \lambda_1''(\bar{s})t}} \left[1 + \frac{\alpha(\bar{s})}{t}\right], \tag{6.40}$$

with $\bar{s}$ determined by the condition

$$t = -\frac{1}{\lambda_1'(\bar{s})} \left[y + \left(\ln \frac{H_1(\bar{s}) D_1(\bar{s})}{\bar{s}}\right)'\right]$$

and

$$\alpha(s) = -\frac{1}{2\lambda_1''(s)} \left(\ln \frac{H_1(s) D_1(s)}{s}\right) + \frac{1}{8} \frac{\lambda_1^{iv}(s)}{[\lambda_1''(s)]^2} - \frac{5}{24} \frac{[\lambda_1'''(s)]^2}{[\lambda_1''(s)]^3}.$$

6.3 Degradation Spectrum Problem

6.3.1 Electron Energy Degradation Spectrum

We end this section with discussion of the *degradation spectra*, this is the name of the energy spectrum integrated over all space:

$$P(E) = \int_0^\infty P(t, E) dt$$

(and so is for photons). Observe that instead of integrating the cascade systems over t and then preparing them for solving by means of the Mellin transformation, it is much easier to just put $\lambda = 0$ in a ready-made solution (6.33):

$$P(\lambda = 0, s) = \frac{E_0^{s+1}}{\beta} \sum_{n=0}^{\infty} \frac{[-f(0)E_0/\beta]^n [\Gamma(s+1)]^2 \Gamma(s+n+1-s_1)}{\Gamma(s+n+1)\Gamma(s+m+2)\Gamma(s+1-s_1)}. \tag{6.41}$$

According to Eq. (6.32), $s_1(0) = 1$, and

$$P(0, s) = \frac{E_0^{s+1}}{\beta} \sum_{n=0}^{\infty} \frac{[-f(0)E_0/\beta]^n s\Gamma(s+1)}{(s+n)\Gamma(s+n+2)}. \tag{6.42}$$

By using the property

$$\frac{\Gamma(s+1)}{\Gamma(s+n+2)} = \frac{1}{n!} \int_0^1 x^s (1-x)^n dx,$$

substitute (6.42) into (6.34):

$$N(E) = \sum_{n=0}^{\infty} \frac{1}{2\pi i} \int_{d-i\infty}^{d+i\infty} ds \int_0^1 dx \frac{[-f(0)(1-x)E_0/\beta]^n}{(s+n)n!} \left(\frac{E_0}{\beta}\right) \left(\frac{E_0 x}{E}\right)^s.$$

Consider the integration with respect to the complex variable s. According to Jordan's lemma, if $\left(\frac{E_0 x}{E}\right) > 1$, then one should close a semicircular integration contour of an infinite radius to the left and take the residues at poles $s = -n$. Vice versa, one should close the contour to the right, where poles are absent, and the integral vanishes. Consequently,

$$N^{\mathrm{TB}}(E) = \frac{E_0}{\beta} \int_{E/E_0}^{1} \exp\left(-f(0)\frac{E}{\beta}\frac{1-x}{x}\right) dx. \tag{6.43}$$

This result is known as the *Tamm–Belenky formula* for the equilibrium electron spectrum in EPC and often but inaccurately[1] is interpreted as a spectrum at the cascade maximum. For photons in light matters, say, for air, where total cross section of pair production and Compton effect (σ_0) is approximately constant, their differential spectrum is up to 20% accuracy linked with electron integral one (6.43):

$$\Gamma^{\mathrm{TB}}(E) = \frac{N^{\mathrm{TB}}(E)}{\sigma_0(E)}.$$

In dimensionless energy variables, $\varepsilon_0 = \frac{E_0 f(0)}{\beta}$ and $\varepsilon = \frac{E f(0)}{\beta}$, the TB formula reads

[1] The physical difference between them is first of all in fluctuations which at maximum are quite large, while in the equilibrium spectrum they are practically zero due to the law of conservation of energy. One more note: we avoid to use the term the equilibrium spectrum, which is often used in cascade theory, in order not to associate it with the thermodynamic equilibrium: a single cascade is not an equilibrium system.

$$N^{TB}(E_0, E) = \frac{E_0}{\beta}\left\{1 - \varepsilon e^{\varepsilon}\int_{\varepsilon}^{\varepsilon_0}\frac{e^{-x}}{x}dx - \frac{\varepsilon e^{\varepsilon}}{\varepsilon_0 e^{\varepsilon_0}}\right\}.$$

For $\varepsilon_0 \gg \varepsilon$

$$N(E) \cong \frac{E_0}{\beta}\chi(\varepsilon), \quad \chi(x) = 1 + xe^x E_i(-x),$$

where $E_i(-x) = -\int_x^{\infty}\frac{e^{-t}}{t}dt$.

If $\varepsilon \gg 1$,

$$N(E) = \frac{E_0}{\beta}\frac{1}{\varepsilon}\left(1 - \frac{2!}{\varepsilon} + \frac{3!}{\varepsilon^2}\right) \cong \frac{E_0}{f(0)E}\left(1 - \frac{2}{\varepsilon}\right), \quad f(0) = 2.29.$$

Finally, for small ε

$$N(E) = \frac{E_0}{\beta}(1 + \varepsilon\ln\varepsilon + \varepsilon C + \cdots), \quad C = 0.577.$$

We see, that the differential spectrum diverges at small energies, that half of total particle number have energies less than $E_{1/2}$, 80% of particles have energies lower than the critical energy, which is about 20 MeV for air and 1.7 MeV for lead. The mean energy per electron is given by the formula $\bar{E}_e = \frac{\beta}{f(0)}[\ln\varepsilon_0 - 0.42]$. Beyond the cascade maximum, the following expression is valid:

$$N^{B}(t, E) = \frac{H_1(\bar{s})D(\bar{s})G(\bar{s}, \varepsilon)}{s\sqrt{2\pi\lambda_1''(\bar{s})t}}e^{ys+\lambda_1(\bar{s})t},$$

where $\bar{s}$ obeys the equation $\lambda_1'(\bar{s})t = -y$, $\varepsilon = f(\lambda_1(\bar{s}))E/\beta$ and

$$G(s, \varepsilon) = e^{\varepsilon}\int_{\varepsilon}^{\infty} e^{-x}\left(1 - \frac{\varepsilon}{x}\right)^s dx.$$

6.3.2 *Photon Energy Degradation Spectrum*

Let us pass to consideration of the photon energy degradation problem. Photons are just as important as component of the cascade as electrons, significantly in Compton scattering processes. In terms of the degradation concept, this process is identical to absorption: the photon disappears from the energy element dE and does not return here anymore, continuing to descend lower and lower on the energy scale. To take it into account, we introduce the photon absorption coefficient into $\sigma_\gamma(E)$.

Taking formula (5.66) for Compton scattering and neglecting the second term in it, playing a significant role only at energies $\sim mc^2$. We derive from the degradation cascade system the following equation:

$$\Phi(E) - \sigma(E)\Gamma(E) + \frac{g}{E}\int_E^\infty \Gamma(E')dE'/E' = 0, \tag{6.44}$$

where

$$\Phi(E) = \int_E^\infty dE' P(E') W_e(E' \to E)$$

is the average number of photons emitted in a unit interval around E by all electrons in the cascade. Introducing function

$$R(E) = \int_E^\infty dE' \Gamma(E') dE''/E',$$

we reduce the integral equation to differential one relatively to $R(E)$,

$$\Phi(E) + \frac{dR}{dE} E\sigma(E) + \frac{g}{E} R(E) = 0$$

which is easy for solving:

$$R(E) = \int_E^\infty \frac{dE_1 \Phi(E_1)}{E_1 \sigma(E_1)} \exp\left\{-g \int_{E_1}^\infty \frac{dE_2}{E_2^2 \sigma(E_2)}\right\}.$$

For air, the total absorption coefficient (pair production + Compton scattering) weakly depends on the photon energy and can be taken to be constant σ_0. On his assumption, the further consideration leads to representation of the photon degradation spectrum $\Gamma(E)$ in the form a sum of two terms,

$$\Gamma(E) = \Gamma_1(E) + \Gamma_2(E),$$

where

$$\Gamma_1(E) = \frac{N_e(E)}{\sigma_0 E}, \tag{6.45}$$

and

$$\Gamma_2(E) = \frac{k}{E\sigma_0} \int_{\varepsilon}^{\infty} e^{k(1/\varepsilon - 1/\varepsilon_1)} N_e(\varepsilon_1) d\varepsilon_1 / \varepsilon_1^2$$

with $k = 0.0543$ (for air).

6.3.3 On Accuracy of Degradation Spectra

To study of degradation problems, i.e., consideration of EPC equations integrated over depth (called often the equilibrium spectra), a lot of works are dedicated.

The interest in these problems is explained by the fact that they are much easier to solve than the original spatial-energy equations, as well as the fact that for finding many one-dimensional and three-dimensional characteristics of the EPC, an effective moment method has been developed [3–5].

Referring here to the article [6] we note that authors undertaken investigation of the accuracy of the degradation Tamm–Belenky spectra

$$P^{\mathrm{TB}}(E) = \frac{\varepsilon_0}{\beta} \left\{ (\varepsilon + 1) e^{\varepsilon} \left[\int_{\varepsilon}^{\varepsilon_0} \frac{e^{-x}}{x} dx + \frac{e^{-\varepsilon_0}}{\varepsilon_0} \right] - 1 \right\},$$

$$N_e^{\mathrm{TB}}(E) = \frac{E_0}{\beta} \left\{ 1 - \varepsilon e^{\varepsilon} \left[\int_{\varepsilon}^{\varepsilon_0} \frac{e^{-x}}{x} dx + \frac{e^{-\varepsilon_0}}{\varepsilon_0} \right] \right\},$$

$$\Gamma^{\mathrm{TB}}(E) = \frac{N^{\mathrm{TB}}(E)}{0.77E},$$

$$N_\gamma^{\mathrm{TB}}(E) = \frac{E_0}{0.77\beta} \left\{ \frac{1}{\varepsilon_0} + e^{\varepsilon} \left[\int_{\varepsilon}^{\varepsilon_0} \frac{e^{-x}}{x} dx + \frac{e^{-\varepsilon_0}}{\varepsilon_0} \right] \right\},$$

where (recall) P and Γ are differential spectra of electrons and photons, respectively, and $N_{e(\gamma)}$ are corresponding integral spectra, and $\varepsilon = 2.29E/\beta$.

The numerical solutions of the forward degradation equations were built sequentially on segments ΔE_i (beginning from E_0 toward decreasing energy). The desired spectra $P(E)$ and $\Gamma(E)$ were presented as products of the analytically obtained spectra and correction functions $p_i(E)$, $\gamma_i(E)$, obtained with the use of linear approximations for each segment separately:

$$P(E) = \frac{E_0}{2.29E^2} p_i(E), \quad \Gamma(E) = \frac{E_0}{2.29 \cdot 0.77E^2} \gamma_i(E).$$

Accuracy at each segment was estimated by analyzing the dependence of the result on the step size, step value (segment length) was chosen automatically, depending on the achieved accuracy (at the level of tenths of a percent, by authors' estimation). Formulating the goals of their calculations, the authors decomposed the concept of accuracy into two components: the actual computational (*arithmetic*) component and conceptual (*physical*) one, arising due to the incomplete adequacy of physical data on elementary processes, introduced into the model. In their conclusions, the authors focused on the second aspect of the problem.

The comparison of analytical and numerical results shows that in the region $E < E_0/2$ numerically calculated spectra are close to analytical TB approximations characterizing its mathematical accuracy. Taking into account the rest mass of electrons affects the cascade characteristics mainly through the pair production process. Because of its threshold property, this decreases the electron number. At the same time, decreasing of photo-absorption cross section leads to the divergence in the electron spectra. Screening refinement also leads to decrease in the pair production cross section, which further increases the deviation of the calculated spectra from the TB formulas. Accounting for the photon absorption due to photo- effect strongly changes the photon spectrum in the region $E \leq 1.1$ MeV.

In general, it can be noted that the electron spectrum is relatively insensitive to taking into account some "subtle" effects in elementary events and differs from the value obtained by "rough" description by no more than $30 \div 40\%$, while the corresponding values of the photon spectrum can differ by two or more times.

The original article cited here contains some other useful information [7, 8] (Figs. 6.1 and 6.2).

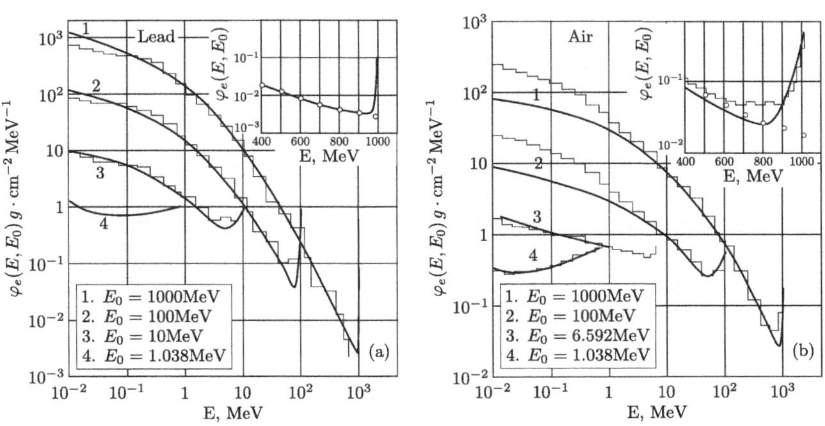

Fig. 6.1 Differential equilibrium electron spectra: **a** in lead; **b** in air. Histograms, calculation by Monte Carlo method [7, 8]; continuous lines, Belenky and MacGinnies spectra; points, equilibrium spectra calculated in A cascade theory approximation

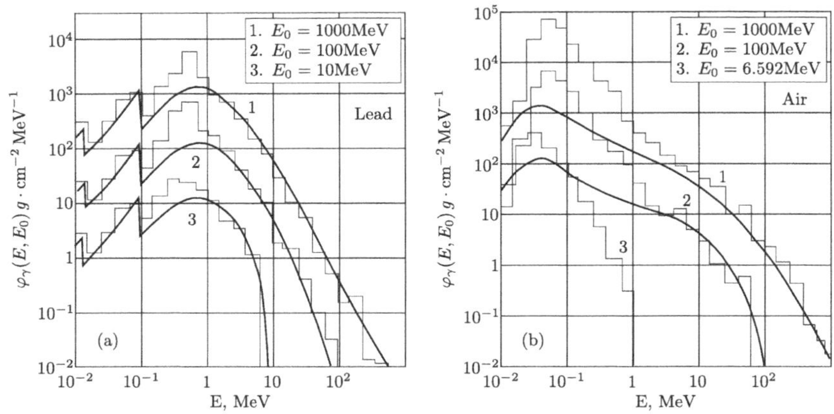

Fig. 6.2 Differential equilibrium photon spectra: **a** in lead; **b** in air. Histograms, calculation by Monte Carlo method [7, 8]; continuous lines, Belenky spectra

6.4 Longitudinal Moment Method

6.4.1 Cascade Curve Phenomenology

From the first works on the cascade theory of the development of electron–photon showers in matter [9–12] is known, that the cascade curve $N(t, E)$(the number of charged particles with energies exceeding E at the depth t) may be represented by means of equation $N(t, E) = Ae^{\varphi(y,t)}$, where t is expressed in special cascade units and each one is responsible (on average) for the doubling of a cascade particle (i.e., pair production $\gamma \to e^+ + e^-$ and bremsstrahlung $e^- \to e^- + \gamma$). Therefore, after passing through this thickness, 2^t particles are obtained, $\ln N(t) \propto t$ and $\varphi(y, t) \propto t \ln(E/\varepsilon)$ with ε some energy value, playing a role of a detector threshold energy (in A-approximation) or a critical energy, below which multiplying event probabilities fade out (B-approximation). Elementary example of such kind is the Furry process, where the number of particles doubles in each new generation. Identifying the number of particles leaving a layer of thickness t with the number N of descendants of the t-th generation from one ancestor, we find that $N(t) = 2^t$, i.e., it exponentially increases in multiplicity and becomes the greatest at: $t_m \equiv t_{max} = \log_2(E_0/\varepsilon)$, which roughly imitates the initial part of the cascade. However, not all particles are detected by the detector: it usually has an energy sensitivity threshold ε and particles with energies below the threshold are not detected. The decisive role in the deviation of the EPC from Furry is played by statistical fluctuations in the number and position of doubling points, and, first of all, fluctuations in the photon paths. Unlike electrons, paths of which, as of other charged particles, are limited by their energies, photon paths are free from such restrictions, and have wide exponential distributions. These fluctuations (together with energy and direction fluctuations in the processes of scattering of both electrons and photons) smear the dangling exponentially growing Furry-type curve, and give the EPC cascade curve its inherent form of a gamma-distribution. As the cascade develops, more and more of its particles find themselves below the ε level, and the period of rapid increasing changes with slow decay. As a result, one can hope to find a long-t asymptotic solution in the form

$$N(t) \sim Ae^{\varphi(t,y)}, \quad \varphi(t, y) = ys + \lambda_1(s)t$$

with $y = \ln(E_0/\varepsilon)$ and $\lambda_1(s)$ being a function of a special parameter s, characterizing the level of the cascade development and called the *cascade parameter* or the *cascade age*. When $s < 1$, the cascade is young, its curve grows with t and energy spectrum $e^{ys} = (E/\varepsilon)^s$ becomes harder (the fraction of high-energy particles increases), whereas, when $s > 1$, the cascade is old, its curve comes down, and spectrum becomes softer. At $s = 1$, the cascade curve reaches the maximum value N_{m} at point t_{m}, where $\partial_t\varphi = 0$ (the derivative is taken under $y = \mathrm{const}$, or, equivalently, $t\lambda'(s) \equiv d\lambda/ds$). It is convenient to expand the exponent $\varphi(t, y)$ in a series by $t - t_{\mathrm{m}}$ up to terms of the second order. Assuming dependence $s = s(t)$, we get

$$\varphi = \varphi_{\mathrm{m}} + 0 + \frac{1}{2}(\partial_t^2\varphi)_{\mathrm{m}}(t - t_{\mathrm{m}})^2.$$

In the framework of B-approximation, the cascade continuously loses its energy in such a way that each electron on path dt loses βdt of energy, and according to the conservation energy law,

$$\int_0^\infty N(t)\beta dt = E_0.$$

Let us comment on the physical aspect of the process in more detail.

The successful use of the Mellin–Laplace transformation in analytical solving EPC problems is easily explained by two important facts: (1) the main quality property of interesting investigators is the multiplicity of elementary particles, growing with their energies and (2) at high-energy region, cross sections of electron–photon interactions responsible for first property are expressed in terms of the homogeneous Bethe–Heitler asymptotics, which give sufficient accuracy in case of cascades developing in light substances (air, plastic, aluminum) but fail in heavy matter such as iron, copper, and lead. Roughly speaking, the conditions for applicability of the conclusions of the cascade theory are estimated as

$$E \gg E_{\mathrm{eff}} \sim 68Z^{-1/3}\mathrm{MeV} \wedge E \gg \beta. \tag{6.46}$$

Under these conditions, cascade equations expressed in terms of *cascade units* (c.u.) (*t-units*) become approximately the same for all substances. As the cascade develops, energy degradation occurs as a result of the particle multiplication in processes of bremsstrahlung, pair production, and Compton scattering. Multiple successive repetition of these events turns a primary particle into a large number of photons and electrons of both signs. A more adequate description of the low-energy part of the cascade process is achieved by numerical methods.

It is known that the less of argument number in the solution, the more effective become numerical methods. In solving differential equations, the number of arguments is reduced by separating variables. This method is convenient due to the general property of differentiating products. Integration does not have this property, and therefore, in solving integral equations, the separation of variables is carried out by integration over "extra" variables. In this case, integration with a weight function is convenient, similar to what is performed in integral transformations. If a certain power of a real argument (or their superposition) is taken as such a weight function, such an integral is called a moment, and the method itself is called the method of moments. In the case of homogeneous media, the derivation of equations for the moments of difficulties is not difficult, the main problem of this method is the inverse problem—the reconstruction of the function from the moments.

The presence of connections between variables complicates the problem: the moments of different variables are also connected, and when solving a system of equations for the moments of different arguments, this connection has to be taken into account. There are, however, favorable cases when this connection turns out to be weak due to objective physical laws. High-energy cascades are among such phenomena. Due to the relativistic features of elementary processes, they deflect particles only by small (of the order of mc^2/E) angles, and the cascade equations themselves contain both differential (along the longitudinal coordinate) and integral (in energy and angular variables). It is this circumstance that gives meaning to solving the degradation problems. The degradation of the cascade energy is represented by the integral of the solution over depth, and this is nothing but the zero moment of the cascade curve.

The very idea of developing the method of moments for the numerical analysis of the longitudinal development of the EPC (that is, for calculating the cascade curve) beyond the limits of applicability of the standard approximations of the cascade theory was very clearly formulated in the Belenky book [13] and then derived by Ivanenko [2]. Referring to Landau's article [14], Belenky represented the cascade curve $N(t)$ (the average number of electrons at a cascade depth t) in the form

$$N(t) = e^{\varphi(t,E_0/\beta)} \approx e^{\varphi_\mathrm{m}-(a/2)(t-t_\mathrm{m})^2}, \quad a = -\left(\partial_t^2\varphi\right)_{t=t_\mathrm{m}},$$

linked with moments $N^{(k)} \equiv \int\limits_0^\infty N(t)t^k dt$:

$$N^{(0)} = N_\mathrm{m}\sqrt{2\pi/a} = \frac{E_0}{\beta},$$

$$\bar{t} \equiv \frac{\int_0^\infty N(t)tdt}{\int_0^\infty N(t)dt} = t_{\rm m},$$

$$\overline{t^2} \equiv \frac{\int_0^\infty N(t)t^2dt}{\int_0^\infty N(t)dt} = \frac{1}{a} + t_{\rm m}^2.$$

Thus, determining cascade curve maximum $N_{\rm m}$

$$N_{\rm m} = \frac{1}{\sqrt{2\pi(\overline{t^2} - t_{\rm m}^2)}} \frac{E_0}{\beta}$$

with its position $t_{\rm m} = \bar{t}$ is reduced to two normalized moment calculations $\bar{t}$ and $\overline{t^2}$.

Notice that $N(t)$ in the above expressions can be replaced by $N(t, E)$, where E is the energy of observed particles, or the lower threshold energy, above which the particles are registered. The distributions are called the *differential* and *integral* spectra, respectively. Their t-dependence can also be characterized by corresponding (differential or integral) longitudinal moments:

$$N^{(k)}(E) = \int_0^\infty N(t, E)t^k dt, \; k = 0, 1, 2, \ldots.$$

Observe that the degradation spectra are nothing but zeroth moments. One should stress also that the cascade curve $N(t)$ here is not a random curve but the deterministic one, representing the averaged over the ensemble curve. The same sense is prescribed to the longitudinal moments: they are not statistical moments (which should be deterministic ones by definition), they are integrals from productions $N(t)t^k, k = 0, 1, 2, \ldots$, and may be used both for averaged cascade curves and their random realizations (see Chap. 8).

6.4.2 Cascade Curves' Reconstruction

Let us describe the regular method of constructing cascade curves suggested by Ivanenko [2].

According to the standard approximation theory, the function $f(x),\ x \in (0, \infty)$, is approximated by superposition of the Laguerre polynomials

$$L_n^i(x) = \sum_{k=0}^{n} \frac{\Gamma(n+i+1)}{\Gamma(k+i+1)} \frac{(-x)^k}{r!(n-k)!} :$$

$$f(x) = A_0 + \sum_n A_n L_n^i(x). \tag{6.47}$$

They are orthogonal in the interval $(0, \infty)$ with weight $e^{-x}x^i$:

$$\int_0^\infty e^{-x} x^i L_m^i(x) L_n^i(x) dx = \begin{cases} 0, & \text{for } m \neq n,\ i > -1; \\ \frac{\Gamma(n+i+1)}{n!}, & \text{for } m = n,\ i > -1. \end{cases} \quad n = 0, 1, 2, \ldots$$

This orthogonality plays a key role in finding the expansion coefficients. The approximate improves as the mean-square error

$$M = \int_0^\infty e^{-x} x^i (A_0 + A_1 L_1^i(x) + A_2 L_2^i(x) + A_3 L_3^i(x) + \cdots - f(x))^2 dx$$

grows smaller. Coefficients A_n, determined from the condition of minimum mean-square error, are calculated by integration

$$A_n = \frac{n!}{\Gamma(n+i+1)} \int_0^\infty e^{-x} x^i f(x) L_n^i(x) dx.$$

Replacing x in Eq. (6.47) by γt, where γ is an arbitrary positive number, multiplying both parts by $(\gamma t)^i e^{-\gamma t}$ and introducing the notation

$$(\gamma t)^i e^{-\gamma t} f(\gamma t) = N(t), \tag{6.48}$$

one can get

$$N(t) = (\gamma t)^i e^{-\gamma t} \sum_{n=0}^{\infty} A_n L_n^i(\gamma t) dt. \tag{6.49}$$

The coefficients A_n are in this case equal to

$$A_n = \frac{\gamma n!}{\Gamma(n+i+1)} \int_0^\infty N(t) L_n^1(\gamma t). \tag{6.50}$$

The cascade curves from primary photons were approximated via expansion over the Laguerre polynomials $L_n^1(x)$,

$$N(t) = \gamma t e^{-\gamma t} \sum_{n=0}^{k} A_n L_n^1(\gamma t), \tag{6.51}$$

where

$$L_0^1(x) = 1,\ L_1^1(x) = 2 - x,\ L_2^1(x) = 3 - 3x + \frac{x^2}{2},\ L_3^1(x) = 4 - 6x + 2x^2 - \frac{x^3}{6}.$$

In accordance with Eq. (6.50)

$$A_n = \frac{\gamma}{n+1} \int_0^{\infty} N(t) L_n^1(\gamma t) dt.$$

For approximation by means of the polynomials, condition $N(0)I = 0$ is required. For this, one can insert the additional term

$$N(t) = e^{-\gamma t} \left(\sum_{n=0}^{k} A_n L_n^0 + C L_{k+1}^0(\gamma t) \right), \tag{6.52}$$

into approximating formula.

The coefficient γ is taken to be equal to the overall absorption coefficient of the most penetrating part of the radiation (photons). Observe that the cascade theory makes it possible to calculate the dependence of the total number of particles in the shower on the depth t for particles of large energy $\ln(E_0/\beta) > 1$. On the other hand, according to the approximation Eq. (6.51), one can compute the cascade curves from primary particles of high energy by making use of the first two moments ($k = 2$).

Figure 6.3 demonstrates two cascade curves, one of which (1) is an exact solution of Eq. (6.40), another (2) shows the approximate formula (6.51) result. The curves differ after the maximum by no more than 5%, before the maximum by no more than 15%, but in this latter region Eq. (6.40) has already ceased to be applicable (condition of its applicability is $t > 1$) (Figs. 6.3 and 6.4).

To move from general reasoning to computational problems, we return to the basic cascade equations for EPC, writing them in the operator form:

$$\partial_t P(t, E) = \mathscr{L}_1[P(t, E), \Gamma(t, E)] + \beta \partial_E P(t, E), \tag{6.53}$$

$$\partial_t \Gamma(t, E) = \mathscr{L}_2[P(t, E), \Gamma(t, E)]. \tag{6.54}$$

Let

$$P(0, E) = \delta(E - E_0), \quad \Gamma(0, E) = 0.$$

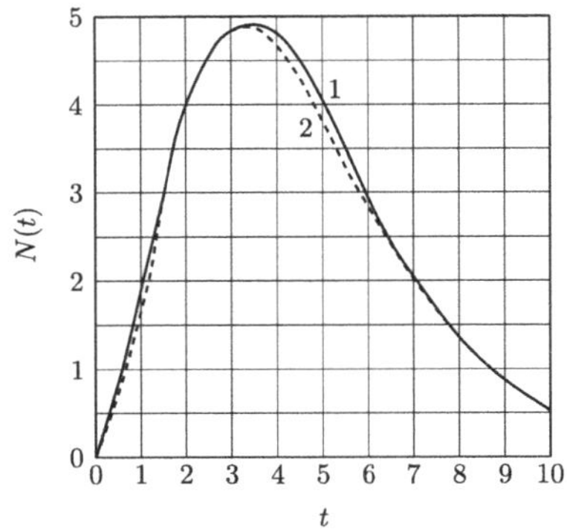

Fig. 6.3 Cascade curves from a primary photon of energy $\varepsilon_0 = 63.62$. Curve 1 is drawn according to Eq. (6.40); curve 2, according to Eq. (6.51)

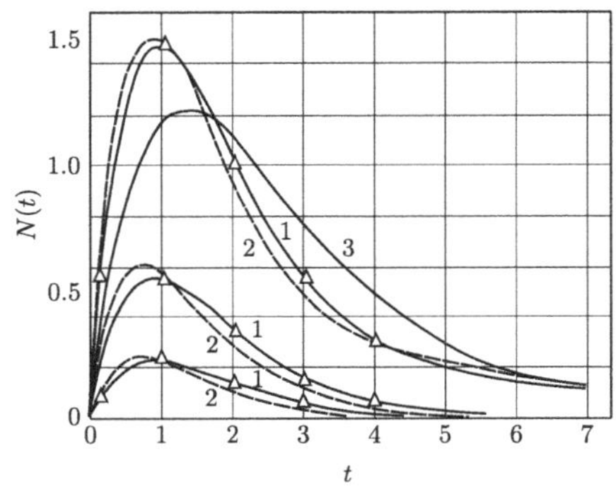

Fig. 6.4 Cascade curves from a primary photon of energy $\varepsilon_0 = 1, 3, 10$. The curves are drawn: 1—according to Eq. (6.51) for $k = 2$ [triangular points computed by Eq. (6.52)]; 2—according to Eq. (6.51) for $k = 3$; 3—according to Eq. (6.51) for $k = 1$

Integrating (6.53)–(6.54) along the entire length of the cascade with these conditions yields

$$\begin{aligned} \mathscr{L}_1[P^{(0)}(E), \Gamma^{(0)}(E)] + \beta\partial_E P^{(0)}(E) &= -\delta(E - E_0), \\ \mathscr{L}_2[P^{(0)}(E), \Gamma^{(0)}(E)] &= 0. \end{aligned} \tag{6.55}$$

The equations for zero-order moments form a known equation system for degradation problem, written this time in the operator form.

To pass to the first-order moment equations, we multiply (6.53)–(6.54) by t and integrate all terms as before:

$$\begin{aligned} \mathscr{L}_1[P^{(1)}(E), \Gamma^{(1)}(E)] + \beta\partial_E P^{(1)}(E) &= -P^{(0)}(E), \\ \mathscr{L}_2[P^{(1)}(E), \Gamma^{(1)}(E)] &= -\Gamma^{(0)}(E). \end{aligned} \tag{6.56}$$

Finally, multiplying those equations by t^2, we obtain after integration over the depth:

$$\begin{aligned} \mathscr{L}_1[P^{(2)}(E), \Gamma^{(2)}(E)] + \beta\partial_E P^{(2)}(E) &= -2P^{(1)}(E), \\ \mathscr{L}_2[P^{(1)}(E), \Gamma^{(1)}(E)] &= -2\Gamma^{(1)}(E). \end{aligned} \tag{6.57}$$

So, one can make from this sufficiently simplified example, at least three significant conclusion. First: the way of derivation of equations for longitudinal moments the derivation is based on the same balance considerations as the derivation of cascade equations, without requiring fundamentally new information. Second: the equations for higher moments have the same mathematical structure as the equation for zero moment, which describes the degradation of energy in a cascade. Therefore, to solve them, you can use the same (analytical or numerical) methods that are used to solve the problem of degradation. Third: a fundamentally new stage of this method is the restoration of the cascade curve from its moments. In the artificially simplified example under consideration, this problem does not look as difficult, but a real problem statement may require much more effort if we are interested in the distribution of cascade particles over all variables (coordinates, directions, energy, and even time). The technique for solving such problems is connected with the concepts of stability, correctness of the formulation, and regularization of the solution, which make up a complete set of tools for the theory of solving inverse problems.

6.5 Including Electron Scattering

In order to pass to a more adequate model of EPC, one should take deviations of particles from the original direction into account deviations of particles from their original direction due to the processes of scattering of electrons on nuclei (Rutherford scattering) and photons on electrons (Compton scattering). They are responsible for the broadening of the cascade process. At high particle energies, most of them are scattered at small angles. According to the relativistic Rutherford formula, the differential cross section for scattering of an electron (positron) with energy E through the angle Θ is

$$d\sigma = 8\pi \left(\frac{Ze^2}{E}\right)^2 \frac{N}{Z}\frac{d\Theta}{\Theta^3},$$

where N is the electron number per unit volume of the substance, and N/Z equals to the number of nuclei in the atom.

To describe effects of scattering in cascade processes, Landau added to the number of active variables of the cascade theory the angle θ between the direction of cascade particle motion and the axis of the shower, in a homogeneous medium given by the direction of the momentum of the primary particle [14]. The deviation of particles in a shower from their original direction is mainly due to Rutherford scattering by heavy nuclei. As is known, this scattering is distinguished by the property that it occurs predominantly at small angles. This fact was used by Landau to involve the small-angle approximation in the cascade theory called often the *Landau approximation*.

Let $P(t, E, \theta)$ be the distribution function of charged particles with respect to energies E and scattering angle θ at the depth t. In small-angle approximation, the motion directions are conveniently characterized by two-dimensional vectors $\boldsymbol{\theta}$ with projections $\theta_x = \theta \cos\chi, \theta_y = \theta \sin\chi$. The change in the distribution function per unit depth of cascade development due to scattering is characterized by the partial derivative

$$(\partial_t P)_{\text{scatt}} = \int \{P(\boldsymbol{\theta} + \boldsymbol{\Theta}) - P(\boldsymbol{\theta})\} d\sigma.$$

Here $d\sigma$ is the cross section for scattering through the angle $\boldsymbol{\Theta}$; the first term under the integral is the number of electrons that acquire an angle of $\boldsymbol{\Theta}$ as a result of scattering, and the second is the number of particles leaving the given element of the solid angle. Expanding the first term in powers of $\boldsymbol{\Theta}$ yields

$$(\partial_t P)_{\text{scatt}} = \frac{1}{4} \int \Theta^2 \cdot d\sigma \, \Delta_\theta P. \tag{6.58}$$

According to relativistic theory

$$d\sigma = 8\pi \left(\frac{Ze^2}{E}\right)^2 \frac{N}{Z} \frac{d\Theta}{\Theta^3}$$

(N stands for the electron number per matter volume unit). When calculating the divergent integral, one should take as the upper limit the average opening angle of the particle beam (of the order of unity) and for the lower limit put $\hbar/a$ (a is of the order of screening distance):

$$\Theta_{\min} \sim \frac{\hbar c^2}{aE} \sim \frac{mc^2}{E} \frac{1}{a} \frac{\hbar}{mc}.$$

As a result, Eq. (6.58) becomes

$$(\partial_t P)_{\text{scatt}} = \left(\frac{K}{2E}\right)^2 \Delta_\theta P, \tag{6.59}$$

whose solution coincides in form with the Fermi distribution.

Observe that Eq. (6.59) has a form of two-dimensional diffusion equation with diffusivity $D_\theta(E) = (K/2E)^2$, $K = 21$ MeV, rapidly increasing as the electron energy decreases in the cascade process.

The resulting expression must be added to the right side of Eq. (6.24), in which P (as and Γ in the second cascade equation) is now a function not only of t, E, but also of θ:

$$\partial_t P = \mathscr{L}_1[P, \Gamma] + \beta \partial_E P(t, E, \theta) + D_\theta(E) \Delta_\theta P(t, E, \theta), \tag{6.60}$$

$$\partial_t \Gamma = \mathscr{L}_2[P, \Gamma]. \tag{6.61}$$

Multiplying the system by $\theta^2 d\Omega$ ($d\Omega = 2\pi \sin\theta d\theta$) and integrating over the total solid angle yield a system for the second angular momenta

$$P^{(2n)} = \int P(t, E, \theta)\theta^{2n} d\Omega, \quad \Gamma^{(2n)} = \int \Gamma(t, E, \theta)\theta^{2n} d\Omega, \quad n = 0, 1, 2, \ldots$$

$$\partial_t P^{(2)} = \mathscr{L}_1[P^{(2)}, \Gamma^{(2)}] + \beta \partial_E P^{(2)}(t, E) + 4D_\theta P^{(0)}(t, E, \theta), \tag{6.62}$$

$$\partial_t \Gamma = \mathscr{L}_2[P^{(2)}, \Gamma^{(2)}] \tag{6.63}$$

involving the zero moment

$$P^{(0)}(t, E) \equiv P(t, E).$$

Investigating (in A-approximation, when $\beta = 0$) the mean-square values

$$\overline{\theta_e^2}(t, E) = \frac{P^{(2)}(t, E)}{P^{(0)}(t, E)} = \frac{D_\theta(\sigma_0 - \lambda(s))}{(A(s+2) - \lambda(s))(\sigma_0 - \lambda(s)) - B(s+2)C(s+2)}, \tag{6.64}$$

$$\overline{\theta_\gamma^2}(t, E) = \frac{\Gamma^{(2)}(t, E)}{\Gamma^{(0)}(t, E)} = \frac{D_\theta C(s+2)B(s)}{(A(s) - \lambda(s))[(A(s+2) - \lambda(s)) - B(s+2)C(s+2)]}, \tag{6.65}$$

of the distribution angles, Landau [14] concluded that these ratios are weakly changing along the cascade, so can approximately be taken constant. Equation (6.64), Landau wrote, shows that $\overline{\theta_e^2}$ in t-interval from $0.1y$ to $3.8y$ changes only 1.7 times. He explained this by the fact that Rutherford scattering has a stronger effect on low-energy particles, scattering has a stronger effect on low-energy particles, and any particle is obtained from particles of higher energy at a distance of the order of one cascade unit (more precisely, 0.65) from a given place. Further, for the same reasons, the scattering of particles does not depend on the energy of the particle that caused the shower.

Based on these conclusions, it is quite logical to assume that the degradation (equilibrium) distributions in θ (as in E) are similar to their counterparts in the cascade maximum.

At the shower maximum point ($s = 1$), formulas (6.64) and (6.65) give

$$\overline{\theta_e^2} = 2.38 D_\theta, \quad \overline{\theta_\gamma^2} = 0.72 D_\theta.$$

Inserting numerical value of D_θ yields

$$\overline{\theta_e^2} = \frac{250}{E^2} \tag{6.66}$$

(E in MeV).

Already at $E = 30$ MeV, the angle $\theta_e \sim 1$. Moreover, owing to the weak dependence of δ on Z, the opening angles of particles of a given energy are practically independent of the material in which the shower propagates. The mean opening angle of most particles, corresponding to $E \sim \varepsilon$ e, apparently increases with atomic number. However, it can be seen from (6.66) that even for air it reaches about 1, while for lead, for example, most of the particles are already all scattered over directions. This circumstance indicates the accuracy limit of formula (6.14) for heavy elements.

Of interest is one important remark by Belenky concerning the multiple scattering of shower particles. In the simplest case, he notes, without ionization and radiation energy loss, the distribution of particles is determined by the equation

$$\partial_t P = D_\theta \Delta_\theta P(t,\theta) \equiv D_\theta \left(\partial^2_{\theta_x} + \partial^2_{\theta_y}\right) P(t,\theta_x,\theta_y).$$

However, the situation changes as soon as bremsstrahlung and pair formation are taken into account. Indeed, the two-dimensional (with respect to θ_x, θ_y projections) Gaussian distribution is directly related to the independence of these mutually orthogonal components,

$$P(t,\theta_x,\theta_y) = P(t,\theta_x)P(t,\theta_y) \tag{6.67}$$

and vice versa. Belenky had proved that such representation is incompatible with cascade equation. We reduce the proof to A-approximation, omitting ionization term (it doesn't play any role here).

Inserting (6.67) into (6.60) with $\beta = 0$ yields

$$P(t,E,\theta_x)\partial_t P(t,E,\theta_y) + P(t,E,\theta_y)\partial_t P(t,E,\theta_x)$$

$$= \mathscr{L}_1[P(t,E,\theta_x)P(t,E,\theta_y)]$$

$$+ \; D_\theta P(t,E,\theta_x)(\partial^2_{\theta_y})P(t,E,\theta_y) + P(t,E,\theta_y)D_\theta(\partial^2_{\theta_x})P(t,E,\theta_x).$$

Thus, in order solution $(t, E, \theta_x, \theta_y)$ to be factorized like (6.67), operator $\mathscr{L}_1$ should possess the property

$$\mathscr{L}_1[P(t,E,\theta_x)P(t,E,\theta_y)] = P(t,E,\theta_x)[\mathscr{L}_1 P(t,E,\theta_y)] + P(t,E,\theta_y)[\mathscr{L}_1 P(t,E,\theta_x)].$$

But this is not the case, so that such factorization of solution $P(t, E, \theta_x, \theta_y)$ is impossible.

The answer to the question, why the seemingly ubiquitous Gaussian law does not work here, is very simple: this law is based on the central limit theorem, which is valid for statistically independent terms. The branching of trajectories in cascade processes includes strong correlations between segments of the same trajectory, thereby violating the CLT conditions.

To consider the angular distributions of EPC particles beyond qualitative moment estimations but in the framework of small-angle version of A-approximation, we apply the Fourier–Bessel representations:

$$\widetilde{P}(t,E,k) = \int_0^\infty P(t,E,\theta)J_0(k\theta)\theta d\theta,$$

and so for photon distribution. The transformed equations join the system

$$\partial_t \widetilde{P}(t, E, k) = \mathscr{L}_1[\widetilde{P}, \widetilde{\Gamma}] - D_\theta k^2 \widetilde{P}(t, E, k), \tag{6.68}$$

$$\partial_t \widetilde{\Gamma}(t, E, k) = \mathscr{L}_2[\widetilde{P}, \widetilde{\Gamma}].$$

Representing solutions of the system in the form of series,

$$\widetilde{P}(t, E, k) = \sum_{n=0}^{\infty} (D_\theta(E_r)k^2)^n M_n^e(t, E, k),$$

where E_r is some energy larger than E_κ, and so for photons, we insert them in the above equations and equate the factors at equal powers of the parameter k, we obtain

$$\partial_t M_0^e(t, E) - \mathscr{L}_1[M_0^e(t, E), M_0^\gamma(t, E)] = 0, \tag{6.69}$$

$$\partial_t M_0^\gamma(t, E) - \mathscr{L}_2[M_0^e(t, E), M_0^\gamma(t, E)] = 0,$$

$$\ldots$$

$$\partial_t M_n^e(t, E) - \mathscr{L}_1[M_n^e(t, E), M_n^\gamma(t, E)] + M_{n-1}^e(t, E)(E_r/E)^2 = 0,$$

$$\partial_t M_n^\gamma(t, E) - \mathscr{L}_2[M_n^e(t, E), M_n^\gamma(t, E)] = 0, \tag{6.70}$$

where n runs all integers from 1 to ∞.

Substituting trial solutions in the form

$$M_1^e \sim M_1^\gamma \sim E^{-s-3} e^{\lambda_1(s)t}$$

we find

$$M_1^e = -\frac{1}{2\pi} \frac{M_0^e}{\psi(\lambda_1, s+2)} \left(\frac{E_r}{E}\right)^2,$$

where

$$\psi(\lambda, s) = A(s) + \lambda - \frac{B(s)C(s)}{\lambda + \sigma_0} = \frac{[\lambda - \lambda_1(s)][\lambda - \lambda_2(s)]}{\lambda + \sigma_0}.$$

Subsequent transformations associated with the transition to the number of electrons $N_e(t, E, \theta)$ with energies above E at a cascade depth corresponding to the age of s had led Belenky to the following solution:

$$\widetilde{N}_e(t, E, k) = \frac{N_e(t, E)}{2\pi[1 + (Pk/E)^2]^{s/2}},$$

where $P = \frac{K}{2\sqrt{f(\lambda_1)}}$. Inverting this transform yields

$$N_e(t, E, \theta) = \frac{N_e(t, E)}{2\pi\theta^{2-s}} \int_0^\infty \frac{J_0(u)udu}{[(E\theta/P)^2 + u^2]^{s/2}}.$$

6.6 Beyond Small-Angle Approximation

Now, we go beyond the small-angle approximation with replacing the small-angle version of the angular Laplacian by its total form (still assuming the scattering to be axially symmetric):

$$\Delta_\theta = \frac{1}{\sin\theta}\partial_\theta\left(\sin\theta\,\partial_\theta\right),$$

and ∂_t in the first terms of cascade equations by $\cos\theta \cdot \partial_t$:

$$\cos\theta \cdot \partial_t P = \mathscr{L}_2[P, \Gamma] + D_\theta(E)\Delta_\theta P, \tag{6.71}$$

$$\cos\theta \cdot \partial_t \Gamma = \mathscr{L}_1[P, \Gamma]. \tag{6.72}$$

Observe, we omit the term with β, so our results will be valid at high values of E only.

Taking boundary condition in the form

$$P(0, E, \theta) = \delta(E - E_0)\delta(\theta), \quad \Gamma(0, E, \theta) = 0$$

and integrating Eq. (6.71)–(6.72) over semi-axis $t > 0$, we arrive at the *angle-energy degradation* cascade system $P(E, \theta) \equiv \int_0^\infty P(E, \theta, t)dt$ and so $\Gamma(E, \theta)$:

$$-\cos\theta\delta(E - E_0)\delta(t) = \mathscr{L}_1[P(E, \theta), \Gamma(E, \theta)],$$

and

$$\mathscr{L}_2[P(E, \theta), \Gamma(E, \theta)] = 0,$$

or eliminating Γ:

$$\mathscr{L}P(E, \theta) + D_\theta(E)\Delta_\theta = -\cos\theta\delta(E - E_0)\delta(\theta). \tag{6.73}$$

Seeking out its solution by expansion in Legendre polynomials,

$$P(E, \theta) = \sum_{n=0}^{\infty} C_n(E)P_n(\cos\theta),$$

we insert it into (6.73), and after known procedures (multiplying by $P_m(\cos\theta)d\Omega$, integrating over directions and using mutual orthogonality of P_n) we arrive at the equation

$$\mathscr{L}[C_n(E)] - n(n+1)(E_\kappa/E)^2 C_n(E) + \frac{2n+1}{4\pi}\delta(E - E_0) = 0.$$

This equation is similar to Eq. (6.68) and is solved in the same way. The angular distributions of electrons with energies higher E in EPC generated by primary electron with energy E_0 have the form [13, p. 173]

$$N_e(E,\theta) = \frac{1}{4\pi f(0)}\frac{E_0}{E}\sum_{n=0}^{\infty}\frac{2n+1}{[1+n(n+1)(P/E)^2)]^{1/2}}P_n(\cos\theta),$$

whereas for electrons with energy equal to E

$$P(E,\theta) = \frac{1}{4\pi f(0)}\frac{E_0}{E^2}\sum_{n=0}^{\infty}\frac{2n+1}{[1+n(n+1)(P/E)^2)]^{3/2}}P_n(\cos\theta).$$

As one can find from the latter distribution,

$$\langle\cos(\theta_e)\rangle(E) = \left[1+2\left(\frac{P}{E}\right)^2\right]^{-3/2}. \qquad (6.74)$$

This formula can be applied to estimation of the mean scattering angle of cascade electron with a given energy. For small angles,

$$\langle\cos(\theta_e)\rangle(E) \approx 1 - \frac{\overline{\theta_e^2}}{2} = 1 - 3\left(\frac{P}{E}\right)^2,$$

whence

$$\overline{\theta_e^2} = 6\left(\frac{P}{E}\right)^2, \qquad (6.75)$$

or, with numerical factor,

$$\overline{\theta_e^2} = 0.65(K/E)^2, \qquad (6.76)$$

$E_\kappa = 21.2$ MeV [13]. This estimation, obtained without taking in account ionization loss, is called the *Landau approximation*. It is valid in the range of the small-angle applicability ($\theta < 1$), for larger ones, the expression (6.74) should be used. The latter, however, is valid only for $E > \beta$, since it was derived without taking into account ionization losses. Therefore, in light substances (air, water), where ionization losses begin to affect before the deflection angles approach 1, the application of formula (6.74) does not make sense. For light elements, it is expedient to take into account, first of all, the effect of ionization losses on scattering, and one can restrict oneself to a small-angle approximation. For heavy elements, for example, lead, on the contrary, scattering becomes intense at energies several times higher than the critical one. Therefore, the application of formula (6.74) in this case makes sense.

Formula (6.74) shows that for an energy of 70 MeV the average scattering angle is 14°, for an energy of 35 MeV it is 27° , and for 6.4 MeV (critical energy in lead) it is already 78°. Thus, the scattering in lead becomes almost limiting already up to the critical energy. Since most of the electrons in the shower have energies less than the critical one, it is obvious that their angular distribution is very wide.

This circumstance is taken into account when measuring the soft (low energy) component under lead. The results of the experiment, in which the measuring device (counter, ionization chamber, scintillation detector) is located under the lead, essentially cannot be compared with the theoretical results obtained without scattering. To interpret such experiments, it is necessary not only to take scattering into account in the basic equations of the theory, but also to re-formulate the boundary conditions, since boundary effects can play an exceptional role here.

6.7 Nishimura–Kamata Theory

Lateral structure of an EPC is formed mainly due to multiple Coulomb scattering which occurred during its traversal through the matter. Based on the features of that cross section, a relation between the shower particle energy E and its lateral distance from the shower center r holds such as

$$E \simeq K/r,$$

where $K \simeq 20$ MeV is the *scattering constant*. Under A-approximation, it is well known that number of shower particles with energy greater than E is given as a function of only the energy ratio E_0/E and the depth t. This means that the lateral distribution function of shower particles is a function of only a factor $(E_0 r/K)$ and the depth t. This approximation is known as the small radius approximation, it is very useful in emulsion chamber data processing.

Kamata and Nishimura [15] constructed the three-dimensional cascade theory without Landau's approximation, taking into account of single and plural scatterings. The theory gives the cascade function II in a power series of a parameter Q, like in the scattering theory of Moliere. See [32] Chap. 5.

Considering the joint lateral-angular distributions $P(\mathbf{r}, \boldsymbol{\theta})$ and $\Gamma(\mathbf{r}, \boldsymbol{\theta})$ in the frame of small-angle approximation, J. Nishimura converted them into the correspondent Fourier images, $P(\mathbf{r}, \boldsymbol{\theta}) \mapsto f(\mathbf{q}, \mathbf{k})$, $\Gamma(\mathbf{r}, \boldsymbol{\theta}) \mapsto g(\mathbf{q}, \mathbf{k})$ and derived equations

$$\partial_t f = -\mathscr{A} f + \mathscr{B} g - D_\theta k^2 f + (\mathbf{q}\partial_\mathbf{k}) f + \beta \partial_E f, \tag{6.77}$$

$$\partial_t g = \mathscr{C} f - \sigma_0 g + (\mathbf{q}\partial_\mathbf{k}) g \tag{6.78}$$

(arguments E and t, not participating in these transformations, temporarily are omitted).

Elimination of g yields

$$[(\partial_t - \mathbf{q}\partial_\mathbf{k}) + \sigma_0][(\partial_t - \mathbf{q}\partial_\mathbf{k})]f = -[(\partial_t - \mathbf{q}\partial_\mathbf{k}) + \sigma_0][\mathscr{A} + D_\theta k^2 - \beta\partial_E]f + \mathscr{B}\mathscr{C} f.$$

For the comparison with experiments, the marginal (lateral or angular) distribution functions are only important:

$$p_1(\boldsymbol{\theta}) = \int P(\boldsymbol{\theta}, \mathbf{r}) d\mathbf{r}, \quad p_2(\mathbf{r}) = \int P(\boldsymbol{\theta}, \mathbf{r}) d\boldsymbol{\theta}$$

and

$$\gamma_1(\boldsymbol{\theta}) = \int \Gamma(\boldsymbol{\theta}, \mathbf{r}) d\mathbf{r}, \quad \gamma_2(\mathbf{r}) = \int \Gamma(\boldsymbol{\theta}, \mathbf{r}) d\boldsymbol{\theta}.$$

Their Fourier transforms are easily obtained from joint ones:

$$f_1(\mathbf{k}) = 2\pi f(\mathbf{0}, \mathbf{k}), \quad f_2(\mathbf{q}) = 2\pi f(\mathbf{q}, \mathbf{0}),$$

and so are for g.

In order to get these functions from Eqs. (6.77)–(6.78), J. Nishimura first computed the solution for $K/E \ll 1$ and $\beta/E \ll 1$,

$$f_1(\mathbf{k}) = \frac{1}{4\pi^2 i} \int\limits_{c-i\infty}^{c+i\infty} \frac{ds}{E} \left(\frac{E_0}{E}\right)^s \sum_{m,n=0}^{\infty} (-D_\theta k^2)^m (-\beta/E)^n \psi_{mn}(s, t), \tag{6.79}$$

expressing it through the function $\psi_{mn}(s, t)$ defined by the difference–differential equation system

$$\{\partial_t^2 + [A(s+2m+n) + \sigma_0]\partial_t + [A(s+2m+n)\sigma_0 - B(s+2m+n)C(s+2m+n)]\}\psi_{mn}$$

$$= (\partial_t + \sigma_0)[\psi_{m-1,n} + (s+2m+n)\psi_{m,n-1}] \tag{6.80}$$

under condition

$$\psi_{00} = H_1(s) e^{\lambda_1(s)t} + H_2(s) e^{\lambda_2(s)t}.$$

It is easy to recognize in the latter Mellin's transform of the cascade curve, and its further development in the Nishimura theory consisted in supplementing the list of arguments with angular and space variables (in the small-angle approximation), including a term describing the Coulomb scattering of charged particles (in the Landau approximation) and ionization losses (in the continuous slowing-down approximation) [16]. The theory is based on the analytic continuation of the psi-function

$$\psi_{pq} = \frac{\mathfrak{M}(p, q, s, t)}{\Gamma(p+1)\Gamma(q+1)},$$

where $\mathfrak{M}(p, q, s, t)$ is the key function of the NK theory, Mellin transform of which

$$F(\xi, \eta, s, t) = -\frac{1}{4\pi^2} \int\limits_{-i\infty}^{+i\infty} dp \int\limits_{-i\infty}^{+i\infty} dq \frac{\mathfrak{M}(p, q, s, t)}{\xi^{p+1}\eta^{q+1}}$$

is determined by equation

$$\sum_{mn=0}^{\infty} (-D_\theta k^2)^m (-\beta/E)^n \psi_{mn}(s, t) = \int\limits_0^{\infty} d\xi \int\limits_0^{\infty} d\eta e^{-\xi D_\theta k^2 - \eta\beta/E} F(\xi, \eta, s, t).$$

It is worth noting that the above formulas were derived for the angular problem, but we are going to focus here on the lateral distribution, considered by Nishimura–Kamata separately. The solution of the lateral distribution problem was obtained according to the same scheme, and we will rewrite here the lateral counterparts of the angular equations in their original notations, replacing $\mathfrak{M} \to \mathfrak{M}_2$ and $\psi \to \phi$ [15]:

$$\mathfrak{M}_2(m, n, s, t) = \frac{\phi_{mn}(s, t)}{\Gamma(n+1)\Gamma(m+1)},$$

$$\sum_{mn=0}^{\infty} \left(-D_\theta q^2\right)^m (-\beta/E)^n \phi_{mn}(s, t) = \int\limits_0^{\infty} d\xi \int\limits_0^{\infty} d\eta e^{-\xi(q^2 D_\theta) - \eta\beta/E} F(\xi, \eta, s, t). \tag{6.81}$$

Moreover, it follows by the inverse Mellin transformation that Eq. (6.81) becomes

$$\sum_{mn=0}^{\infty} \left(-D_\theta q^2\right)^m (-\beta/E)^n \phi_{mn}(s, t)$$

$$= -\frac{1}{4\pi^2} \int\limits_0^{\infty} d\xi \int\limits_0^{\infty} d\eta e^{-\xi(q^2 D_\theta) - \eta\beta/E} \int\limits_{-i\infty}^{i\infty} dp \int\limits_{-i\infty}^{i\infty} dq \frac{\mathfrak{M}_2(p, q, s, t)}{\xi^{p+1}\eta^{q+1}}. \tag{6.82}$$

Since the double series is uniformly and absolutely convergent for $E > Kqt^{3/2}/3$ and $E > 8\beta t$ [15], the order of the integrations can be changed (under these conditions), and finally

$$p_2(\mathbf{r}) = -\frac{1}{8\pi^4 i}\int_{-i\infty}^{i\infty}\int_{-i\infty}^{i\infty}\int_{-i\infty}^{i\infty} dsdpdq\left(\frac{E_0}{E}\right)^s \frac{1}{E}\left(\frac{E}{K}\right)^2\left(\frac{\beta}{E}\right)^q\left(\frac{E^2r^2}{K^2}\right)^{-p-1}$$
$$\times\ \Gamma(p+1)\Gamma(-q)\mathfrak{M}_2(p, q, s, t). \tag{6.83}$$

The above restrictions can be ignored now, because (6.83) exists for any value of E and r. Observe that (6.83) is the exact solution of the differential lateral structure functions for any value of E and r derived by the *principle of analytic continuation* (see Appendix A).

The expression for the integral lateral distribution function, i.e., the lateral structure function for the electrons with energy higher than E, is obtained by the integration with respect to energy, and given by

$$N_e(E_0, E, r, t) = -\frac{1}{8\pi^4 i}\int_{-i\infty}^{i\infty}\int_{-i\infty}^{i\infty}\int_{-i\infty}^{i\infty} \frac{dsdpdq}{s+2p+q}\left(\frac{E_0}{E}\right)^s\left(\frac{E}{K}\right)^2\left(\frac{\beta}{E}\right)^q$$
$$\times\left(\frac{E^2r^2}{K^2}\right)^{-p-1}\Gamma(p+1)\Gamma(-q)\mathfrak{M}_2(p, q, s, t). \tag{6.84}$$

At the limit for $E = 0$, the integration with respect to q is equal to the residues of the integrands at $q = -s - 2p$, so that

$$N_e(E_0, 0, r, t) = -\frac{1}{4\pi^3}\int_{-i\infty}^{i\infty}\int_{-i\infty}^{i\infty} dsdp\left(\frac{E_0}{\beta}\right)^s\left(\frac{\beta}{K}\right)\left(\frac{\beta^2r^2}{K^2}\right)^{-p-1}$$
$$\times\ \Gamma(p+1)\Gamma(s+2p)\mathfrak{M}_2(p, -s-2p, s, t). \tag{6.85}$$

The total number $N_e(E_0, 0, t)$ of all electrons crossing the plane $t = \text{const}$ is obtained by integrating $N_e(E_0, 0, r, t)$ over the plane:

$$N_e(E_0, 0, t) = 2\pi\int_0^\infty rdrN_e(E_0, 0, r, t) = \frac{1}{2\pi i}\int ds\left(\frac{E_0}{\beta}\right)^s\Gamma(s)\mathfrak{M}_2(0, -s, s, t). \tag{6.86}$$

This is the same expression as that obtained by Snyder [17] in one-dimensional shower theory.

Similar treatments are also possible for the analytical solutions of the structure functions for photons, which are sometimes important for the analysis of the spread of extensive air showers detected by the apparatus under a thick absorber.

Dealing with experimental investigation of the lateral shower structure, it is convenient to have information in some standard form, for example, normalized to 1. It is the structure function, defined as

$$P(E_0, 0, r, t) = \frac{N_e(E_0, 0, r, t)}{N_e(E_0, 0, t)}, \tag{6.87}$$

so that

$$\int_0^\infty P(E_0, 0, r, t) 2\pi r dr = 1. \tag{6.88}$$

Using Eqs. (6.85)–(6.86), Kamata and Nishimura numerically evaluated the asymptotic case $E_0/\beta \to \infty$, admitting the saddle-point method application for integration with respect to s. As a result, they obtained

$$P(\infty, 0, r, t) = \int_{-i\infty}^{i\infty} dp \left(\frac{\beta}{K}\right)^2 \left(\frac{\beta^2 r^2}{K^2}\right)^{-p-1} \frac{\Gamma(p+1)\Gamma(\bar{s}+2p)}{2\pi^2 i \Gamma(\bar{s})} \frac{\mathfrak{M}_2(p, -\bar{s}-2p, \bar{s}, t)}{\mathfrak{M}_2(0, -\bar{s}, \bar{s}, t)}, \tag{6.89}$$

where $\bar{s}$ is the cascade age parameter defined by equation

$$\ln(E_0/\beta) + \lambda_1(\bar{s})t = 0. \tag{6.90}$$

The integral with respect to p can also be evaluated by the saddle-point method. Writing

$$e^{u(\bar{s},p,t)} = \Gamma(p+1)\Gamma(\bar{s}+2p)\mathfrak{M}_2(p, -\bar{s}-2p, \bar{s}, t),$$

the saddle point $\bar{p}$ was determined by the equation

$$-2\ln(\beta r/K) + \left.\frac{\partial u(s, p, t)}{\partial p}\right|_{p=\bar{p}} = 0,$$

so the result is represented by expression

$$f(\infty, 0, r, t) = \frac{e^{u(\bar{s},\bar{p},\, t)}}{\pi\sqrt{2\pi u''(\bar{s}, \bar{p}, t)}\Gamma(\bar{s})\mathfrak{M}_2(0, -\bar{s}, \bar{s}, t)} \left(\frac{\beta}{K}\right)^2 \left(\frac{\beta^2 r^2}{K^2}\right)^{-\bar{p}-1}. \tag{6.91}$$

The numerical data are shown in Table 6.1.

Seeing that $\mathfrak{M}_2(p, -\bar{s}-2p, \bar{s}, t)/\mathfrak{M}_2(0, -\bar{s}, \bar{s}, t)$ and $\partial u/\partial p$ do not explicitly depend on t, the number of independent arguments in the structure function may be reduced by involving age parameter s. One more parameter, r_1, appears as a radial

Table 6.1 The lateral structure function $Q(x, s)$ for $s = 0.6, 1.0, 1.4$ and 2.0

x	0.6	1.0	1.4	2.0
0.001	$1.95 \cdot 10^{-1}$	$4.4 \cdot 10^{-1}$	$8.5 \cdot 10^{-1}$	2.4
0.003	$1.90 \cdot 10^{-1}$	$4.2 \cdot 10^{-1}$	$8.0 \cdot 10^{-1}$	1.92
0.01	$1.81 \cdot 10^{-1}$	$3.9 \cdot 10^{-1}$	$7.1 \cdot 10^{-1}$	1.40
0.03	$1.58 \cdot 10^{-1}$	$3.4 \cdot 10^{-1}$	$5.8 \cdot 10^{-1}$	$9.3 \cdot 10^{-1}$
0.1	$1.34 \cdot 10^{-1}$	$2.65 \cdot 10^{-1}$	$3.8 \cdot 10^{-1}$	$5.3 \cdot 10^{-1}$
0.2	$1.08 \cdot 10^{-1}$	$2.03 \cdot 10^{-1}$	$2.6 \cdot 10^{-1}$	$3.2 \cdot 10^{-1}$
0.5	$4.9 \cdot 10^{-2}$	$1.09 \cdot 10^{-1}$	$1.22 \cdot 10^{-1}$	$1.21 \cdot 10^{-1}$
1.0	$1.35 \cdot 10^{-2}$	$4.1 \cdot 10^{-2}$	$5.0 \cdot 10^{-2}$	$4.4 \cdot 10^{-2}$
2.0	$2.64 \cdot 10^{-3}$	$8.8 \cdot 10^{-3}$	$1.48 \cdot 10^{-3}$	$1.18 \cdot 10^{-2}$
5.0	$2.16 \cdot 10^{-4}$	$7.7 \cdot 10^{-4}$	$1.55 \cdot 10^{-4}$	$1.32 \cdot 10^{-3}$

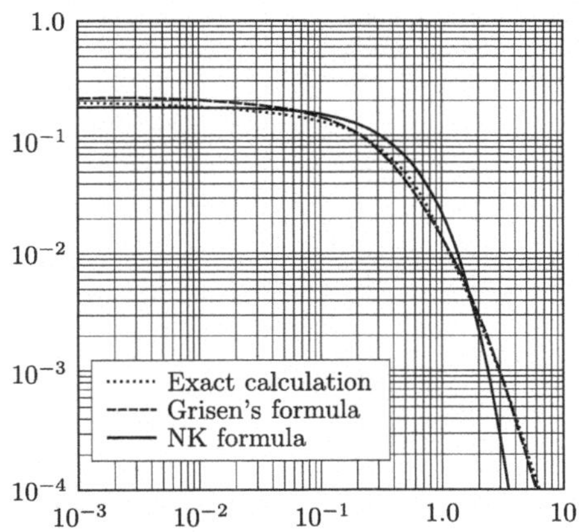

Fig. 6.5 Asymptotic formula of the structure function for $s = 0.6$. Dotted line represents the exact calculation, dashed line Greisen's formula, and solid line NK formula [15]

scale, $f(\infty, 0, r, t) = Q(r/r_1, \bar{s})$. As such, the cascade unit K/β, the Molière radius r_M, or the root-mean-square radius $r_{\text{r.m.s.}}$ of the lateral distribution of shower particles may be chosen, depending on the conditions of the problem (Figs. 6.5, 6.6, and 6.7).

The Nishimura–Kamata theory of electron–photon cascades has made a great contribution to the cosmic ray study, especially to the study of extensive air showers. The original result by Kamata and Nishimura is given in a short numerical table, so that application of the NK function to cosmic ray phenomena used to be made by means of various approximation formulas. Among them a formula obtained by Kamata and Nishimura themselves [15, 18] and that by Greisen [19] are well known.

However, the practical use of the theory by astrophysicists requires the long and tedious numerical calculations and the results' application to experimental data is inconvenient. To make the results more handy for experimenters, a few simple approximation formulae were elaborated. We consider here the way of derivation of the most known of them, the Nishimura–Kamata–Greisen (NKG) formula for the lateral structure function of electrons in EPC in a homogeneous medium.

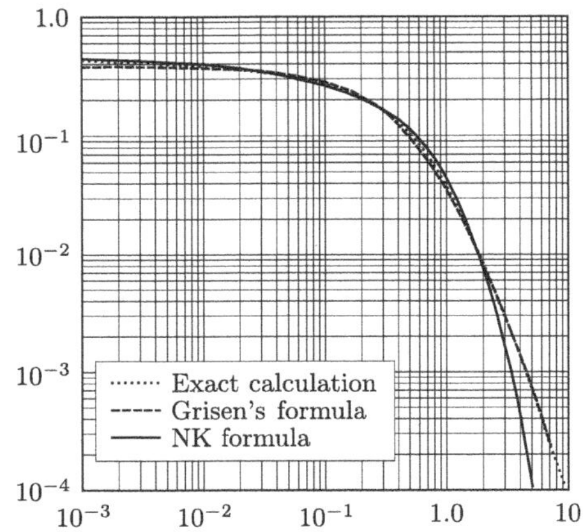

Fig. 6.6 Asymptotic formula of the structure function for $s = 1.0$ [15]

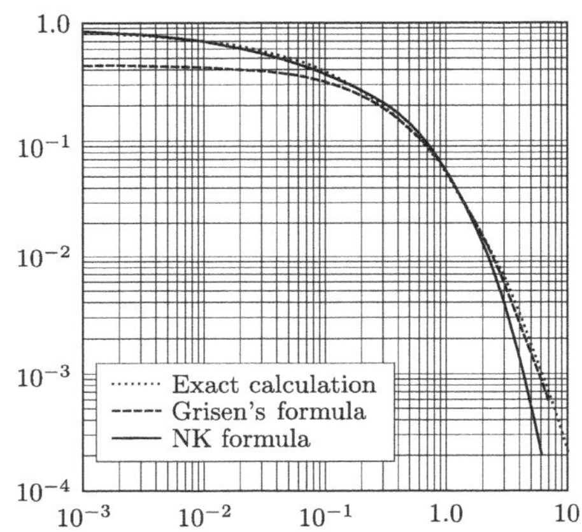

Fig. 6.7 Asymptotic formula of the structure function for $s = 1.4$ [15]

Let us represent the Nishimura–Kamata lateral structure function for EPC electron [15] in the form

$$Q(x, s) = x^{2-s}P(x, s) = \frac{1}{\pi}\frac{1}{2\pi i}\int dp x^{-2p-s}\frac{\Gamma(p+1)\Gamma(2p+s)}{\Gamma(s)}\mu(p, s, t), \tag{6.92}$$

where $x = r/r_1$ and

$$\mu(p, s, t) = \frac{\mathfrak{M}_2(p, -2p - s, s, t)}{\mathfrak{M}_2(0, -s, s, t)}. \tag{6.93}$$

The first analytic approximation was offered by Nishimura–Kamata themselves [20] in the form

$$Q_{\rm NK}(x) = c(s)[1 + (4/s)x]e^{-a(s)x^{b(s)}}$$

with

$$a(s) = \frac{4}{s}e^{0.915(s-1)}, \quad b(s) = 0.15 + \frac{1}{1+s}, \quad c(s) = \frac{a^{s/b}}{2\pi\left[\Gamma(\frac{s}{b}) + \frac{4\Gamma(\frac{s+1}{b})}{sa^{1/b}}\right]}.$$

H. Oda, in [20], remarked that $\mu(p, s, t)$ is a convex function of p with values

$$\mu(0, s, t) = 1, \tag{6.94}$$

$$\mu(1, s, t) = r_{\mathrm{ms}}^2/s(s+1), \tag{6.95}$$

and singularity at $p = 2$. As a function of t, $\mu(p, s, t)$ can be considered as a constant, if t is not too small (say, $t > 5$ c.u.). Then, H.Oda examined various approximations for $\mu(p, s, t)$ and compared the resultant $P(r, s)$ with NK function. Starting with *zeroth-approximation*

$$\mu_0(p, s, t) = 1$$

and found it as a very crude representation, H.Oda obtained

$$Q_0(x, s) = \phi(x, s) = \frac{1}{\pi\Gamma(s)}\int_0^\infty e^{-t^2-x/t}t^{-s+1}dt.$$

Then, consequently continuing elaboration in such a way, that it would give correct value not only when $p \cong 0$ but $p \cong 1$ as well, by choosing

$$\mu_1(p, s, t) = b^{2p}$$

with a parameter b determined, say, by root-mean-square (rms) radius r_{rms}, as

$$b = r_{\mathrm{rms}}^2/s(s+1)$$

led to the next (*1-approximation*),

$$Q_1(x, s) = b^{-s}\phi(x/b, s). \tag{6.96}$$

This equation says that the 1-approximation result $Q_1(x, s)$ can be obtained by shifting $Q_0(x, s)$ parallel to the abscissa by the factor b and to the ordinate by b^s. It was found that with a proper fitting of $b(s)$, namely, by function

$$b(s) = 0.25s + 0.45 \tag{6.97}$$

(second approximation in Oda's terminology) $Q_1(x, s)$ agrees with Q_{NK} fairly well.

There were considered multiscale representations of the lateral structure function, the simplest of which is

$$Q_2(x, s) = (1 - w)b_1^{-s}\psi(x/b_1, s) + wb_2^{-s}\psi(x/b_2, s).$$

Putting $w = 1/2$, b_1 and b_2 were determined from NK calculations empirically:

$$b_1 = 2^{s-1}, \quad b_2 = (13/9)^{s-3.17},$$

but no significant refinement of the approximation was achieved.

On the same NK calculation basis, K.Greisen proposed a more simple approximation formula [19]:

$$P_{\mathrm{NKG}}(x, s) = x^{s-2}Q(x, s), \quad Q(x, s) = C(s)(1 + x)^{s-4.5}$$

with the normalizing coefficient

$$C(s) = \frac{1}{2\pi}\frac{\Gamma(4.5 - s)}{\Gamma(4.5 - 2s)},$$

and $x = r/r_{\mathrm{M}}$ (the distance from the EPC axis, expressed in Moliere's units r_{M}). Observe that the exponent of the power law changes from $s - 2$ near the axis to $4.5 - s$ far from it.

However, this simple form presents just a tolerable agreement (as long as s is far from 1) with the numerical results of the cascade theory [18]. Furthermore, the coherence of the numerical values of the age parameter coming from experimental lateral densities in EAS with the values derived from the longitudinal development has raised numerous problems [20]. The experimenters introduced several reductions of Moliere radii, giving steeper profiles, as observed in the EAS at mountain altitude. The *local age parameter* (LAP) introduced by [31] has an experimental hint of these difficulties, as follows [21, 22].

The consequences of the LAP conception were underlined by the rapporteurs of various ICRC conferences [23–25]. The following quote from the Capdevielle and Cohen article is appropriate to place in the context of our discussion. It was then pointed out [26] that at small distances (< 200 m) and at lower energy (< 107 GeV) a reduction of Moliere radius by the scaling factor

$$m = 0.78 - 0.21s,$$

derived from the resolution of the diffusion equations by the *adjoint equations method* [27], *the best agreement with the local age parameter experimental and simulated densities. This modified NKG function was employed recently to calculate the synchrotron radiation at radio frequencies in giant EAS above* 10^9 *GeV. We have checked that between 10 and 100 m this modified NKG function is close to our simulation with CoRSiKa and their approach (as the radio emission is mainly generated by highest densities) remains acceptable.*

We will come back to the discussion in Conclusion (see 14.3–14.4).

References

1. Tamm I, Belenky S (1939) On the soft component of cosmic rays at sea level. J Phys USSR 1:177
2. Ivanenko IP (1956) Cascade showers produced by electrons and photons in light and heavy elements. Zh Exper Teor Fiz 31:86–104 (in Russian)
3. Ivanenko IP, Syzov VV (1988) Moments of longitudinal distributions of electron-photon cascades. Moscow State University Bulletion, ser 3, Fizika. Astronomia 29(2):21–25 (in Russian)
4. Ivanenko IP, Syzov VV (1988) Calculation of longitudinal distributions of electron-photon cascades by moment method, ibid, pp 22–28 (in Russian)
5. Ivanenko IP, Syzov VV (1988) On the accuracy of angular distribution moments in electron-photon cascades, ibid, pp 28–33 (1988) (in Russian)
6. Ivanenko IP, Syzov VV (1985) On the accuracy of equilibrium characteristics of electron-photon cascades. Moscow State University Bulletion, ser 3, Fizika. Astronomia 26(4):28–32 (1985) (in Russian)
7. Kol'chuzhkin AM, Uchaikin VV, Bespalov VI (1972) Equilibrium electron and photon spectra in lead and air. Atom Energia 32:224 (in Russian)
8. Kol'chuzhkin AM, Bespalov VI (1974) Equilibrium electron and photon spectra in lead and air. Soviet Phys J 741–743 (1974) (in Russian)
9. Carlson JF, Oppenheimer JR (1937) On multiplicative showers. Phys Rev 51:220–231
10. Landau LD, Rumer G (1938) The cascade theory of electronic showers. Proc R Soc London 213–228
11. Greisen K, Rossi B (1941) Cosmic-ray theory. Rev Mod Phys 13(4):240–309
12. Bhabha HJ, Chakrabarty SK (1942) Calculations on the cascade theory with collision loss. Proc Indian Acad Sci A 15:464–476
13. Belenky SZ (1948) Avalanche processes in cosmic rays. Gostekhizdat, Moscow (in Russian)
14. Landau LD (1940) Angular distribution of particles in showers. J Phys USSR 10(3):237
15. Kamata K, Nishimura J (1958) The lateral and the angular structure functions of electron showers. Prog Theor Phys Suppl 6:93–155
16. Nishimura J, Kamata K (1952) On the theory of cascade showers, I. Prog Theor Phys 7(2):185–192
17. Snyder HS (1949) Comparison of calculations on cascade theory. Phys Rev 76:1563
18. Nishimura J (1967) Theory of cascade showers. In: Handbuch der Physik, vol XL VI/2. Springer, Berlin
19. Greisen K (1956) In: Wilson JG (ed) Progress in cosmic ray physics, vol III
20. Hiroshi O (1976) Approximate derivation of the lateral structure function of electrons in a cascade shower. Prog Theor Phys 56(6):1863–1870
21. Hillas AM, Lapikens J, Electron-photon cascades in the atmosphere and in detectors. In: Proceedings of the 15th ICRC, Plovdiv, vol 8, pp 460–465
22. Hillas AM, Results of Monte Carlo simulation of electron-photon cascades in the atmosphere and in detectors. In: Proceedings of the 17th ICRC, Paris, vol 6, pp 244–247
23. Capdevielle JN, Cohen F (2005) The relation between the lateral profile of giant extensive air showers and the age parameter. J Phys G: Nucl Part Phys 31:507–524
24. Tonwar S (1981) Cosmic rays at earth. In: Proceedings of the 17th ICRC, Paris, vol 13, p 330

25. Rao MVS (1983) Behavior of high-energy interaction, from structure of small and medium air showers. In: Proceedings of the 18th ICRC, Bangalore, vol 12, pp 449–474
26. Capdevielle JN, Procureur J, Gawin J (1983) Possible analytical descriptions for the lateral electron distribution in EAS. In: Proceedings of the 18th ICRC, Bangalore, vol 11, pp 307–310
27. Bourdeau MF, Capdevielle JN, Procureur J (1980) The lateral age parameter in extensive air showers. J Phys G: Nucl Phys 6:901–908

For Further Reading

28. Lagutin AA, Plyasheshnikov Uchaikin VV (1979) The radial distribution of electromagnetic cascade particles in the air. In: Proceedings of the 16th ICRC, Kyoto, vol 7, pp 18–23
29. Capdevielle JN, Le Gall C, Gawin J et al (2002) Lateral-distribution functions for giant air showers. Il Nuovo Cimento 25C(4):393–424
30. Nishimura J (1967) Handbuch der Physik, 46/2. Springer, Berlin
31. Bourdeau MF, Capdevielle JN (1980) The lateral age parameter in extensive air showers. J Phys G: Nucl Phys 6(7):901

Chapter 7
SBCE Method

7.1 Lagrange Polynomial Interpolation

The important direction in the cascade theory development is elaboration of more efficient numerical methods free from simplifying assumptions of the analytical theory and slowly decreasing statistical errors with calculation time, inherent in Monte Carlo simulations. Among such methods are finite-difference method, multigroup method, moment method, decomposition in orthogonal functions, and their combinations with Monte Carlo and numerical methods of recovery of desired solutions from decomposition coefficients. This book focuses on one of these methods, the spline backward cascade equation (SBCE), for two reasons. The first is that, unlike other numerical methods, it is not yet sufficiently reflected in the literature, and the second is that the author directly (in collaboration with his colleagues—A. V. Plyasheshnikov and A. A. Lagutin) participated in SBCE development.

For numerical solution of EPC problem including statistical fluctuations, using the backward equations seemed to us more convenient than direct ones. The solution of the latter is complicated by larger number of independent variables, which eventually have to calculate multidimensional integral. The backward equations can be derived directly for moments of the studied characteristic, which depend on a single-particle set of variables and have exactly the same structure (with the exception of the right parts containing information about lower moments). For this reason, their solution can be fulfilled within the same universal program, based on piecewise polynomial representation of the energy dependence of the desired functions.

The first publications about this method appeared in 1977–79 [1–4].

To reveal the main idea of the method, consider first solving integral equation

$$\sigma(E)f(E) = \int\limits_{E^*}^{E} w(E \to E')f(E')dE' + F(E), \tag{7.1}$$

© Higher Education Press Limited Company 2024

V. V. Uchaikin, *Electron-Photon Cascades*, Nonlinear Physical Science,
https://doi.org/10.1007/978-981-99-7524-2_7

similar to those that describe, for example, averages and fluctuations of particle ranges in an infinite medium, covariance matrix $C^{(k,l)}$, and other characteristics of the cascade, depending only from primary energy. Let us choose on the energy axis an increasing sequence of points $E_0 = E^*, E_1, \dots, E_k \dots$ and introducing the notation

$$f_k = f(E_k), \qquad \sigma_k = \sigma(E_k), \qquad I_{ki} = \int_{E_{i-1}}^{E_i} w(E_k \to E)dE, \tag{7.2}$$

rewrite (7.1) as

$$\sigma_k f_k = \sum_{i=1}^{k} I_{ki} + F_k. \tag{7.3}$$

As $k = 0$, the integral in (7.1) vanishes, and we have $f_0 = F_0/\sigma_0$. The solution at the next point f_1 can be found from equations $\sigma_1 f_1 = I_{11} + F_1$, if $f(E)$ is approximated under the integral I_{11} by a linear function given by the values f_0 and f_1. The next value can be found in a similar way, based on the values of f_0 and f_1 obtained above and applying now quadratic interpolation of the function $f(E)$ in each of integrals I_{21} and I_{22}. When calculating f_3, one can use third-degree polynomial interpolation, etc. Note that, however, what to describe the solution (7.1) with a single polynomial in a wide range of changes in the energy variable is practically impossible. A simpler and more stable approximation is *spline method* providing piecewise polynomial approximation with a limited number continuous derivatives. Applying it to Eq. (7.1) looks like as follows: the maximum degree of polynomials N is given and for $k \geqslant N$ only $N + 1$ are used for calculation node points $E_{i-N}, \dots, E_N$ if $i \geqslant N$. If $k < N$, then there is no solution at the points $k + 1, \dots, N$, and the degree of the interpolation polynomial is reduced to k, and as nodal points are $E_0, \dots, E_k$. So the degree interpolation polynomial approximating the function $f(E)$ under integral I_{ki} is defined by the formula

$$n(k, i) = \min\{k, N\},$$

and the number of the rightmost interpolation node

$$m(k, i) = \begin{cases} n(k, i), & i < N, \\ i, & i \geqslant N. \end{cases}$$

Using Lagrange polynomials for interpolation

$$L_{ij}^{n}(E) = \prod_{\substack{r=0 \\ r \neq m-j}}^{n} (E - E_{m-r})(E_j - E_{m-r})^{-1}, \qquad E \in \Delta E_i,$$

represent the solution $f(E)$ on the i-th interval as

$$f(E) = \sum_{j=m-n}^{m} L_{ij}^{n}(E) f_j, \quad E \in \Delta E_i. \tag{7.4}$$

Inserting (7.4) into (7.2), we obtain

$$I_{ki} = \sum_{j=m-n}^{m} \Delta_{kij} f_j, \tag{7.5}$$

where

$$\Delta_{kij} = \int_{E_{i-1}}^{E_i} w(E_k \to E) L_{ij}^{n}(E) dE. \tag{7.6}$$

Further, using (7.5) in (7.3), arrive at the algebraic equation system

$$(\sigma_k - a_{kk}) f_k = F_k', \tag{7.7}$$

$$F_k' = F_k + \sum_{j=0}^{k-1} a_{kj} f_j, \tag{7.8}$$

where

$$a_{kj} = \sum_{i=1}^{k} \Delta_{kij} \sum_{s=m-n}^{m} \delta_{sj} \tag{7.9}$$

are coefficients depending on the cross sections used (a kind of *group constants*).
According to (7.8), the right side (7.7) contains f_j for all $j < k$, so (7.7) gives a recursive way calculation of unknowns:

$$f_k = (\sigma_k - a_{kk})^{-1} F_k'. \tag{7.10}$$

7.2 Straight-Ahead Equation System

In the *straight-ahead approximation* (see the explanation before (5.147)), we are dealing with integro-differential equation

$$[\partial_t + \sigma(E)] f(t, E) - \int_{E^*}^{E} w(E \to E') f(t, E') dE' = F(t, E), \tag{7.11}$$

which is reduced by the method described above to the system of differential equations $[\partial_t + (\sigma_k\ a_{kk})]f_k = F'_k$, where the right side has a former view (7.8) cascades with a few types of particles $(a, b, \dots)$. Eq. (7.11) becomes the system:

$$[\partial_t + \sigma_a(E)]f_a(t, E) - \sum_b \int_{E^*}^{E} w_{ab}(E \to E')f_b(t, E')dE' + F_a(t, E).$$

EPC includes two types of particles, electrons ($\sigma_e = \sigma$, $f_e = f$, $F_e = F$), and photons ($\sigma_\gamma = \mu$, $f_\gamma = g$, $F_g = G$), so we have

$$[\partial_t + \sigma(E)]f(t, E) - \int_{E^*}^{E} w_{\gamma e}(E \to E')f(t, E')dE' - \int_{E^*}^{E} w_{e\gamma}(E \to E')g(t, E')dE' = F(t, E), \tag{7.12}$$

$$[\partial_t + \mu(E)]g(t, E) - \int_{E^*}^{E} w_{ee}(E \to E')f(t, E')dE'$$

$$- \int_{E^*}^{E} w_{\gamma\gamma}(E \to E')g(t, E')dE' = G(t, E). \tag{7.13}$$

The detector response to the cascade process is determined by functions F and G in Eqs. (7.12) and (7.13) and boundary conditions $f(0, E)$, $f(0, E)$.

For example, if the device measures electrons above some threshold energy E^*, then

$$F(t, E) = G(t, E) = 0; \quad f(0, E) = 1(E - E^*), \quad g(0, E) = 0.$$

For solving the adjoint (backward) cascade equations, a special numerical method SBCE (Spline Backward Cascade Equations) was suggested in Ref. [1] and then was developed for application to solving other problems in cascade (both EPC and EAS) theory [2–5, 7, 8, 25, 29, 45, 49, 63]. To use this method for solving Eqs. (7.12) and (7.13), an increasing subsequence $\{E_k\}$, $k = 1, 2, 3, \dots$ should be chosen and corresponding values of functions $f_k(t) \equiv f(t, E_k)$, $g_k(t) \equiv g(t, E_k)$. The updated description of the SBCE method can be found in article [6].

Let us approximate functions f and g inside the energy intervals $\Delta(E_k) = (E_{k-1}, E_k)$ by the Lagrange polynomial interpolation through their values at the support points E_k,

$$f(E, t) \approx \tilde{f}(E, t) = \sum_{j=k-n}^{k} f_j(t)L_{kj}^n(E), \tag{7.14}$$

$$g(E,t) \approx \tilde{g}(E,t) = \sum_{j=k-n}^{k} g_j(t) L_{kj}^n(E), \tag{7.15}$$

where

$$L_{kj}^n(E) = \prod_{r \neq k-j}^{n} (E - E_{k-r})(E_j - E_{k-r})^{-1},\ E \in (\Delta E_k), \tag{7.16}$$

$$n \equiv n(k) = \min(N, k),$$

and N is the maximum available power of the Lagrange polynomial.

For the first k energy intervals, the reference pairs $f_0, g_0; \ldots; f_k, g_k$ are used, in this case the polynomial power $n = k$. For intervals with $k \geq N$ the pairs $f_{k-N}, g_{k-N}; f_{k-N+1}, g_{k-N+1}, \ldots, f_k, g_k$ are involved in the interpolation with the use of term with $n = N$.

Further, a simple algebra leads to equations

$$\partial_t f_k + A_k f_k - B_k g_k = F_k', \quad \partial_t g_k - C_k f_k + D_k g_k = G_k', \tag{7.17}$$

where

$$A_k = 1/\lambda_k - a_{kk},\ B_k = b_{kk},\ C_k = c_{kk},\ D_k = 1/\mu_k - d_{kk}, \tag{7.18}$$

and

$$F_k' = F_k + \sum_{j=0}^{k-1} [a_{jk} f_j + b_{jk} g_j], \quad G_k' = G_k + \sum_{j=0}^{k-1} [c_{jk} f_j + d_{jk} g_j]. \tag{7.19}$$

The coefficients a_{jk}, b_{jk}, c_{jk}, and d_{jk} can be expressed through cross sections of relevant processes:

$$a_{jk} = \sum_{i=1}^{k} \int_{E_{i-1}}^{E_i} w_{ee}(E_k, E) L_{ij}^n(E) \sum_{s=i-n}^{i} \delta_{js} dE,$$

$$b_{jk} = \sum_{i=1}^{k} \int_{E_{i-1}}^{E_i} w_{e\gamma}(E_k, E) L_{ij}^n(E) \sum_{s=i-n}^{i} \delta_{js} dE,$$

with $\delta_{js} = 1$ if $j = s$, and $\delta_{js} = 0$ if $j \neq s$. Solving Eq. (7.17) begins with the next two steps:

(1) substituting $E = E_0 = E^*$ into Eqs. (7.12)–(7.13) converts them to the pair

$$\partial_t f_0 + f_0(t)/\lambda_e(E_0) = F_0(t),\ \partial_t g_0 + g_0(t)/\lambda_\gamma(E_0) = G_0(t); \tag{7.20}$$

(2) having solutions of Eqs. (7.20) opens way to find the next pair $f_2(t)$, $g_2(t)$, and so on.

7.3 Solving Straight-Ahead System

For a fixed value of k, an increasing subsequence of depth values t_l ($t_0 = 0, t_1 = \tau_1, \ldots, t_l = t_{l-1} + \tau_l$), and relevant $f_k(t_l) = f_{k,l}$, $g_k(t_l) = g_{k,l}$ are introduced. In each interval (t_{l-1}, t_l), Eq. (7.17) are solved with the use of $f_{k,l-1}$, $g_{k,l-1}$ as boundary conditions. Passing to the next boundary, $f_k(t) \to f_k(t_l) = f_{k,l}$, $g_k(t) \to g_k(t_l) = g_{k,l}$ yields the system

$$f_{k,l} = (\lambda_{0k} - \lambda_{1k})^{-1} \sum_{\nu=0}^{1} (-1)^{\nu} \Bigg\{ \exp(\lambda_{\nu k}\tau_l)[(D_k + \lambda_{\nu k}) f_{k,l-1}$$

$$+ B_k g_{k,l-1}] \int_{t_{l-1}}^{t_l} \exp[\lambda_{\nu k}(t_l - \tau)][(D_k + \lambda_{\nu k}) F_k'(\tau) + B_k G_k'(\tau)] \Bigg\} d\tau, \tag{7.21}$$

$$g_{k,l} = (\lambda_{0k} - \lambda_{1k})^{-1} \sum_{\nu=0}^{1} (-1)^{\nu} \Bigg\{ \exp(\lambda_{\nu k}\tau_l)[(A_k + \lambda_{\nu k}) g_{k,l-1}$$

$$+ C_k f_{k,l-1}] \int_{t_{l-1}}^{t_l} \exp[\lambda_{\nu k}(t_l - \tau)][(A_k + \lambda_{\nu k}) G_k'(\tau) + C_k F_k'(\tau)] \Bigg\} d\tau, \tag{7.22}$$

where

$$\lambda_{\nu k} = \frac{1}{2} \left\{ -(A_k + D_k) + (-1)^{\nu} [(A_k - D_k)^2 + 4 B_k C_k]^{1/2} \right\}. \tag{7.23}$$

Note that in case of the homogeneous medium these relations are valid for an arbitrary values of τ_l, otherwise they can be applied only for $\tau_l \ll t_l$.

The knowledge of $f_{k,0}$ and $f_{k,0}$ (these quantities can be calculated on the basis of boundary conditions like Eqs. (7.14) or (7.15)) and the multiple application of Eqs. (7.21)–(7.23) allow us to calculate quantities $f_{k,l}$, $g_{k,l}$ for $l = 1, 2$, and so on.

As noticed by authors who used the SBCE approach in considerations of astroparticle and other problems [5], it provides an accuracy better than a few percent and gives results for a wide region of the primary energy in just one run of calculations. Also they notice that the computational time consumed by this method does not exceed on average a few minutes for a 1 GHz PC-type computer and increases only weakly (logarithmically) with the primary energy.

In summary, the important feature of this technique is its flexibility to describe the cascade processes in three different substances. It allows detailed numerical calculations with an adequate accuracy throughout very large energy intervals of both primary and secondary energies. This is an important condition for the quantitative analysis as well as for clear understanding of similarities and differences in cascade development in environments dominated by matter, photon gas, or magnetic field. The method allows accurate calculations in inhomogeneous environments. For example, authors of [6] have applied this method [7] for calculations of characteristics of air showers initiated by extremely high-energy γ-rays in a "sandwich"-type structure consisting of the geomagnetic field and the Earth atmosphere, both the magnetic field and the gas density being treated as functions of coordinates. Moreover, the method is not limited by the geometrical size or by the optical depth of the region where the cascade develops. In particular, recently this method has been used for study of propagation of multi-TeV c-rays from the TeV blazar Mkn 501 over huge cosmological distances [8].

7.4 Integral Terms in SBCE for Statistical Moments

Now, we recall that $f_a(t, E)$ in above equation denotes the average reading $\langle Q\rangle$ of detector from the one-dimensional EP cascade, generated by a-particle born at point t with energy E: $f_a(t, E) = \langle Q\rangle$. It is the first statistical moment of the random value Q. Further, we will consider statistical moments of higher orders, $\langle Q^n\rangle$, $n = 2, 3, \ldots$ under the same conditions, using for them designations $f_a^{(n)}(t, E)$. In particular, $f_a^{(2)}(t, E) - f_a^2(t, E)$ is the variance of the random reading, the measure of cascade-to-cascade fluctuations in Q. The corresponding equation system is of the form:

$$\partial_t f_a^{(n)} + \mathscr{L}^+_{e\to e} f_e^{(n)} + \mathscr{L}^+_{e\to\gamma} f_\gamma^{(n)} = F_e^{(n)};$$

$$\partial_t f_\gamma^{(n)} + \mathscr{L}^+_{\gamma\to\gamma} f_\gamma^{(n)} + \mathscr{L}^+_{\gamma\to e} f_e^{(n)} = F_\gamma^{(n)}.$$

Here $f_a^{(n)}(t, E)$ is nth statistical moment of a measured functional produced by EPC from a-particle with energy E [2, 9],

$$\mathscr{L}^+_{e\to e} f_e^{(n)} = [\sigma_e^{\text{brems}} + \sigma_e^{\text{ion}}] f_e^{(n)}(t, E)$$

$$- \int_{E^*}^{E} dE' [w_{e\to e}^{\text{brems}}(E \to E') + w_{e\to e}^{\text{ion}}(E \to E')] f_e^{(n)}(t, E'),$$

$$\mathscr{L}^+_{e\to\gamma} f_\gamma^{(n)} = - \int_{E^*}^{E} dE' w_{e\to\gamma}^{\text{brems}}(E \to E') f_\gamma^{(n)}(t, E'),$$

$$\mathscr{L}^{+}_{\gamma\to\gamma} f^{(n)}_{\gamma} = [\sigma^{\text{pair}}_{\gamma} + \sigma^{\text{Comp}}_{\gamma} + \sigma^{\text{photo}}_{\gamma}] f^{(n)}_{\gamma}(t, E)$$

$$- \int\limits_{\max\{E^*, \alpha(E)\}}^{E} dE' w^{\text{Compt}}_{\gamma\to\gamma}(E \to E') f^{(n)}_{\gamma}(t, E'),$$

$$\mathscr{L}^{+}_{\gamma\to e} f^{(n)}_{e} = - \int\limits_{E^*}^{E-2mc^2} dE' w^{\text{pair}}_{\gamma\to e}(E \to E') f^{(n)}_{e}(t, E')$$

$$- \int\limits_{E^*}^{E-\alpha(E)} dE' w^{\text{Compt}}_{\gamma\to e}(E \to E') f^{(n)}_{e}(t, E'),\ \alpha(E) = \frac{E}{1 + 2E/(mc^2)}.$$

For trace detectors, the right parts have the form

$$F^{(n)}_{e}(t, E) = n b_e(t, E) f^{(n-1)}_{e}(t, E) + (1 - \delta_{n1}) 1(E - 2E^*)$$

$$\times \sum_{i=1}^{n-1} \binom{n}{i} \left\{ \int\limits_{E^*}^{E-E^*} dE' w^{\text{brems}}_{e\to\gamma}(E \to E')\ f^{(n-i)}_{\gamma}(t, E') f^{(i)}_{e}(t, E - E') \right.$$

$$\left. + \int\limits_{E^*}^{E-E^*} dE' w^{\text{ion}}_{e\to e}(E \to E') f^{(n-i)}_{e}(t, E') f^{(i)}_{e}(t, E - E') \right\},$$

$$f^{(n)}_{\gamma}(t, E) = n b_{\gamma}(t, E) f^{(n-1)}_{\gamma}(t, E) + (1 - \delta_{n1}) 1(E - 2E^*)$$

$$\times \sum_{i=1}^{n-1} \binom{n}{i} \left\{ \int\limits_{E^*}^{E-E^*-2mc^2} dE' w^{\text{pair}}_{\gamma\to e}(E \to E')\ f^{(n-i)}_{e}(t, E') f^{(i)}_{e}(t, E - E' - 2mc^2) \right.$$

$$\left. + \int\limits_{\max\{E^*, \alpha(E)\}}^{E-E^*} dE' w^{\text{Compt}}_{\gamma\to\gamma}(E \to E')\ f^{(n-i)}_{\gamma}(t, E') f^{(i)}_{e}(t, E - E') \right\}.$$

For mixed moments of readings of various detectors on the right parts of the corresponding equations will be symmetrized products of the lowest moments of the readings of these detectors. Yes, for mixed moment $\overline{q_1 q_2}$ of readings from two detectors $\overline{q_1}$ and $\overline{q_2}$ whose first moments are equal $f^{(1)}_{1a}(t, E)$ and $f^{(1)}_{2a}(t, E)$, we get (see (4.82))

$$F_e^{(2)}(t,E) = b_{1e}(t,E) f_{2e}^{(1)}(t,E) + b_{2e}(t,E) f_{1e}^{(1)}(t,E) + 1(E-2E^*)$$

$$\times \left\{ \int\limits_{E^*}^{E-E^*} dE' w_{e\to\gamma}^{\text{brems}}(E \to E') \left[f_{1\gamma}^{(1)}(t,E') f_{2e}^{(1)}(t,E-E') \right.\right.$$

$$\left.\left. + f_{2\gamma}^{(1)}(t,E') f_{1e}^{(1)}(t,E-E') \right] \int\limits_{E^*}^{E-E^*} dE' w_{e\to e}^{\text{ion}}(E \to E') f_{1e}^{(1)}(t,E') f_{2e}^{(1)}(t,E-E') \right\}.$$

7.5 Some Details in Bremsstrahlung Term

The next step in constructing the SBCE working scheme is involving in the code befitting cross sections. Some problems may arise here. Let's take, for example, the Bethe–Heitler formula for bremsstrahlung, which has a singularity at low photon energies:

$$w_{e\to\gamma}^{\text{brems}}(E \to E') \sim E'^{-1}, \quad E' \to 0.$$

Let us represent the radiation component of the coefficient

$$A_k \equiv [\sigma_k - a_{kk}]_{ee}$$

in the form

$$A_k^{\text{brems}} = \int\limits_{E_{k-1}}^{E_k} dE w_{e\to\gamma}^{\text{brems}}(E_k \to E_k - E) \int\limits_{E_{k-1}}^{E_k} dE w_{e\to\gamma}^{\text{brems}}(E_k \to E_k - E) L_{kk}^n(E) + \ldots,$$

where the integrals are omitted, whose convergence is beyond doubt. Substituting here the decomposition

$$L_{kk}^n(E) = 1 + (E - E_k) \sum_{r=1}^{n} (E_k - E_{k-r})^{-1} + O((E - E_k)^2),$$

make sure that A_k^{brems} are finite. The Meller cross section has a characteristic feature $w_{e\to e}^{\text{ion}}(E \to E') \sim E'^{-2}$, $E' \to 0$.

$$A_k^{\text{ion}} = \int\limits_{E_k/2}^{E_{k-1}} dE w_{e\to e}^{\text{ion}}(E_k \to E)$$

$$-\left[\int\limits_{E_k-E_{k-1}}^{E_k/2} dE w_{e\to e}^{\text{ion}}(E_k \to E)E - \beta_k\right]\sum_{r=1}^{n}(E_k - E_{k-r})^{-1} + \ldots,$$

where $\beta_k = \beta(E_k) = \int\limits_0^{E_k/2} dE w_{e\to e}^{\text{ion}}(E_k \to E)E$ are ionization losses energy per unit path length. Therefore, the ionization component A_k is finite. The complete formula for A_k is

$$A_k = \int\limits_0^{E_{k-1}} dE w_{e\to\gamma}^{\text{brems}}(E_k \to E_k - E) + \int\limits_{E_k/2}^{E_{k-1}} dE w_{e\to e}^{\text{ion}}(E_k \to E)$$

$$+\left[\beta_k - \int\limits_{E_k-E_{k-1}}^{E_k/2} dE w_{e\to e}^{\text{ion}}(E_k \to E)E\right]\sum_{i=1}^{n}(E_k - E_{k-i})^{-1}$$

$$+\int\limits_{E_{k-1}}^{E_k} dE\{w_{e\to\gamma}^{\text{brems}}(E_k \to E_k - E)\left[1 - L_{kk}^n(E)\right] + w_{e\to e}^{\text{ion}}(E_k \to E)(E - E_k)^2\widetilde{L}_{kk}^n(E)\}$$

$$-\sum_{i=1}^{k-1}\int\limits_{E_{i-1}}^{E_i} dE[w_{e\to\gamma}^{\text{brems}}(E_k \to E_k - E) + w_{e\to e}^{\text{ion}}(E_k \to E)]L_{ik}^n(E)\sum_{s=m-n}^{m}\delta_{sk},$$

where

$$\widetilde{L}_{kk}^n(E) = (E - E_k)^{-2}\left[1 - L_{kk}^n(E) + (E - E_k)\sum_{r=1}^{m}(E_k - E_{k-r})^{-1}\right].$$

7.6 The SBCE Results Review

The EPC method made it possible for the first time theoretical study of the statistical characteristics of EPC with taking into account the real energy dependence of the cross sections of all elementary processes in an extremely wide range: from 10^4 up to 10^{20} eV.

Carrying out a large series of numerical calculations of average values and fluctuations of various EPC characteristics pursued the following goals:

1. Testing SBCE method by comparison with the results of other numerical calculations and experiments to verify the accuracy and reliability of the developed method, its applicability to the calculation of various EPC characteristics in a wide range of energies.

2. A numerical analysis of averages and fluctuations of various measured characteristics of EPC. Comparison with analytical results to determine their accuracy in problems of fluctuations.

3. Calculation of the characteristics of EPC ultrahigh energies in air for their subsequent use in calculations of wide atmospheric showers.

Table 7.1 gives a general idea of the scope of work under item 1. Quite satisfactory agreement has been obtained between our results and all the works listed in the table.

Let us give a brief review of the results obtained.

Table 7.1 SBCE testing

Characteristics	Matter	Primary energy E, GeV	Threshold energy E^* MeV	Method
Electron path in infinite medium	Lead	0.1–1	1	Monte Carlo [12]
	Lead	0.15–0.55	5	Experiment Path fluctuations [13]
	Xenon	0.025–1	3.066	Monte Carlo [14]
Path fluctuations in infinite medium	Lead	0.1–1	1	Monte Carlo [12]
	Xenon	0.025–1	3.066	Monte Carlo [14]
Cascade curves	Lead	$1–10^3$	1; 10	Multigroup [15]
	Lead	10^3	1	Multigroup [16]
	Lead	6	10	Monte Carlo [17]
	Lead	10^6	10^6	Semianalyt
				Monte Carlo [18, 19]
	Iron	10; 10^2	0.3; 1	Multigroup [15]
	Air	10; 10^2	0.316	Multigroup [15]
	Air	10	0.1–4	Monte Carlo [20]
	Air	10; 10^2	0;4	Monte Carlo [21, 22]
	Air	$10–10^6$	0	Semianalyt
				Monte Carlo [23]
Electron number fluctuations	Lead	6	10	Monte Carlo [17]
	Lead	10^6	10^6	Monte Carlo [24]
	Lead	1;2;6	10	Monte Carlo [25]
	Lead	1;2	10	Monte Carlo [24]
	Lead	103	6.7	Experim. [26]
	Iron	140	0.3	Experim. [26]

(continued)

Table 7.1 (continued)

Electron long. correlat.	Iron	400	0	Monte Carlo [25]
Cherenkov radiat. and fluct.	Atmosphere	1–300		Monte Carlo [26]
Electron path in layer	Xenon	1	3.066	Monte Carlo [12]
Electron path in layer (with fixed photon interact.)	Lead	1	3.066	Monte Carlo [12]
Electron number sum in diff. depths	Lead	2–15	0.5	Experim. [27]
	Xenon	1.5	1.5	Experim. [28]
Electron path fluct.in layer	Xenon	0.154	3.066	Monte Carlo [12]
El. path with fixed point fluct.	Xenon	0.154	3.066	Monte Carlo [12]
Angular distribution	Air	10	1	Monte Carlo [18]
	Air	$10–10^5$	25–80	Semianalyt.
				Monte Carlo [29]
RMS radius	Air	10	4	Monte Carlo [20]
	Air	10	0.1; 1;4	Monte Carlo [18]
Lateral distribution	Air	$10; 10^2$	0	Monte Carlo [20]
	Air	10	0.25; 4	Monte Carlo [18]
	Air	10^3	0	Monte Carlo [30]
	Air	$10–10^6$	0	Semianalyt.
				Monte Carlo [31]

7.6.1 The Total Electron Trace Length in an Infinite Medium

The SBCE calculations were performed for $\overline{R}(E)$ and $\sigma = \sqrt{\mathsf{Var} R}$ of the EPC electron path in air, xenon, and lead for threshold energies $E^* = 1.10$, and 50 MeV and energies of primary particles (electron or photon) E up to $100E^*$ [32]. The results showed that predicted by the theory asymptotic behavior

$$\overline{R}(E) = aE, \qquad \sigma(E) = b\sqrt{E} \tag{7.24}$$

is reached in the interval $10E^* < E < 100E^*$ (Fig. 7.1). As seen from Table 7.2, this is quite consistent with Monte Carlo calculations (MC) according to the formula of Tamm–Belenky (TB).

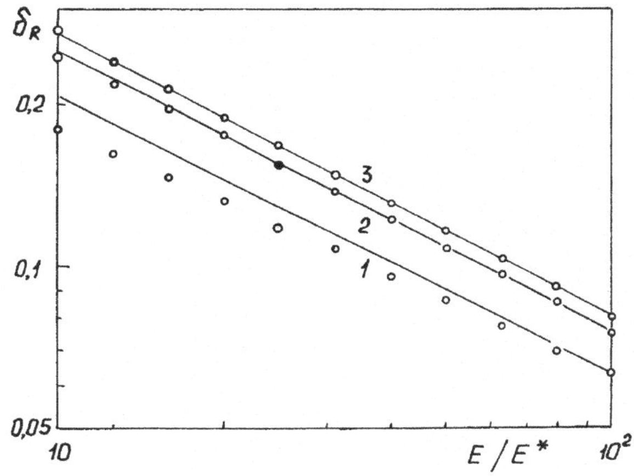

Fig. 7.1 Relative fluctuations of the total range (with energy above E^*) of EPC in xenon: 1—$E^* = 1\,\text{MeV}$, 2—$E^* = 10\,\text{MeV}$, 3—$E^* = 50\,\text{MeV}$. Points—SBCE calculation, curves—analytical approximation

Table 7.2 Approximation coefficients a (rad.un.MeV^{-1}) and b (rad.un.MeV$^{-1/2}$)

Matter, t_0, g/cm^2	E^*, MeV	Calcul. [32]		Other data		
		a	b	a	b	Method
Lead, 5.81	1	0.091	0.062	0.091	0.053	Monte Carlo [10]
				0.099		Tamm–Belenky [12]
	5	0.046		0.048		Experim. [11]
	10	0.029	0.074			
	50	0.0077	0.046			
Xenon, 8.13	1	0.068	0.043			
	3.066	0.046	0.055	0.046	0.050	Monte Carlo [12]
				0.047		Tamm-Belenky [12]
	10	0.025	0.062			
	50	0.0074	0.044			
Air, 37.0	1	0.012				
	10	0.0079	0.016			

7.6.2 Cascade Curves

Testing of SBCE method began with calculations of average cascade curves in air ($E^* = 0.1 \div 25$ MeV, $E = 1 \div 10^6$ GeV), iron ($E^* = 0.1 \div 35$ MeV, $E = 1 \div 10^4$ GeV), and lead ($E^* = 0.1 \div 10^6$ MeV, $E = 1 \div 10^6$ GeV) [1, 9] (Fig. 7.2).

The calculations showed, that the number of electrons stops changing as the threshold energy decreases below 0.25 MeV and at $E^* = 0.1$ MeV it can be con-

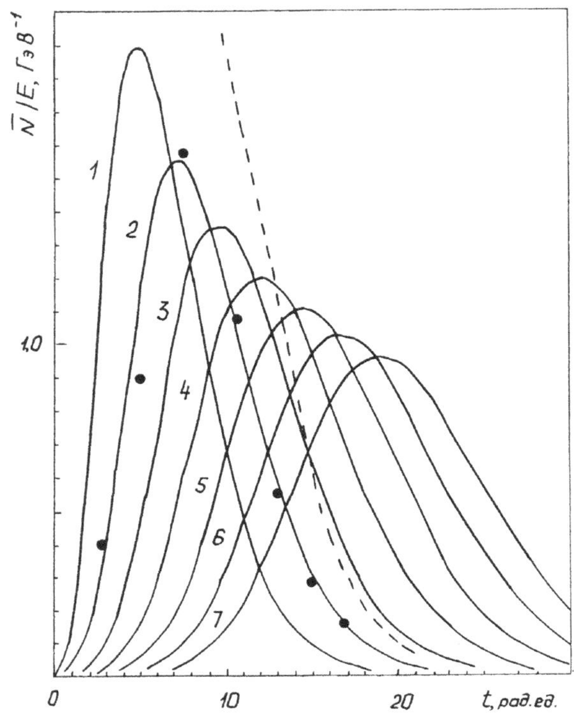

Fig. 7.2 Average number of EPC electrons in air from primary electron (1, 3) and photon (2, 4) with energy E. 1—$E = 10\,\text{GeV}$, 2—$E = 10^2\,\text{GeV}$, ..., 7—$E = 10^7\,\text{GeV}$; $E^* = 0.25\,\text{MeV}$. Dotted line—data [28] ($E^* = 0.316\,\text{MeV}$, $E = 10^2\,\text{GeV}$), lines—SBCE calculation, •—Monte Carlo results

sidered as a number of all electrons. This fact, in agreement with the conclusions of [33], allows us to interpret these results as the total number of particles in the shower. Calculation results of the total number of electrons in B-approximation of the cascade theory were approximated by Greisen using the formula

$$N_G(t, E) = \frac{0.31}{\sqrt{\log(E/\beta)}} \exp\left\{\left(1 - \frac{3}{2}\ln s\right) t\right\},$$

where t is the depth in rad. units, $s = 3(1 + 2t_{max}/t)^{-1}$, $t_{max} = \ln(E/\beta)$ for the primary photon. Hillas showed [19] that the Greisen approximation agrees with the data Monte Carlo method at $E = 10.10^2$ GeV for depths up to 2.5 t_{max}. Having generally confirmed this approximation, our calculations showed that at great depths (and in the $E > 10^4$ GeV region and at shallow depths), its error noticeably increases. To compensate for it, a correction is introduced into the approximating formula multiplier $\gamma(t, E)$:

$$\overline{N}(t, E) = \gamma(t, E) N_G(t, E). \quad (7.25)$$

Table 7.3 Correction factor $\gamma(t, E)$ for air

t, glad. units	Primary photon energy (GeV)						
	10	10^2	10^3	10^4	10^5	10^6	10^7
1	1.02	1.01	1.06	1.14	1.24	1.35	1.47
2	1.06	1.04	1.07	1.13	1.22	1.32	1.45
4	1.09	1.06	1.05	1.06	1.12	1.19	1.28
6	1.10	1.06	1.03	1.02	1.07	1.11	1.16
8	1.15	1.08	1.04	1.01	1.05	1.06	1.09
10	1.20	1.11	1.05	1.01	1.04	1.02	1.06
12	1.36	1.18	1.09	1.03	1.05	1.04	1.04
14	1.56	1.28	1.15	1.07	1.06	1.05	1.04
16	1.76	1.39	1.21	1.11	1.09	1.07	1.04
18	2.06	1.52	1.29	1.19	1.13	1.09	1.06
20	2.40	1.70	1.41	1.25	1.18	1.12	1.08
22	2.81	1.92	1.56	1.33	1.24	1.17	1.11
24	3.32	2.22	1.72	1.44	1.30	1.22	1.16
26	3.93	2.54	1.91	1.57	1.38	1.28	1.21
28	4.65	2.92	2.14	1.72	1.48	1.35	1.26

Given in Table 7.3 values of this multiplier give understanding of the accuracy of the Greisen formula in the region of ultrahigh energies.

7.6.3 Particle Number Fluctuations

SBCE calculations of fluctuations in particle number at different depths of EPC were carried out for air ($E^* = 1$ MeV, $E = 1 \div 10^6$ GeV), iron ($E^* = 0.1 \div 35$ MeV, $E = 1 \div 10^3$ GeV), and lead ($E^* = 0.1 \div 10$ MeV, $E = 1 \div 10^6$ GeV) [9, 34].

A specific feature of the dependence of relative fluctuations of number of electrons $\delta N = \sqrt{DN}/\overline{N}$ versus depth is the presence of a minimum near the maximum of the average cascade curve. Another feature is observed at the beginning of the cascade (in the case primary photon $\delta N(t)$ increases, and in the case of an electron—passes through a sharp maximum and tends to zero at $t \to 0$). The results of SBCE calculations coincided with the data Monte Carlo method (with an accuracy of ~5%–10%), experiment (in within its error), and confirmed by later calculations [22] carried out specifically to test our method (Figs. 7.3 and 7.4).

Comparison with results of analytical theory shows that the latter gives a correct qualitative description fluctuations, but in the quantitative relation differs from the exact ones by 20% − 40% and more (Figs. 7.5 and 7.6). When discussing the results,

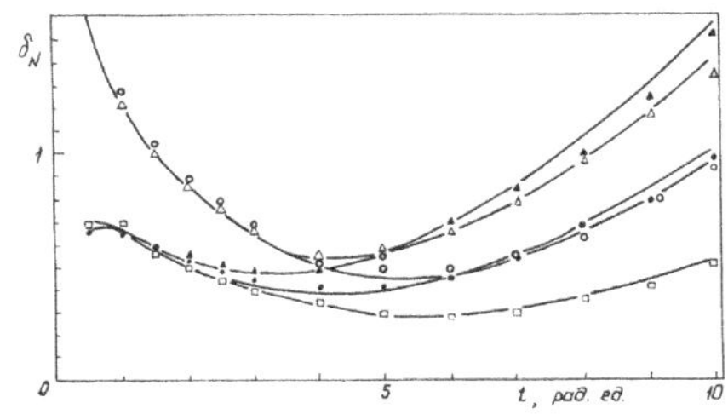

Fig. 7.3 Electron number fluctuations in lead ($E^* = 10$ MeV). Dots mark Monte Carlo results △, ▲ $- E = 1$ GeV [23], o, •—2 GeV [25]; ×—6 GeV [15]. Solid line SBCE-results [9], ×, •, ▲—primary electron, o, △—primary photon

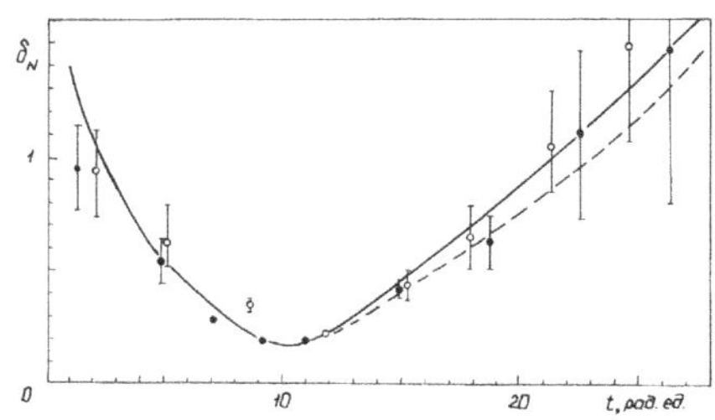

Fig. 7.4 Electron number fluctuations. (1) in lead ($E = 10^3$ GeV, $E^* = 6.7$ MeV), (2) in iron ($E = 140$ GeV, $E^* = 0.3$ MeV). Dots are experim. data [24], curves are SBCE results [9]

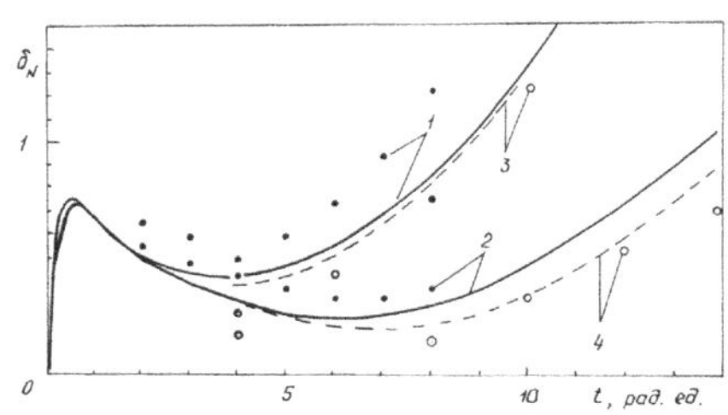

Fig. 7.5 Electron number fluctuations. ⊙ show A-approximation [60], • show B-approximation [61]; 1 relates to $y = 4$, 2 to $y = 6$, 3 to $E/E^* = 10^2$, 4 to $E/E^* = 10^3$. Curves show SBCE results[9]

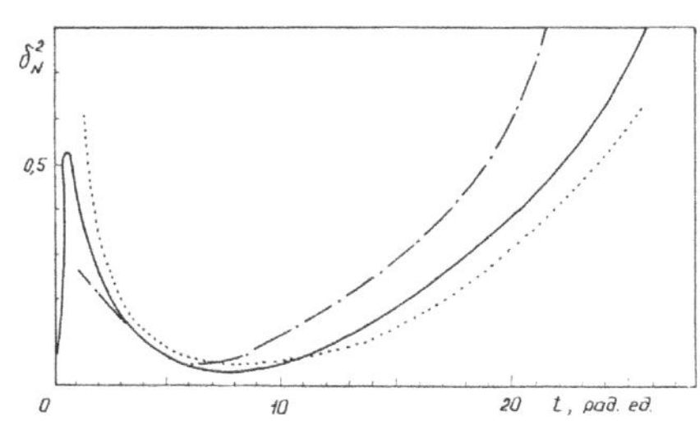

Fig. 7.6 Electron number fluctuations in lead ($E = 10$ GeV, $E^* = 1$ MeV): 1—[9] , 2—[36], 3—refined method, 4—[62]

Prof. I. P. Ivanenko proposed to consider fluctuations in depending on the dimensionless variables $t/t_{\max}(E)$. As seen from work [34], such a representation reduces the mutual scatter relative fluctuations of results related to different primary energies ($10^3 \div 10^6$ GeV). Further analysis led to the conclusion that the transition to a variable $t/t_{\min}$ ($t_{\min}$—depth of minimum relative fluctuations) practically eliminates the difference between fluctuations in a fairly wide range of depths, excluding the vicinity of the maximum. At the cascade maximum, the fluctuation dependence on

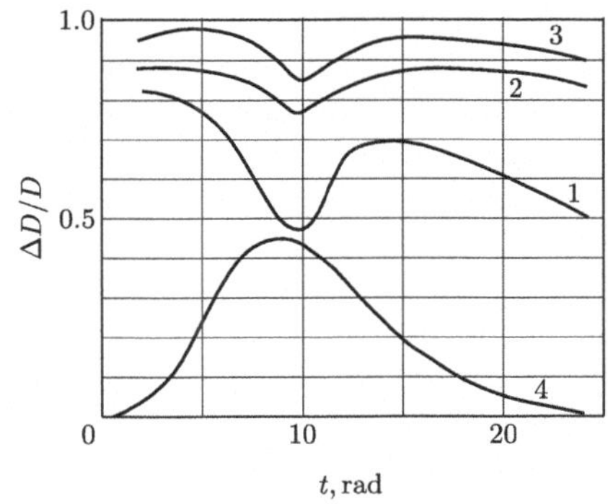

Fig. 7.7 Electron number fluctuations in iron from primary photon: obtained by SBCE calculation ($E = 10^2$ GeV, $E^* = 0.1$ MeV): A relative contribution into the variance: (1) of primary photon path, (2) of total path of all photons, (3) of part of EPC in range $1 \div 10^2$ GeV, (4) the average cascade curve

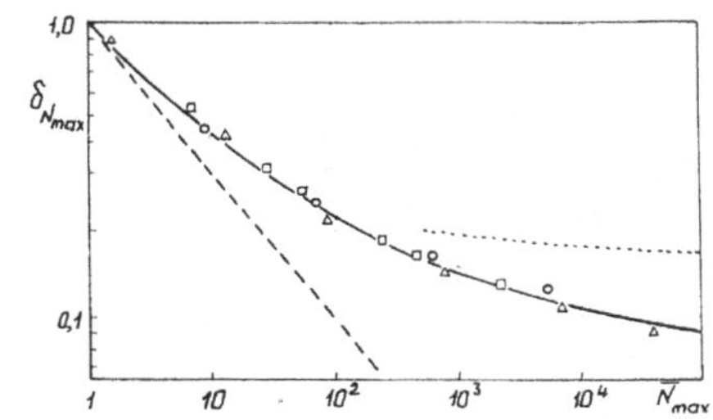

Fig. 7.8 Electron number fluctuations in lead $\bar{N}_{max}$ (primary photon). Dots – Monte Carlo results, △—Fe ($E^* = 0.1$MeV), ×—Pb ($E^* = 1$MeV); dashed line shows the Poisson law, dotted line represents analytical results from [62], solid line relates to SBCE calculations

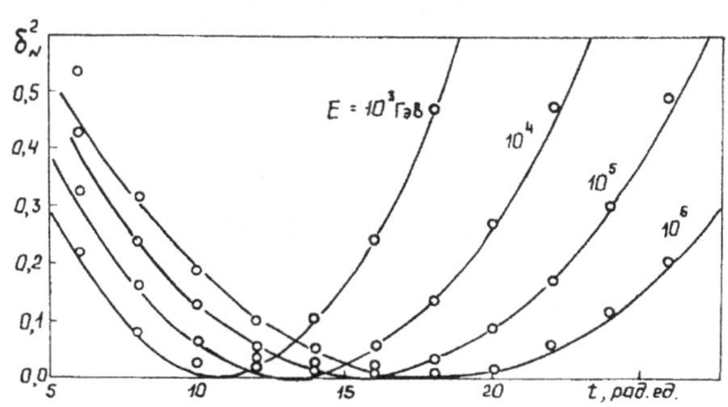

Fig. 7.9 Fluctuation of the number of electrons in air from the primary photon ($E^* = 1$ MeV): Points—SBCE calculation, curves—approximation (7.26)

N_{max} is described by the same curve regardless of the primary energy and absorbent substances (Fig. 7.7).

The square of the relative fluctuations in the number of electrons with energies above 1 MeV in air from primary photon in the energy range $10^3 \div 10^6$ GeV can be approximated (Fig. 7.9) by the expression

$$\delta^2 N = (1 - t/t_{min})^2; \ 0.3 < t/t_{min} < 2; \ t_{min} \approx 1 + t_{max}, \tag{7.26}$$

sufficient for EAS calculations taking into account EPC fluctuations (obvious the error near the minimum does not matter due to the smallness fluctuations in this region at high energies) (Fig. 7.8).

Using the equations derived in Sect. 4.6, there were analyzed sources of fluctuations in the cascade curves. The calculation results showed (Fig. 7.7) that taking into account the fluctuations of only the first free path is explicitly not enough to describe fluctuations in general, while taking into account fluctuations of all paths of all photons in the cascade make a decisive contribution ($\gtrsim$80%) into the resulting variance. It can be said that *the main source of fluctuations in the EPC cascade curves is fluctuations of photon ranges.* Note that when studying fluctuations full range in an infinite medium range fluctuations between elementary interactions "turned off" ($\partial_t \overline{R} = 0$) and the only fluctuation source is the random energy distribution of produced particles.

If the right parts of equations for variances to put zero for $E < E^{**}$ ($E^{**} > E^{*}$), then we get contribution to the fluctuation of the high-energy ($E > E^{**}$) part of cascade. Such a calculation would correspond to the incomplete method Monte Carlo, in which particles are modeled only when $E > E^{**}$, and for smaller values their contributions to the result are replaced by average values (deterministic value). The calculation results showed (see Fig. 7.7) that the main contribution to the dispersion is introduced by the energy interval $(0.01E, E)$. This result can be considered as a justification for the applicability of incomplete Monte Carlo method for the study of fluctuations. Another important conclusion following from this is that the effects associated with the low-energy part of the cascade (scattering, transient effects in inhomogeneous absorbers, etc.), in calculations, it is sufficient to produce fluctuations only in the right part of the inverse equation for the variance (i.e., in the average cascade characteristics). The equation itself can be solved with only those processes that are essential in the above energy range.

Correlations in Cascade Curves

Correlations are one of the most important characteristics of cascade processes. They are determined statistical links between fluctuations in the number of interactions in the cascade, fluctuations of free paths and energy losses of particles, fluctuations in reading ionization calorimeters and Cherenkov telescopes, etc. Correlations are used to separate cascades formed by primary electrons from cascades generated by primary π-mesons in experiments on accelerators [35].

In [25], some results of SBCE calculations of the particle number correlations are represented. They are related to EPC in the air produced by photons with primary energies $10 \div 10^5$ GeV. Results for autocorrelation moments (Sect. 4.6) obtained by SBCE method are consistent both with the Monte Carlo results [25] and the method of Green's functions [36].

Let us note the main regularities in the behavior of the coefficient correlations (Tables 7.4 and 7.5).

If both planes t_1 and t_2 of electron registration are placed before the cascade curve maximum, or after it, the correlations are positive. Otherwise (if t_1 and t_2 far enough apart), we observe anti-correlations with coefficient close to—1 (this is

Table 7.4 Electron number correlations matrix at different depths in air from primary photon with energy $E = 10^3$ GeV)

t	2	4	6	8	10	12	14	16	18	20
2	1	0.92	0.75	0.50	−0.09	−0.69	−0.66	−0.56	−0.49	−0.40
4		1	0.93	0.70	0.04	−0.80	−0.85	−0.76	−0.66	−0.56
6			1	0.90	0.29	−0.76	−0.94	−0.90	−0.82	−0.71
8				1	0.65	−0.51	−0.90	−0.97	−0.94	−0.86
10					1	0.30	−0.33	−0.61	−0.75	−0.80
12						1	0.80	0.53	0.30	0.12
14							1	0.93	0.79	0.63
16								1	0.95	0.85
18									1	0.97
20										1

Table 7.5 Electron number correlations matrix at different depths in air from primary photon with energy $E = 10^4$ GeV)

t	2	4	6	8	10	12	14	16	18	20
2	1	0.93	0.82	0.68	−0.48	−0.06	−0.65	−0.68	−0.59	−0.50
4		1	0.96	0.85	0.65	0.17	−0.75	−0.85	−0.76	−0.67
6			1	0.96	0.81	0.32	−0.76	−0.96	−0.89	−0.81
8				1	0.94	0.54	−0.64	−0.99	−0.98	−0.92
10					1	0.79	−0.36	−0.92	−0.99	−0.99
12						1	0.27	−0.48	−0.76	−0.89
14							1	0.71	0.41	0.20
16								1	0.93	0.82
18									1	0.97
20										1

especially noticeable in case of a larger energy—10^4 GeV). The change in the sign of the correlation is easy to understand: due to the law of conservation of energy, the integral of the cascade curve (at high energies) practically does not fluctuate, therefore an increase in the number in one area must be compensated reducing it to another. The larger absolute value correlation coefficient together with the characteristic dependence fluctuations with depth and a significant contribution to them from the first run suggests the possibility of separating fluctuations into fluctuations shear (for example, fluctuations in the center of gravity) and fluctuations in shape cascade curve (i.e., the dependence of the number of particles on the distance to center of gravity). In [37], where this idea was put forward and substantiated, it was shown that the coincidence of the centers of gravity of the cascade curves sharply reduces fluctuations, which indicates the stability of the EPC form in wide area around the center of gravity.

Further development of this idea is to increase the number moments characterizing the shape of the curve, and the development of a method calculation of the probabilistic characteristics of the cascade curve based on using the characteristics of random moments.

Total Path of Electrons in a Layer of Finite Thickness

In Ref. [34], the mean values and fluctuations of the total range electrons in a layer of iron ($E = 10 \div 300$ GeV), xenon ($0.1 \div 3.4$ GeV), and lead ($0.1 \div 10$ GeV) are provided. General nature of addiction relative fluctuations on the layer thickness is seen from Fig. 7.10: the values of $\delta R(t, E)$ decrease and approach as t increases to the limit $\delta R(\infty, E)$. Such behavior can be explained as follows. According to Sect. 8.3, $\delta R(\infty, E) \backsim E^{-1/2}$ and higher energies correspond to smaller fluctuations. However, as t decreases, first of all fluctuations in the ranges of high-energy cascades begin to increase, lower energy cascades are shorter and "feel" the influence of the limb layers at much smaller thicknesses. Especially clearly this is seen in Fig. 7.11.

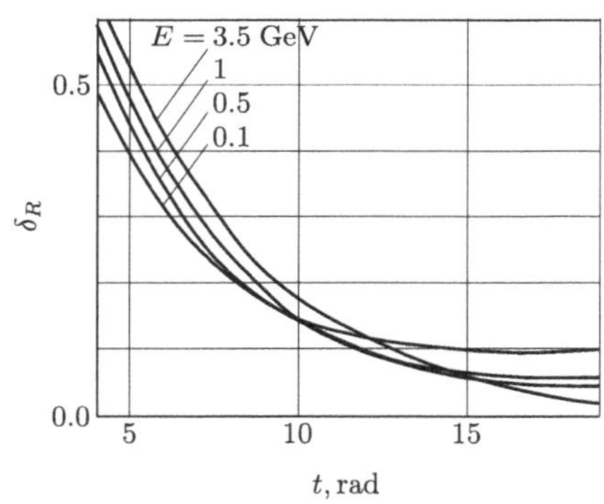

Fig. 7.10 Electrons' path fluctuations in a xenon layer (primary photon, $E^* = 1.5$ MeV)

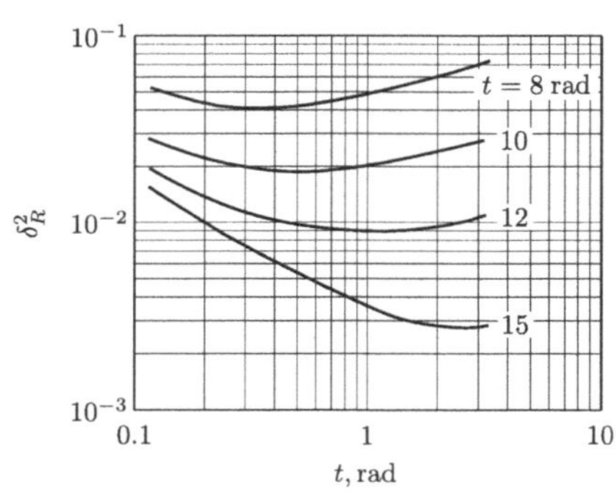

Fig. 7.11 Electrons' path fluctuations in a xenon layer (primary photon, $E^* = 3.066$ MeV)

Table 7.6 Fluctuations of electrons path in iron

Method	$s = 1.1$	1.2	1.3	1.4	1.5	1.6
SBCE-result	0.16	0.12	0.088	0.055	0.034	0.022
Equation (8.33)	0.25	0.11	0.076	0.050	0.032	0.016

Table 7.7 Electron paths in xenon

Quantity	Method	t, rad					
		1	2	3	5	7	10
$\overline{R}$	SBCE	0.53	1.75	3.09	5.06	6.10	6.76
$\overline{R}'$	Equation (4.75)	2.07	3.93	5.09	6.20	6.69	7.00
δR	Monte Carlo	2.15	4.08	5.16	6.23	6.65	6.95
	SBCE	1.37	0.90	0.65	0.38	0.25	0.14
$\delta R'$	Equation (4.75)	0.25	0.27	0.23	0.18	0.14	0.11
	Monte Carlo	0.26	0.27	0.25	0.19	0.15	0.10

Comparison of the results of numerical calculations $\overline{R}$ and δR with N and δN for the number of electrons with energies above 0.1 MeV in iron from a primary electron with an energy of 100 GeV showed that approximate relation (8.33) works satisfactorily in $1.2 < s < 1.6$ (Table 7.6).

When determining the energy of photons using a xenon bubble chamber, the thickness of the substance traversed by the cascade is measured not from point of birth (entry) of a photon, but from the point of its first interactions [39]. The relation (4.75) makes it easy to find moments q' at a fixed point of the first interaction in calculated moments q at a fixed birth point. The characteristics of the electron range obtained in this way in the layer xenon are given in Table 7.7.

It shows that the average values agree with the data of the Monte Carlo method with an accuracy of 4%, and relative fluctuations—up to 10%. It can also be seen that the first run plays the greatest role in the region of small thicknesses (where its contribution to fluctuations reaches 80%). With increasing thickness, the contribution of the first mileage is decreasing. In the limit $t \to \infty$ as $E \to \infty$ fluctuations in the total range of EPC electrons are due to, as can be seen from the derivation of the asymptotics in Sect. 8.3, fluctuations in distribution of energy between an electron and a bremsstrahlung photon radiation.

7.6.4 Cherenkov Radiation in Atmosphere

The photon flux calculation in the spectrum section (0.3 m $< \lambda <$ 0.6 m) of the Cherenkov EPC radiation in the atmosphere was calculated for primary photon's energy range $0.3 \div 10^6$ GeV. The atmosphere characteristics depending on high dependences of temperature and other characteristics of the atmosphere depending on the height were taken from ARDC data [40] and included in the coefficient $a(t, E)$. Obtained dependencies $\overline{N}/E$ and δN from energy it is easy to give physical explanation (Fig. 7.12). As the energy of the primary photon increases, the range of all shower electrons grows in proportion to the energy. However, the range of "luminous" electrons increases faster, since with an increase in energy, the shower reaches denser layer atmospheres where the threshold energy of Cherenkov radiation is lower. Relative fluctuations decrease in this case. As soon as a significant proportion range of electrons is below the observation level, there is reverse trend: as E increases, the Cherenkov flux shower radiation at the observation level decreases, and fluctuations increase.

7.6.5 LPM *Effect*

Differential sections of the brake radiation and pair production calculated taking into account the Landau–Pomeranchuk–Migdal effect. In the region of high energies (for the lead—above 10^3 GeV) it significantly changes the Bethe–Heitler cross sections—In view of the fact that different authors [16, 17, 38] obtained various quantitative estimates of the influence of this effect on the cascade curves in heavy substances, we applied to the solution of this problem the numerical method described above. Calculation results (lead $E^* = 10^3$, $E = 10^6$ GeV) confirmed the conclusion works [16, 17] on the much stronger influence of the CVL effect onto cascade curves, which was obtained in [38] (Fig. 7.13).

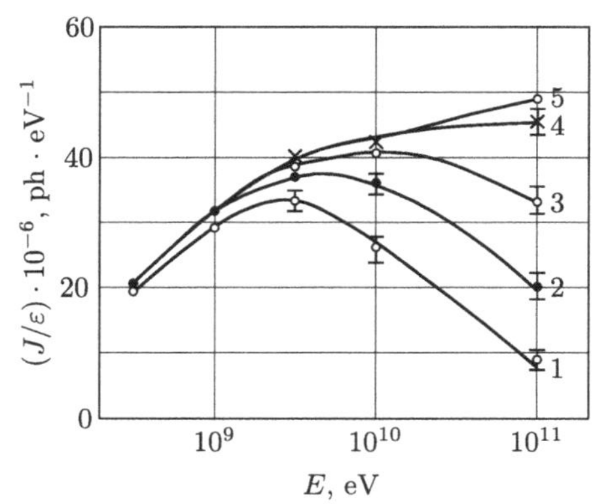

Fig. 7.12 The mean flux of Cherenkov radiation photons at various depths in atmosphere (primary photon) (1) 200, (2) 300, (3) 400, (4) 616 g/cm^2, (5) at sea level [26]

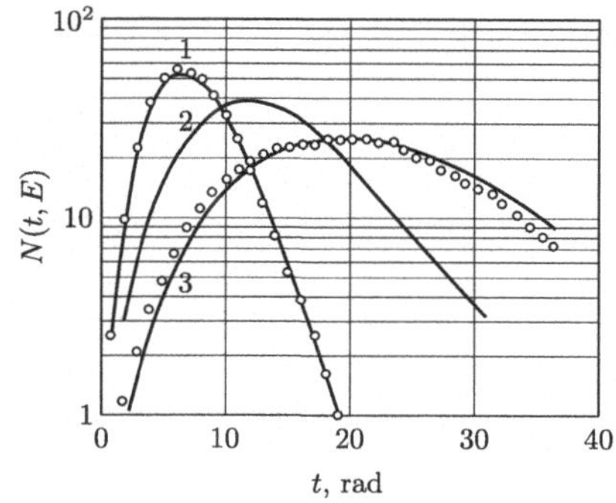

Fig. 7.13 Cascade curves in lead from primary photon ($E = 10^6$ GeV, $E^* = 10^3$ GeV): 1—without LPM effect, 2—[38], and 3—with LPM effect, dots are hybrid method result [16], solid line is SBCE result

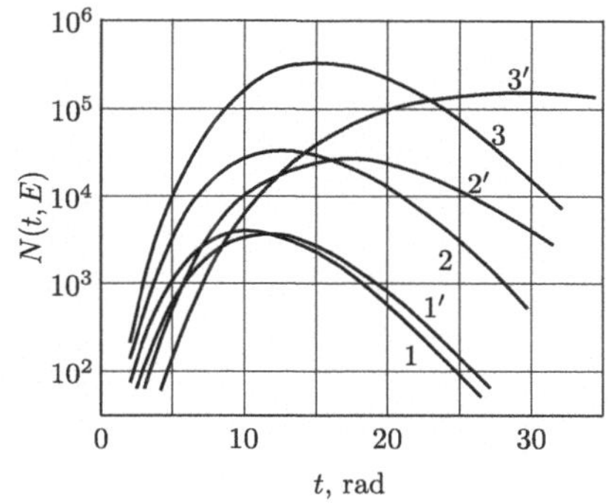

Fig. 7.14 Cascade curves in lead from primary electron ($E^* = 0.1$ GeV): 1—$E = 10^4$, 2—10^5, 3—10^6 GeV, 1′, 2′, 3′—for the same energies with taking into account LPM

Figure 7.14 demonstrates comparison of cascade curves in lead, calculated with the Bethe–Heitler cross sections with and without taking into account the LPM effect. At $E = 10^4$ GeV these curves do not yet differ too much from each other, but at $E = 10^6$ GeV the depth of the maximum of the curve increases due to this effect almost two times and its height decreases in the same proportion maximum. Systematic results at low threshold energies are given in Table 7.8. In accordance with the change in the cascade curves their fluctuations follow as well (Fig. 7.15).

7.7 Small-Angle Approximation

When deriving direct equations in the small-angle approximation transverse coordinates $\mathbf{r}$ and $\boldsymbol{\vartheta} = \dot{\mathbf{r}}$ secondary particles are counted from shower axis defined by the birth point and the direction of motion primary particle, which are the parameters of the problem. In reverse (adjoint) equations, the variables $\mathbf{r}$ and $\boldsymbol{\vartheta}$ refer to the primary particle, and the parameter tasks are the characteristics of the detector, to which it is necessary tie down to perform a physically adequate staging tasks. The symmetry of the problem plays an important role in this. Let's show it on specific examples. Let it be necessary to find the mean value and fluctuations in the number of particles

Table 7.8 Electron cascade curves in lead ($E^* = 1\text{MeV}$)

t	E, GeV						
	1	10	10^2	10^3	10^4	10^5	10^6
2	0.735+1	0.197+2	0.430+2	0.806+2	0.495+2	0.182+2	0.435+1
	0.476+1	0.112+2	0.224+2	0.402+2	0.315+2	0.119+2	0.265+1
4	0.113+2	0.643+2	0.249+3	0.760+3	0.984+3	0.546+3	0.126+3
	0.990+1	0.477+2	0.166+3	0.474+3	0.686+3	0.374+3	0.840+2
6	0.900+1	0.902+2	0.567+3	0.262+4	0.595+4	0.518+4	0.151+4
	0.965+1	0.803+	0.446+3	0.190+4	0.448+4	0.371+4	0.104+4
8	0.582+1	0.833+2	0.770+3	0.507+4	0.184+5	0.246+5	0.930+4
	0.699+1	0.849+2	0.691+3	0.415+4	0.148+5	0.184+5	0.652+4
10	0.357+1	0.622+2	0.766+3	0.679+4	0.368+5	0.744+5	0.369+5
	0.452+1	0.697+2	0.762+3	0.612+4	0.315+5	0.581+5	0.264+5
12	0.217+1	0.418+2	0.627+3	0.707+4	0.539+5	0.163+6	0.107+6
	0.281+1	0.497+2	0.676+3	0.693+4	0.489+5	0.132+6	0.783+5
14	0.133+1	0.266+2	0.456+3	0.619+4	0.631+5	0.278+6	0.248+6
	0.173+1	0.329+2	0.522+3	0.649+4	0.602+5	0.235+6	0.185+6
16	0.833	0.167+2	0.308+3	0.480+4	0.623+5	0.394+6	0.480+6
	0.107+1	0.209+2	0.368+3	0.532+4	0.622+5	0.347+6	0.366+6
18	0.532	0.104+2	0.200+3	0.343+4	0.541+5	0.480+6	0.808+6
	0.676	0.132+2	0.245+3	0.397+4	0.563+5	0.440+6	0.630+6
20	0.346	0.660+1	0.128+3	0.232+4	0.427+5	0.517+6	0.122+7
	0.435	0.830+1	0.159+3	0.277+4	0.460+5	0.493+6	0.971+6
22	0.230	0.424+1	0.810+2	0.152+4	0.314+5	0.504+6	0.167+7
	0.285	0.529+1	0.101+3	0.185+4	0.348+5	0.499+6	0.137+7
24	0.155	0.277+1	0.518+2	0.983+3	0.219+5	0.453+6	0.212+7
	0.190	0.343+1	0.647+2	0.121+4	0.249+5	0.465+6	0.178+7
26	0.105	0.184+1	0.336+2	0.632+3	0.148+5	0.380+6	0.253+7
	0.129	0.226+1	0.417+2	0.782+3	0.171+5	0.404+6	0.218+7
28	0.727-1	0.124+1	0.221+2	0.408+3	0.972+4	0.301+6	0.285+7
	0.879-1	0.151+1	0.273+2	0.504+3	0.114+5	0.331+6	0.252+7

The top number corresponds to the primary to an electron, bottom—to a primary photon; 0.735+1 means $0.735 \cdot 10^1$ etc.

crossing a certain plane in interval of angles $(0, \vartheta^*)$ with respect to the normal to it (regardless of the point of intersection). Direction of movement of the primary particle, which is an argument for solving the conjugate equation, cannot be tied to a fixed direction. Considering, however, that the dependence of the desired moments on the angle of incidence of the primary particle is described by a rapidly decreasing function ϑ, it suffices to confine ourselves to considering small ones that make the main contribution to result. The corresponding equations for the moments have the form

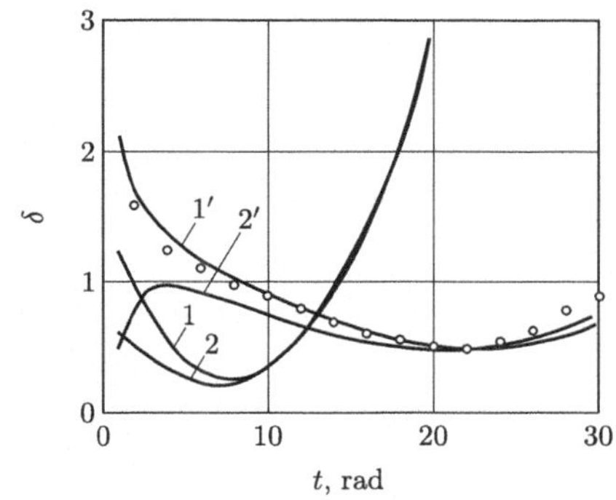

Fig. 7.15 Electron ($E^* = 10^3$GeV) number fluctuations in EPC from primary photon (1) and electron (2) with energy $E = 10^6$GeV. 1′, 2′—the same with the LPM effect accounting. Dots are Monte Carlo results [38]

$$[\partial_t + \sigma(E)] f(t, \boldsymbol{\vartheta}, E)$$

$$= \int\limits_{E^*}^{E} dE' \int d\boldsymbol{\vartheta}' w(E \to E', \boldsymbol{\vartheta} \to \boldsymbol{\vartheta}') f(t, \boldsymbol{\vartheta}', E') + F(t, \boldsymbol{\vartheta}, E), \tag{7.27}$$

where t is the distance from the primary particle to the observation plane (for obvious reasons, there is no dependency on $\mathbf{r}$). To find the number of particles in the case of normal incidence of the primary particle, you need to solve Eq. (7.27) and then put in the solution $\boldsymbol{\vartheta} = 0$.

It is easy to show, however, that if the detector (as in this case) has azimuthal symmetry and the same the differential cross section has symmetry

$$w(E \to E', \boldsymbol{\vartheta} \to \boldsymbol{\vartheta}') = w(E \to E', |\boldsymbol{\vartheta} - \boldsymbol{\vartheta}'|), \tag{7.28}$$

then the solution of Eq. (7.27) depends only on $\theta = |\boldsymbol{\vartheta} - \boldsymbol{\vartheta}'|$ and the Fourier transform of angular variable

$$\widetilde{f}(t, k, E) = \int e^{ik\theta} f(t, \theta, E) d\boldsymbol{\theta} = 2\pi \int\limits_0^\infty J_0(k\theta) f(t, \theta, E) \theta d\theta,$$

this equation is reduced to the form

$$[\partial_t + \sigma(E)] \widetilde{f}(t, k, E)$$

$$= \int\limits_{E^*}^{E} dE' \int d\vartheta' \widetilde{w}(E \to E', k) \widetilde{f}(t, k, E') + \widetilde{F}(t, k, E),$$

coinciding with (7.11). Here

$$\widetilde{w}(E \to E', k) = 2\pi \int_0^\infty J_0(k\theta) w(E \to E', \theta)\theta d\theta.$$

The desired solution is found using the inverse *Fourier–Bessel (Hankel) transform*:

$$f(t, \theta, E) = \frac{1}{2\pi} \int_0^\infty J_0(k, \theta) \widetilde{f}(t, k, E) k dk. \tag{7.29}$$

In the problem of the angular distribution of the average number of particles $\widetilde{F}(t, k, E) = 0$, and $\widetilde{f}(0, k, E) = 1(E - E^*)$. Twice differentiating equation for $\widetilde{f}$ twice in k and setting $k = 0$, we obtain a corresponding equation for the second moment of angular distribution,

$$\theta^{(2)}(t, E) = 2\pi \int_0^\infty f(t, \theta, E)\theta^3 d\theta = -2[\partial_k^2 \widetilde{f}(t, k, E)]_{k=0}, \tag{7.30}$$

solving which yields the root-mean-square angle

$$\theta_{rms}(t, E) = [\theta^{(2)}(t, E)/\overline{N}(t, E)]^{1/2}.$$

If the sensitivity function depends on the transverse coordinates, then instead of (7.27) we have

$$[\partial_t + \boldsymbol{\vartheta}\partial_{\mathbf{r}} + \sigma(E)] f(t, \mathbf{r}, \boldsymbol{\vartheta}, E)$$

$$= \int_{E^*}^{E} dE' \int d\boldsymbol{\vartheta}' w(E \to E', \boldsymbol{\vartheta} \to \boldsymbol{\vartheta}') f(t, \mathbf{r}, \boldsymbol{\vartheta}', E') + F(t, \mathbf{r}, \boldsymbol{\vartheta}, E).$$

If the detector is spherically symmetrical with respect to, for example, origin ($t = 0$, $\mathbf{r} = 0$), isotropic, placed in homogeneous environment and holds (7.28), then the solution of the latter equation depends only on the distance $|\mathbf{t} + \mathbf{r}|$ from the particle to the origin and the angle ψ between the direction particle motion and direction to the origin. Such property of inverse equations in this problem gives them a large advantage over straight lines even when calculating the first moment: the number of independent variables is smaller [41]. Selecting as the "reference" direction the direction to the beginning coordinates, in the small-angle approximation ($|\mathbf{t} + \mathbf{r}| \approx t$, $\psi \approx r/t$) reads

$$[\partial_t - (\psi/t)\partial_\psi + \sigma(E)] f(t, \psi, E)$$

$$= \int_{E^*}^{E} dE' \int_0^{2\pi} d\varphi \int_0^\infty d\psi' \psi' w(E \to E', \theta) f(t, \psi', E') + F(t, \psi, E),$$

with $\theta = [\psi^2 + \psi'^2 - 2\psi\psi' \cos\varphi]^{1/2}$. Denoting

$$\widetilde{f}(t, kt, E) = 2\pi t^2 \int\limits_0^\infty J_0(kt\psi) f(t, \psi, E)\psi d\psi,$$

write the following equation for transform:

$$[\partial_t + \sigma(E)]\widetilde{f}(t, kt, E) = \int\limits_{E^*}^{E} dE'\widetilde{w}(E \to E', kt)\widetilde{f}(t, kt, E') + \widetilde{F}(t, kt, E).$$

Inverse transformation yields

$$f(t, \psi, E) = \frac{1}{2\pi} \int\limits_0^\infty J_0(kt\psi)\widetilde{f}(t, kt, E)kdk. \tag{7.31}$$

According to (7.29) and (7.31), the problem of restoring the angular and spatial distributions over known Fourier transforms—Bessel is reduced to the calculation of an improper integral of the form

$$I = \int\limits_0^\infty \varphi(x) J_0(x)dx. \tag{7.32}$$

In numerical inversion, this integral is replaced by its own (i.e., the infinite upper limit is replaced by such a finite value $x_{\max}$ so that the resulting error in the integral was small enough) and is presented as a sum integrals over segments Δx:

$$I = \sum_i \int\limits_{\Delta x_i} \varphi(x) J_0(x)dx. \tag{7.33}$$

Note, however, that in the case of narrow distributions (especially concerns the spatial problem) when the transformant decreases very slowly, recovery results at large values x begin to behave irregularly: a characteristic incorrect task "saw". Reason for this solution behavior lies in the fact that the contributions of two neighboring sections integrands with different signs turn out to be very close in absolute value, their algebraic sum becomes sensitive to errors in the approximation transformant, and the cumulative result—to the value of the upper limit. Increasing the latter while maintaining the necessary approximation accuracy of transformants leads to an increase in time computing. To overcome this difficulty, the method regularization, which consists in replacing

$$\int_0^\infty \varphi(x) J_0(x) dx \to \int_0^\infty h(x, \varepsilon)\varphi(x) J_0(x) dx, \tag{7.34}$$

where $h(x, \varepsilon)$ is a decreasing function of x depending on a small parameter $\varepsilon(h(x, \varepsilon) \to 1,\ \varepsilon \to 0)$, which is chosen empirically. Consistent application of this method to calculating the transform of the estimated distribution and then to recovery from this transformant of the original distribution allows you to verify the accuracy of the result gave a good effect the simplest stabilizing function $h(x, \varepsilon) = e^{-\varepsilon x}$, allowing three to four times increase in the maximum radius of sure recovery spatial distribution.

7.8 Angular Electron Distributions

Angular distribution electron, reflecting the influence mainly of the Coulomb scattering, is an important characteristic of EPC and is used in calculations of the Cherenkov radiation of showers in the atmosphere and other optically transparent media ("luminous point" approximation [42]).

Calculations of angular distributions of electrons and rms EPC angles in air were drawn for threshold energies $E^* = 0.25 \div 80$ MeV, primary energies $E = 1 \div 10^7$ GeV, and depths t up to 28 rad. units. Matching results with method data Monte Carlo for $E = 10$ and $E = 10^2$ GeV shows a satisfactory agreement over the entire range of changes in the angular variable, thus confirming the considerations expressed in [43] about that the small-angle approximation (more precisely, the Landau approximation) may be applicable for all angles θ from 0 to π, and also the conclusions of the later work [42]: "... the function of the angular distribution in the small-angle approximation is close to exact in a wide range of angles $\theta \lesssim 1$ and threshold energies $E^* \lesssim \beta$" (Fig. 7.16).

It was shown in [29] that the distribution of electrons over variable $x = \theta/\theta_{\text{rms}}$ (E, E^*, s) in the region $0/05 < x < 5$ depends only on the age of s and (with an error of $15 - 20\%$) does not depend on the primary ($10 < E < 10^7$ GeV) and threshold ($E^* < 80$ GeV) energies:

$$\rho(\theta; E, E^*, s)\theta d\theta = \rho_0(x; s) x dx.$$

The function ρ_0 is given in Table 7.9. Comparing it with the approximation

$$\rho_0(x; s) = C(s) x^{s-1} \exp\{-\sqrt{s(s+1)}x\},$$

obtained in [42], shows that the largest error (higher 25%) at $s = 0.6$ is observed in the region $x \gtrsim 5$, at $s = 1$ in the region $x = 2 \div 3$, at $s = 1.4$ in the region $x \lesssim 0.1 \cup x \gtrsim 4$.

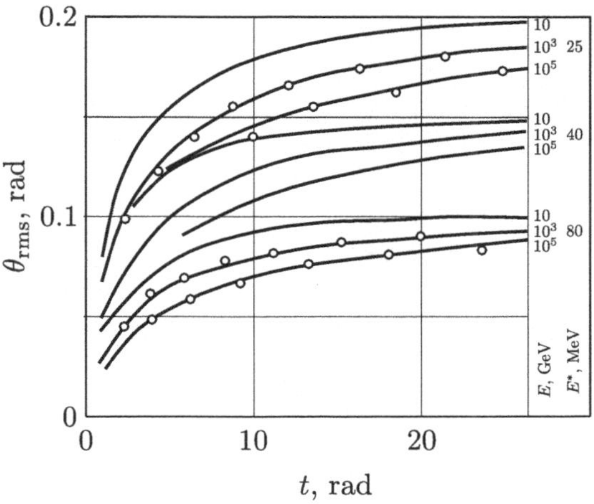

Fig. 7.16 The root-mean-square angle of distribution of EPC electrons in air from the primary photon (in radians), curves—results of numerical solution of equations, points—the Monte Carlo method. On the right are the primary and threshold energies

Table 7.9 EFC electron angular distribution function $\rho_0(x, s)$, $x = \theta/\theta_{rms}$

x	$s = 0.6$		$s = 1.0$		$s = 1.4$	
	num. [29]*	appr. [42]	num. [29]	appr. [42]	num. [29]	appr. [42]
0.05	3.35-1	3.35-1	2.10-1	2.10-1	1.68-1	1.15-1
0.1	2.53-1	2.42-1	2.10-1	1.95-1	1.81-1	1.39-1
0.2	1.56-1	1.61-1	1.75-1	1.69-1	1.70-1	1.52-1
0.5	9.70-2	8.62-2	1.08-1	1.11-1	1.35-1	1.27-1
1	3.67-2	3.98-2	5.32-2	5.47-2	6.41-2	6.69-2
2	1.08-2	1.12-2	1.00-2	1.33-2	1.35-2	1.42-2
3	3.81-3	3.61-2	2.44-3	3.24-3	2.83-3	2.68-3
4	1.35-3	1.21-3	6.41-4	7.83-4	3.81-4	4.75-4
5	3.29-4	4.17-4	1.68-4	1.90-4	1.03-4	8.40-5

7.9 Lateral Electron Distribution

The width of the spatial distribution of particles in a shower is usually characterized by the root-mean-square radius r_{rms}, but this is a poor parameter. It is brought here from those problems of statistical physics, where two-dimensional bell-shaped distributions decrease rapidly with distance from the center, like in the Gaussian distribution $e^{-(r/r_0)^2}$. But the lateral distribution in the EPC is characterized by a slow of a power-kind law decrease at the margins, which makes it absolutely impossible to reliably conclude anything from comparing the sample estimates of the r_{rms} cor-

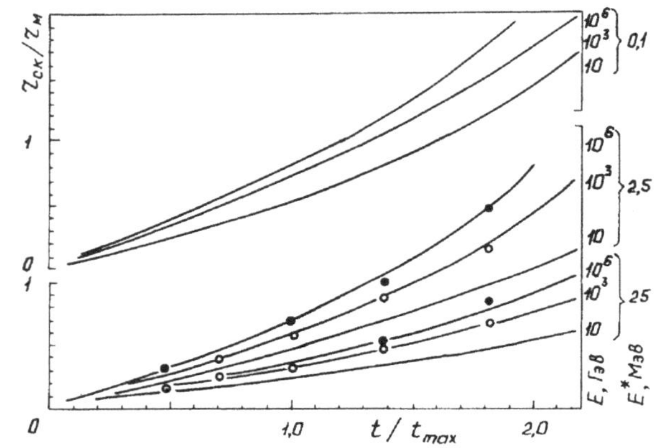

Fig. 7.17 Root-mean-square radius (in Moliere units) of electron distribution in EPC in air from a primary photon, curves are SBCE results, dots are Monte Carlo data

responding with theoretical data.[1] Interest to r_{rms} had been increased in connection with the discussion of 1975–1983 about the width of the EFC in air. A thorough introduction to the subject matter was given in [44], at the beginning of which it was noted that the reason for the discussion was presented at the 14th–15th International Conferences on cosmic rays calculations by Allan; A. A. Belyaev, V. V. Guzhavin and I. P. Ivanenko, Hillas and Lapikens, who gave about twice narrower distribution than Nishimura–Kamata's (NK) known distribution. In this regard, the authors of [44] returned to the discussion made by Nelson back in 1966, the assumption of independence spatial distribution of EPC particles from primary energy and atomic number of a substance, provided that the level of observation characterized by a cascade parameter, and the radius is expressed in Moliere units r_M. J. Nishimura, I. Ivanenko, and some took a different point of view, believing that root cause discrepancies between Monte Carlo data and analytical results are in the difference in primary energies: $10 \div 10^2$ GeV—in Monte Carlo calculations and $E = \infty$ in the analytic theory.

In view of this issue importance for the study of EAS, r_{rms} were calculated in air (threshold energies 0.1; 0.25; 1.4; 25; 80; 10^3 MeV) and carbon (1; 10^2 MeV) in primary energy range $1 \div 10^7$ GeV (Fig. 7.17). The results Table 7.10 ($E^* = 0.1$ MeV) were published in [45] ($E^* = 2.5$ and 25 MeV). They were in good agreement with the data Monte Carlo and multigroup method data.

To compare with B-approximation, the SBCE results for $E^* \to 0$ and $E \to \infty$ were obtained. We restricted ourselves here to the analysis of r_{rms} in maximum of the cascade ($s = 1$), where there is a dependence $r_{rms}(\infty, E^*)$ on the threshold energy, calculated in the framework of the analytic theory [46] in the domain $E^* > 4$ MeV. Presenting the numerical SBCE—results r_{rms} for a fixed E^* as a function of the variable $1/\ln E$, one can see that behavior of r_{rms} in the region $E = 10 \div 10^7$ GeV

[1] Discarding typically large values of r_i (*outliers*) only exacerbates uncertainty (experienced statisticians use this technique only in conditions of homogeneous and uncorrelated data, which obviously contradicts the nature of the observed distributions in the EPC).

Table 7.10 RMS radius r_{rms} (in rad units) of electron distribution in EPC in air ($E^* = 0.1$ MeV) from a primary photon with energy E

s	E, GeV					
	1	10	10^2	10^3	10^4	10^5
0.5	0.441-1	0.555-1	0.644-1	0.700-1	0.734-1	0.754-1
0.6	0.558-1	0.694-1	0.801-1	0.867-1	0.908-1	0.934-1
0.7	0.681-1	0.844-1	0.976-1	0.106	0.111	0.114
0.8	0.813-1	0.102	0.118	0.128	0.136	0.139
0.9	0.960-1	0.123	0.143	0.156	0.165	0.169
1.0	0.013	0.147	0.173	0.189	0.200	0.206
1.1	0.133	0.177	0.209	0.229	0.243	0.251
1.2	0.156	0.214	0.253	0.278	0.296	0.306
1.3	0.186	0.258	0.307	0.338	0.361	0.373
1.4	0.221	0.313	0.373	0.411	0.440	0.457
1.5	0.264	0.378	0.452	0.501	0.536	0.559
1.6	0.315	0.455	0.547	0.609	0.656	0.687
1.7	0.375	0.544	0.662	0.739	0.801	
1.8	0.445	0.655	0.800	0.896		

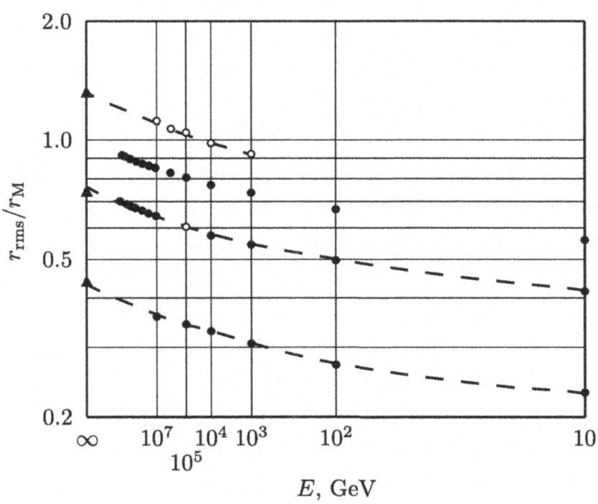

Fig. 7.18 The dependence r_{rms} (in Molier units) on primary energy at the maximum cascade. 1—$E^* = 25$ MeV, 2—$E^* = 4$ MeV, 3—$E^* = 0.1$ MeV, 1—$E^* \approx 0$ MeV, •—SBCE-results, ▲—the limit (as $E \to \infty$) values of r_{rms} for finite threshold energies (after Nishimura), o—Oda–Misaki analytical results for $E^* = 0$ and $E \to \infty$

agrees with limit ($E \to \infty$) value, given by analytic theory (0, 73 r_M at $E^* = 4$ MeV, 0.43 r_M at $E^* = 25$ MeV, and so on) (Fig. 7.18). At the same time, $r_{rms}^{NK}(E, 0)$, found in B-approximation from Nishimura–Kamata transforms [47], turned out to be 30% larger; in particular $r_{rms}^{NK}(\infty, 0) = 1.3$; $r_{rms}^{SBCE}(\infty, 0) \approx 1$. The cause of discrepancies is incorrect accounting of elementary processes in low-energy region ($0.1 \div 1$ MeV) in the frame of analytical theory. Assumption that energy losses $\beta(E) =$ const, identification of the kinetic energy with total one and extension down to zero the electron–positron pair generation increase the number and total path of low-energy electrons, which, due to their wide spatial distribution, contribute more to root-mean-square radius than in the total number of particles determining the cascade curve.

At the same time, SBCE analysis of the dependence $r_{ms}(E)$ confirmed the point of view of J. Nishimura, I. P. Ivanenko et al. on the significant influence of the finiteness of the primary energy E_0 on the width of the distribution of electrons in the EPC. During the discussion continued in [48] were eliminated our inaccuracies. It was noted, in particular, that conclusion about the constancy of r_{rms} in the $E \gtrsim 10^6$ GeV region does not follow from the data given in [45]. This remark was completely fair: subsequent calculations showed that the finite E effect affects up to very high energies (10% proximity to the r_{rms} limit reaches at $E \gtrsim 10^{10}$ GeV, and 5% for $E \gtrsim 10^{20}$ GeV, which is of no practical importance).

As noted above, RMS-radius studies have only methodological value. To solve practical problems of EAS analysis, detailed data on the shape of the spatial distribution of electrons are needed. The shape is represented by normalized to 1 radial density $\rho(r; E, E^*, s)$:

$$2\pi \int_0^\infty \rho(r; E, E^*, s) r dr = 1.$$

The electron LDF in air was computed with SBCE for threshold energies $E^* = 0.25$; 2.5; 4; 25 MeV and primary photon energy from 10 to 10^6 GeV up to $t = 23.5$ rad. units. The results were obtained for distances 100–150 m from the shower axis at small depths ($s < 1$) and $200 \div 300$ m at largest ones ($s \geqslant 1$) [31, 45, 49]. Obtaining LDF at large radii was hindered by the "saw effect", which is a consequence of instability inherent in inverse problem-solving and overcome by using the regularization method which made it possible to advance up to 500–1000 m (A. I. Goncharov).

Analysis of the results showed that, like the angular distribution, its lateral counterpart expressed in dimensionless variable $x = r/r_{\text{rms}}(E, E^*, s)$ depends only on cascade age s but does not depend (except for the region of small x) on the primary and threshold energies in the range $E^* \ll \beta$, $E > 10^2$ GeV (Fig. 7.19). Thus, the statement made in [42, 50] that in a certain range of distances the RDF is determined by the value of its second moment, received the exact quantitative expression:

$$\rho(r; E, E^*, s) = [r_{\text{rms}}(E, E^*, s)]^{-2} \rho_0(x, s).$$

It is interesting to note that Nishimura–Kamata LDF and corresponding $r_{\text{rms}}^{\text{NK}}(\infty, 0, s)$ with good accuracy satisfies the same relation: $\rho^{\text{NK}}(r; \infty, 0, s) = [r_{\text{rms}}^{\text{NK}}(\infty, 0, s)]^{-2}$ $\rho_0(x, s)$, which gives grounds for approximating the function $\rho_0(x, s)$ formula of the Nishimura–Kamata–Greisen type.

The above considerations led to the following picture of LDF evolution with primary energy increasing: graph of the function $r^2 \rho(r; E, E^*, s)$ plotted in logarithmic (along the abscissa axis) scale is shifted to the right by an amount proportional to logarithm of root-mean-square radius (Fig. 7.20). If $r_{\text{rms}}(E, 0, s)$ had

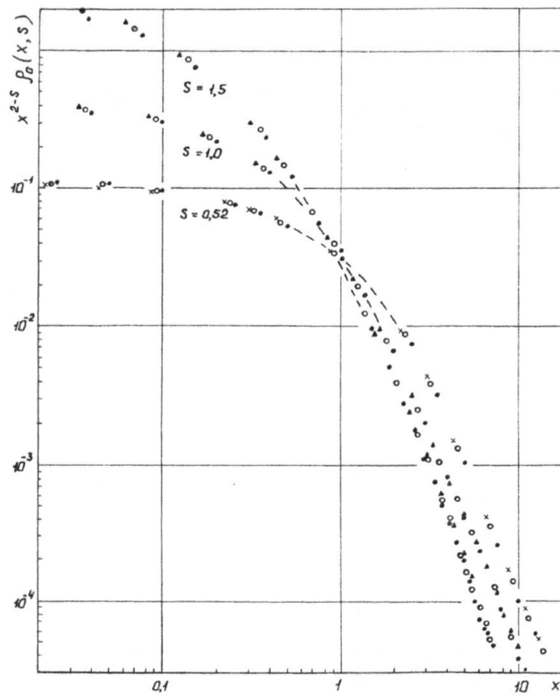

Fig. 7.19 Lateral distribution electron density in dimensionless variable $x = r/r_{c.k}(E, E^*, s)$ in air. $E^* = 0.1$ MeV, $E = 10^2$ (•), 10^3 (∘), 10^4 (×) 10^5(▲)GeV

as its limit $r^{NK}_{rms}(\infty, 0, s)$, RDF would be distribution of NK. Since, however, $\lim r_{rms}(E, 0, s) < r_{rms}(\infty, 0, s)$, the limit distribution turns out to be substantially narrower than the NK distribution, but it is not achieved in the energy field, of practical interest.

The least change in LDF is observed at $1 \lesssim r \lesssim 100$ m, so the region $10^2 < E < 10^5$ GeV can be recommended for such an approximation [2] energy-dependent function [49]:

$$\rho(r, s) = \frac{C(s)}{(mr_M)^2}\left(\frac{r}{mr_M}\right)^{s-2}\left(\frac{r}{mr_M}+1\right)^{s-4,5} \tag{7.35}$$

$$m(s) = 0.78 - 0.21s. \tag{7.36}$$

It should be noted that the SBCE method developed by our team for the first time gave the ability to calculate the FPR EFC within the framework of one program with

[2] The need introducing a scale factor into the formula according to the data Tien Shan experiment was pointed out in [51].

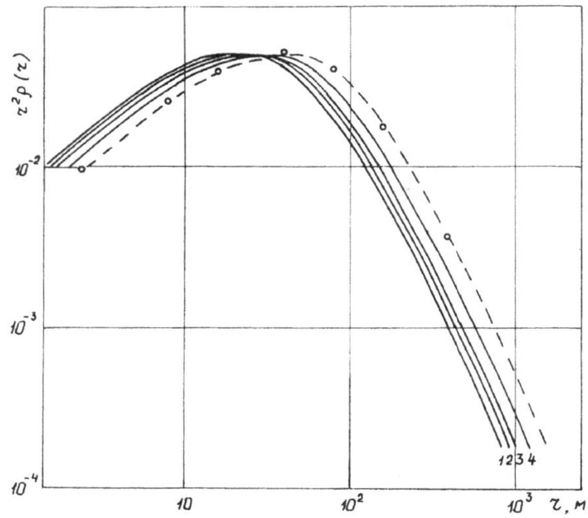

Fig. 7.20 Total electron number LDF evolution at EPC maximum in air with increasing primary photon energy 10^3, 10^5, 10^7 GeV and ∞ (solid curves 1–4). The dashed curve relates to LDF with $r_{\rm rms}(\infty, 0, 1) = 1, 3$, obtained by Oda–Misaki in B-approximation. Dots represent the Nishimura–Kamata distribution

taking into account the real energy dependence of the cross sections of elementary processes in an extremely wide energy range—from 10^{-4} to 10^7 GeV. At the end of these calculations the maximum energy of the EPC calculated by the Monte Carlo method with detailed account of processes in the low-energy region was 10^2 GeV [18, 20]. Two years later, the energy of 10^3 GeV was achieved by the analog Monte Carlo (AMC) method (Fig. 7.20, [30]). Let us stress that neither SBCE nor AMC can reach high primary energy with accuracy taking into account low-energy processes in EPC. Only a combination of the SBCE method with the modified Monte Carlo (MMC) method carried out in [21] made it possible to go through this energy range [31, 45] with accurate calculation of low-energy processes. The hybrid SBCE+MMC code produced a wide spectrum of results, confirming most obtained to that time results and corrected some of them. Thus, due to the inaccuracy of modeling the scattering electrons and low statistics at $s \geqslant 1.3$ in [31] were represented wider distribution; after correcting the algorithm and increase in statistics, the results coincided with the numerical data before distances $\lesssim 200$ m. The refined conditions of the pass [48] in significantly narrowed the gap between the LDF of the analytical theory taking into account the finite value of the primary energy and the results of numerical method. The subsequent discussion and application of these results are given in [30, 52–58] and others.

It should also be noted that Lagutin—Plyasheshnikov–Uchaikin approximation (7.35) was included as a special option into EGS4 (**E**lectron **G**amma **S**hower system version **4**) for calculation of EP subcascades in program CORSIKA, replacing the previous Nishimura–Kamata–Greisen anzatz.

The electron densities are calculated for 80 reference points centered around the shower axis on a circular grid extending in 8 directions spaced by 45° and with 10 radial distances in each direction, covering the range from 1 m to a maximum radius which has to be specified (see Ref. [20]) in logarithmic steps. We are aware of the fact that Eq. (7.35) is correct only in the Landau approximation, which restricts (besides others) the distance r between the considered grid point and the subshower axis to $|r/r_M| > |E_{cr}/E$, the authors of [63] inform.

It is interesting to see traces of this discussion in the publications of those years. So, in the article [53] we read "It is now generally well known that the NKG (Nishimura–Kamata–Greisen) function does not give a good fit to the lateral distribution of electrons for a single value of the age parameter. This is admitted on the basis of analysis of experimental results over a large range of distances, from the shower core at mountain altitude as well as at sea level, and from sizes of 10^4 up to 10' particles. In previous work [52] we considered the consequences of the conflicting propositions of solutions for three-dimensional transport equations of pure electromagnetic (EM) cascades [4, 59] in extensive air shower simulation and, in the same way, the effects of results of Monte Carlo simulation [20]. The remarkable agreement between Uchaikin's and Hillas' formulae was underlined and a typical variation of the local age parameter with distance with a general decrease was noticed".

References

1. Pljasheshnikov AV, Lagutin AA, Uchaikin VV (1979) The numerical method of solution of one-dimension cascade theory adjoint equations. In: Proceedings of the 16-th ICRC, Kyoto, vol. 7, pp. 1–6
2. Uchaikin VV, Pljasheshnikov AV, Lagutin AA, et al (1977) The calculational techniques of the fluctuations of the electron range corresponding to finite thickness layer. In: Proceedings of 15-th ICRC, Plovdiv, vol. 7, pp. 502–507
3. Uchaikin VV, Plyasheshnikov AV, Lagutin AA, Vetoshkin VV (1978) Numerical method for calculating fluctuations in the path of charged particles of an electromagnetic cascade in an infinite medium. Izv Vyssh Uchebn Zaved Fizica 4:27–30 (in Russian)
4. Pljasheshnikov AV, Lagutin AA, Uchaikin VV (1979) The numerical method of the radial distribution of electromagnetic cascade particles calculation. In: Proceedings of 16-th ICRC, Kyoto, vol 7, pp 13–17
5. Lagutin AA, Uchaikin VV, Chernyaev GV, Adjoint approach to EAS problem, ibid, pp 260–263
6. Aharonian FA, Plyasheshnikov AV (2003) Similarities and differences between relativistic electron-photon cascades developed in matter, photon gas and magnetic field. Astroparticle Phys 19:525–548
7. Chernyaev GV, Lagutin AA, Uchaikin VV, The model of hadron-nuclear interaction in the super-high energy range. In: Proceedings of 18th ICRC, Bombay, vol. 6, pp. 17–20
8. Lagutin AA, Litvinov VA, Uchaikin VV (1987) Mathematical aspects of applied cosmic ray physics. In: Proceedings of the 20th ICRC, Moscow, vol 4, pp 276–279
9. Uchaikin VV, Lagutin AA, Plyasheshnikov AV (1979) On fluctuations in the number of particles in an electromagnetic cascade. Yader. Physics 30:429–436 (in Russian)
10. Nagel HH (1965) Electron—Phonon—Cascaden in Blei. Zeitschrift fur Physik 186:319–346
11. Belousov AS, Malinovsky EI, Rusakov SV et al (1973) Study of electron-photon showers in lead in the energy range 150–550 MeV. Yader. Physics 17:1028–1036 (in Russian)
12. Borkovsky MY, Kruglov SP (1976) Calculations of the released energy and total ranges of charged particles during the development of showers in xenon. Atom Energy 40:243 (in Russian)

13. Adler D, Fuchs B, Thielheim KO (1977) Numerical integration of electromagnetic cascade equations. discussion of results for air, copper, iron and lead. In: Proceedings of 15th ICRC, Plovdiv, vol 7, pp 466–471
14. Belyaev AA, Guzhavin VV, Ivanenko IP (1975) Multigroup method in the theory of electron-photon showers. Moscow, (Preprint/Phys. Institute of the Academy of Sciences of the USSR; No 34) (in Russian)
15. Varfolomeev AA, Drabkii LB (1970) Calculation of electromagnetic showers in lead with an initial energy of 6 GeV. Cosmic Rays 12:32–59 (in Russian)
16. Ellsworth RW, Streitmatter RE, Bowen T (1979) Hybrid Monte-Carlo and analityc calculation of electromagnetic cascades in Pb Including the Landau—Pomeranchuk—Migdal effects. In: Proceedings of 16-th ICRC, Kyoto 7:55–59
17. Konishi E, Misaki A, Fujimaki N (1978) Three-dimensional cascade showers in lead taking account of the Landau—Pomeranchuk—Migdal effect. Nuovo Cimento 44A(4):509–530
18. Kolchuzhkin AM, Bespalov VI (1979) The investigation of electron-photon showers by Monte-Carlo Method. In: Proceedings of 16-lh ICRC, Kyoto, vol 9, pp 222–227
19. Hillas AM (1982) Angular and energy distribution of charged particles in electron-photon cascade in air. J Phys G: Nucl Phys 8(10):1461–1473
20. Hillas AM, Lapikens J (1977) Electron-photon cascades in the atmosphere and in detectors. In: Proceedings of 15th ICRC, Plovdiv, vol 8, pp 460–465
21. Plyasheshnikov AV, Vorobjev KV (1981) The Algorithm of the calculation of extreme high energy electromagnetic cascade parameters by the Monte-Carlo method. In: Proceedings of 17th ICRC, Paris, vol 5, pp 206–209
22. Vankov Ch, Petrov S, Stanev T (1981) Fluctuation in the longitudinal development of electron-photon cascade in lead. In: Proceedings of 17th ICRC, Paris 5:166–169
23. Crawford DF, Messel H (1965) The electron-photon cascade in lead, emulsion and copper absorbers. Nucl Phys 61:145–172
24. Emeljanov YA (1977) Fluctuations in high-energy electromagnetic cascade showers. In: Proceedings of 15th ICRC, Plovdiv, vol 7, pp 472–477
25. Lagutin AA, Plyasheshnikov AV, Uchaikin VV et al (1981) The calculation of correlations in electron-photon cascades. In: Proceedings of 17th ICRG, Paris, vol 5, pp 198–201
26. Kolchuzhkin AM, Uchaikin VV, Bespalov VI (1973) Calculation of fluctuations of the Cherenkov radiation of an electron-photon shower in the atmosphere, Academy of Sciences of the USSR Bull. Ser Phys 7:1442 (in Russian)
27. Muller D (1972) Electron showers of high energy in lead. Phys Rev D 5:2677–2683
28. Barmin VV, Varylov VG, Golubchikov VM et al (1972) Measurement of electron energy in a xenon bubble chamber by the full length shower track method, Moscow (Proprint/Institute of Theoretical and Experimental Physics; No 938) (in Russian)
29. Lagutin AA, Pljasheshnikov AV, Uchaikin VV et al (1981) The calculation of the angular distributions of electron-photon cascades electrons. In: Proceedings of 17th ICRC, Paris, vol 5, pp 202–205
30. Hillas AM (1981) Results of monte-carlo simulations of electron-photon cascades in the atmosphere and in detectors. In: Proceedings of 17th ICRC, Paris 6:244–247
31. Vorobyov KV, Plyasheshnikov AV, Uchaikin VV (1982) Radial distribution of electrons of an electromagnetic cascade in the atmosphere. Izv AN USSR Ser phys 46:2437 (in Russian)
32. Uchaikin VV, Plyasheshnikov AV, Lagutin AA, Vetoshkin VV (1978) Numerical method for calculating fluctuations in the path of charged particles of an electromagnetic cascade in an infinite medium. Izv Vyssh Uchebn Zaved Fizica 4:27–30 (in Russian)
33. Linde IA, Kanevsky BL (1981) Method of solution of cascade theory non-linear ecmation for function of distribution over the number of particles in nuclear-electromagnetic shower. In: Proceedings of 17th ICRC, Paris, vol 5, pp 268–271
34. Lagutin AA, Plyasheshnikov AV, Uchaikin VV (1979) The calculation of the electromagnetic cascade characteristics fluctuations. In: Proceedings of 16th ICRC, Kyoto, vol 7, pp 7–12
35. Borkovsky MY (1980) On correlations in particle numbers at different depths in electron-photon lines during the extraction of electrons from meson beams (in Russian)

36. Betman RG, Gedalin EV (1976) The distribution function of the number of electrons in an electromagnetic avalanche. Nuclear Interactions at high energies, Tbilisi, vol 2, pp 13–18 (in Russian)
37. Kokoulin RP, Petrukhin AA (1980) On the possibility of a new approach to cascade theory. Yader. Physics 32:1030–1038 (in Russian)
38. Stanev T, Vankov Ch (1979) The Landau—Pomeranchuk—Migdal effect and the development of electron-photon cascade in lead. In: Proceedings of 16th ICRC, Kyoto, vol 7, pp 47–52
39. Okhrimenko LS, Slovinsky B, Strugalsky Z (1975) Characteristics of the longitudinal development of electron-photon showers caused by γ-quanta with an energy of 30–120 MeV in liquid xenon. Yader. Physics 22:110–115 (in Russian)
40. Zuev VE (1970) Propagation of visible and infrared waves in the atmosphere. Moscow, Sov Radio (in Russian)
41. Kolchuzhkin AM, Uchaikin VV (1978) Introduction to the theory of particle passage through matter. Atomizdat, Moscow (in Russian)
42. Belyaev AA, Ivanenko IP, Kanevsky BL et al (1980) Electron-photon cascades in cosmic rays at superhigh energies. Moscow, Science (in Russian)
43. Belenky SZ (1948) Avalanche processes in cosmic rays. Gostekhizdat, Moscow (in Russian)
44. Crannell CJ, Allan HR, Sun MP (1978) Widths of electron—photon cascades. Phys Rev D 18(1):216–219
45. Lagutin AA, Plyasheshnikov AV, Uchaikin VV (1981) The spatial distribution of electron-photon cascade particles in air with the threshold energy above zero II. In: Proceedings of 17th ICRC, Paris, vol 5, pp 194–197
46. Nishimura J (1967) Theory of cascade showers. Handbuch der Physik. Springer, Berlin, pp 3–114
47. Oda H, Misaki A (1979) Mean square lateral spread of electrons in cascade shower under approximation B. In: Proceedings of 16th ICRC, Kyoto, vol 13, pp 305–309
48. Ivanenko IP, Roganova TM (1983) Cascade showers caused by ultrahigh energy particles. Nauka, Moscow (in Russian)
49. Lagutin AA, Plyasheshnikov AV, Uchaikin VV (1979) The radial distribution of electromagnetic cascade particles in the air. In: Proceedings of 16th ICRC, Kyoto, vol 7, pp 18–23
50. Ivanenko IP (1972) Electromagnetic cascade processes. Moscow State University (in Russian)
51. Dedenko LG, Nikolsky SI, Stamenov IN (1976) On the spatial distribution of the electron flux in extensive air showers. Brief Commun Phys 1:30 (in Russian)
52. Bourdeau MF, Capdevielle JN, Gawin G, Procureur J (1980) Consequences of radical electron distribution of primary energy spectrum, cosmic rays. In: Proceedings of the 7th European symposium, Leningrad, pp 351–355
53. Capdevielle JN, Gawin G (1982) The radial electron distribution in extensive air showers. J Phys G: Nucl Phys 8:1317–1336
54. Danilova TV, Erlykin AD (1983) Influence of the finite primary energy on the properties of EAS electromagnetic component. Proceedings of 18th ICRC, Bombay 6:37–40
55. Sivaprasad K (1983) The Electron lateral distribution in EAS. In: Proceedings of 18th ICRC, Bombay, vol 8, p 12
56. Rao MVS (1983) Behavior of high energy interactions from structure of small and medium air showers. In: Proceedings of 18th ICRC, Bombay, vol 12, pp 449–474
57. Krys E, Wasilewski A (1984) Charged particle distributions in a photon-initiated cascades. Acta UL. Folia Phys 7:111–123
58. Nagano M, Natano Y, Hara T et al (1984) The lateral distribution of electrons of extensive air shower observed at Akeno (920 g/sm^2). J Phys Soc Jpn 53(5):1667–1681
59. Uchaikin VV (1979) The theory of cascade process nonlinear measurements and the statement of backward problems. In: Proceedings of 16th ICRC, Kyoto, vol 7, pp 24–29
60. Gerasimova NM (1964) Fluctuations in the number of particles in electron-photon Lithium. In: Proceedings of the physical institute of the academy of sciences of the USSR, vol 26, pp 192–206 (in Russian)
61. Ramakrishnan A (1965) Elementary particles and cosmic rays. Mir, Moscow (in Russian)

62. Gerasimova NM (1979) Analysis of the results of the studies of fluctuations in electron-photon shower. In: Proceedings of the 16th ICRC, Kyoto, vol 9, pp 216–221
63. Konopelko AK, Litvinov VA, Plyasheshnikov AV, Uchaikin VV et al (1987) The lateral development of extremely high energy EPC in the atmosphere. In: Proceedings of 20th ICRC, Moscow, vol 5, pp 395–402
64. Heck D, Knapp J, Capdevielle JN, Schatz G, Thouw T (1998) CORSIKA: a monte carlo code to simulate extensive air showers. Forschungszentrum Karlsruhe GmbH, Karlsruhe

Chapter 8
Statistical Fluctuations in EPC

8.1 Fluctuations in Electron Number

The main physical feature of the cascade process is statistical fluctuations due to the corpuscular nature of the medium in which the high-energy medium moves. It is for this reason that the trajectories of cosmic ray particles in the Earth's atmosphere are not continuous, like the trajectories of raindrops, but appear broken, consisting of almost rectilinear segments of random length and random directions. At high energies, the processes of particle production also take place, so that the trajectory of the cascade in the atmosphere becomes similar to an inverted tree.

The characteristics of cascades (cascade curves, energy, angular and spatial distributions), which we have considered so far, are characteristics averaged over the statistical ensemble, while real measurements give us information only about individual members of this ensemble. To correlate experimental data with a theoretical construction and thereby justify or reject a theoretical model, statistical analysis is necessary. This chapter contains a theoretical description of some aspects of this, in general, rather complex problem.

For calculating the particles number variance, one should find also the second moment, i.e., the mean square of the number. However, it is more convenient to deal with the corresponding factorial moment

$$N_e^{[2]}(t,x) = \int\limits_x^{x_1+x_2<1} dx_1 \int\limits_x dx_2 \bar{\Phi}_{ee}(t,x_1,x_2).$$

Under $x \to 0$, correlations in upper limits can be neglected [1] and the moment may be expressed through two-particle transform as follows:

$$N_e^{[2]}(t,x) = \frac{1}{(2\pi i)^2} \int\int \frac{ds_1 ds_2}{s_1 s_2} x^{-s_1-s_2} \bar{\Phi}_{ee}(t,s_1,s_2) \tag{8.1}$$

© Higher Education Press Limited Company 2024

V. V. Uchaikin, *Electron-Photon Cascades*, Nonlinear Physical Science,
https://doi.org/10.1007/978-981-99-7524-2_8

(recall, t means here the layer thickness). Taking into account the symmetrization agreement (2.41), we represent transform (5.146) in the form

$$\bar{\Phi}_{ee}(t, s_1, s_2) = \int_0^t dt' \sum_{c=e,\gamma} [2\bar{G}_{a\to c}(t', s_1 + s_2)\bar{V}_c(s_1, s_2)$$
$$\times \bar{G}_{e\to e}(t - t', s_1)\bar{G}_{\bar{c}\to e}(t - t', s_2)]. \tag{8.2}$$

Here

$$\bar{c} = \begin{cases} \gamma, & c = e, \\ e, & c = \gamma. \end{cases}$$

Using (5.141), (5.142), we obtain

$$\bar{V}_e(s_1, s_2) = \left[\frac{s_2 + 1}{s_1 + s_2 + 2} + \delta\frac{s_1 + 1}{s_2}\right] \frac{\Gamma(s_1 + 1)\Gamma(s_2 + 1)}{\Gamma(s_2 + s_2 + 2)},$$

$$\bar{V}_\gamma(s_1, s_2) = \left[1 - \delta\frac{(s_1 + 1)(s_2 + 1)}{(s_1 + s_2 + 2)(s_1 + s_2 + 3)}\right] \frac{\Gamma(s_1 + 1)\Gamma(s_2 + 1)}{\Gamma(s_1 + s_2 + 2)}.$$

Inserting (6.5) into (8.2) and integrating with respect to t' yield

$$\bar{\Phi}_{ee}(t, s_1, s_2) = \sum_{ijk=1}^{2} \sum_c 2[H^{(i)}_{a\to c}(s_1 + s_2)\bar{V}_c(s_1, s_2)H^{(j)}_{e\to e}(s_1)H^{(k)}_{\bar{c}\to e}(s_2)]$$
$$\times \frac{\exp\{\lambda_i(s_1 + s_2)t\} - \exp\{[\lambda_j(s_1) + \lambda_k(s_2)]t\}}{\lambda_i(s_1 + s_2) - \lambda_j(s_1) - \lambda_k(s_2)}.$$

Since the terms with λ_1 in the exponents are much more than terms with λ_2, the latter can be neglected at large depths. The corresponding result is given in work [1].

By inserting (8.2) into (8.1) and neglecting terms containing at least one factor of the form $\exp\{\lambda_2(s_i)t\}$, we rewrite the second factorial moment as

$$N^{[2]}_{a\to b}(t, x) = \frac{1}{(2\pi i)^2} \sum_{k=1}^{2} \int ds_1 \int ds_2 \int_0^t ds_3$$
$$\times M^k_{a\to b}(s_1, s_2) \exp[\nu^k(s_1, s_2, s_3)], \tag{8.3}$$

where

$$\nu^k(s_1, s_2, s_3) = \lambda_k(s_1 + s_2)(z - s_3) + [\lambda_1(s_1) + \lambda_2(s_2)]s_3 + (s_1 + s_2)y, \tag{8.4}$$

$$M^k_{a\to b}(s_1, s_2) = 2\sum_c H^{(k)}_{a\to c}(s_1 + s_2)\bar{V}_c(s_1, s_2)H^{(1)}_{e\to b}(s_1)H^{(1)}_{c\to b}(s_2)/(s_1 s_2). \quad (8.5)$$

Following [2], calculate integrals in Eq. (8.3) by using the multivariate saddle-point method. Due to the obvious symmetry of the integrand, the first two coordinates of the saddle point are equal: $\bar{s}_1 = \bar{s}_2$. Since the integration along the third coordinate is carried out over a finite interval $(0, t)$, the position of maximum of $v^k(s_1, s_2, s_3)$ may fall both inside this strip and outside it. Begin with the first integral in (8.3). Three cases are possible here.

1. $0 < s_{3\,\max} < t$ $(2y < t \lesssim 12y)$. Saddle point coordinates are determined by the equations

$$\partial v^1/\partial s_i = 0, \qquad i = 1, 2, 3. \quad (8.6)$$

The exponent (8.4) expands in the series

$$v^1(s_1, s_2, s_3) = v^1(\bar{s}_1, \bar{s}_2, \bar{s}_3)$$

$$+\frac{1}{2}\sum_{ij=1}^{3} v^1_{ij}(\bar{s}_1, \bar{s}_2, \bar{s}_3)(s_i - \bar{s}_i)(s_j - \bar{s}_j), \ldots$$

$$v^1_{ij}(\bar{s}_1, \bar{s}_2, \bar{s}_3) = \partial^2 v^1(\bar{s}_1, \bar{s}_2, \bar{s}_3)/d\bar{s}_i d\bar{s}_j.$$

2. $s_{3\,\max} \geqslant z$ $(t \leqslant 2y)$. In this case $\bar{s}_3 = z$, coordinates $\bar{s}_1$ and $\bar{s}_2$ are determined by equations

$$\partial v^1/\partial s_i = 0, \; i = 1, 2, \quad (8.7)$$

and the expansion of the exponent has the form

$$v^1(s_1, s_2, s_3) = v^1(\bar{s}_1, \bar{s}_2, \bar{s}_3) + v^1_3(\bar{s}_1, \bar{s}_2, \bar{s}_3)(s_3 - \bar{s}_3)$$

$$+\frac{1}{2}\sum_{ij=1}^{2} v^1_{ij}(\bar{s}_1, \bar{s}_2, \bar{s}_3)(s_i - \bar{s}_i)(s_j - \bar{s}_j).$$

3. $s_{3\,\max} \leqslant 0$. This case meets the condition $t \geqslant 12y$ that practically is of no interest.

The exponent in the second term (8.3) monotonically increases with growing s_3, so we should put $\bar{s}_3 = t$. As a result, we arrive at formula (8.7) for the function v^2:

$$\partial v^2/\partial s_i = 0, \qquad i = 1, 2, \quad (8.8)$$

and corresponding expansion of the exponent.

Reducing quadratic forms to diagonal shape via sequential change of variables and integrating in appropriated limits, we get

$$N_{a\to b}^{[2]}(t,x)=\sum_{k=1}^{2}S_{a\to b}^{k[2]}(t,x,\mathbf{s}), \tag{8.9}$$

where $\mathbf{s}$ means a set of three saddle-point coordinates.

The multivariate saddle-point method was used by Gedalin [2], Betman and Gedalin [3] in calculations of ESP fluctuations and correlations in frame of the B-approximation. Unfortunately, the cumbersome shape of the formulas obtained in this case deprives the analytical approach of its main advantage—transparency and compactness. The results satisfying such requirements were obtained by N. M. Gerasimova based on the use of inverse equations (in one of her publications she thanks L. D. Landau for advising her this approach) [4]:

$$N_{e\to e}^{[2]}(t,x)=2\{[2\lambda_1(s)+\sigma_0]\bar{V}_e(s,s)\overline{N}_{\gamma\to e}(t,x)\overline{N}_{e\to e}(t,x)$$

$$+C(2s)\bar{V}_\gamma(s,s)\overline{N}_{e\to e}^2(t,x)\}/\varphi_2(s),$$

$$N_{\gamma\to e}^{[2]}(t,x)=2[B(2s)\bar{V}_e(s,s)\overline{N}_{\gamma\to e}(t,x)\overline{N}_{e\to e}(t,x)$$

$$+[2\lambda_1(s)+A(2s)]\bar{V}_\gamma(s,s)\overline{N}_{e\to e}^2(t,x)]/\varphi_2(s),$$

$$\varphi_2(s)=[2\lambda_1(s)-\lambda_1(2s)][2\lambda_1(s)-\lambda_2(2s)].$$

They were found by solving backward equations with the use of right-hand side approximations in the form

$$F_e^{(2)}(t,x)\approx 2\bar{V}_e(s,s)\overline{N}_e(t,x)\overline{N}_\gamma(t,x)$$

$$F_\gamma^{(2)}(t,x)\approx 2\bar{V}_\gamma(s,s)\overline{N}_e^2(t,x),$$

following from the power-kind approximations to integrand:

$$\overline{N}_{a\to e}(t,x/x')\approx(x/x')^{-s}\overline{N}_{a\to e}(t,x),$$

$$\overline{N}_{a\to e}(t,x/(1-x'))\approx[x/(1-x')]^{-s}\overline{N}_{a\to e}(t,x).$$

The relative fluctuations found using these formulas are of two terms:

$$\delta^2 N_{a\to e}=\delta^{[2]}N_{a\to e}+1/\overline{N}_{a\to e}.$$

The first of them, $\delta^{[2]}N_{a\to e}=[N_{a\to e}^{[2]}/\overline{N}_{a\to e}^2]-1$, depends only on the cascade parameter s and makes the basic contribution in fluctuations, except for the region of the cascade curve maximum ($s=1$). The similar conclusion is made in [5]. This gave the reason to believe that fluctuations at the maximum have Poisson character (see the discussion to the report [1]), but the refined method [4, 6] showed the first term in this region gives a slower decrease in fluctuations at the maximum with the primary energy increasing: $\delta^{[2]} N_{a\to e}(t_{\max}) \sim t_{\max}^{-1}$. Although quantitatively given result and does not quite agree with more accurate numerical calculations, the general conclusion about the non-Poissonian nature of fluctuations at the maximum turns out to be valid.

For $s = 1.47$, the above expressions for the second factorial moments lose their physical meaning (turn to infinity). Region $s > 1.47$ $(t > 2y)$ was considered separately in [4] using another saddle point, and the obtained moments in the vicinity of $s = 1.47$ stitched with the above. It was shown [7] that the discontinuity of the solution at the point $s = 1.47$ may be avoided if in the expansion of the exponent to keep the terms with v_{13}^1 and v_{23}^1, neglecting of which is equivalent to approximations, used in [8].

If we exclude from consideration the region of the maximum of the cascade, the analytical theory gives a generally correct qualitative description fluctuations. Its most important conclusion is that the fluctuations are weakly dependent on the primary energy of the incident particle (at a fixed value of the cascade parameter) from the energy of the primary particle. This fact, indicating the nature of fluctuations as a whole far from Poisson, is confirmed by numerical calculations.

In conclusion, we note several works in which attempts to analytically solve Eq. (4.72) for the function distribution of the number of particles in EPC [9–11]. They use two approaches: one is based on calculating a few first moments and restoring the distribution function from them, the other is on direct solution of the backward equation for the function distribution by using iterative methods. Most of the numerical results were obtained by the Monte Carlo method [11, etc.].

8.2 The Electron Total Trace Length

Unlike fluctuations in the number of particles, fluctuations in the total electron path and in proportional to them ionization losses of EPCs initially were investigated only by the method Monte Carlo [12, 13] and experimentally [14–16].

To investigate statistical property of the total path of EPC electron in the layer of a finite thickness t, use the backward equations (5.147) for the moments $R_a^{(1)}(t, x)$ and $R_a^{(2)}(t, x)$ setting there $a_a(t, x) = \delta_{ae}$ and boundary conditions $R_a^{(n)}(0, x) = 0$. Because of obvious symmetry, one can conclude that the characteristics of the total path in an infinite homogeneous medium do not depend on the point of birth of the primary particle. This means that for $z \to \infty$ the moments cease depend on t and the time derivative can be omitted. Remaining system of equations for transforms

$$A(s)\bar{R}_e^{(n)}(s) - C(s)\bar{R}_\gamma^{(n)}(s) = \bar{F}_e^{(n)}(s),$$

$$-B(s)\bar{R}_e^{(n)}(s) + \sigma_0 \bar{R}_\gamma^{(n)}(s) = \bar{F}_\gamma^{(n)}(s)$$

is easy for solving:

$$\bar{R}_e^{(n)}(s) = [\sigma_0 \bar{F}_e^{(n)}(s) + C(s)\bar{F}_\gamma^{(n)}(s)]/\varphi(s), \tag{8.10}$$

$$\bar{R}_\gamma^{(n)}(s) = [A(s)\bar{F}_\gamma^{(n)}(s) + B(s)\bar{F}_e^{(n)}(s)]/\varphi(s), \tag{8.11}$$

where

$$\varphi(s) = A(s)\sigma_0 - B(s)C(s) = \lambda_1(s)\lambda_2(s). \tag{8.12}$$

The right-hand sides of equations for the first moments are transformed into $\bar{F}_e^{(1)}(s) = 1/s$, and $\bar{F}_\gamma^{(1)}(s) = 0$, respectively, and we get

$$\bar{R}_j^{(1)}(s) = \sigma_0/[s\varphi(s)] \tag{8.13}$$

and

$$\bar{R}_\gamma^{(1)}(s) = B(s)/[s\varphi(s)]. \tag{8.14}$$

The average values of the path are found by inverse transformation:

$$R_e^{(1)}(x) = \frac{1}{2\pi i}\int x^{-s}\sigma_0/[s\varphi(s)]ds \tag{8.15}$$

and

$$R_\gamma^{(1)}(x) = \frac{1}{2\pi i}\int x^{-s}B(s)/[s\varphi(s)]ds. \tag{8.16}$$

As one can see from Eqs. (8.12) and (6.2)–(6.4), the poles of integrand in (8.15), (8.16) coincide and are determined by zeros s_k of the function $\varphi(s)$ ($s_1 = 1$, $\mathrm{Re}\, s_k < 0$ as $k > 1$). As $s \to 0$, $s\varphi(s) \to -2\sigma_0\delta$, therefore the point $s = 0$ is not a pole. Applying the residue method, we get

$$R_a^{(1)}(x) = \sum_k c_k^{(a)} x^{-s_k}, \tag{8.17}$$

where $c_k^{(e)} = \sigma_0/[s_k\varphi'(s_k)]$, $c_k^{(\gamma)} = B(s_k)/[s_k\varphi'(s_k)]$. According to (8.17), first coefficients $c_1^{(e)} = c_1^{(\gamma)} \equiv \alpha = 0.437$ determine asymptotic behavior of first moments as $x \to 0$ ($E/E^* \to \infty$):

$$R_e^{(1)}(x) \sim \alpha x^{-1}, \qquad R_\gamma^{(1)}(x) \sim \alpha x^{-1}, \qquad x \to 0. \tag{8.18}$$

Transforms of right-hand sides of equations for second moments according to (5.153) have form

$$\bar{F}_a^{(2)}(s) = \frac{2}{(2\pi i)^2} \int\int \frac{ds_1 ds_2}{s - s_1 - s_2}$$

$$\times \left[\frac{\delta_{ae}}{s_1} \bar{R}_e^{(1)}(s_2) + \widehat{V}_a(s_1, s_2; s) \bar{R}_e^{(1)}(s_1) \bar{R}_{\bar{a}}^{(1)}(s_2) \right], \tag{8.19}$$

where

$$\widehat{V}_a(s_1, s_2; s) = \widehat{V}_a'(s - s_2, s_2) + \widehat{V}_a''(s_1, s - s_1),$$

$$\widehat{V}_a'(s - s_2, s_2) = \int_0^{1/2} x^{s-s_2}(1-x)^{s_2} V_a(x) dx,$$

$$\widehat{V}_a''(s_1, s - s_1) = \int_{1/2}^{1} x^{s_1}(1-x)^{s-s_1} V_a(x) dx. \tag{8.20}$$

Inserting (8.13) and (8.14) into (8.19) and integrating over both variables s_1, s_2 by means of the residue theorem, we get

$$\bar{F}_e^{(2)}(s) = 2\bar{R}_e^{(1)}(x) + 2\sum_{ij} \frac{c_i^{(e)} c_j^{(\gamma)}}{s - s_i - s_j}$$

$$\times \left[\widehat{V}_e'(s - s_j, s_j) \widehat{V}_e''(s_i, s - s_i) \right], \tag{8.21}$$

$$\bar{F}_\gamma^{(2)}(s) = 4\sum_{ij} \frac{c_i^{(e)} c_j^{(\gamma)}}{s - s_i - s_j} \widehat{V}_\gamma'(s - s_j, s_j), \tag{8.22}$$

where summing is assumed over all zeros of the function $\varphi(s)$.

Using expressions for poles of second moment transforms, write inverse transformations in the form of residue sum:

$$R_e^{(2)}(x) = \frac{1}{2\pi i} \int \bar{R}_e^{(2)}(s) x^{-s} ds$$

$$= \sum \text{res}\{x^{-s}[\sigma_0 F_e^{(2)}(s) + C(s) F_\gamma^{(2)}(s)]/\varphi(s)\}, \tag{8.23}$$

$$R_\gamma^{(2)}(x) = \frac{1}{2\pi i} \int \bar{R}_\gamma^{(2)}(s) x^{-s} ds$$

$$= \sum \text{res}\{x^{-s}[A(s) F_\gamma^{(2)}(s) + B(s) F_e^{(2)}(s)]/\varphi(s)\}. \tag{8.24}$$

Asymptotical behavior of second moments at $x \to 0$ is determined by a simple simple rightmost pole $s = 2$, generated by terms with $i = j = 1$ of double sums in Eqs. (8.21), (8.22):

$$R_e^{(2)}(x) \sim \alpha^2 x^{-2}\{2\sigma_0[\widehat{V}_e'(1,1) + \widehat{V}_e''(1,1)] + 4C(2)\widehat{V}_\gamma'(1,1)\}/\varphi(2), \quad (8.25)$$

$$R_\gamma^{(2)}(x) \sim \alpha^2 x^{-2}\{4A(2)\widehat{V}_\gamma'(1,1) + 2B(2)[\widehat{V}_e'(1,1) + \widehat{V}_e''(1,1)]\}/\varphi(2). \quad (8.26)$$

Using easily verifiable relations

$$\widehat{V}_e'(1,1) + \widehat{V}_e''(1,1) = C(1) - C(2), 2V_\gamma'(1,1) = [B(1) - B(2)]/2$$

in Eqs. (8.25), (8.26), and taking into account that $B(1) = \sigma_0, 2C(1) - C(2) = A(2)$, we find out, that the contents of the curly braces in each of the expressions (8.25), (8.26) equals $\varphi(2)$ (see (8.12)), so that

$$R_e^{(2)}(x) \sim R_\gamma^{(2)}(x) \sim \alpha^2 x^{-2}, \qquad x \to 0.$$

Using Eq. (8.17), we conclude that the relative fluctuations of the total path of all electrons with energies higher E^* in EPC, generated by an incident photon with energy E in a homogeneous boundless medium tend n to zero, as $E/E^* \to \infty$.

To clarify the law of the decreasing one need to find the next (closest to $s = 2$) pole. Such a pole is the point $s = 1$, at which it has singularity function (8.21) and vanishes $\varphi(s)$, standing in the denominator of the integrands. Feature in (8.21) is generated by the transform of the first moment (i.e., the averaged total path) and the function

$$\widehat{V}_e''(1, s-1) = \delta\frac{(1/2)^{s-1}}{s-1} - 2\delta\frac{(1/2)^s}{s} + (1+\delta)\frac{(1/2)^{s+1}}{s+1} - \frac{(1/2)^{s+2}}{s+2}.$$

Thus, the point $s = 1$ is a pole of the second order, whose contribution to integrals (8.23), (8.24) is given by the expressions

$$\operatorname*{res}_{s=1} R_e^2(s)x^{-s} = -f(1)x^{-1}\ln x + f'(1)x^{-1},$$

$$\operatorname*{res}_{s=1} R_\gamma^2(s)x^{-s} = -g(1)x^{-1}\ln x + g'(1)x^{-1},$$

where $f(s) = (s-1)^2 R_e^2(s)$, $g(s) == (s-1)^2 R_\gamma^2(s)$. Inserting (8.21), (8.22) into (8.10), (8.11) and taking into account the features noted above, we find

$$f(1) = g(1) = 2\alpha^2\left[1 - \delta\sum_j c_j^{(\gamma)}/s_j\right]. \quad (8.27)$$

According to (8.14), (8.17)

$$B(s)/[s\varphi(s)] = -\sum_j c_j^{(\gamma)}/(s+s_j) \rightarrow -\sum_j c_j^{(\gamma)}/s_j,\ s \rightarrow 0.$$

On the other hand, explicit expressions (6.2)–(6.5) and (8.12) lead to conclusion that $\lim_{s\to 0} B(s)/[s\varphi(s)] = -1/\delta$. Comparison of these equalities shows that $f(1) = g(1) = 0$. Numerical calculations show that $f'(1) = g'(1) \equiv \gamma = 0,16$, so that for variances, we obtain

$$\mathrm{Var}R_e(x) \sim \mathrm{Var}R_\gamma(x) \sim \gamma x^{-1}, \qquad x \rightarrow 0.$$

Thus, the relative fluctuations of the total path of EPC electrons in an infinite homogeneous medium decrease with increasing the primary energy proportionally to $E^{-1/2}$ [17]. Numerical calculations confirm this behavior of fluctuations with increasing energy, and inclusion of ionization losses into consideration does not change its character, which is in good agreement with experiment. Of course if we put $\beta = \text{const}$, ignore photo-absorption and consider the range of all electrons with $E > 0$, then, by virtue of the law of conservation of energy, fluctuations of this path in infinite environment will be zero. In reality, $\beta(E)$ grows with decreasing E, and the total path fluctuates. Moreover, talking about the energy, one should keep in mind the heat component of the energy fluctuations. Fortunately, the phenomena studied in ionization calorimeters are caused by particles of such high energies that thermal fluctuations play no role [8].

Note the importance of taking into account all the poles in sum (8.27) to determine correct asymptotical term in fluctuations. When solving a similar problem in [18] (the fluctuations of the number of electrons produced in infinite environment) the first moment was presented only by the main expansion term (8.17). As a result of this and other errors from this work followed, that the relative fluctuations of the total number the particles produced tend to a nonzero constant at $E \rightarrow \infty$. The correct calculations lead to the conclusion that the fluctuations of this characteristic of the cascade have the same asymptotic behavior $E^{-1/2}$, as fluctuations of the path [19].

8.3 Fluctuations of Electron Path in a Layer

To the end of this chapter, we consider not the total path of all electrons of EPC, but only a part of it, belonging to the homogeneous layer of a finite depth t. According to (5.150), the transforms of its moments are expressed in terms of Green's function transforms via relation

$$\bar{R}_a^{(n)}(t,s) = \sum_b \int_0^t \bar{G}_{a\to b}(t-t',s)\bar{F}_b^{(n)}(t',s)dt', \tag{8.28}$$

where

$$\bar{F}_b^{(1)}(z, s) = \delta_{be}/s,$$

$$\bar{F}_b^{(2)}(z, s) = \frac{2}{(2\pi i)^2} \iint \frac{ds_1 ds_2}{s - s_1 - s_2}$$

$$\times \left[\frac{\delta_{be}}{s_1} \bar{R}_e^{(1)}(t, s_2) + \widehat{V}_b(s_1, s_2; s) \bar{R}_e^{(1)}(t, s_1) \bar{R}_{\bar{b}}^{(1)}(t, s_2) \right].$$

We will get for first moments

$$\bar{R}_a^{(1)}(t, s) = \sum_{i=1}^{2} H_{a\to e}^{(i)}(s) \frac{\exp\{\lambda_i(s)t\} - 1}{s\lambda_i(s)} = \bar{R}_a^{(1)}(\infty, s) + \Delta \widetilde{R}_a^{(1)}(t, s), \qquad (8.29)$$

where

$$\bar{R}_a^{(1)}(t, s) = -\left[H_{a\to e}^{(1)}/\lambda_1(s) - H_{a\to e}^{(2)}/\lambda_2(s) \right]/s$$

is in agreement with formulas (8.13), (8.14), and

$$\Delta \bar{R}_a^{(1)}(t, s) \approx \frac{H_{a\to e}^{(1)}(s)}{s\lambda_1(s)} e^{\lambda_1(s)t} \qquad (8.30)$$

for not very small depths. Specific for this task is the presence of a pole of the transform (8.30) at the point $s = 1$. In this regard, the multiplier $\lambda_1(s)$ during the reverse transformation by the saddle method should be raised to the exponent:

$$\Delta R_a^{(1)}(t, x) = \frac{1}{2\pi i} \int H_{a\to e}^{(1)}(s) e^{\psi(s)+sy} ds/s$$

$$\approx \frac{1}{p\sqrt{2\pi|\psi''(p)|}} \frac{H_{a\to e}^{(1)}(p)}{\lambda_1(p)} e^{\lambda_1(p)t} x^{-p},$$

where $\psi(s) = \lambda_1(s)z - \ln|\lambda_1(s)|$, and the saddle-point p satisfies the equation

$$t = \frac{\ln x}{\lambda_1'(p)} + \frac{1}{\lambda_1(p)}, \quad p > 1, \qquad (8.31)$$

being different from the usual saddle-point condition.

The result obtained can be represented also as

$$\Delta R_a^{(1)}(t, x) = \overset{*}{S}_{a\to e}(t, x, p),$$

where

$$\overset{*}{S}_{a\to e}(t, x, p) = \sqrt{\lambda_1''(p)t/|\psi''(p)|}\, S_{a\to e}(t, x, p)/\lambda_1(p).$$

Under $t \to \infty$, $p \to s$, $\psi''(p) \to \lambda_1''(s)t$ and we get the easily explainable expression

$$\Delta R_a^{(1)}(t,x) \sim \overline{N}_{a\to e}(t,x)/\lambda_1(s).$$

Inserting (8.29) into (5.153) yields

$$\bar{F}_a^{(2)}(t,s) = \bar{F}_a^{(2)}(\infty,s) + \frac{2}{(2\pi i)^2}\int\int \frac{ds_1 ds_2}{s-s_1-s_2}\left\{\frac{\delta_{ae}}{s_1}\Delta\bar{R}_e^{(1)}(t,s_2) + \widehat{V}_a(s_1,s_2;s)\right.$$

$$\left.\times\left[\bar{R}_e^{(1)}(\infty,s_1)\Delta\bar{R}_{\bar{a}}^{(1)}(t,s_2) + \Delta\bar{R}_e^{(1)}(z,s_1)\bar{R}_{\bar{a}}^{(1)}(\infty,s_2) + \Delta\bar{R}_e^{(1)}(z,s_1)\Delta\bar{R}_{\bar{a}}^{(1)}(t,s_2)\right]\right\}, \tag{8.32}$$

where $\bar{F}_a^{(2)}(\infty,s)$ coincides with corresponding formula (8.19) in the previous problem. Substituting the first term (8.32) into (8.28) and integrating with respect to t', we find

$$\left[\bar{R}_a^{(2)}(t,s)\right]_1 = \sum_k\sum_b \frac{H_{a\to b}^{(k)}(s)\bar{F}_b^{(2)}(\infty,s)}{[-\lambda_k(s)]} + \sum_k\sum_b \frac{H_{a\to b}^{(k)}(s)\bar{F}_b^{(2)}(\infty,s)}{[-\lambda_k(s)]}e^{\lambda_k(s)t}.$$

Here, the first term coincides with transforms obtained above $\bar{R}_a^{(2)}(\infty,s)$. Consequently,

$$\left[R_a^{(2)}(t,x)\right]_1 = R_a^{(2)}(\infty,x) + \frac{1}{2\pi i}\int ds x^{-s}\sum_k\sum_b \frac{H_{a\to b}^{(k)}(s)\bar{F}_b^{(2)}(\infty,s)}{[-\lambda_k(s)]}e^{\lambda_k(s)t}.$$

At large t and $x \to 0$, the behavior of this integral is governed by the multiplier x^{-p}. Neglecting such terms, we get

$$\left[R_a^{(2)}(z,x)\right]_1 \sim R_a^{(2)}(\infty,x), \qquad x \to 0.$$

For the same reason, we omit in (5.64) the term with $(\delta_{ae}/s_1)\Delta\bar{R}_e^{(1)}(t,s_2)$.

Computing next component of the second moment,

$$\left[R_a^{(2)}(t,x)\right]_2 = \frac{2}{(2\pi i)^3}\int\int\int \frac{ds ds_1 ds_2}{(s-s_1-s_2)}$$

$$\int_0^t dt' \sum_b \bar{G}_{a\to b}(t-t',s)\bar{V}_b(s_1,s_2;s)$$

$$\times\left[\frac{\bar{R}_e^{(1)}(\infty,s_1)H_{\bar{b}\to e}(s_2)}{s_2\lambda_1(s_2)}e^{\lambda_1(s_2)t'} + \frac{\bar{R}_{\bar{b}}^{(1)}(\infty,s_2)H_{e\to e}(s_1)}{s_1\lambda_1(s_1)}e^{\lambda_1(s_1)t'}\right]$$

we will take into account only principal poles in $\bar{R}_a^{(1)}(\infty, s)$ and apply the saddle-point method with respect to another variable:

$$\left[R_a^{(2)}(t,x)\right]_2 = -\frac{2\alpha x^{-p-1}e^{\lambda_1(p)t}}{p\sqrt{2\pi|\psi''(p)|}}\sum_{k,b}\frac{H_{a\to b}^{(k)}(p+1)}{\lambda_1(p)-\lambda_k(p+1)}$$

$$\times\left[\widehat{V}_b(p,1;p+1)H_{e\to e}^{(1)}(p)+\widehat{V}_b(1,p;p+1)H_{b\to e}^{(1)}(p)\right].$$

Finally, write the last term

$$\left[R_a^{(2)}(t,x)\right]_3 = \frac{1}{(2\pi i)^2}\sum_{k=1}^{2}\int ds_1\int ds_2\int ds_3 = \frac{M_{a\to e}^k(s_1,s_2)}{\lambda_1(s_1)\lambda_1(s_2)}\exp\left\{v^k(s_1,s_2,s_3)\right\},$$

where $M_{a\to e}^k$ and v^k are functions determined above. By using multi-variable saddle-point method (taking into account what had been said above regarding function $\lambda_1(s)$ in the denominator), we get

$$\left[R_a^{(2)}(t,x)\right]_3 = \sum_{k=1}^{2}\overset{*}{S}{}_{a\to b}^{k[2]}(t,x,\mathbf{p}),$$

where $\mathbf{p} = (p_1, p_2, p_3)$ is the saddle point determined by relations (8.6)–(8.8) with replacement $v^1 \to\to v^1 - \ln|\lambda_1(s_1)\lambda_1(s_2)|$, and $\overset{*}{S}{}_{a\to b}^{k[2]}$ is the function$S_{a\to b}^{k[2]}$, computed with the same replacement. Thus, calculation of the last component has been reduced to calculation of the second factorial moment of particle number (with saddle-point shifted).

As in case with first moments, under $t \to \infty$, $\mathbf{p} \to \mathbf{s}$ and, as a result,

$$\left[R_a^{(2)}(t,x)\right]_3 \sim N_{a\to b}^{[2]}(t,x)/[\lambda_1(s)]^2.$$

Ignoring term of smaller order of smallness under $x \to 0$, then others, one can obtain a formula

$$\delta^2 R_a(t,x) \approx \delta^2 R_a(\infty,x) + \left[\frac{\overline{N}_{a\to e}(t,x)}{\lambda_1(s)\overline{R}_a(z,x)}\right]^2 \delta^2 N_{a\to e}(t,x) \qquad (8.33)$$

for the square of relative fluctuations of total path in a thick layer. It shows the general character of the dependence of path fluctuations (losses energy) from the energy of the primary particle. When the energy is not sufficiently high, the ratio $\bar{N}/\bar{R}$ is small (cascades practically do not reach layer boundaries), and the relative fluctuations decrease with increasing E as in an infinite environment ($\sim E^{-1/2}$). When a noticeable part of the cascade turns out to be outside the layer, a significant contribution to fluctuations is introduced by the second term, which increases with the increase in E (mainly due to an increase in the ratio $\bar{N}/\bar{R} \sim E^{s-1}$, other factors change more slowly). In general, the dependence $\delta R(E)$ under a fixed t has the form of a curve with a minimum at $E = E_{\min}$, defined by the thickness of the layer t and the value of the threshold energy E^*. A similar result is obtained in the B-approximation (subject to replacement (8.2)). In this case, the threshold energy can be considered as a zero level, and (providing $\beta =$ const) for each random cascade sample $\beta R_a(\infty, E) = E$,, i.e., the total path stops fluctuating and the first term in (8.33) disappears (this is what the model says; in reality $\beta(E) \neq$ const and total path fluctuate, whereas the energy loss becomes constant):

$$\delta R_a(t, x) = \frac{1}{|\lambda_1(s)|} \frac{\overline{N}_{a\to e}(t, x)}{\overline{R}_a(t, x)} \delta N_{a\to e}(t, x). \tag{8.34}$$

This relation becomes clear, if to assume, that near boundary, the cascade curve fluctuates in such a way that its slope changes are small (in other words, take place the shift-fluctuations), then for each random realization can be written

$$R_a(t, x) \approx R_a(\infty, x) + \frac{1}{\lambda_1(s)} N_{a\to e}(t, x). \tag{8.35}$$

Computing the variance of this variable (subject to determinate variables $R_a(\infty, E)$) and $\lambda_1(s)$ yields (8.34). By using (8.35) Eq. (8.33) can be rewritten as

$$\delta^2 R_a(t, x) \approx \delta^2 R_a(\infty, x) + \left[\frac{\overline{R}_a(\infty, x) - \overline{R}_a(t, x)}{\overline{R}_a(t, x)}\right]^2 \delta^2 N_{a\to e}(t, x). \tag{8.36}$$

The results presented here were obtained in the works [20, 21].

8.4 On the First Free Path in EPC Fluctuations

In the standard formulation of the cascade theory, the beginning of the cascade is the point of appearance (birth) of the particle initiating the cascade. In EPC, it can be one of these particles: electron or photon, but using ionization camera or other trace-kind detector, an experimenter cannot see photon traces, if the primary particle is a photon, then this point will be the point of the trajectory of the parent electron, in which it emits this photon. In experiments with ionization calorimeters, scintillation spectrometers, Cherenkov detectors, bubble chambers, and X-ray films, photons are registered only as a result of their conversion into charged particles, the trajectories of which, due to ionization, leave "tracks" accessible for optical registration, while the photons themselves do not have such traces. For this reason, it is more convenient to analyze this kind of experimental data by taking the primary photon conversion point as the beginning of the cascade process. Longitudinal is with respect to the direction of development of the cascade (the coordinates of these points are shifted relative to each other by the free path of the primary photon). This shift is a random variable, and therefore there is a difference between the characteristics (both average and fluctuations) of the cascades in these two cases. In [22], fluctuations in the number of particles in EPC were calculated under assumption that the primary photon free path plays the main role in fluctuations of the cascade as a whole (at a fixed point of the first interaction) while the contributions of other events may be neglected. However, in experimental determination of photon energy by using xenon cameras [14, 23] such an approximation should be refined taking into account other sources of fluctuations. We will consider here the relationship between fluctuations in the characteristics of the cascade at random and fixed points of the first interaction. Without limiting ourselves to the study of fluctuations of only the number of particles, we write the studied functional of a cascade random realization

$$q_{z,E} = \int_{z}^{z^*} \sum_{i=1}^{\nu(z')} a_i(z', E_i(z'))dz', \quad z < z^*, \tag{8.37}$$

$$q_{z,E} = 0, \quad z > z^*,$$

where z is the initial longitudinal coordinate of the primary particle (point of its birth); E is its energy; $E_1(z'), \ldots, E_\nu(z')$ are the energies of the cascade electrons at the point $z' > z$; $a(z, E_i)dz$ is the detector response to path dz of the particle with energy E_i; and z^* is the boundary coordinate of the sensitive layer. In particular, for a detector measuring the number of electrons at a point $z = z^*$, $a_e(z, E) = \delta(z - z^*)$; for a detector that measures the total path of the cascade electrons in the region $z < z' < z^*$, $a_e(z, E) = 1$; etc. Denoting by z_1 the first collision point, from (8.37), we obtain

$$q_{z,E} = \int_{z}^{z_1} a(z', E(z'))dz' + q'_{z_1,E_1},$$

where q'_{z_1,E_1} is the response of detector to the cascade with the first collision point z_1 and the energy of the primary particle at the moment of collision E_1, $a = a_e$ for the primary electron and $a = 0$ for the primary photon. Considering the interval $(z, z + dz)$ and taking into account that the probability of a collision of a primary particle in this interval (i.e., the probability of an event $z < z_1 < z + dz$) is equal to σdz, we write the following relationship:

$$q_{z,E} = \begin{cases} a(z, E)dz + q_{z+dz,E-\beta dz} + o(dz) & \text{with probability} \quad 1 - \sigma dz \\ q'_{z,E} + O(dz) & \text{with probability} \quad \sigma dz \end{cases},$$

where $\beta = \beta_e$ for the primary electron (ionization loss) and $\beta = 0$ for the primary photon. Averaging q and q^2 by using the above relation, we get the following expressions:

$$\Big[-\partial_z+\beta\partial_E+\sigma\Big]\overline{q}=a+\sigma\overline{q}', \tag{8.38}$$

$$\Big[-\partial_z+\beta\partial_E+\sigma\Big]\overline{q^2}=2a\overline{q}+\sigma\overline{q'^2}. \tag{8.39}$$

Multiplying the first of them by $2\overline{q}$ and subtracting from the second, we arrive at the corresponding relation for the variances:

$$\Big[-\partial_z+\beta\partial_E+\sigma\Big]D=\sigma(\overline{q}-\overline{q}')^2+\sigma D'. \tag{8.40}$$

In particular, for a primary photon, it follows from (8.38) that

$$\sigma_\gamma(\overline{q}_\gamma-\overline{q}'_\gamma)=\partial_z\bar{q}_\gamma$$

therefore (8.40) can be reduced to the form

$$\partial_t D_\gamma+\sigma_\gamma D_\gamma=\frac{1}{\sigma_\gamma}\left(\partial_t\bar{q}_\gamma\right)^2+\sigma_\gamma D'_\gamma, \tag{8.41}$$

where $t=z^*-z$. Integrating (8.41), taking into account the boundary condition $D_\gamma(0)=0$, which follows from (8.37), we obtain

$$D_\gamma=\int\limits_0^t\exp[-\sigma_\gamma(t-t')]\Big[\left(\frac{1}{\sigma_\gamma}\partial_{t'}q_\gamma\right)^2+D'_\gamma\Big]\sigma_\gamma dt'=D^{\rm I}_\gamma(t)+D^{\rm II}_\gamma(t). \tag{8.42}$$

According to [22], the first term $D^{\rm I}_\gamma$ relates to the first path fluctuations. The second term $D^{\rm II}_\gamma$ reflects the effect of fluctuations in the rest of the cascade and is expressed in terms of the variance D'_γ of the cascade with a fixed point of the first collision.

Formula (8.41) allows us to calculate D'_γ from the available function $D_\gamma(t)$. The calculation results shown in the figure show good agreement with the data of the Monte Carlo method [24]. Formula (8.42) allows us to analyze the relationship between $D^{\rm I}_\gamma$ and $D^{\rm II}_\gamma$. It can be seen from Table 8.1 that $D^{\rm II}_\gamma$ makes a significant contribution to the total dispersion (8.42) and there is no reason to neglect this term.

Here, ($E_{th}\geqslant 0.1$ MeV, $E_\gamma=10^6$ MeV) (Fig. 8.1).

From formula (8.39) one can obtain a system of equations for the variances D_e and D_γ

$$\Big[-\partial_z+\sigma_\gamma\Big]D_\gamma(t,E)-2\int W_p(E,E')D_e(t,E')dE'$$

$$=\int W_p(E,E')[\overline{q}_\gamma(t,E)-\overline{q}_e(t,E')-\overline{q}_e(t,E-E')]^2dE'. \tag{8.43}$$

Table 8.1 Different components of the total electron path variance in a layer of thickness t

t	D_γ	D_γ^{I}	D_γ^{II}	D'_γ
2	$0.172 \cdot 10^4$	$0.835 \cdot 10^3$	$0.885 \cdot 10^3$	$0.193 \cdot 10^4$
4	$0.855 \cdot 10^6$	$0.398 \cdot 10^5$	$0.457 \cdot 10^5$	$0.510 \cdot 10^5$
8	$0.916 \cdot 10^6$	$0.369 \cdot 10^6$	$0.547 \cdot 10^6$	$0.616 \cdot 10^6$
14	$0.261 \cdot 10^6$	$0.833 \cdot 10^5$	$0.178 \cdot 10^6$	$0.134 \cdot 10^6$
22	$0.128 \cdot 10^6$	$0.422 \cdot 10^5$	$0.860 \cdot 10^5$	$0.288 \cdot 10^5$
26	$0.244 \cdot 10^4$	$0.941 \cdot 10^4$	$0.150 \cdot 10^5$	$0.492 \cdot 10^4$

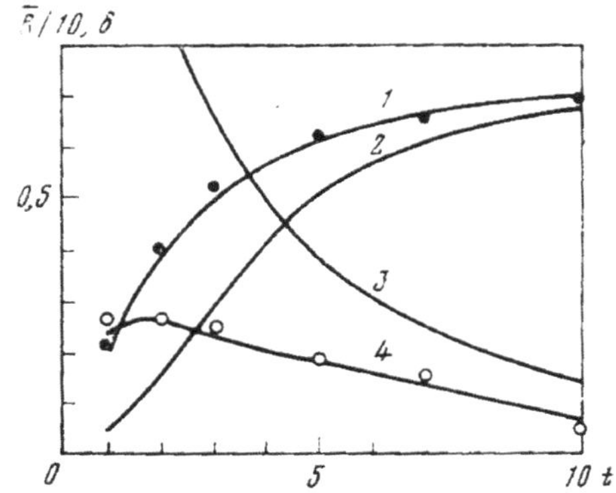

Fig. 8.1 Average values $\overline{R}$ (1, 2) and relative fluctuations δ (3, 4) of electron trace length in xenon ($E_e \geqslant 3.066$ MeV, $E_\gamma = 154$ MeV): 1, 4–for EPC with a fixed point of the first interaction; 2, 3–for EPC with a fixed point of birth. Dots are data from [4]

$$\Big[-\partial_z + \beta_e \partial_E + \sigma_e \Big] D_e - \int W_e(E, E')[D_\gamma(t, E') + D_e(t, E - E')] dE'$$

$$= \int W_e(E, E')[\overline{q}_e(t, E) - \overline{q}_\gamma(t, E') - \overline{q}_e(t, E - E')]^2 dE'. \tag{8.44}$$

The use of this system of equations, where the unknown functions are directly the variances, has certain advantages over the usual approach to the fluctuation problem, which consists in solving the equations for the moments $\overline{q}$ and $\overline{q}^2$ and then calculating variances according to the formula $D = \overline{q^2} - \overline{q}^2$ [4, 21]. These advantages are especially significant in cases where the dispersion is small, and $\overline{q^2}$ and $\overline{q}^2$ are large, so that the magnitude of the difference $\overline{q^2} - \overline{q}^2$ becomes commensurate with the error in calculating the moments.

As a result of viewing about 15,000 xeno-bubble chambers irradiated in a beam of π^- mesons with a momentum of 3.5 GeV/c, 232 cases were selected electron–photon showers from gamma rays with energies 30–120 MeV. In order to determine the energy of gamma quanta, a complete identification of all traces of particles related to the shower caused by these gamma quanta. The shower energy was determined from the lengths electron ranges as the sum of the energies of all belonging electrons. For this, the link between path lengths of electrons and their energies, recalculated for liquid xenon. Minimum of the recorded trace length was about 1.5 mm.

In conclusion, the authors emphasized that the parameters characterizing the longitudinal development of showers: average length of shower development $\langle d \rangle$, the length d_A of its partial development, the share of the total path of cascade electrons A, as well as the relative variance $\sigma_A/\bar{A}$ and the length of partial development showers σ_d/d_A are very convenient when studying the characteristics of longitudinal development of electron–photon avalanches caused by gamma quanta in homogeneous media. Cascade characteristics and fluctuations expressed through these parameters exhibit useful and interesting scale properties.

The results are presented of investigations of the development lengths, the total ranges of shower electrons, and corresponding fluctuations in the showers induced by the γ-quanta of 30, 50, 70, 100, and 120 MeV in liquid xenon. The scaling property is revealed in the integral cascade curves and in a studied dependence of fluctuations on the part of the shower electron total range.

8.5 Random Moments Method (Longitudinal Development)

8.5.1 Fluctuations in Cascade Curves

The moment method is one of the most popular methods in the theory of radiation transfer in general [25–28] and in the EPC, in particular [28, 29]. Usually, it was applied to computing averaged EPC characteristics, whereas fluctuation analysis left the Monte Carlo method. Meanwhile, it was noticed that the random cascade examples are similar in shape to the averaged cascade curves, differing from them in height and position of the maximum (Fig. 8.2). But the mean characteristics of cascades in our articles [30–32] was successful. In our work [10], a new approach to solving this problem is proposed, developed on the basis of spatial moments

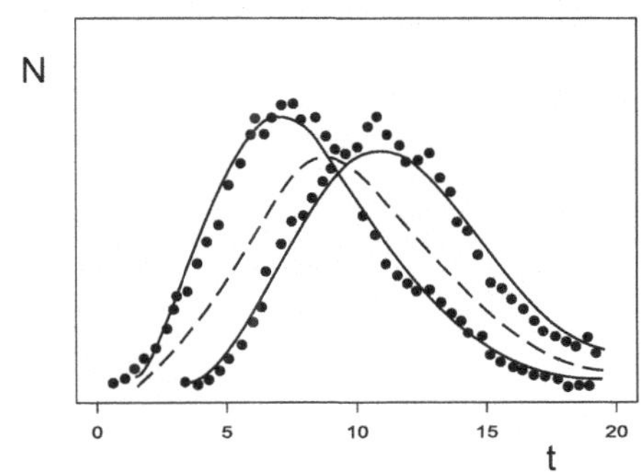

Fig. 8.2 Cascade curves: •—three-moment approximations; dashed line—averaged cascade curve

of random cascade curves (for the sake of brevity, they will be called the *random moments*). For their mean values, mean squares and higher moments (here the word "moment" is used in the probabilistic sense as the expectation of some degree of a random quantity), a closed system of backward cascade equations which is suitable for SBCE solving may be written.

Suppose that $N_{a\to d}(E, z \to E^*, z^*)$ is the random number of d-particles crossing the plane $z = z^*$ with an energy $E > E^*$ in a cascade produced by a primary a-particle with energy E, created at the point $z < z^*$ with momentum along the positive direction of z-axis). In a homogeneous medium, $N_{a\to d}$ depends on the distance $t = z^* - z$. The kth longitudinal moment of the cascade curve is expressed in the form:

$$N_a^{(k)}(E) = \int\limits_0^\infty t^k N_{a\to d}(E, z \to E^*, z^*)dt. \tag{8.45}$$

Observe that $N_a^{(0)}$ is equal to the area under the cascade curve or, in other words, the total path of d-particles; $N_a^{(1)}/N_a^{(0)}$ is the center of gravity of the cascade curve means the position g of shifting them along the cascade axis until c.g. coincide. As shown in [33], significantly reduces the fluctuations in the number of particles, making the random curves ensemble to be more homogeneous. One can expect that more general transformation allows to reconcile the zeroth, first, and second moments of random curves. As our MC calculations in A-approximation for $E_\gamma/E^* = 10^4$, the level of fluctuations proves even lower here (Fig. 8.3). Since the complete set of moments (under definite conditions) uniquely determines the curve, the remaining fluctuations may be regarded as the influence of fluctuations of higher moments and on the basis of the calculations it may be concluded that their contribution to the total fluctuation is relatively small. In connection with this, it is appropriate to ask how accurately the first random moments reflect the statistical characteristics of the cascade curves.

To answer this question, the first few moments are found for each random cascade curve, and the usual cascade theory technique of derivation by means of Laguerre polynomials is used, in the form

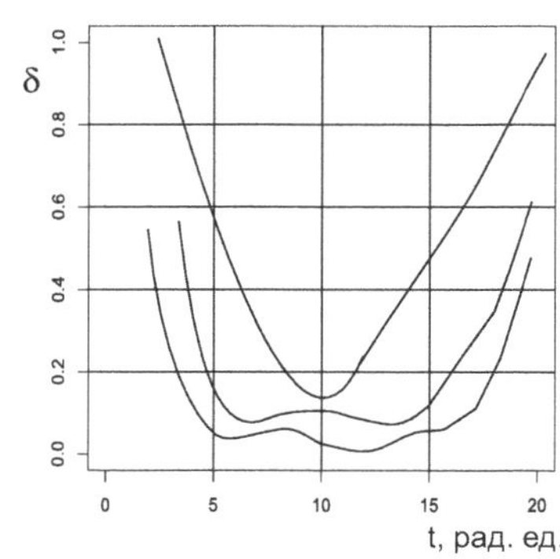

Fig. 8.3 Fluctuations in cascade curves (top down): reduced to the common point of the primary photon collision; reduced to three first common moments, averaged over the whole ensemble

$$N_{a\to d}(E,t) \approx F(E,t;N_a^{(0)},\dots,N_a^{(k)}). \tag{8.46}$$

It is found that even the first three moments satisfactorily approximate the random samples of the cascade curves (Fig. 8.2), and four or more moments are sufficient to reproduce both the mean values and the fluctuations in the number of particles. In other words, the contribution to the fluctuations from cascade curves of more complex form, which must be described by moments of higher order, is negligibly small.

Suppose that

$$q(E,t) = \int_0^\infty h(t)N_{a\to d}(E,t)dt \tag{8.47}$$

is the cascade characteristics under investigation, where $h(t)$ is the detector response to a single electron at point t. For instance, $h(t) = \delta(t-t^*)$, if the number of electrons at depth t^* is measured, or

$$h(t) = 1(t) = \begin{cases} 1, \ t > 0 \\ 0, \ t < 0, \end{cases}$$

if the total path of cascade electrons in the layer of thickness t is of interest and so on.

8.5.2 Correlations in Cascade Curves

As shown in Sect. 8.2 (see also [32]), the relative fluctuations of the zero moment decrease with increase in energy as $E^{-1/2}$. The higher order moment fluctuations also decrease with the primary energy increasing, although less rapidly (Table 8.2). It is explained in that the basic contribution to the integral in Eq. (8.45) is made by the region close to the maximum of the integrand, and therefore its value correlates with the number of particles in this region. With increase in the order of the moment,

Table 8.2 Relative fluctuations of electron numbers random moments in EPC from photon with energy E_γ

k	E_γ/E^*		
	10^2	10^3	10^4
0	0.0934	0.0294	0.00931
1	0.304	0.199	0.152
2	0.596	0.400	0.305
3	0.957	0.626	0.469
4	1.44	0.896	0.655
5	2.14	1.24	0.874

the maximum is shifted to greater depths, where the fluctuations of the cascade curve are higher. Taking into account that a small number of moments is sufficient for the derivation of the high-energy cascade curves, the kth moment is written in the form

$$N_a^{(k)} = \overline{N}_a^{(k)} + \Delta N_a^{(k)} \tag{8.48}$$

and substituting Eq. (8.46) into Eq. (8.47), the function F is expanded in Taylor series with respect to $\Delta N_a^{(k)}$, retaining terms of second-order smallness

$$q = Q + \sum_k Q_k \Delta N_a^{(k)} + \frac{1}{2} \sum_{k,m} Q_{km} C_a^{(k,m)}, \tag{8.49}$$

where

$$Q = \int_0^\infty F(E, t; \overline{N}_a^{(0)}, \ldots, \overline{N}_a^{(k)}) h(t) dt, \tag{8.50}$$

$$Q_k = \int_0^\infty \frac{\partial F(E, t; \overline{N}_a^{(0)}, \ldots, \overline{N}_a^{(k)})}{\partial \overline{N}_a^{(k)}} h(t) dt, \tag{8.51}$$

$$Q_{km} = \int_0^\infty \frac{\partial^2 F(E, t; \overline{N}_a^{(0)}, \ldots, \overline{N}_a^{(k)})}{\partial \overline{N}_a^{(k)} \partial \overline{N}_a^{(m)}} h(t) dt, \tag{8.52}$$

$$C_a^{(k,m)} = \langle \Delta N_a^{(k)} \cdot \Delta N_a^{(m)} \rangle.$$

So the mean value

$$\bar{q} = Q + \frac{1}{2} \sum_{k,m} Q_{km} \overline{C}_a^{(k,m)}, \tag{8.53}$$

the variance

$$\sigma_q^2 = \sum_{k,m} Q_k Q_m \overline{C}_a^{(k,m)},$$

and the relative fluctuations

$$\delta_q = \frac{\sigma_q}{\bar{q}} \tag{8.54}$$

for the functional under consideration may be found.

The expression for the correlation of the readings of two detectors measuring the characteristics q_1 and q_2, respectively, is derived analogously

$$\overline{(q_1 - \overline{q}_1)(q_2 - \overline{q}_2)} = \sum_{k,m} Q_{1;k} Q_{2;m} \overline{C}_a^{(k,m)}. \tag{8.55}$$

The set of $\bar{C}_a^{(k,m)}$ in terms of which and the known quantities in Eqs. (8.49) and (8.50) the fluctuations of various characteristics of the electromagnetic cascade in Eqs. (8.52) and (8.54) may be expressed, forms a *covariational matrix* of random moments, and may be found by solving the corresponding equations.

8.5.3 Moment Covariance Matrix Equation

To derive equations, describing correlations in EPC, we again use the stochastic interrelation approach, considering the motion of a primary particle in the interval $(z, z + \Delta z)$. With a probability $1 - \sigma_a(E)\Delta z$, the particle passes it without interactions (event A), but with probability $\sigma_a(E)\Delta z$ it interacts with the matter, producing an arbitrary number ν of daughter particles (event B). Denoting indicators of these random events by 1_A and 1_B, respectively, we obtain stochastic interrelation

$$N_{a\to d}(E, z \to E_d, z_d) = 1_A\{\delta_{a,d} 1(E - E_d) 1(z + \Delta z - z_d)$$

$$+ N_{a\to d}(E, z + \Delta z \to E_d, z_d) 1(z_d - z - \Delta z)\} + 1_B \sum_{i=1}^{\nu} N_{a_i\to d}(E_i, z \to E_d, z_d). \tag{8.56}$$

Multiplying Eq. (8.55) by $(z_d - z)^k$ and integrating over z_d from z to ∞ yield

$$N_a^{(k)}(E) = 1_A\{\delta_{a,d} 1(E - E^*)\delta_{k,0}\Delta z$$

$$+ k\Delta z N_a^{(k-1)}(E) + N_a^{(k)}(E)\} + 1_B \sum_{i=1}^{\nu} N_{a_i}^{(k)}(E_i). \tag{8.57}$$

Noting that the mean values of the random moments coincide with the moments of the mean cascade curves usually employed, let us subtract them from both sides of Eq. (8.56):

$$\Delta N_a^{(k)}(E) = 1_A\{\delta_{a,d} 1(E - E_d)\delta_{k,0}\Delta z + k\Delta N_a^{(k-1)}(E) + \Delta N_a^{(k)}(E)$$

$$+ k\overline{N}_a^{(k-1)}(E) + \overline{N}_a^{(k)}(E)\} + 1_B\Big\{\sum_{i=1}^{\nu}[\Delta N_{a_i}^{(k)} + \overline{N}_{a_i}^{(k)}(E)] - \overline{N}_a^{(k)}(E)\Big\}. \tag{8.58}$$

Writing analogous expressions for mth moment, multiplying them, and averaging over the curve ensemble yield, a system of integral equations for the elements of the covariance matrix is obtained in the following:

$$\sigma_a(E)\overline{C}_a^{(k,m)}(E) - \sum_b \int dE' W_{a\to b}(E \to E')\overline{C}_b^{(k,m)}(E')$$

$$-\sum_{b,c}\int dE'W_{a\to b,c}(E\to E')[\overline{C}_b^{(k,m)}(E')+\overline{C}_c^{(k,m)}(E-E')]=R_a^{(k,m)}(E), \tag{8.59}$$

where

$$R_a^{(k,m)}(E)=k\overline{C}_a^{(k,m-1)}(E)+m\overline{C}_a^{(k-1,m)}(E)$$

$$+\sum_{b}\int dE'W_{a\to b}(E\to E')\Big[\overline{N}_b^{(k)}(E')-\overline{N}_a^{(k)}(E)\Big]\Big[\overline{N}_b^{(m)}(E')-\overline{N}_a^{(m)}(E)\Big]$$

$$+\sum_{b,c}\int dE'W_{a\to b,c}(E\to E')\Big[\overline{N}_b^{(k)}(E')+\overline{N}_c^{(k)}(E-E')-\overline{N}_a^{(k)}(E)\Big]$$

$$\times\Big[\overline{N}_b^{(m)}(E')+\overline{N}_c^{(m)}(E-E')-\overline{N}_a^{(m)}(E)\Big].$$

In the particular case $k=m=0$, $C_a^{(0,0)}$ has the meaning of the variance of the total path of the secondary particles in an infinite homogeneous medium. This problem was considered in [32], where the SBCE method for solving equations of the type of Eq. (8.58) applied. Substituting the result for $C_a^{(0,0)}$ into the right-hand side of Eq. (8.58), we find the possibility of solving the system for $C_a^{(0,1)}$. Using this result, one can obtain $C_a^{(1,1)}$, and so on.

Numerical results of calculations performed by SBCE method are given in Tables 8.3 and 8.4, the correlation matrix

$$\rho^{(k,l)}=C^{(k,l)}[C^{(k,k)}C^{(l,l)}]^{-1/2}$$

are represented in Table 8.5.

Table 8.3 Covariance matrix $C^{(k,l)}$ for the EPC from a primary photon in the A-approximation $E_\gamma/E^*=10^3$ (top numbers are MC results, lower numbers are SBCE results)

k	l					
	0	1	2	3	4	5
0	1.62+2	1.66+3	2.06+4	2.82+5	4.13+6	6.38+7
	1.63+2	1.46+3	1.51+4	1.76+5	2.06+6	2.57+7
1		4.60+5	8.41+6	1.34+8	2.13+9	3.52+10
		4.56+5	8.38+6	1.35+8	2.19+9	3.72+10
2			1.58+8	2.57+9	4.18+10	7.01+11
			1.59+8	2.64+9	4.42+10	7.72+11
3				4.27+10	7.05+11	1.20+13
				4.52+10	7.77+11	1.40+13
4					1.18+13	2.05+14
					1.37+13	2.54+14
5						3.58+15
						4.85+15

Table 8.4 Covariance matrix $C^{(k,l)}$ for the EPC from a primary photon in the A-approximation $E_\gamma/E^* = 10^4$ (the arrangement is the same as in Table 8.3)

k	l					
	0	1	2	3	4	5
0	1.57+3	2.27+4	3.79+5	6.64+6	1.20+8	2.28+9
	1.63+3	1.74+4	2.11+5	2.84+6	3.98+7	5.80+8
1		4.32+7	9.98+8	1.95+10	3.77+11	7.49+12
		4.44+7	1.03+9	2.03+10	3.94+11	7.88+12
2			2.34+10	4.66+11	9.12+12	1.83+14
			2.44+10	4.88+11	9.67+12	1.97+14
3				9.39+12	1.86+14	3.79+15
				9.97+12	2.01+14	4.17+15
4					3.75+15	7.72+16
					4.13+15	8.72+16
5						1.61+18
						1.88+18

Table 8.5 Characteristics of the moments of the spatial distribution of electrons EPC

t, rad.	$\delta R^{(0)}$		$C^{(1,1)}$		$\delta R^{(2)}$		$\rho^{(0,2)}$	
	1	2	1	2	1	2	1	2
2	0.84	0.85	$4.79 \cdot 10^{-3}$	$9.54 \cdot 10^{-3}$	1.93	2.66	0.55	0.41
4	0.49	0.50	$3.86 \cdot 10^{-2}$	$3.92 \cdot 10^{-2}$	1.23	1.70	0.40	0.30
6	0.50	0.51	$7.16 \cdot 10^{-2}$	$7.49 \cdot 10^{-2}$	1.34	2.00	0.13	0.12
8	0.74	0.79	$8.27 \cdot 10^{-2}$	$8.31 \cdot 10^{-2}$	2.26	2.80	0.13	0.10

Notes Primary photon. $E^* = 25$ MeV, $E_\gamma = 10$ GeV. 1—Monte Carlo method, 2—SBCE method

8.5.4 Some Parameter Estimations

The efficiency and accuracy of the proposed method are tested by means of trial calculations of the fluctuations of EPC in A-approximation of cascade theory, the results of which are tested by direct modeling of cascades by the Monte Carlo method in the same approximation (for $E_\gamma/E_1 = 10^2$, 3500 case histories; for $E_\gamma/E_1 = 10^3$ and 10^4 700 each). Numerical solution of Eq. (8.59) for the covariational matrix continues until $k = m = 5$.

Comparison of the results shows that the mean values of the longitudinal moments and elements of the covariational matrix agree within the limits of statistical error of the Monte Carlo method. The non-diagonal elements have a large statistical error here, reaching 70% at $k = 0$ and $m = 5$.

By analogy with the center of gravity and longitudinal dimension (extent) of the mean cascade curve [34]

$$t_0 = \bar{N}^{(1)}/\bar{N}^{(0)}, \tag{8.60}$$

$$\tau_0 = \bar{N}^{(2)}/\bar{N}^{(0)} - t_0^2, \tag{8.61}$$

the corresponding characteristics of the random curve are introduced as follows:

$$t_0 = N^{(1)}/N^{(0)}, \tag{8.62}$$

$$\tau_0 = N^{(2)}/N^{(0)} - t^2. \tag{8.63}$$

Since the fluctuations of $N^{(0)}$ are small and decrease rapidly with increase in E, this quantity may be replaced by its mean value in the denominator of Eqs. (8.62) and (8.63)

$$t' = N^{(1)}/\bar{N}^{(0)}, \qquad \tau' = N^{(2)}/\bar{N}^{(0)} - t'^2.$$

In this approximation

$$\bar{t} \approx \bar{t'} = t_0, \qquad \delta_t \approx \delta_{t'} = \delta_{N^{(1)}}, \qquad \bar{\tau} \approx \bar{\tau'} = \tau_0 - t'^2\delta_t^2.$$

More accurate results may be obtained if the denominator in Eqs. (8.62) and (8.63) is written in the form

$$N^{(0)} = \bar{N}^{(0)}(1 + \Delta N^{(0)}/\bar{N}^{(0)})$$

and the ratio is expanded in series in powers of $\Delta N^{(0)}/\bar{N}^{(0)}$. In this (covariational) approximation, it is found that

$$\bar{t} \approx \bar{t''} = t_0 + t_0\delta^2_{V^{(0)}} - \bar{C}^{(0,1)}/\bar{N}^{(0)2},$$

$$\delta_t^2 \approx \delta_{t''}^2 = \delta^2_{N^{(1)}} + (2t_0\bar{C}^{(0,1)} - t_0^2\bar{C}^{(0,0)})/(\bar{N}^{(0)})\bar{t''})^2,$$

$$\bar{\tau} \approx \bar{\tau}'' = \tau_0 + (\tau_0 + 2t_0^2)\delta^2_{N^{(0)}} + (\bar{C}^{(1,1)} - \bar{C}^{(0,2)} - 4t_0\bar{C}^{(0,1)})/\bar{N}^{(0)2}.$$

The probabilistic moments of higher order required for calculation of the fluctuations of τ are not considered here.

It is evident from Tables 8.6 and 8.7 that the results of numerical calculations in the above approximations agree with the results of direct modeling by the Monte Carlo method. Significant difference in τ_0 and τ is seen, whereas t_0 and t practically coincide. Note also the fairly large value of the fluctuations of t and τ and their slow decrease with increase in E_γ/E_1.

The results of calculating the relative fluctuations in the number of electrons according to Eq. (8.53) using a covariational matrix obtained numerically are shown in Fig. 8.4. Completely satisfactory agreement is seen both with direct modeling by

Table 8.6 Parameters of the average cascade curve (in rad. units)

E_γ/E^*	$\overline{\theta}$		$\overline{\tau}$	
	Analytical method	Numerical	Analytical method	Numerical
10^2	5.45	5.48	2.83	2.85
10^3	7.78	7.81	3.42	3.43
10^4	10.1	10.1	3.92	3.92

Table 8.7 Longitudinal size

E_γ/E_1	τ_0		$\overline{\tau'}$	$\overline{\tau''}$	$\overline{\tau}$	δ_τ
	Analit [35]	SBCE	SBCE	SBCE	MC	MC
10^2	8.01	8.10	5.33	5.89	–	–
10^3	11.7	11.8	9.40	9.48	9.62	0.33
10^4	15.4	15.5	13.1	13.2	13.3	0.24

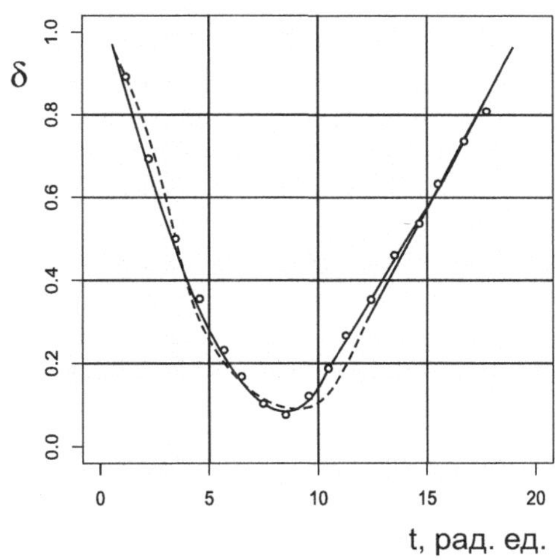

Fig. 8.4 Cascade curve fluctuations: o—Monte Carlo data; solid—[37]; dotted line—method of random moments (formula (8.52))

the Monte Carlo method and with the results obtained by numerical solution of the conjugate equations for $\bar{N}$ and $\bar{N}_2$ [36].

8.5.5 On the Detector Reading Probability Distribution

Let's return to the covariance matrix of moments (Tables 8.3 and 8.4) and calculate the correlation coefficients of different moments:

$$\rho^{(k,l)} = C^{(k,l)}[C^{(k,k)}C^{(l,l)}]^{-1/2}.$$

Table 8.8 Moment correlation matrix

k	l					
	0	1	2	3	4	5
0	1.00	0.170	0.091	0.065	0.044	0.029
	1.00	0.065	0.034	0.022	0.015	0.011
1		1.00	0.983	0.940	0.875	0.791
		1.00	0.990	0.963	0.920	0.863
2			1.00	0.986	0.944	0.878
			1.00	0.991	0.964	0.920
3				1.00	0.986	0.943
				1.00	0.991	0.963
4					1.00	0.985
					1.00	0.990
5						1.00
						1.00

Note Top numbers for–$E_\gamma E^* = 10^2$, lower ones for 10^4

The calculation results given in Table 8.8 show that the highest ($n \geqslant 1$) moments practically do not correlate with zeroth one but are closely related to each other.

Statistical modeling showed that the zeroth and first moments are not only weakly correlated, but also practically independent (Fig. 8.5a). As for the higher moments, with sufficient accuracy for applications, they can be considered functionally related to the first moment, and for the second moment this relationship is close to linear, and with an increase in the order of the moment it becomes more nonlinear (Fig. 8.5b, c).

Observe that having written the functional connection in the form $N^{(k)} = f_k(N^{(1)})$, we arrive at the following approximation of the joint distribution density of the set of moments:

$$\psi(N^{(0)}, \ldots, N^{(n)}) \approx \Psi^{(0)}(N^{(0)})\Psi^{(1)}(N^{(1)})$$

$$\times\ \delta[N^{(2)} - f_2(N^{(1)})] \ldots \delta[N^{(n)} - f_n(N^{(1)})]. \tag{8.64}$$

Then, the distribution of (8.64) reads

$$\begin{aligned}\psi(q) &= \int dN^{(0)} \ldots \int dN^{(n)} \Psi(N^{(0)}, \ldots, N^{(n)}) \delta[q - Q(N^{(0)}, \ldots, N^{(n)})] \\ &= \int dN^{(0)} \int dN^{(1)} \Psi^{(0)}(N^{(0)}) \Psi^{(1)}(N^{(1)}) \\ &\times\ \delta[q - Q(N^{(0)}, N^{(1)}), f_2(N^{(1)}), \ldots, f_n(N^{(1)}))].\end{aligned} \tag{8.65}$$

The distribution functions $\Psi^{(0)}$ and $\Psi^{(1)}$ can be obtained by Monte Carlo or numerical methods. In the latter case, it is convenient to first find the moments of the distributions

Fig. 8.5 Joint distributions of the moments of the cascade curve (Monte Carlo method, $E_\gamma/E^* = 10^4$, 50 cascades, the length is expressed in radiation units 0)

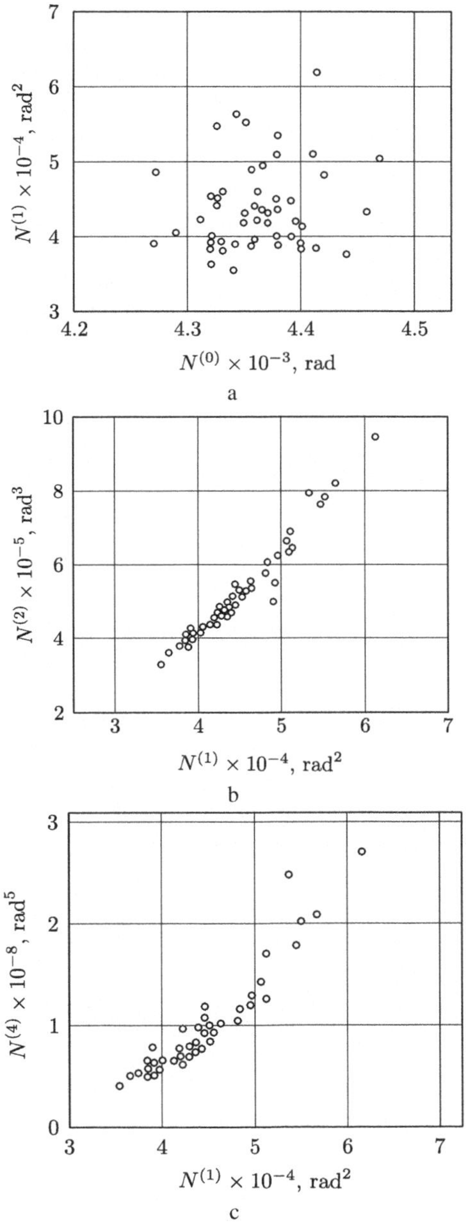

$N^{(0)}$, $N^{(1)}$, and use them to reconstruct the distributions themselves. The equations for the moments $\langle N^{(k)n}\rangle = \mathsf{M}N^{(k)n}$ can be easily obtained from the stochastic relation for random variables (4.75) $N_a^{(k)}$. After raising to the nth power, it becomes

$$[N_a^{(k)}(E)]^n = 1_A \left\{[1(E - E^*)\delta_{ad}\delta_{k0} + kN_a^{(k-1)}(E)]\Delta t + N_a^{(k)}(E)\right\}^n$$

$$+ (1 - 1_A)\left[\sum_{i=1}^{\nu} N_{a_i}^{(k)}(E_i)\right]^n 1_A \left\{[N_a^{(k)}(E)]^n + \right.$$

$$\left. + n[N_a^{(k)}(E)]^{n-1}[1(E - E^*)\delta_{ad}\delta_{k0} + kN_a^{(k-1)}(E)]\Delta t\right\}$$

$$+ (1 - 1_A)\left[\sum_{i-1}^{\nu} N_{a_i}^{(k)}(E_i)\right]^n + o(\Delta t).$$

After averaging, we get (for $\Delta t \to 0$)

$$\sigma_a(E)\langle N_a^{(k)n}\rangle = 1(E - E^*)\delta_{ad}\delta_{k0}n\langle N_a^{(k)(n-1)}\rangle$$

$$+ kn\langle N_a^{(k-1)}N_a^{(k)(n-1)}\rangle + \sigma_a(E)\mathsf{M}\left[\sum_{i=1}^{\nu} N_{a_i}^{(k)}(E_I)\right]^n. \tag{8.66}$$

The zeroth moment in accordance with the normalization condition is equal to 1. The equation for the first moment ($n = 1$) is consistent with (8.66) and does not require additional information to solve. The equation for the second moment contains the element $\langle N_a^{(k-1)}N_a^{(k)}\rangle = C^{(k-1,k)}$ in its right side. When analyzing equations for higher moments, we take into account that we need to find solutions only for two values of the index k: 0 and 1. According to (8.64) $\langle N_a^{(0)}N_a^{(1)(n-1)}\rangle \approx \langle N_a^{(0)}\rangle\langle N_a^{(1)(n-1)}\rangle$, so Eq. (8.66) turns out to be closed again. Note also that for the EPC $\nu = 2$, so the last term on the right-hand side (8.66) is written as

$$\sigma_a(E)\mathsf{M}[N_{a_1}^{(k)}(E_1) + N_{a_2}^{(k)}(E_2)]^n \equiv \sigma(x)\mathsf{M}[N^{(k)}(x_1) + N^{(k)}(x_2)]^n$$

$$= \sigma(x)\mathsf{M}\left[\sum_{m=0}^{n}\binom{n}{m}N^{(k)m}(x_1)N^{(k)(n-m)}(x_2)\right] = \int dx' w(x \to x')\langle N^{(k)n}(x')\rangle$$

$$+ \sigma(x)\sum_{m=1}^{n-1}\binom{n}{m}\frac{1}{2!}\int dx'\int dx'' v_2(x \to x', x'')\langle N^{(k)m}(x')\rangle\langle N^{(k)(n-m)}(x'')\rangle.$$

Monte Carlo calculations have shown that the distribution of the zero moment is satisfactorily described by a normal law with mean value and variance consistent with both the numerical solution of the inverse equations and the analytical results

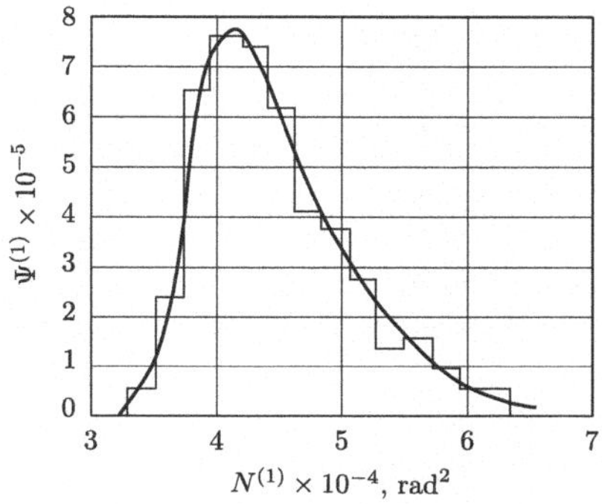

Fig. 8.6 Distribution of the first moment of the random cascade curve (A-approximation, $E_\gamma/E^* = 10^4$). Histogram—Monte Carlo method, curve—four-moment approximation

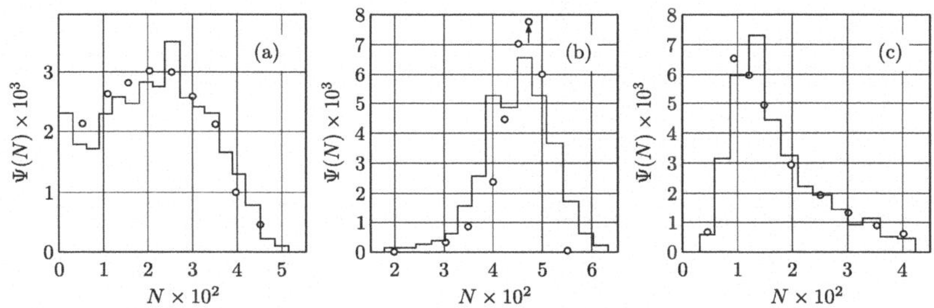

Fig. 8.7 Distribution of electron random number at different EPC depths from the primary photon (approximation $E_\gamma/E^* = 10^4$). Histograms show the MC results, dots are RMM results—Monte Carlo method, dots—moment method

for the asymptotics. The $N_a^{(1)}$ distribution was found by numerically solving the inverse equations for the first four moments of this quantity and approximating the distribution of the Pearson curve (Fig. 8.6). The distribution of the number of electrons calculated by formula (7.27) at different depths of EPC development also agrees with the results of the Monte Carlo method (Fig. 8.7). The formula (8.65) also clarifies the cause for the difference between the fluctuations in the cascade maximum and the fluctuations in the degradation spectrum, whereas spectrum the average values of these quantities coincide). The difference is due to the fluctuations of the first moment $N^{(1)}$: if its fluctuations are excluded or the dependence of Q on $N^{(1)}$ is neglected, the mentioned difference disappears.

8.6 Transverse Random Moments

The concept of a random moment can also be extended to transverse distributions. Let $\mu_{a\to d}(t, E \to \mathbf{r})$ be a random density of particles of type d with energy above E^* at point $\mathbf{r}$ (the 2D cross vector) of the plane perpendicular to the direction of motion of the primary particle and located at a distance t from the point of its birth in a homogeneous medium. We confine ourselves to considering the moments

$$R_a^{(k)}(t, E) = \int \mu_{a\to d}(t, E \to \mathbf{r})\mathbf{r}^k d\mathbf{r} = \sum_i \mathbf{r}_i^k, \qquad k = 0, 1, 2. \tag{8.67}$$

For $k = 0$ the moment (8.67) is equal to the random number of particles at depth t, the ratio

$$\mathbf{R}_a^{(1)}/R_a^{(0)} = g_\perp \tag{8.68}$$

is the random position of the center of gravity of the transverse distribution, and

$$\{R_a^{(2)}/R_a^{(0)} - [\mathbf{R}_a^{(1)}/R_a^{(0)}]^2\}^{1/2} = \tau_\perp \tag{8.69}$$

describes the width of the distribution. These characteristics were experimentally studied in [38–41].

It follows from symmetry considerations that the centroid of the mean distribution and the mean position of the centroid of the random distribution coincide: $\overline{\boldsymbol{\rho}} = \langle \boldsymbol{\rho} \rangle = 0$. For the same reasons $\langle \mathbf{R}_a^{(1)} \rangle = 0$. Further,

$$\overline{\tau} = [\overline{R}_a^{(2)}/\overline{R}_a^{(0)}]^{1/2} \equiv (\overline{r}_{\text{rms}})_a$$

is RMS radius of the average distribution of electrons in EPC, generated by a-particle,

$$\langle \tau^2 \rangle = \langle R_a^{(2)}/R_a^{(0)} \rangle - \langle [\mathbf{R}_a^{(1)}/R_a^{(0)}]^2 \rangle,$$

and

$$C_a^{(k,l)} = \operatorname{cov}(R_a^{(k)}, R_a^{(l)})$$

means the covariance matrix of transverse moments,

$$\rho_a^{(k,l)}(t, E) = \frac{C_a^{(k,l)}}{\sqrt{\mathsf{D}R_a^{(k)}\mathsf{D}R_a^{(l)}}}$$

—correlation coefficients. The parameters of the transverse distribution of particles (8.68), (8.69) coincide in shape with the corresponding parameters of the longitudinal development, and their probabilistic characteristics in the covariance approximation are given by the expressions, in which $N^{(k)}$ should be replaced by $R^{(k)}$. The difference is that the elements of the covariant matrix expend on the depth and naturally satisfy another system of equations.

To derive this system, it is convenient to use the generating functional form

$$K_a(t, E \to s(\cdot)) = \ln \mathsf{M} \exp\left\{i \int s(\mathbf{r})\mu_{a\to d}(t, E \to \mathbf{r})d\mathbf{r}\right\}. \qquad (8.70)$$

Put here

$$s(\mathbf{r}) = \sum_j s_j \mathbf{r}^j, \qquad j = 0, 1, 2,$$

then

$$\int s(\mathbf{r})\mu_{a\to d}(t, E \to \mathbf{r})d\mathbf{r} = \sum_j s_j R_a^{(j)}$$

and functional (8.70) becomes a function of the vector $\mathbf{s} = (s_0, s_1, s_2)$:

$$K_a(t, E \to \mathbf{s}) = \ln \mathsf{M} \exp\left\{i \sum_j s_j R_a^{(j)}\right\}.$$

According to (4.6) this function obeys the equation

$$\frac{\partial K_a}{\partial t} + \sigma_a = e^{-K_a(t,E\to\mathbf{s})} W_a(t, E \to e^{K.(t,\cdot\to\mathbf{s})}).$$

The further calculations are based on the equality

$$R_a^{(j)}(\boldsymbol{\vartheta}) = \int (\mathbf{r} + t\boldsymbol{\vartheta})^j \mu_{a\to d}(t, E \to \mathbf{r})d\mathbf{r} = \sum_{k=0}^{j} \binom{j}{k} (t\boldsymbol{\vartheta})^k R_a^{(j-k)}. \qquad (8.71)$$

To explain this term, let us consider the process of Coulomb scattering in more detail.

Recall that the coordinate system we have chosen is related to the direction of motion of the primary particle. Let the direction of motion of the particle change by the two-dimensional vector $\boldsymbol{\vartheta}$ (Fig. 8.8) during the process of Coulomb scattering.

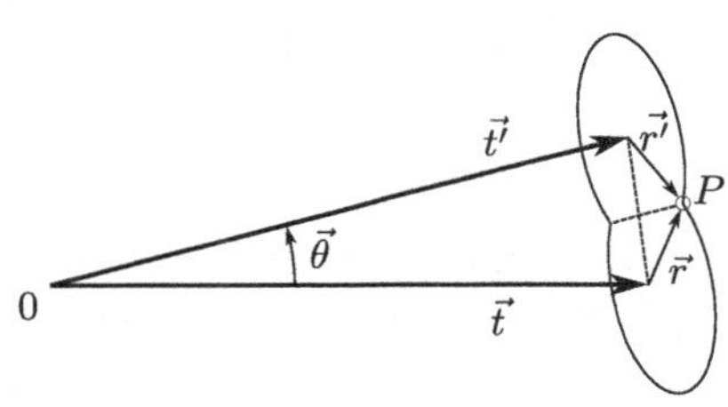

Fig. 8.8 To the discussion of the term in Eq. (8.71)

Table 8.9 RMS of the radius of the spatial distribution

t, rad un.	$\bar{r}^2_{ms}$		$\langle r^2_{ms}\rangle$	
	1	2	1	2
2	$7.91 \cdot 10^{-4}$	$8.14 \cdot 10^{-4}$	$5.05 \cdot 10^{-4}$	$6.43 \cdot 10^{-4}$
4	$2.96 \cdot 10^{-3}$	$3.17 \cdot 10^{-3}$	$2.71 \cdot 10^{-3}$	$2.94 \cdot 10^{-3}$
6	$7.24 \cdot 10^{-3}$	$7.64 \cdot 10^{-3}$	$8.43 \cdot 10^{-3}$	$8.40 \cdot 10^{-3}$
8	$1.52 \cdot 10^{-2}$	$1.49 \cdot 10^{-2}$	$1.87 \cdot 10^{-3}$	$1.87 \cdot 10^{-2}$

Notes Primary photon. $E^* = 25$ MeV, $E_\gamma = 10$ GeV. 1—Monte Carlo method, 2—SBCE method

Marking with a prime the coordinates of the observation point P in the rotated coordinate system, we obtain $\mathbf{t} + \mathbf{r} = \mathbf{t}' + \mathbf{r}'$. Projecting this equality onto the $\mathbf{t}'$ axis, we see that up to terms of the first order of smallness in the scattering angle $t \approx t'$, $\mathbf{t}' - \mathbf{t} = t\boldsymbol{\vartheta}$. For the transverse vector $\mathbf{r}' = \mathbf{r} - t\boldsymbol{\vartheta}$, so the density of the random measure in the rotated coordinate system is expressed in terms of the density in the original coordinate system by the relation $\mu'_{a\to d}(t', E \to \mathbf{r}) = \mu_{a\to d}(t, E \to \mathbf{r} - t\boldsymbol{\vartheta})$, from which follows the expression (8.71).

It remains to use the well-known properties of the generating function K:

$$\left.\frac{\partial K}{\partial s_k}\right|_{s=0} = \overline{R}^{(k)},$$

$$\left.\frac{\partial^2 K}{\partial s_k \partial s_l}\right|_{s=0} = \operatorname{cov}(R^{(k)}, R^{(l)}) \equiv C^{(k,l)},$$

to obtain from (8.71) a system of equations for the means and the covariance matrix of transverse (spatial) moments.

By SBCE solving equations for the mean values and the covariance matrix of transverse moments are found ($k = 0, 1, 2$) EPC in air ($E^* = 25$ MeV, $E \leqslant 10^5$ GeV) and iron ($E^* = 0.1$ MeV , $E = 10^3$ GeV). The results agree satisfactorily with the data of direct simulation by the Monte Carlo method according to the program described in [42] (Table 8.5). The exception is fluctuations of the second moment. The observed discrepancy can be explained by a large statistical error in estimating the dispersion of the second moment and by the well-known uncertainty of the fourth moment of the Coulomb cross section in the small-angle approximation [34, 43, 44].

The calculation results also showed that the ratio $\varkappa_a^2 = C_a^{(1,1)}/\overline{R}_a^{(2)}$ in the region of maximum cascade is practically independent of depth and primary energy (Fig. 8.9). Hence it follows that the dispersion of the first moment in this region has the form

$$\mathsf{D}R_a^{(1)} = \varkappa_a^2 (\bar{r}_{\text{rms}})_a^2 \overline{N}_a.$$

At high energies, particle fluctuations at the maximum can be neglected, and for the standard deviation of the center of gravity $\boldsymbol{\rho}_\perp$ in transversal plane, we obtain

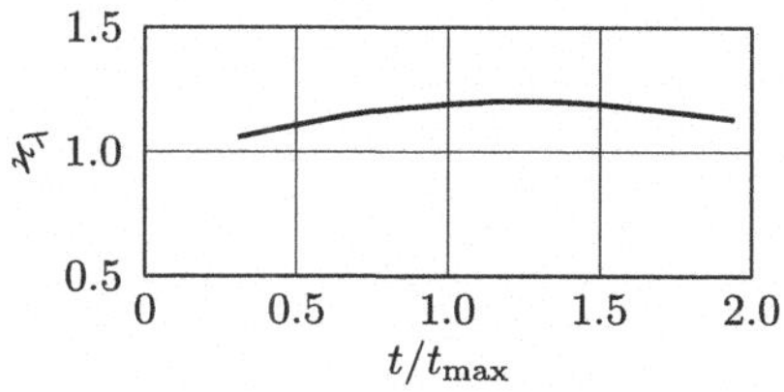

Fig. 8.9 Dependence of $\varkappa$ on the EPC depth in air (primary photon, $E^* = 25$ MeV)

$$\sigma_{c.g.} = \sqrt{\mathsf{D}\rho_\perp} = \varkappa_a(\bar{r}_{c.g.})_a \overline{N}_a^{-1/2}. \tag{8.72}$$

At high energies, particle fluctuations at the maximum can be neglected, and for the standard deviation of the center of gravity g, we obtain

$$\sigma_g = \text{const} \cdot \overline{N}^{-1/2}. \tag{8.73}$$

From (8.72), in addition, it is clear that the coefficient of proportionality in (8.73) contains a characteristic value—the root-mean-square radius of the distribution. The presence of this coefficient is an additional argument in favor of the assumption made in [39] about the statistical independence of the transverse coordinates of peripheral electrons. Indeed, in this case,

$$\mathsf{D}\rho = \mathsf{D}\left[\frac{1}{N}\sum_{i=1}^{N}\mathbf{r}_i\right] = \frac{1}{\overline{N}_a}[1 + (\delta N_a)^2](\bar{r}_{\text{ms}})_a^2,$$

which, up to a factor of $\sim$1, agrees with (8.72).

For a quantitative comparison of the calculation results with experiment [39], it was taken into account that the distributions were measured along one of the Cartesian coordinates of the observation plane. In addition, when calculating the fluctuations, the uncertainty of the shower axis introduced by the finite dimensions of the hodoscope elements is taken into account. The calculation results are in satisfactory agreement with the experimental data (Fig. 8.10). In Table 8.9, the results of calculations of the average value

$$\langle R_a^{(2)}/R_a^{(0)}\rangle \approx \left[1 + (\delta N_a)^2 - \rho_a^{(0,2)}\delta N_a \delta R_a^{(2)}\right](\bar{r}_{\text{ms}})_a^2. \tag{8.74}$$

Note that in the region of the maximum, both characteristics practically coincide, up to the maximum, random showers are the same, and behind the maximum, they are wider than the "average" cascade. From the data in Table 8.9, it can also be seen that the calculations by the formula (8.74) are in good agreement with the simulation results in the region of the maximum, where fluctuations in the number of particles are small.

On Figs. 8.11 and 8.12 the fluctuations of r_{rms}^2 and its correlation with the number of particles are given. It can be seen, in particular, that beyond the maximum of

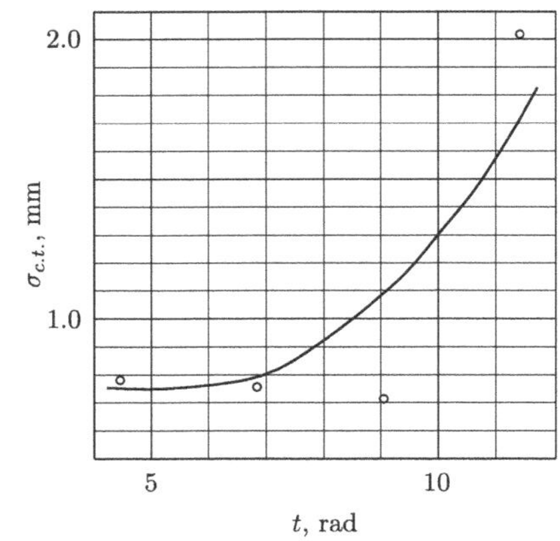

Fig. 8.10 Fluctuations of center of gravity of the transverse distribution of EPA electrons in iron, from the primary electron (approximation $E_e = 40$ GeV, $E^* = 0.1$ MeV). Points—experiment [39], curve—calculation by the formula (8.72), $\varkappa = 1.2$

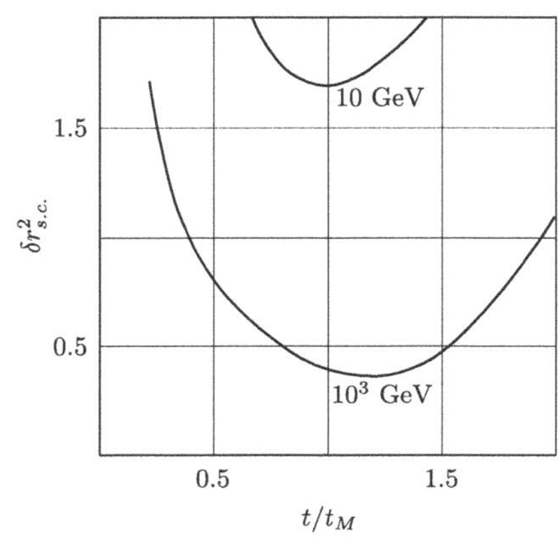

Fig. 8.11 Relative fluctuations of the EPC width from the primary photon in air ($E^* = 25$ MeV)

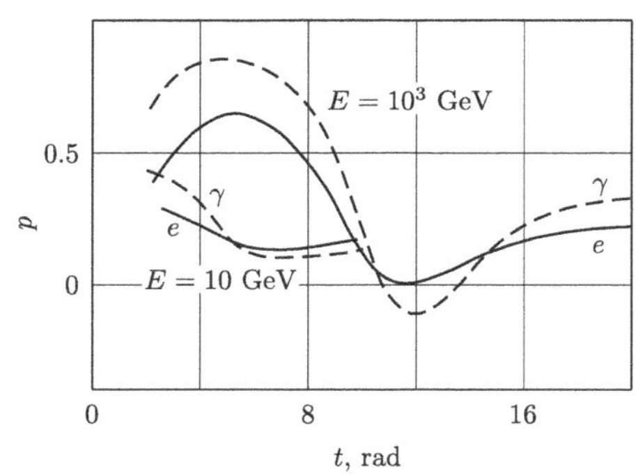

Fig. 8.12 Correlation coefficient between N and $R^2_{c.k}$ in a shower from primary photons and electrons in air ($E^* = 25$ MeV)

the cascade curve, the width of the distribution along the radius and the number of particles practically do not correlate, while up to the maximum, a positive correlation is noticeable, which increases with increasing primary energy: a larger number of particles (i.e., a more "old" shower) corresponds to a wider distribution.

The least studied is the problem of fluctuations in the spatial distribution of shower particles. For practical consideration of these fluctuations in EAS, for example, various approximate estimates are used. In [45], the random distribution density is represented as $\Phi(r) = Nf(r)$, where N is a random value of the total number of particles at the observation level, $f(r)$ is a deterministic PDF corresponding to the average distribution. In other papers, the number of particles arriving at the ΔS area is assumed to be distributed according to the Poisson law with an average value $\overline{N}(\Delta S)$. It is obvious, however, that none of these representations fully reflects the actual situation. In the first case, the relative fluctuations $N(\Delta S)$ do not depend on the position of ΔS, nor on its size, which obviously contradicts reality. The either second case leads to Poisson fluctuations of the total number of particles at the observation level, which also does not correspond to reality.

The common term "fluctuations in the density of the spatial distribution of particles" is also unsuitable. The point is that the random realization of the density is described by a generalized function (the sum of delta functions), the square of which does not exist. This seems quite obvious from a practical point of view: the density of the number of particles is the limit of the ratio $N(\Delta S)/\Delta S$ at $\Delta S \to 0$, the relative fluctuations of this quantity at a finite ΔS coincide with fluctuations $N(\Delta S)$, while the latter go to infinity as $\Delta S \to 0$.

In our paper [46], a phenomenological description of fluctuations in the spatial distribution of particles is given, which leads to agreement with the results of statistical modeling. It is based on the assumption that the transverse coordinates of particles are independent for a fixed value of their total number. If we also assume that the distribution of the transverse radius-vector of each particle $f(r)$ is independent of the total number, then we arrive at the formula $\delta^2_{N(\Delta S)} = \delta^2_N - 1/\overline{N} + 1/\overline{N}(\Delta S)$ published in [46]. At $\Delta S \to 0$ it gives Poisson fluctuations, at $\Delta S \to \infty$ (when the detector occupies the entire observation plane) it gives fluctuations of the total number of particles δ^2_N. While quite satisfactorily describing fluctuations in the muon distribution in EASs [46], this formula cannot explain the observed dependence of fluctuations of the EPC electronic component on the distance to the shower axis, in particular, in which the observed regions $\delta_{N(\Delta S)}$ (fluctuations in the number of particles arriving at ΔS) are less than fluctuations in the total number of particles at the observation level. A similar situation is also observed in EAS. We assumed that this behavior of fluctuations can be explained by fluctuations in the shape of the spatial distribution. The bulk of the shape fluctuation can be obtained by relating it to the total number at the observational level:

$$\delta^2_{N(\Delta S)} = \langle N^{2(1+\gamma)} \rangle / \langle N^{(1+\gamma)} \rangle^2 + 1/\overline{N}(\Delta S) - 1.$$

For $\gamma = 0$ this formula coincides with the previous one. They lead to the same result at the maximum of the cascade, where the fluctuations of the total number are negligibly small. At great depths $\gamma(r)$ is a monotonically decreasing function that intersects the abscissa axis and passes into the region of negative values at large r. At shallow depths (up to a maximum), $\gamma(r)$, on the contrary, increases from negative to positive values. The interpretation of this phenomenon is quite simple. Since the main source of EPC fluctuations is photon path fluctuations that shift and stretch or compress the cascade curve, then a random increase in N after the maximum indicates stretching of the cascade curve and a corresponding decrease in the S parameter, which leads to a narrowing of the spatial distribution, which partially compensates for the effect of fluctuations N at medium and large radii (the total range of electrons above the observation level decreases and the effect of electron scattering becomes smaller). A negative fluctuation, on the contrary, indicates an increase in the random age, which leads to a broadening of the spatial distribution, which again compensates for the decrease in N at medium and large distances. The result of this is a decrease in the $\delta_{N(\Delta S)}$ fluctuation compared to δ_N beyond the cascade maximum if the Poisson component of the fluctuation is small.

In the initial stage of the cascade (up to the maximum), the relationship between N and random age has the opposite character: an increase in N means an increase in age and a broadening of the spatial distribution. Here the $\delta_{N(\Delta S)}$ fluctuations significantly exceed the δ_N fluctuations in the entire region, except for small distances from the axis.

An analysis of the results of calculations by the Monte Carlo method confirmed the validity of the above considerations (Fig. 8.13).

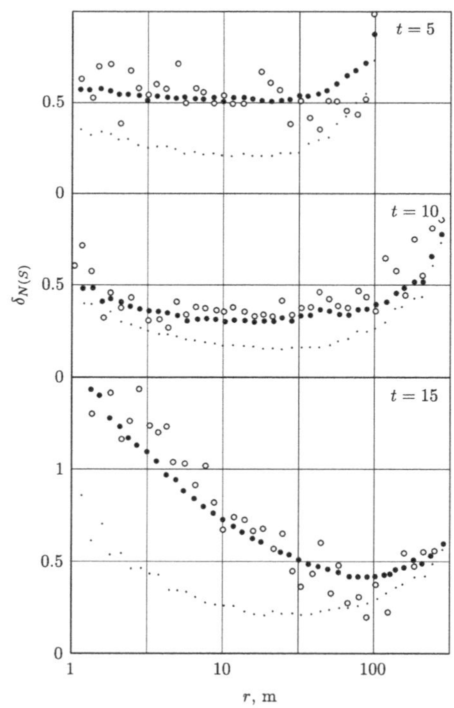

Fig. 8.13 Relative fluctuations in electron numbers in rings $r \in (10^{n/16}, 10^{(n+1)/16})$. Air, primary photon, $E = 10^3$GeV, $E^* = 1$MeV. o—Monte Carlo calculation (9 showers), •—approximation, · · · · ·—Poisson law

References

1. Dunaytsev AF, Ivanshin YI, Kakauridze DB et al (1970) Measurement of electron energy using a shower spectrometer. In: Conference on equipment in high energy physics, Dubna, pp 414–418 (in Russian)
2. Gedalin EV (1966) Cascade theory of showers. In: Nuclear interactions at high energies, Tbilisi, vol 2, pp 5–12 (in Russian)
3. Betman RG, Gedalin EV (1976) Distribution function of the number of electrons in an electromagnetic avalanche. In: Nuclear interactions at high energies, Tbilisi, vol 2, pp 13–18 (in Russian)
4. Gerasimova NM (1964) Fluctuations in the number of particles in an electron-photon shower. Proc Phys Inst Acad Sci USSR 26:192–206 (in Russian)
5. Janossy L (1955) Research on the theory of cascades. Zh Exper Teor Fiz 20(386–404):518–535 (in Russian)
6. Gerasimova NM (1979) Analysis of the results of the studies of fluctuations in electron-photon shower. In: Proceedings of the 16-th ICRC, vol 9. Kyoto, pp 216–221
7. Kolchuzhkin AM, Uchaikin VV (1975) On statistical fluctuations of detector readings in a radiation field. Izv Vyssh Uchebn Zaved Fizika 10:90–96 (in Russian)
8. Barmin VV, Varylov VG, Golubchikov VM et al (1972) Measurement of electron energy in a xenon bubble chamber by the full length shower track method, Moscow, p 12. (Preprint/Institute of Theoretical and Experimental Physics; No. 938) (in Russian)
9. Lagutin AA, Vetoshkin VV (1978) Numerical method of calculation of charged particles path fluctuations in electromagnetic cascade in infinite medium. Izv Vyssh Uchebn Zaved Fizika 4:27–30
10. Lagutin AA, Plyasheshnikov AV, Uchaikin VV (1979) On fluctuations in particle number in electromagnetic cascade. Yadernaja Fizika 30:429–436
11. Hayakawa S (1969) Cosmic ray physics. Nuclear and astrophysical aspects. Interscience Monographs and Texts in Physics and Astronomy
12. Nagel HH (1965) Electron Cascaden in Blei. Zeitschrift fur Physik Bd 186:319–346
13. Borkovsky M. Ya., Kruglov SP (1976) Calculations of the released energy and total ranges of charged particles during the development of showers in xenon. At Energy 40:243 (in Russian)
14. Okhrimenko LS, Slovinsky V, Strugalsky Z (1974) Characteristics of the longitudinal development of the electron-photon showers induced by the gamma-quanta of 30,50,70,100 and 120 MeV in liquid xenon. Preprint P1-8199. Joint Institute for Nuclear Research, Dubna (in Russian)
15. Okhrimenko LS, Slovinsky V, Strugalsky Z et al (1978) Longitudinal development of electron-photon showers caused by gamma rays with energy γ = 875–2625 MeV in xenon, Preprint R1–11980. Joint Institute for Nuclear Research, Dubna (in Russian)
16. Slovinsky B (1978) Fluctuations in electron-photon showers produced by gamma rays with an energy of 30-3400 MeV, Dubna, p 10. Preprint R1-11890. Joint Institute for Nuclear Research, Dubna (in Russian)
17. Uchaikin VV, Lagutin AA (1978) Fluctuations in the path of particles of an electron-photon shower in an infinite medium. Bull Russ Acad Sci 42:1458–1461 (in Russian)
18. Srinivasan SK et al (1961) Fluctuation problem in electromagnetic cascades. Zeitschrift fur Physik Bd 161:346–352, 1964; Bd 177:164–173
19. Uchaikin VV, Lagutin AA (1977) About fluctuations of the number of the particles produced in electron-photon shower. In: Proceedings of 15-th ICRC, vol 7. Plovdiv, pp 490–495
20. Uchaikin VV, Lagutin AA (1975) Calculation of the fluctuation of the Cherenkov radiation of an electron-photon shower. Yadernaya Fizika 21:1257–1264 (in Russian)
21. Uchaikin VV, Lagutin AA (1975) Fluctuations in the energy loss of an electron-photon shower. Bull Russ Acad Sci 39:1255 (in Russian)
22. Kokoulin RP, Petrukhin AA (1980) On possibility of a new approach to the cascade theory. Yadernaya Fizika 32:1030–1038

23. Slovinsky B, Strugalsky Z (1972) Investigation of the process of interaction of fast pions with xenon nuclei. Yadernaya Fizika 16:734 (in Russian)
24. Borkovsky MJ (1981) On the possibility of using the correlations existing in electron-photon cascades for determination of the π-meson beam composition. Nucl Instrum Methods Phys Res 189(2–3):405–407
25. Goldstein H, Wilkins Jr JE (1954) Calculations of the penetrations of gamma rays. Final Report (No. NYO-3075; NDA-15C-41). Nuclear Development Associates, Inc
26. Fano U, Spencer LV, Berger MJ (1959) Penetration and diffusion of X rays. Handbuch der Physik, Bd XXXVIII/2
27. Davison B, Sykes JB (1957) Neutron transport theory. Clarendon Press, Oxford
28. Belenkiy SZ (1948) Shower processes in cosmic rays. Gostechizdat, Moscow (in Russian)
29. Ivanenko IP (1972) Electromagnetic cascade processes. Moscow State University Press (in Russian)
30. Uchaikin VV (1978) Fluctuations in cascade processes of particle propagation through the matter. Izv Vyssh Uchebn Zaved Fizika 2:30–34 (in Russian)
31. Uchaikin VV (1978) Fluctuations in cascade processes of particle propagation through the matter II. Izv Vyssh Uchebn Zaved Fizika 2:7–12 (in Russian)
32. Uchaikin VV (1978) Fluctuations in cascade processes of particle propagation through the matter III. Izv Vyssh Uchebn Zaved Fizika 2:14–21 (in Russian)
33. Kokoulin RP, Petrukhin AA (1980) On the possibility of a new approach to cascade theory. Yadernaja Fizika 32:1030–1038 (in Russian)
34. Hayakawa S (1969) Cosmic ray physics. Nuclear and astrophysical aspects. Wiley-Interscience, New York
35. Varfolomeev AA, Drabkin LB (1970) Calculation of electromagnetic showers in lead with an initial energy of 6 GeV. Cosmic Rays 12:32–59 (in Russian)
36. Uchaikin VV, Lagutin AA, Plyasheshnikov AV (1979) On fluctuations in the number of particles in an electromagnetic cascade. Yadernaja Fizika 30:429–436 (in Russian)
37. Uchaikin VV, Lagutin AA, Vetoshkin VV, Pljasheshnikov AV (1977) The calculational techniques of the fluctionations of the electron range corresponding to finite thickness layer. In: Proceedings of 15th ICRC, vol 5. Plovdiv, pp 496–501
38. Vishnevsky NK, Lavkin AG, Lapshin VG (1978) Study of the profile of an electromagnetic shower in a gamma spectrometer on gas-discharge chambers with automatic data retrieval, Serpukhov, p 16 (Preprint/Institute of High Energy Physics, No. 78-169) (in Russian)
39. Melnikov NS, Razuvaev EI, Semenov VK (1978) The distribution of coordinates of the centers of gravity of electromagnetic showers in hodoscope spectrometers. Serpukhov, p 11. (Preprint/Institute of High Energy Physics; No. 78-144) (in Russian)
40. Budagov YA, Omelyanenko AA, Semenov AA (1984) Measuring the coordinates of the axis of an electromagnetic shower using an electrodeless drift chamber. Dubna, p 7. Preprint 13-84-757. Joint Institute for Nuclear Research, Dubna (in Russian)
41. Blocker C, Brandenburg GW, Carey R et al (1983) Intellegent towers for electromagnetic showers. Nucl Instrum Methods Phys Res 216(1–2):71–78
42. Akimov VV, Blokhintsev ID, Vorobyov KV et al (1981) Method for calculating the physical characteristics of the Gamma-1 gamma-ray telescope, Moscow, p 55 (Preprint/Inst. Space Research; Pr-684)
43. Rossi B, Greisen K (1941) Cosmic-ray theory. Rev Mod Phys 13:240
44. Rossi B (1952) High-energy particles. Prentice-Hall, New York
45. Khristiansen GV, Kulikov GV, Fomin YuA (1975) Cosmic radiation of superhigh energy. Atomizdat, Moscow (in Russian)
46. Konopelko AK, Litvinov VA, Plyasheshnikov AV et al (1986) On fluctuations in extensive air showers, Izv. Acad Sci USSR. Ser Phys 50(11):2211–2213 (in Russian)
47. Gerasimova NM (1981) Function of distribution over the particle number in electron-photon shower theory. In: Proceedings of 17-th ICRC, vol 5. Paris, pp 248–251
48. Hinkley SW, Tsokos CP (1975) Stochastic processes in particle-number fluctuations in an electron-photon shower. Acta Biotheoretica 24:58–74

49. Linde IA, Kanevsky BL (1981) Method of solution of cascade theory non-linear estimation for function of distribution over the number of particles in nuclear-electromagnetic shower. In: Proceedings of 17-th ICRC, vol 5. Paris, pp 268–271
50. Adachi A, Fujimoto Y, Ogia N et al (1964) A Monte Carlo analysis of electromagnetic cascade showers. Suppl Progr Theor Phys 32:154–192
51. Ivanenko IP, Kirillov AA, Lyutov YG (1979) Investigating the fluctuations in electron-photon cascade development. In: Proceedings of 16-th ICRC, vol 5. Kyoto, pp 252–257
52. Vankov Ch, Petrov S, Stanev T (1981) Fluctuation in the longitudinal development of electron-photon cascade in lead. In: Proceedings of 17-th ICRC, vol 5. Paris, pp 166–169
53. Fano U, Spencer LV, Berger MJ (1959) Penetration and diffusion of X rays. Springer, Berlin, Heidelberg
54. Goldstein H, Wilkins Jr JE (1954) Calculations of the penetrations of gamma rays. Final Report (No. NYO-3075; NDA-15C-41), Nuclear Development Associates, Inc
55. Davison B, Sykes JB (1957) Neutron transport theory. Clarendon Press, Oxford
56. Belenky SZ (1948) Avalanche processes in cosmic rays. Gostekhizdat, Moscow (in Russian)
57. Ivanenko IP (1972) Electromagnetic cascade processes. Press Moscow State University, Moscow (in Russian)
58. Vetoshkin VV, Uchaikin VV (1987) Moment method in the theory of cascade-process fluctuations. Soviet Phys J 30(3):195–200
59. Lagutin AA, Uchaikin VV (1998) Adjoint cascade theory in astroparticle physics. Izv Altai State Univ Press 9:4–32

For Further Reading

60. Hayakawa S (1969) Cosmic ray physics. Nuclear and astrophysical aspects. Interscience Monographs and Texts in Physics and Astronomy

Chapter 9
Cascades in Calorimeters

9.1 The Transition Effects

The most important feature of EPC development in a multilayer absorber used in calorimeters, is a sharp change in the flux of particles (electrons) near the layer boundaries—the *transition effect*. Transient effects play a significant role in a number of methods determination of particle energy. The readings of ionization chambers and calorimeters, Cherenkov and scintillation detectors are distorted by transient effects [27].

The experimental results show that the error in determining the particle energy due to the transition effect can reach 50%.

Qualitative explanation of the transition effect was given by S.N. Vernov: when a shower passes from one medium to another, ionization losses abruptly increase, that leads to a sharp change in number of electrons, mainly of low energy Experiments by S. Vernov and O. Vavilov confirmed the conclusion about the determining role of low-energetic electrons in the phenomenon [28]. The first quantitative estimates of the transition effect due to Christy and Kusaka [29] were based on assumption that the number of particles behind a two-layer barrier lead-iron is equal to their number in a homogeneous barrier of the thickness expressed in the radiation units. More detailed calculations, based on consideration of cascade equations in inhomogeneous medium, were carried out by S. Belenkiy [4] with the use of an equilibrium spectrum of particles and later by K. Pinkau [30] using energy in B-approximation of the cascade theory.

In these works, the deflection of shower particles from the original direction due to scattering was neglected. However, the results of numerical calculations electromagnetic cascades in heavy substances [5, 6, 18] showed that the low-energy electrons have a wide angular distribution. It was this circumstance that caused the discrepancies between theoretical and experimental data discussed in [14–33].

© Higher Education Press Limited Company 2024

V. V. Uchaikin, *Electron-Photon Cascades*, Nonlinear Physical Science,
https://doi.org/10.1007/978-981-99-7524-2_9

9.2 The Perturbation Theory Formulas

A common idea of the perturbation method family is based on the appropriated choice of a close problem, whose solution can be taken as an initial approximation, and the subsequent refinement is achieved by constructing an expansion in powers of the smallness parameter (in operator terminology, in powers of the perturbation operator). For example, if we should solve the problem of the development of cascade in heterogeneous medium, it would be reasonable to choose for initial approximation some adequate (i.e., with the same source but homogeneous medium). The advantage of the dual approach, described in Chaps. 3–4, is the possibility of two independent initial approximations, relating to basic and adjoint equations).

In the case of an additive detector, the adjoint vector-function contains only one (the first) non-zero component, so only diagonal term $\mathscr{L}_{11}$ left in Eqs. (4.45)–(4.48). We further assume that the properties of the medium do not depend on the physical time t, and the additive detector is integrating in time:

$$J = \int_{\theta}^{\infty} \bar{q}(\theta, T)dT.$$

Integrating (4.45) over T yields

$$\Delta J \equiv \int_{\theta}^{\infty} \Delta\bar{q}(\theta, T)dT = \int \bar{f}'^{+}_{1}(x)\Delta\mathscr{L}_{11}\bar{f}_1(x)dx, \tag{9.1}$$

where $\bar{f}_1(x) = \int_{\theta}^{\infty} f_1(t, x)dt$ and $f'^{+}_{1}(x) = \int_{\theta}^{\infty} f'^{+}_{1}(t, x)dt$ are solutions of stationary (basic and adjoint) equations $-\mathscr{L}_{11}\bar{f}_1 = f_{1\theta}$, $\mathscr{L}'^{+}_{11}\bar{f}'^{+}_{1}(t, x) = g_1$, following from (3.33) and (4.29) after time integration.

If we now pass from the particle density $\bar{f}_1$ to the particle flux density Φ and define the stationary kinetic operator L by the relation $L\Phi = -\mathscr{L}_{11}f_1$, we get from (9.1) the usual formula for the functional change due to the replacement of medium A by medium B:

$$\Delta J^{\mathrm{BA}} = J^{\mathrm{B}} - J^{\mathrm{A}} = -\int \Phi^{\mathrm{B}+}(x)\Delta L^{\mathrm{BA}}\Phi^{\mathrm{A}}(x)dx, \tag{9.2}$$

where Φ^{A} and $\Phi^{\mathrm{B}+} \equiv \bar{f}'^{+}_{1}$ satisfy equations

$$L^{\mathrm{A}}\Phi^{\mathrm{A}} = S, \tag{9.3}$$

$$L^{\mathrm{B}}\Phi^{\mathrm{B}} = D, \tag{9.4}$$

and $S = f_{1\theta}$ and $D = g_1$ are the source density and the detector response function, respectively.

The formulas will also take a similar form (4.46)–(4.48):

$$\Delta J^{\mathrm{BA}} = -\int \Phi^{\mathrm{A}+} \Delta L^{\mathrm{BA}} \Phi^{\mathrm{B}} dx,$$

$$\Delta J^{\mathrm{BA}} = -\int \Phi^{\mathrm{A}} \Delta L^{\mathrm{BA}+} \Phi^{\mathrm{B}+} dx,$$

$$\Delta J^{\mathrm{BA}} = -\int \Phi^{\mathrm{B}+} \Delta L^{\mathrm{BA}+} \Phi^{\mathrm{A}+} dx.$$

If the perturbation is small, then in these formulas one can replace Φ^{B} by Φ^{A} and $\Phi^{\mathrm{B}+}$ by $\Phi^{\mathrm{A}+}$:

$$\Delta J^{\mathrm{BA}} = -\int \Phi^{\mathrm{A}+} \Delta L^{\mathrm{BA}} \Phi^{\mathrm{A}} dx = -\int \Phi^{\mathrm{A}} \Delta L^{\mathrm{BA}+} \Phi^{\mathrm{A}+} dx. \tag{9.5}$$

Comparing the small disturbance formula (9.5) with the reading expression detector through source density $J^{\mathrm{A}} = \int \Phi^{\mathrm{A}+}(x) S(x) dx$, one can interpret increment J^{BA} of the indication given detector as the indication of this detector in environment A, due to "source" $-\Delta L^{\mathrm{BA}} \Phi^{\mathrm{A}}$ equivalent to the perturbation under consideration.

We restrict ourselves further to consideration of the plane geometry, when the flux density depends only on one of spatial coordinates (z), the angle ϑ between the z axis and the direction of particle motion $\boldsymbol{\Omega}$ and its energies E: $\Phi(x) = \Phi_a(z, \cos\vartheta, E)$, $\Phi^+(x) = \Phi_a^+(z, \cos\vartheta, E)$, $a = e, \gamma$. By writing these functions as vector functions

$$\Phi = \begin{pmatrix} \Phi_e \\ \Phi_\gamma \end{pmatrix}, \qquad \Phi^+ = (\Phi_e^+, \Phi_\gamma^+),$$

(9.3), (9.4) can be interpreted as matrix operator equations

$$L = \begin{pmatrix} L_{ee} & L_{e\gamma} \\ L_{\gamma e} & L_{\gamma\gamma} \end{pmatrix} = \begin{pmatrix} \cos\vartheta\,\partial_z - \partial_E \beta + L'_{ee} & L_{e\gamma} \\ L_{\gamma e} & \cos\vartheta\,\partial_z + L'_{\gamma\gamma} \end{pmatrix}.$$

Here L_{ee} is an operator describing the change in the state of electrons in bremsstrahlung processes and elastic scattering; $L_{e\gamma}$ describes the production of electrons in Compton scattering processes and formation of pairs; $L_{\gamma e}$ describes electron bremsstrahlung, and $L'_{\gamma\gamma}$ describes Compton scattering and photoabsorption of photons. In the full screening approximation, the operators L'_{ee} and $L_{\gamma e}$, calculated per radiation unit of length do not depend on the atomic number of the substance so that the difference kinetic operators for two different media A and B have the form

$$\Delta L^{\mathrm{BA}} = \begin{pmatrix} -\partial_E \Delta\beta^{\mathrm{BA}} & \Delta L^{\mathrm{BA}}_{e\gamma} \\ 0 & \Delta L^{\mathrm{BA}}_{\gamma\gamma} \end{pmatrix}. \tag{9.6}$$

9.3 Adjoint Functions for Transition Effect

We consider EPC passage through a plane boundary $t = 0$, separating two media 1 ($t < 0$) and 2 ($t > 0$) (rad units).

Let A be a homogeneous medium of type 1, and B be the inhomogeneous medium under consideration. Then

$$\Delta L^{\mathrm{BA}} = \begin{cases} 0, & t < 0; \\ \Delta L^{21}, & t > 0 \end{cases}$$

and the formula (9.2) for the increment of the detector readings caused by the replacement in right half-space of medium 1 onto medium 2, taking into account (9.6), takes the form

$$\Delta J = \int\limits_0^\infty dt \int d\Omega \int dE \left\{ \Phi_e^{\mathrm{B}+} \left[\partial_E(\Delta\beta\Phi_e^{\mathrm{A}}) - \Delta L_{e\gamma}\Phi_\gamma^{\mathrm{A}} \right] - \Phi_\gamma^{\mathrm{BA}} \Delta L'_{\gamma\gamma}\Phi_\gamma^{\mathrm{A}} \right\}. \tag{9.7}$$

Indices BA at ΔJ and 21 will temporarily be dropped. Thus, the perturbation under consideration is equivalent to the appearance in a heterogeneous B medium electron sources distributed over the right half-space with the density

$$\Delta S_e = \partial_E(\Delta\beta\Phi_e^{\mathrm{A}}) - \Delta L_{e\gamma}\Phi_\gamma^{\mathrm{A}} \tag{9.8}$$

and photon sources with the density

$$\Delta S_\gamma = -\Delta L'_{\gamma\gamma}\Phi_\gamma^{\mathrm{A}}.$$

When studying the energy–angular distribution of electrons near the boundary, we will assume detector measuring differential flux electron density in the plane $t = \mathrm{const}$, i.e., its response function

$$D_a(t', \Omega', E') = \delta_{ae}\delta(t - t')\delta(E - E')\delta(\cos\vartheta - \cos\vartheta')/2\pi.$$

Here and below, the primed variables belong to adjoint functions Φ^+, but without primes to the flux density Φ. Knowing that the main change in shower characteristics occurs at very small ($\sim 10^{-1}-10^{-2}$ rad. units) distances from the boundary, the quantity (9.7) can be calculated in *continuous slowing down* approximation, disregarding scattering . An explicit expression for this function can be obtained both by solving the adjoint equation, and directly, guided by their physical meaning. For greater clarity, we use the latter method.

Since the Coulomb scattering is not taken into account, the adjoint function contains factor

$$\delta(\cos\vartheta - \cos\vartheta')/2\pi$$

and becomes zero, if particles move to side opposite to the observation plane (i.e., for combinations $t' < t$, $\cos\theta < 0$ and $t' > t$, $\cos\vartheta > 0$. This condition can be written as

$$(t - t')/\cos\vartheta > 0 \tag{9.9}$$

and taken into account by factor $1(t - t')/\cos\vartheta)$. Energy distribution of electrons passing through the layer (t', t) at angle ϑ and having energy E' at the point t' is represented via delta-function $\delta[E - \mathscr{E}(t', E'; t, \cos\vartheta)]$ where $\mathscr{E}(t', E'; t, \cos\vartheta)$—electron energy at the point t provided that it moves at angle ϑ to the axis and at the point t' had energy E'. Given that intersection of a plane by one particle at an angle ϑ to the normal gives a contribution to the flow equal to $1/|\cos\vartheta|$, finally, we get

$$\Phi_e^+(t', \cos\vartheta', E') == \delta[E - \mathscr{E}(t', E'; t, \cos\vartheta)]\frac{\delta(\cos\vartheta - \cos\vartheta')}{2\pi|\cos\vartheta|}1_+\left(\frac{t - t'}{\cos\vartheta}\right). \tag{9.10}$$

To connect the function $\mathscr{E}$ with physical characteristics of the medium, we introduce into consideration the total electron path in an infinite medium:

$$\tau = R(E) = \int_0^E \frac{dE'}{\beta(E')}. \tag{9.11}$$

Being a continuous monotonically increasing function, $R(E)$ has an inverse function

$$E = R^{-1}(\tau) \tag{9.12}$$

with derivative

$$\frac{dR^{-1}}{d\tau} = \left(\frac{dR}{dE}\right)^{-1} = \beta(E). \tag{9.13}$$

Equation (9.12) gives the dependence of the total electron energy on its path, and Eq. (9.13) expresses energy loss per unit length. If between the points t' and t there is a layer of homogeneous matter intersected electron at an angle ϑ to the normal, then the relationship between the energy at the input E' and at the output $\mathscr{E}$ is given by the equation

$$\int_{\mathscr{E}}^{E'} \frac{dE}{\beta(E)} = \frac{t - t'}{\cos\vartheta}. \tag{9.14}$$

According to (9.9), this value is positive. According to (9.11), the Eq. (9.14) can be represented as

$$R(E') - R(\mathscr{E}) = \frac{t - t'}{\cos\vartheta} \tag{9.15}$$

or

$$\mathscr{E} = R^{-1}\left[R(E') - \frac{t - t'}{\cos\vartheta}\right]. \tag{9.16}$$

Let now $t' > 0, t \leqslant 0$, and through the beginning coordinates, there is a flat boundary separating medium 1 from medium 2 with different losses β_1 and β_2. Then the energy of a particle crossing the boundary in the opposite ($\cos\vartheta < 0$) direction is

$$\mathscr{E}_0 = R_2^{-1}[R_2(E') + t'/\cos\vartheta], \tag{9.17}$$

and at the point $t < 0$

$$\mathscr{E}_- = R_1^{-1}[R_1(\mathscr{E}_0) - t/\cos\vartheta]. \tag{9.18}$$

Under the same assumptions (i.e., neglecting particle collisions) the conjugate function Φ_γ^+ in calculations of the electronic shower components should be set equal to zero.

9.4 Flux Perturbation by Boundaries

The perturbation of the differential electron flux density caused by replacement in the right half-space of medium 1 by medium 2—transition effect—in the considered approximation is equal to

$$\Delta\Phi_e(t, \cos\vartheta, E) = \int_0^\infty dt' \int_{4\pi} d\Omega' \int_E^{E_0} dE' \Phi_e^+(t', \cos\vartheta', E') \Delta S_e(t', \cos\vartheta', E'),$$

where Φ_e^+ is defined by the formula (9.10); ΔS_e is expression (9.8), and E_0 is the maximum energy in the spectrum source. After integrating over the angles, we get

$$\Delta\Phi_e(t, \cos\vartheta, E) = \frac{1}{|\cos\vartheta|} \int_0^\infty dt' \int_E^{E_0} dE' \delta[E - \mathscr{E}(t', E'; t, \cos\vartheta)]$$
$$\times\ 1_+\left(\frac{t - t'}{|\cos\vartheta|}\right) \Delta S_e(t', \cos\vartheta', E'). \tag{9.19}$$

This shows that for $t < 0$ and $\cos\vartheta > 0$ the integrand the expression is equal to zero, i.e., the transient effect in the direct particle flow is absent in the left half-space, and the angular distribution in this region coincides with the unperturbed one.

In the right half-space for positive directions, the integrand vanishes for $t' > t$, so

$$\Delta\Phi_e(t,\cos\vartheta,E) = \frac{1}{\cos\vartheta}\int_0^1 dt' \int_E^{E_0} dE'\delta[E - \mathscr{E}(t',E';t,\cos\vartheta)]\Delta S_e(t',\cos\vartheta,E').$$

Passing in the inner integral to the variable of integration

$$\mathscr{E} = R_2^{-1}\left[R_2(E') - \frac{t-t'}{\cos\vartheta}\right] \tag{9.20}$$

and finding from (9.13) and (9.15) the differential

$$dE' = \frac{\beta_2(E')}{\beta_2(\mathscr{E})}d\mathscr{E}, \tag{9.21}$$

we get

$$\Delta\Phi_e(t,\cos\vartheta,E) = \frac{1}{\cos\vartheta}\int_0^t dt' \int_{\mathscr{E}_1}^{\mathscr{E}_2} d\mathscr{E}\frac{\beta_2(E'_{\mathscr{E}})}{\beta_2(\mathscr{E})}\delta(E-\mathscr{E})\Delta S_e(t',\cos\vartheta,E'_{\mathscr{E}}), \tag{9.22}$$

where

$$E'_{\mathscr{E}} = R_2^{-1}\left[R_2(\mathscr{E}) + \frac{t-t'}{\cos\vartheta}\right],$$

$$\mathscr{E}_1 = R_2^{-1}\left[R_2(E) - \frac{t-t'}{\cos\vartheta}\right],$$

and

$$\mathscr{E}_2 = R_2^{-1}[R_2(E_0) - \frac{t-t'}{\cos\vartheta}].$$

The inner integral in (9.22) is non-zero and equals $[\beta_2(E'_E)/\beta_2(E)]\Delta S_e(t',\cos\vartheta, E'_E)$ only if the integration segment covers the singular point of the δ-function $\mathscr{E} = E$, i.e., if

$$R_2^{-1}\left[R_2(E) - \frac{t-t'}{\cos\vartheta}\right] < E < R_2^{-1}\left[R_2(E_0) - \frac{t-t'}{\cos\vartheta}\right]. \tag{9.23}$$

Transforming these inequalities into the form

$$0 < \frac{t - t'}{\cos \vartheta} < R_2(E_0) - R_2(E) \equiv R_2(E_0, E),$$

we make sure that the execution of the left part is provided by the condition (9.9), and the right inequality holds only for

$$t' > t - R_2(E_0, E) \cos \vartheta. \tag{9.24}$$

The value $R_2(E_0, E)$ is the path of the particle when it slows down in medium 2 from the energy E_0 (maximum energy in the spectrum of the source) to E. The condition (9.24) means that if the length of the segment $(t - t')/\cos \vartheta$ exceeds the length of the run $R_2(E_0, E)$, then among the particles with energy E and the point t there will be none coming from the point t'. Thus, the lower limit on t' in the integral (9.22) should be replaced by $t_1 = \max\{0, t - R_1(E_0, E) \cos \vartheta\}$

$$\Delta\Phi_e(t, \cos \vartheta, E) = \frac{1}{\cos \vartheta \beta_2(E)} \int_{t_1}^{t_2} \beta_2(E'_E) \Delta S_e(t', \cos \vartheta, E'_E) dt'. \tag{9.25}$$

For particles moving in the opposite direction ($\cos \vartheta < 0$), the integrand of the function in (9.19) is nonzero only for $t' > t$, so

$$\Delta\Phi_e(t, \cos \vartheta, E) = \frac{1}{|\cos \vartheta|} \int_t^{\infty} dt' \int_E^{E_0} dE' \delta E' \delta[E - E(t', E'; t, \cos \vartheta)] \Delta S_e(t', \cos \vartheta, E').$$

As can be seen from this formula, $\Delta\Phi_e$ does not vanish for negative t, i.e., the transition effect in the reverse flow of particles is observed on both sides of the boundary. Consider first the right region ($t > 0$). Passing from E' to a new variable integrating E by the formulas (9.20), (9.21) and taking into account that from (9.23) in this case follows $t' < t - R_2(E_0, E)|\cos \vartheta|$, we get

$$\Delta\Phi_e(t, \cos \vartheta, E) = \frac{1}{|\cos \vartheta| \beta_2(E)} \int_t^{t + R_2(E_0, E)|\cos \vartheta|} \beta_2(E'_E) \Delta S_e(t', \cos \vartheta, E'_E) dt'. \tag{9.26}$$

Formulas (9.25) and (9.26) can be combined into one. Introducing a new variable integration $\tau = (t - t')/\cos \vartheta$, we write the transition effect formula in right half-space for an arbitrary direction ϑ:

$$\Delta\Phi_e(t, \cos \vartheta, E) = \frac{1}{\beta_2(E)} \int_0^{b_1} \beta_2(E_\tau) \Delta S_e(t - \tau \cos \vartheta, \cos \vartheta, E_\tau) d\tau, \tag{9.27}$$

where

$$b_1 = \begin{cases} \min\{t/\cos\vartheta, R - 2(E_0, E)\}, & \cos\vartheta > 0; \\ R_2(E_0, E), & \cos\vartheta < 0; \end{cases}$$

$$E_\tau = R_2^{-1}[R_2(E) + \tau]. \tag{9.28}$$

Finally, substituting (9.8) into (9.27), we arrive at formula (2) from [1]

$$\Delta\Phi_e(t, \cos\vartheta, E) = \frac{1}{\beta_2(E_\tau)} \int\limits_0^{b_1} \beta_2(E_\tau) \left\{\partial_{E_\tau} \left[\Delta\beta(E_\tau)\Phi_e^A(t - \tau\cos\vartheta, \cos\vartheta, E_\tau)\right.\right.$$

$$\left.\left. - \Delta L_{e\gamma}\Phi_\gamma^A(t - \tau\cos\vartheta, \cos\vartheta, E_\tau)\right]\right\} d\tau. \tag{9.29}$$

For $\cos\vartheta > 0$ and $t/\cos\vartheta \to 0$ it becomes the result of [2], describing the effect in the approximation linear in t and valid at very small distances from the boundary, but neglecting scattering and the energy dependence of β and assuming continuous deceleration of the unperturbed electron flow into the formula used by Pinkau [3]. The transition effect to the left of the boundary can be calculated in a similar way, only when passing from E' to $\mathcal{E}$ instead of (9.16), you should use the formulas (9.17), (9.18):

$$\Delta\Phi_e(t, \cos\vartheta, E) = \frac{\beta_1(\mathcal{E}^0)}{\beta_1(E)} \Delta\Phi_e(0, \cos\vartheta, \mathcal{E}^0), \quad t < 0, \quad \cos\vartheta < 0. \tag{9.30}$$

Here $\mathcal{E}^0 = R_1^{-1}[R_1(E) + t/\cos\vartheta]$ is the energy which the particle had when crossing the boundary, if at the point $t < 0$ it has energy E and moves at an angle $\vartheta > \pi/2$ to the axis. Meaning, that the perturbation of the flow at the boundary included in the formula (9.30) can be determined as the limit of the solution in the right half-space

$$\Delta\Phi_e(0, \cos\vartheta, E') = \lim_{t \to +0} \Delta\Phi(t, \cos\vartheta, E').$$

In what follows, we confine ourselves to writing formulas only for the region of perturbation. Disturbance the flow outside this area is found as a flow from the surface "source", equal to the perturbation of the flow at the boundary (the perturbation "leaves" from the perturbing layer and propagates to the left half as ordinary particles, but the sign $\Delta\Phi$ can also be negative).

9.5 Belenky's Comment of the Transition Effect

The more we refine the angular and energy distribution in an undisturbed particle flow, the further we go from the original theory of the transition effect proposed by the

Soviet physicist S. Z. Belenkiy in his book [4]. Naturally, the restrictions he adopted (mainly ignoring the wide angular distribution of shower particles) prevented him from obtaining a correct (in numerical terms) description of this effect, but both the formulation of the problem, the indications of the mathematical way to solve it, and the physical comments made by him in this book retain its significance even now, when the one-dimensional is in the direction of a more adequate reflection of reality, although it has not lost its pedagogical purpose.

Let's turn to chapter 6 of this book, which deals with transition effects. Noting that in cosmic radiation, along with particles of extremely high energies, the author points to a very important role played by particles with energies of $10^6 \div 10^7$ eV, which are the product of degradation of the energy of primary particles, and although in numerical terms the fraction of such particles in the shower is small. Their flows are very sensitive to local changes in the properties of the medium even in relatively thin layers of matter (walls of ionization chambers, counters, scintillation plates, X-ray films, gas gaps in calorimeters).

The chapter begins with consideration of the transition of an electron–photon shower from medium I to medium II through the plane of separation of these media perpendicular to the direction of its axis. It is assumed that equilibrium spectra of both shower components are established in medium I, and with these distributions they enter a plane layer of medium II, keeping the direction of motion parallel to the shower axis. Assuming, the layer is sufficiently thin, the author takes the equilibrium distributions being valid for the infinite boundless medium I as a zero-approximation for electron–photon distributions in the thin layer of a matter II, which may be quite different from the first one in its density, atomic, and order numbers.

The first approximation is found from the same system of cascade equations under assumptions that (1) the spectrum of photons over the considered layer of the second substance does not have time to change significantly, i.e., $t < q/\sigma_\gamma$, and (2) the electron spectrum also does not change noticeably, i.e., $t < 1$. With these, frankly speaking, rather modest assumptions, the author was able to draw some pretty important conclusions from these calculations.

When passing from a layer with a higher atomic number Z to a layer with a lower number, the relative role of ionization losses increases in comparison with radiative drag and pair formation, so the number of particles decreases. On passing from a layer with a smaller Z to a layer with a larger number of particles, the number of particles will increase, since their absorption (per radiation unit) will become smaller.

This example shows the role of ionization losses in the interpretation of transient effects. Indeed, when writing cascade equations in radiation units of length, only ionization losses depend significantly on the physical properties of the substance, the remaining terms of cascade equations make smaller contributions and even opposite signs.

For example, the effect of the Compton effect on the development of a cascade is inversely proportional to Z, and a relative increase in its cross section on going to a substance with a smaller Z will cause, in contrast to ionization losses, an increase in the number of particles. This is understandable, the author writes, since the Compton effect is associated with the appearance of charged particles, and not with their absorption. A change in the pair formation cross section acts in the same direction as the Compton effect, but plays a smaller role. Therefore, if there were no ionization losses (well, and also Compton scattering with the formation of pairs, to a lesser extent related to the atomic number of the medium in which the cascade develops), the transition effect would be imperceptible.

9.6 Transition Effect at Shower Maximum

The formula (9.29) is greatly simplified if we neglect the dependence of the unperturbed functions Φ_e^A and Φ_γ^A on depth. This can be done, in particular, when the interface falls on the maximum of the cascade curve in a homogeneous unperturbed medium A. The first term in (9.29) describing the perturbation of the electron flow due to the difference in ionization losses, is converted taking into account (9.28) as follows:

$$\Delta\Phi_e^{\mathrm{I}}(t, \cos\vartheta, E) = \frac{1}{\beta_2(E)} \int_0^{b_1} \partial_{E_\tau} \left[\Delta\beta(E_\tau)\Phi_e^A(\cos\vartheta, E_\tau)\right] \beta_2(E_\tau) d\tau$$

$$= \frac{1}{\beta_2(E)} \int_E^{E_{b_1}} \partial E' \left[\Delta\beta(E')\Phi_e^A(\cos\vartheta, E')\right] dE'$$

$$= \frac{1}{\beta_2(E)} \left[\Delta\beta(E_{b_1})\Phi_e^A(\cos\vartheta, E_b) - \Delta\beta(E)\Phi_e^A(\cos\vartheta, E)\right]. \tag{9.31}$$

Numerical calculations show that this is the main (and, in approximation B, the only) factor determining the magnitude of the transition effect. Second term

$$\Phi_e^{\mathrm{II}}(t, \cos\vartheta, E) = \frac{1}{\beta_2(E)} \int_E^{E_{b_1}} \Delta L_{e\gamma} \Phi_\gamma^A(\cos\vartheta; E') dE', \tag{9.32}$$

describing the effect of changing the cross section for the production of electrons by photons plays a small role and can be calculated approximately. Summing up (9.31), (9.32), we get

$$\Phi_e(t, \cos\vartheta, E) = \Phi_e^{\mathrm{I}}(t, \cos\vartheta, E) + \Phi_e^{\mathrm{II}}(t, \cos\vartheta, E). \tag{9.33}$$

From Eqs. (9.31) to (9.33), neglecting scattering and the dependence of β on energy follows the formula of S. Z. Belenky for the transition effect in the electron component of the shower [4].

To show the main regularities of the transformation of the energy and angular distributions particles when passing through the interface between two media, we present the results of calculations using the formulas (9.31)–(9.33) for the lead–copper combination in the region of the shower maximum formed by electron with energy $E_0 = 316$ MeV. As unperturbed differential flux density electrons, the data of [5] were used. The corresponding data for photons were taken from [6]. The calculation of the integral (9.32) was carried out under the assumption that the direction of motion secondary electrons coincides with the direction of the photon momentum, and their energy is equal to the average the energy lost by a photon in a collision.

Figure 9.1 shows the evolution of the differential energy spectrum of electrons

$$\Phi_e^B(t, E) = 2\pi \int_{-1}^{1} \Phi_e^B(t, \cos\vartheta, E) d\cos\vartheta$$

when crossing the interface Pb–Cu. At a distance of 0.1 rad. units to the left of the border the spectrum in an inhomogeneous medium is still close to the spectrum of particles at the shower maximum in lead. As approaching the border, the difference between them constantly increases. In the region $-0.01 < t < 0.01$, there is a sharp drop in the low-energy part of the spectrum, caused by a jump change in the value of ionization losses at the boundary. Note that the absolute change in the number of particles with an energy of 0.1 MeV is twice the change in the spectrum at $E = 0.5$ MeV and five times at $E = 1$ MeV. The change in the spectrum at $E = 10$ MeV is already quite small, which is in full agreement with the assumptions about the determining role of low-energy electrons in the transition effect made in [4, 7]. It also follows from this that very important for the analysis of the transition effect is to take into account the dependence β from energy: approximation $\beta = \text{const}$ can lead

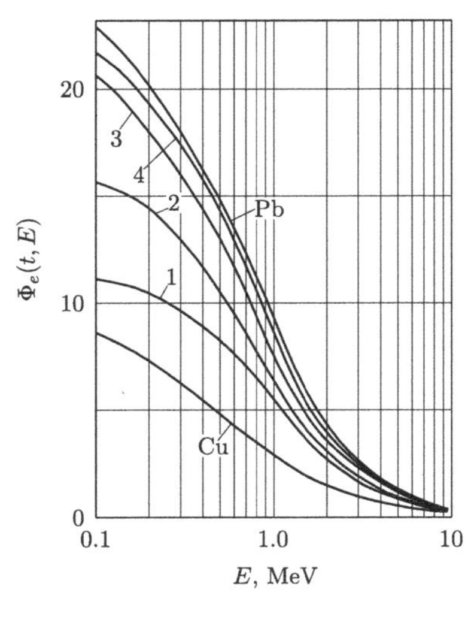

Fig. 9.1 Differential energy spectrum of shower electrons near borders Pb–Cu. 1—$t = 0, 01$; 2—0; 3—$-0, 01$; 4—$t = -0, 1$ [8]

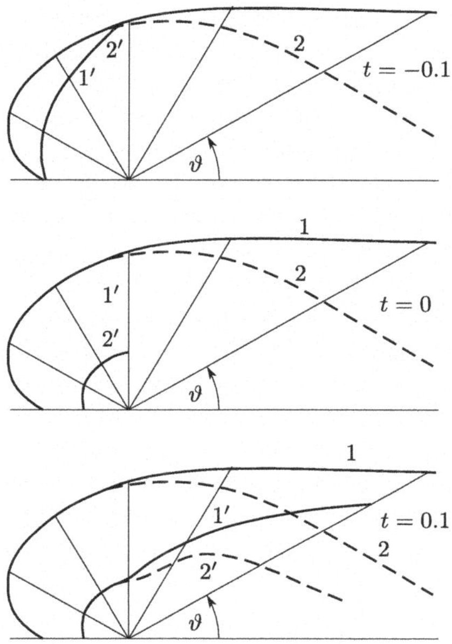

Fig. 9.2 The angular distribution of shower electrons near the Pb–Cu interface in polar coordinates

to significant distortions of the transition effect. Of interest is the dependence of the rate of spectrum change on the distance to the boundary. In particular, the change in the number of particles with an energy of 0.2 MeV at distances of 0.1–0.01 rad. units from the border is 3–5 times less than the change in this number at smaller distances. At the same time, the number of electrons, say, with an energy of 0.5 MeV, changes to a lesser extent degree, i.e., the influence of an equivalent source extends over long distances. A sharp change in the electron flux near the boundary ($t \lesssim 0.01$) and a slow—at large distances is a characteristic feature of the transition effect.

As we move away from the boundary into the right half-space, the electron spectrum approaches to the equilibrium spectrum in copper.

Consider the influence of the boundary on the angular distribution of particles:

$$\Phi_e^{\mathrm{B}}(t, \cos\vartheta) = \int_0^{E_0} \Phi_e^{\mathrm{B}}(t, \cos\vartheta, E)dE. \tag{9.34}$$

Figure 9.2 shows polar diagrams of angular distributions electrons at points $t = -0.1$; 0 and 0.1. Curve 1 shows the angular distribution of electrons (9.34) in homogeneous lead (unperturbed medium), curve 2 is the same, for electrons with energies less than 10 MeV:

$$\Phi_e^{\mathrm{B}}(t, \cos\vartheta, E < 10\,\mathrm{MeV}) = \int_0^{10\mathrm{MeV}} \Phi_e^{\mathrm{B}}(t, \cos\vartheta, E)dE.$$

The indices 1′ and 2′ mark the corresponding curves in an inhomogeneous medium. From the data in Fig. 9.2, as well as directly from Eqs. (9.31) to (9.33), the following regularities emerge. In the left half-space, the perturbation affects only the reverse current of particles ($\cos\vartheta < 0$), forward current in this region does not depend on the medium in the right half-space and coincides with the unperturbed one. At the interface itself, the perturbation in the direct particle current is still absent, and the perturbation in the reverse has reached the limiting value, therefore, for $\vartheta = \pi/2$, the angular distribution at this point has a discontinuity. In the right half-space, the reverse current of particles remains unchanged, and the angular distribution of direct current particles sharply narrows, which leads to the disappearance of the discontinuity. The resulting angular particle distribution in copper is more forward compared to the unperturbed distribution in lead. From this, it is clear that it is an important factor affecting the accuracy of the calculation of the transient effect which is the inclusion of scattering in an unperturbed medium, with a special role played by the angular distribution of the inverse particle current ($\vartheta > \pi/2$). Neglecting scattering, the transition effect in the left region and on the boundary itself turns out to be equal to zero, which is fundamentally wrong.

9.7 Total Particle Flux at a Boundary

The most important characteristic of the transition effect is the behavior near the interface of the total flow of particles (electrons)

$$\Phi_e^{\mathrm{B}}(t) = 2\pi \int_{-1}^{+1} d\cos\vartheta \int_{0}^{E_0} dE\,\Phi_e^{\mathrm{B}}(t, \cos\vartheta, E),$$

which can be (and has been) explored in experiments. Measurement results are usually presented as a ratio

$$K(t) = \Phi_e^{\mathrm{B}}(t)/\Phi_e^{\mathrm{A}}(t),$$

called the transition effect coefficient (or simply the transition effect). In our notation

$$K(t) = 1 + \Delta\Phi_e(t)\Phi_e^{\mathrm{A}}(t). \tag{9.35}$$

Let us now continue the investigation begun in the previous section. Integrating the differential perturbations $\Delta\Phi_e(t, \cos\vartheta, E)$ in angles and energies, we obtain a perturbation of the total electron flow $\Delta\Phi_e(t)$ and we can determine the coefficient (9.35). Recall that, according to (9.33), the perturbation $\Delta\Phi_e$ consists of two terms $\Delta\Phi_e = \Delta\Phi_e^{\mathrm{I}} + \Delta\Phi_e^{\mathrm{II}}$, due to the perturbation of ionization losses and the perturbation of the cross section for the production of electrons by photons, respectively. As noted in [4], these processes compete: an increase in β upon transition from lead to

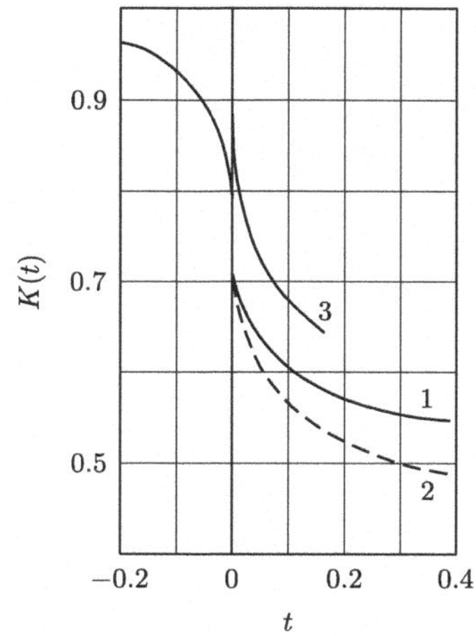

Fig. 9.3 Transition effect in the total electron flow near the boundary Pb–Cu. 1—calculation without taking into account the contribution of photons, 2—taking into account this contribution, 3—according to the Belenky formula [8]

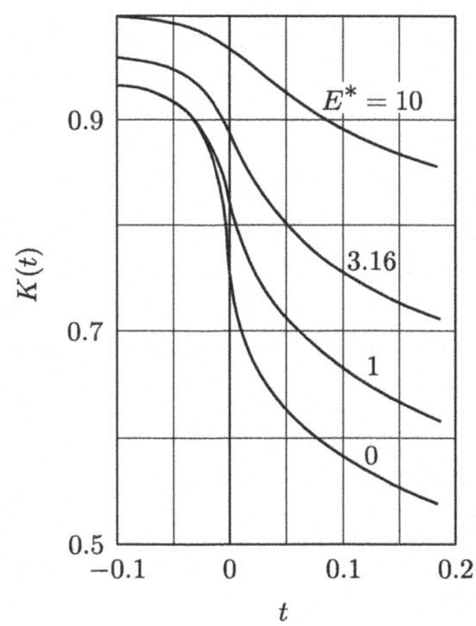

Fig. 9.4 Transition effect in the integrated spectrum near the boundary Pb–Cu

copper leads to a decrease in the electron flux, and an increase in the cross sections for the interaction of photons with matter is accompanied by an increase number of secondary electrons. However, the influence of the latter process at small distances from the boundary is small, it begins affect only when $t \sim 0.05$. The reverse current of photons in lead is small, and their influence on the transient effect to the left of the border does not exceed 1%. Figure 9.3 shows the transient effect in the total electron flow at the lead–copper interface. We note that on the boundary $K(0) = 0.78$—this is the contribution of the reverse particle current to the transition effect. Fig. 9.4 gives the dependence of the transition effect of the threshold energy E^*, if in (9.35) we mean by Φ_e the flux of particles with energies above E^*. It is seen, that low-energy electrons also play the main role here; in particular, the contribution of electrons with energies less than 1 MeV is ∼10%.

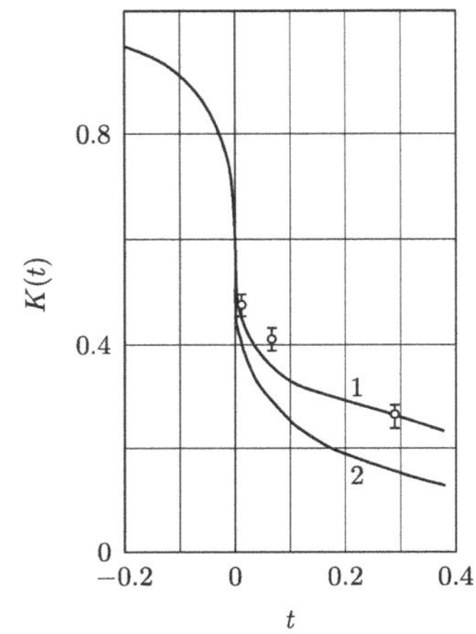

Fig. 9.5 Transition effect to the right of the border Pb–plexiglass. Dots—experiment, solid line—calculation [9]

9.8 Scintillation and Ionization Chambers

Scintillation detector and ionization chamber used for the experimental study development of showers under conditions of flat geometry, are themselves sources of perturbations if they consist of material with different properties from the main medium. To calculate the magnitude of this perturbation, it is necessary to consider the passage of a shower through a thin layer of substance 2 placed in the main medium of substance 1. The picture of the transition effect is somewhat complicated by the second boundary at the point $t = d$ (d is the thickness of the given layer, the origin $t = 0$ coincides with the left boundary). In view of the finite thickness of the perturbing layer, the formulas for the perturbation of the electron flow inside the layer will differ from the corresponding formulas for the case of two semi-infinite environments only by the upper limit in the integral over τ for $\cos\vartheta < 0$: $b = \min\{(t-d)/\cos\vartheta, R_2(E_0, E)\}$. Like that as before the perturbing layer, the perturbation of the flow is observed only for reverse directions (9.30), behind this layer the flow is perturbed only in positive directions:

$$\Delta\Phi_e(t, \cos\vartheta, E) = \frac{\beta_1(E^d)}{\beta_1(E)}\Delta\Phi_e(d, \cos\vartheta, E^d),$$

$$E^d = R_1^{-1}\left[R_1(E) + \frac{t-d}{\cos\vartheta}\right].$$

Of course, as you move away from the equivalent source in one direction or another, the formula for the adjoint functions derived in (9.3) is becoming less and less reliable. However less they give a correct qualitative description of the effect: as you move away from the boundary to the left $\Delta\Phi_e \to 0$ and the Φ_e^B curve in an inhomogeneous medium merges with the curve in homogeneous medium 1. Some-

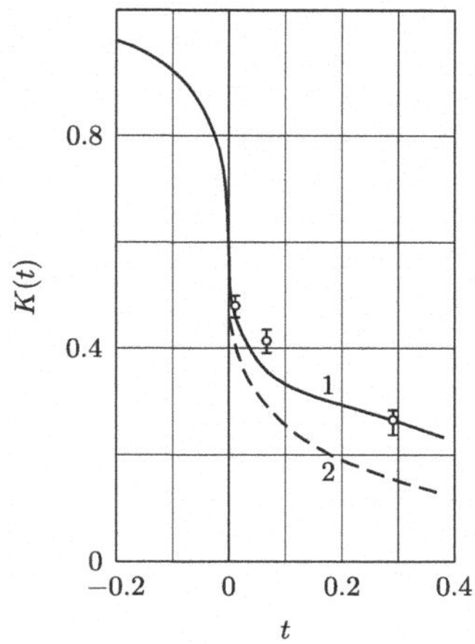

Fig. 9.6 Transition effect to the right of the border Pb–plexiglass. Calculation taking into account (1) and without taking into account (2) the contribution of photons [9]

what less obvious is the similar behavior of $\Delta\Phi_e$ to the right of the perturbing layer: as t increases, the Φ_e^B curve tends to a cascade curve in a homogeneous medium 1, the cascade "forgets" the perturbation and is restored. This explains vanishing the transition effect for high-energy particles: they cross the boundary without noticing the difference of matters on both sides of it and quickly recovering after passing. Such behavior of cascade curves at a small layer thickness d is confirmed by a number of experiments [12, 13]. That fact, that approximate expressions of conjugate functions still "work" (albeit with a smaller accuracy) and in these cases, can be interpreted as follows. First, the main role in the equivalent source is played by low-energy particles, the exchange of energy which is determined mainly by ionization collisions and quite satisfactorily described in the model of continuous deceleration. Second, their angular distribution is close to isotropic, so that the account of scattering in the process of subsequent motion cannot be an appreciable way to change the picture. Finally, with the help of these functions, the correction to the main cascade curve rapidly decreases with distance from the perturbation area, therefore, the accuracy of conjugate functions at large distances is no longer a significant factor (Figs. 9.5 and 9.6).

The boundary peculiarities in EPC listed above are illustrated in Fig. 9.7, which shows the penetration of EPC through a layer of plastic scintillator (plexiglass) immersed into lead. It is seen, that the tendency to recovery depends on the layer thickness: with an increase in d, it noticeably weakens. At the layer boundaries $d = 0.021$, the calculated curve agrees with the experimental values, obtained in [11]. Note that the measurements of the transition effect were carried out in multilayer absorber, while the calculation was performed for a single layer.

Scintillation detector readings (neglecting the internal light absorption) were found by integrating the given curves over the entire layer. Attitude

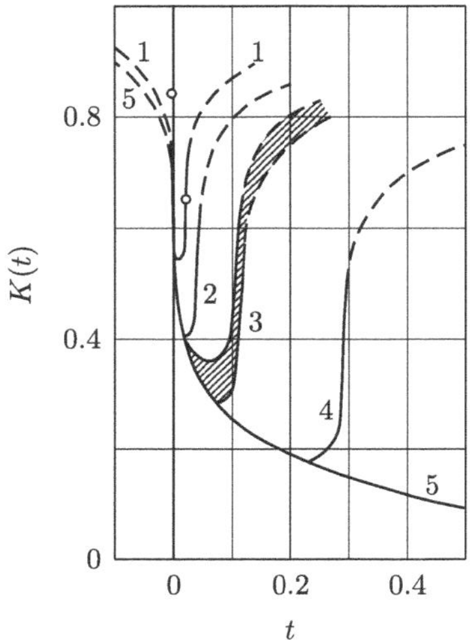

Fig. 9.7 Transition effect in a layer of plexiglass placed in Pb. Layer thickness d (rad): 1—0.021; 2—0.05; 3—0.12; 4—0.3; 5—∞. Shaded area—photon contribution for the layer $d = 0.12$ rad [10], dots—experiment [11]

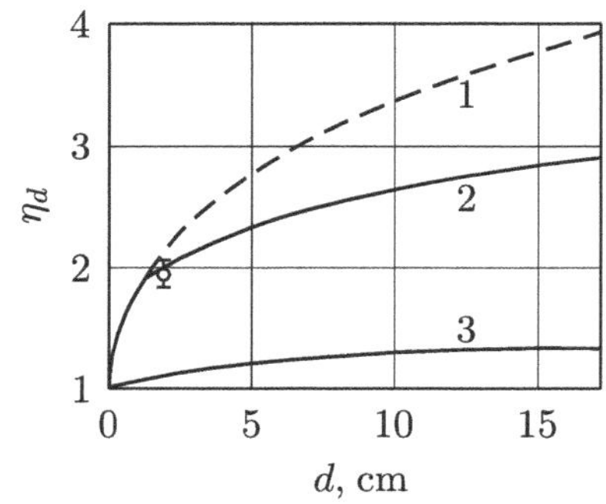

Fig. 9.8 Correction factor for scintillation detector. 1—calculation without taking into account the contribution of photons; 2—taking into account this contribution [10]; 3—data from [50]. dot—experiment [14]

$$\eta_d = \frac{\int_0^d \Phi_e^{(d)}(t)dt}{\int_0^d \Phi_e^{(\mathrm{B})}(t)dt}$$

gives the coefficient that determines the correction for the transient effect in the volume of the scintillator. The dependence of the correction factor η_d on the scintillator thickness d is given in Fig. 9.8. For $d > 3$ cm, a large role is seen in the contribution of photons due to the significant difference in the cross sections Compton scattering in lead and plexiglass. Here are shown results of numerical calculation (neglecting scattering) [9] and experimental data [14].

When studying the influence of the transition effect on the readings of an ionization chamber consisting of two walls of material 2 and the gas gap between them and placed in medium 1, by changing the electron flux in the gap, one can neglect (the gap thickness is usually small: for IK-6 it is ~0.001 rad. units). In this case, the camera can be considered as a continuous layer with a thickness equal to the total thickness of its walls, and its readings as a value proportional to the total electron flow

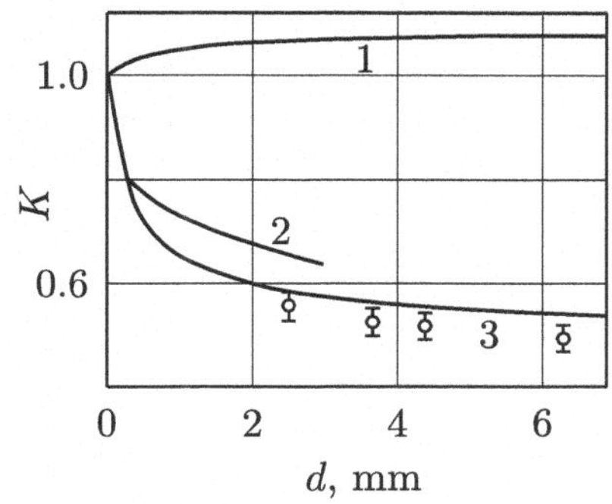

Fig. 9.9 Transition effect in a copper-walled ionization chamber placed in iron (1) and lead (2—calculation [15], 3—calculation [10], dots—experiment [16])

at the point corresponding to the position of the gap. Fig. 9.9 gives the dependence of the transition effect at the shower maximum in lead and iron on the thickness of the copper wall of the ionization chamber. It can be seen that the flux of particles recorded by the camera first changes rapidly with increasing wall thickness, and then remains almost constant, and for lead and iron the perturbation is different. So, with a wall thickness of 2.5 mm, the K coefficient is 0.58 and 1.05 for lead and iron, respectively. Results of comparison with experiment [16] show that the discrepancy between it and the calculation, even for a thickness of $d = 6$ mm, does not exceed 6, and the results of another calculation [15], carried out taking into account (albeit very roughly) the influence of particle scattering, are given. One of the methods for measuring the energy of high-energy particles is based on determining the area $S^{(1)}$ under the cascade curve:

$$S^{(1)} = \int dz \int d\Omega \int dE \Phi_e^{(1)}(z, \cos\vartheta, E) = \int \Phi_e^{(1)}(z) dz, \tag{9.36}$$

which, in the $\beta = \text{const}$ approximation, is related to the energy E_0 by the relation $E_0 = \beta \cdot S^{(1)}$. The dependence $\Phi_e(z)$ can be obtained experimentally by moving the ionization chamber along the depth absorber and fixing its readings at various points z. However, due to the transition effect in the walls camera, the measured area S will differ from $S^{(1)}$. The relation $\bar{\eta} = S^{(1)}/S$ will be called integral transition effect.

Representing (9.36) as

$$S^{(1)} = \int d\Omega \int dE \Phi_e^{(1)}(\cos\vartheta, E),$$

we see that the area under the cascade curve in a homogeneous medium is expressed in terms of the equilibrium angular and energy distribution of electrons:

$$\Phi_e^{(1)}(\cos\vartheta, E) = \int \Phi_e^{(1)}(z, \cos\vartheta, E) dz,$$

the equation for which is easily obtained from the general Eq. (9.3) by integration over space variables. Let us show that the integral coefficient $\bar{\eta}$ can also be expressed in terms of the equilibrium distribution if using the approach developed above. Recalling that in the above formulas t is the coordinate of the point measurement, counted from the left boundary of the perturbing layer, for a chamber with a thickness of the front wall d_1 we set $t = d_1$. Denoting the coordinate of the ionization gap of the chamber, counted from the beginning of the cascade, by z, we obtain

$$\Delta\Phi_e(z, \cos\vartheta, E) = \frac{1}{\beta_2(E)} \int\limits_0^b \beta_2(E_\tau)\Big\{\partial_{E_\tau}[\Delta\beta(E_\tau)\Phi_e^{(1)}(z - \tau\cos\vartheta, \cos\vartheta, E_\tau)]$$
$$- \Delta L_{e\gamma}\Phi_\gamma^{(1)}(z - \tau\cos\vartheta, \cos\vartheta, E_\tau)\Big\}d\tau, \tag{9.37}$$

where

$$b = \begin{cases} \min\{d_1/\cos\vartheta, R_2(E_0, E)\}, & \cos\vartheta > 0; \\ \min\{(d - d_1)/|\cos\vartheta|, R_2(E_0, E)\}, & \cos\vartheta < 0. \end{cases}$$

Integrating (9.37) over z and taking into account that b does not depend on z, we obtain

$$\int\limits_{-\infty}^{\infty} \Delta\Phi_e(z, \cos\vartheta, E)dz = \frac{1}{\beta_2(E)}\Big\{[\Delta\beta(E_b)\Phi_e^{(1)}(\cos\vartheta, E_b)$$
$$- \Delta\beta(E)\Phi_e^{(1)}(\cos\vartheta, E)] - \int\limits_0^b \beta_2(E_\tau)\Delta L_{e\gamma}\Phi_e^{(1)}(\cos\vartheta, E_\tau)d\tau\Big\}.$$

It follows from this that the integral coefficient of the transition effect is really expressed only through the equilibrium distributions of electrons and photons in a homogeneous medium and is numerically equal to the transition effect at the maximum of the shower.

The above analysis has been confirmed by a series of experimental works [12, 13, 17], as well as by calculations Monte Carlo method [18].

9.9 Transition Effect in Electromagnetic Calorimetry

Knowing the average characteristic of a shower in a multilayer calorimeter makes it possible to calculate its averaged (from shower to shower) response to a shower formed by a single primary particle. According to the additivity of this detector, its reading is represented by the sum of all sensitive layers,

$$S = \sum_i S_i.$$

In the absence of transition effect, these contributions would be proportional to corresponding values of the (random) cascade curve in the homogeneous absorber, $S_i = kN_i(t_i)$, and its mean value

$$\bar{S} = k \sum_i \bar{N}(t_i) \approx k \int \bar{N}(t)dt \propto E_0.$$

Since the sensitivity domains (gas gaps, scintillation layers) are usually made from matters with a lower atomic number, than a basic absorber (typically, Pb, U), these gaps reduce the response of the device compared to what would be observed in the case of the same chemical composition of the absorbing and measuring parts, and we have

$$\bar{S} = \sum_i K(t_i)\bar{N}(t_i) \approx \bar{K} E_0$$

where the weighted average (over all measuring layers) value of the transition coefficient

$$\bar{K} = \frac{\sum_i \bar{K}(t_i)\bar{N}(t_i)}{\sum_i \bar{N}(t_i)} \approx \frac{\int \bar{K}(t)\bar{N}(t)dt}{\int \bar{N}(t)dt} < k.$$

Taking into account the influence of the transition effect in the calculations of the energy resolution of an ionization calorimeter at first glance is a more difficult task, since this effect is largely determined by the multiple scattering of particles of both types near the boundaries, while the analytical theory covers problems in which it is sufficient to take into account only scattering at small angles, or ignore it altogether.

There are, however, two circumstances that make it possible to develop an effective method for solving this problem: (1) transition effects are due to low-energy electrons, (2) the main source of statistical fluctuations that limit the energy resolution of the device is the region of intense multiplication (that is, the high-energy part) of the cascade. This separation led to the construction of a hybrid model of the process proposed in the book [19] and later was developed in the theory of Cherenkov radiation under the name *Semi-analytical Monte Carlo Method* (see 12.5).

To include the transition effect in the theory of fluctuations, we again turn to the adjoint equation. Let, with energy E, move at the depth t of a calorimeter of thickness d, producing its own subcascade in (t, d). Let $l = d - t$ be the length of this interval, and $S_a(l, E)$, $D_a(l, E)$ the expectation and dispersion of the calorimeter response to this subcascade. According to Sect. 4.6, the latter obeys the equation

$$\partial_l D_a + \sigma_a D_a - \sum_b \int_{E^*}^{E} dE' w_{a\to b}(E \to E') D_b(l, E')$$

$$= 1(E - E^{**})\sigma_a \mathsf{M}[\bar{S}_b(l, E_b) + \bar{S}_{\bar{b}}(l, E_{\bar{b}}) - \bar{S}_a(l, E)]^2.$$

Here b is a random index of one of the particles, born in the first collision, $\bar{b}$ is the index of other particles, E^{**} separates the high-energy region from low-energy one, and σ_a, $w_{a\to b}$ are taken for a homogeneous absorber. Note that the transition effect is taken into account only on the right-hand sides of these equations by using the instrument readings as the average value

$$\bar{S}_a(t, E) = \sum_i \bar{N}_a(t \to t_i, E)K(s_i)1(t - t_i).$$

In the first case, the results of SBCE calculations for a homogeneous lead absorber [20] were used; in the second case, corrections for scattering and the transition effect found from experimental data [21] were introduced. It can be seen that taking into account the transition effect noticeably increases the relative fluctuations, bringing them closer to the experimental data.

The remaining difference can be explained by fluctuations in the responses of the device to the intersection of the layer by an electron with a fixed energy (loss fluctuations), which include the spread of transverse coordinates, intersection angles, as well as causes not related to the dynamics of the cascade: light collection, fluctuations in photomultipliers, etc.

Neglecting, in accordance with experimental evidence, the energy dependence of fluctuations in individual losses δ_1 on the energy of particles crossing the layers, and assuming these contributions to be independent, we arrive at the following formula:

$$\delta^2 = [\delta S_e]^2 + \delta_1^2/\bar{S}_e,$$

giving satisfactory agreement with the experiment (see Fig. 9.18).

Experimental studies were carried out at the electron synchrotron of the Tomsk Polytechnic Institute. X-ray film (RT-1) was used as a detector and full electron flow at a given depth was found by numerical integration of the radial distributions of blackening density. The value of the integral transition effect $\bar{\eta}$ was determined as the ratio of the areas under the cascade curves in a homogeneous absorber (Pb) and in a multilayer (Pb–Cu–...) absorber simulating an ionization calorimeter. Copper layers 1, 2, 4, 8, and 16 mm thick were used. The front layer of lead in all experiments had a thickness of 1.72 rad. units, the thickness of other layers of lead is 1.8 rad. units. The films were placed in the front layer lead at depths of 0.05; 0.62 and 0.84 rad. units (for normalizing cascade curves) and at the centers of each copper layer. In contrast to [16], where the measurements were carried out using a single camera, moving in depth, this technique made it possible to take into account the mutual influence of the layers on each other, which enhances the integral transition effect (Figs. 9.10 and 9.11).

In Fig. 9.12 the results of measurements of transition effects at $E_0 = 950$ MeV in copper layers of various thicknesses in lead are given. As follows from our calculations, the transient effect at the layer boundary is smaller than at the boundary

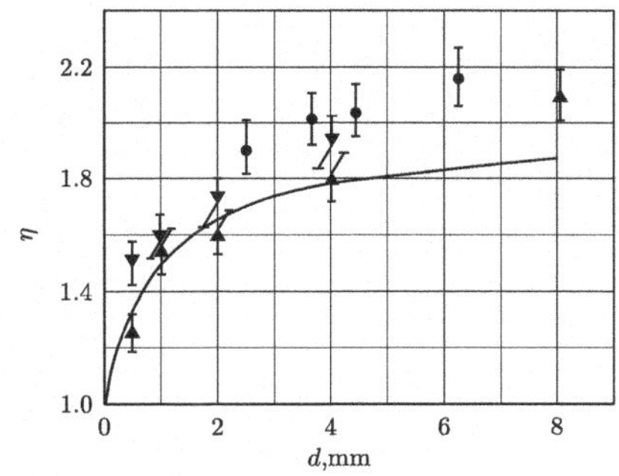

Fig. 9.10 Integral transition effect in an ionized chamber with copper walls placed in lead. Solid line—calculation [10], dots—experimental data: ▼$-E = 600$ MeV, ▲$-E = 900$ MeV [17], •$-E = 100 - 500$ MeV [16]

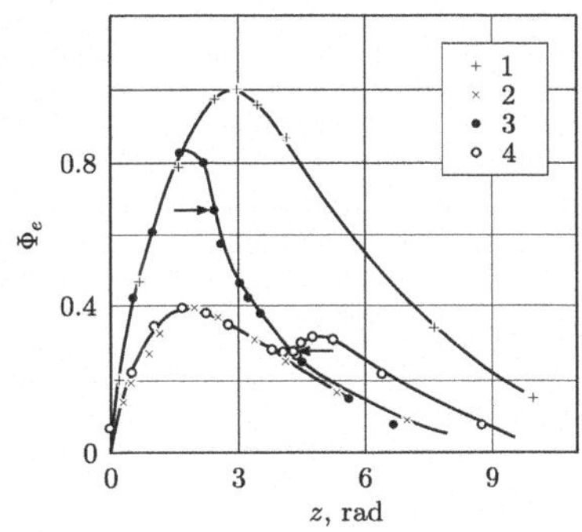

Fig. 9.11 Cascade EPC curves in a two-layer absorber from a bremsstrahlung beam [164] ($E_0 = 600$) MeV. 1—Pb, 2—Cu, 3—Pb-Cu, 4—Cu-Pb. The arrows indicate the values of the electron flux at the interfaces

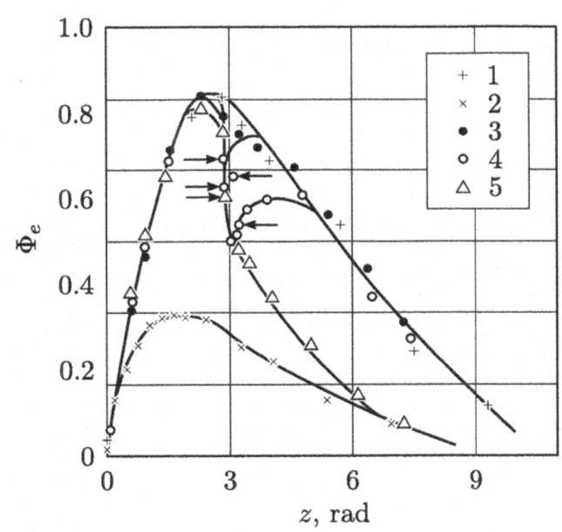

Fig. 9.12 Cascade curves in a three-layer absorber ($E_0 = 950$) MeV. 1—Pb, 2—Cu, 3—Pb-Cu, 4—Pb-Cu-Pb, $d_{Cu} = 2$ mm; 5—Pb-Cu-Pb, $d_{Cu} = 5$ mm

with a semi-infinite medium. In particular, at the Pb–Cu interface, $K = 0.72$, while for layers 2 and 5 mm thick, $K = 0.81$ and 0.74, respectively. In the middle of the layers, the flux is 1.3 and 1.7 times less than at the maximum in lead. It is known [16] that reading of the ionization chamber IC-6 with a total wall thickness of 5 mm, placed in a shower maximum in lead, in 1.8 times less than in homogeneous lead, which is consistent with the results of this experiment. When the shower crosses the second boundary (Cu–Pb) of the layer, the electron flux increases sharply and

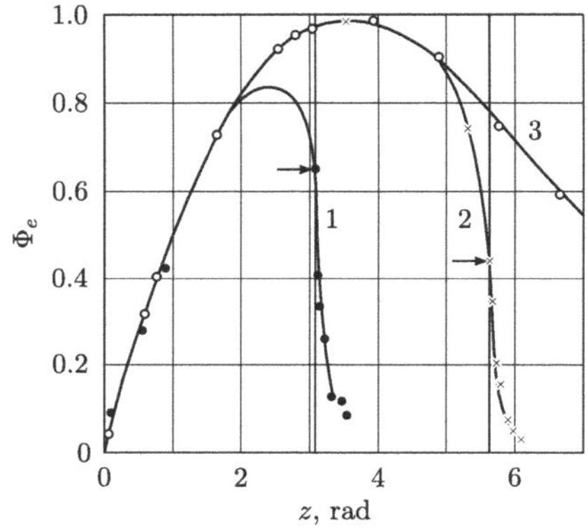

Fig. 9.13 Cascade curves in a two-layer absorber. 1, 2—Pb–plexiglass; 3—Pb [13]. The arrows indicate the values of the electron flux at the interfaces

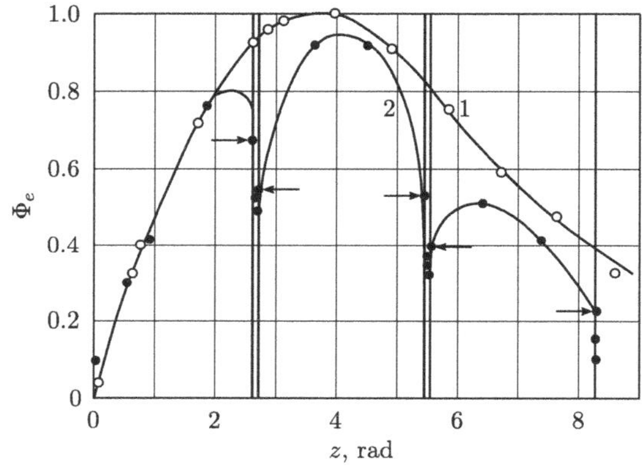

Fig. 9.14 Cascade curves in a multilayer absorber. 1—Pb, 2—lead with layers plexiglass thickness 2 mm [13]

at some distance from the boundary reaches the value of the flux in a homogeneous lead absorber (the shower is restoring). At infinite thickness of the copper layer, the flow approaches to the cascade curve in a homogeneous copper absorber (Figs. 9.13 and 9.14).

$$\bar{S} = \sum_i K(t_i)\bar{N}(t_i) \approx k' E_0, \quad k' < k. \tag{9.38}$$

Let us consider as an example the calorimeter described in [21]. It is made up of eight layers of lead 2.5 rad. thick. units each (except the first, whose thickness is 1 rad. units) and eight plates scintillator with a thickness of 0.025 rad. units placed between them (Fig. 9.15). Comparing the experimental cascade curves presented in this work with the calculated curves for a homogeneous lead [18], one can find the magnitude of the transient effect $K(t_i)$ at different stages of shower development. It was shown above that for $s = 1$ (at the maximum of the cascade) the magnitude of the transition effect does not depend on the energy and type of the primary particle. The results of the experiment show that on other at depths, the transition effect depends only

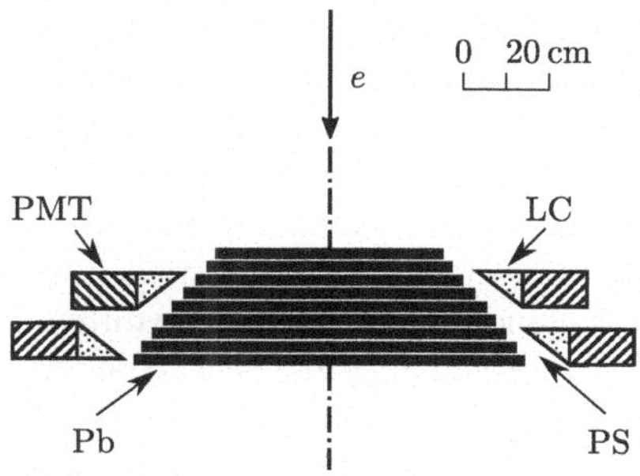

Fig. 9.15 Scintillation calorimeter (sandwich) [21]. LC—luminous LED, PS—plastic scintillator

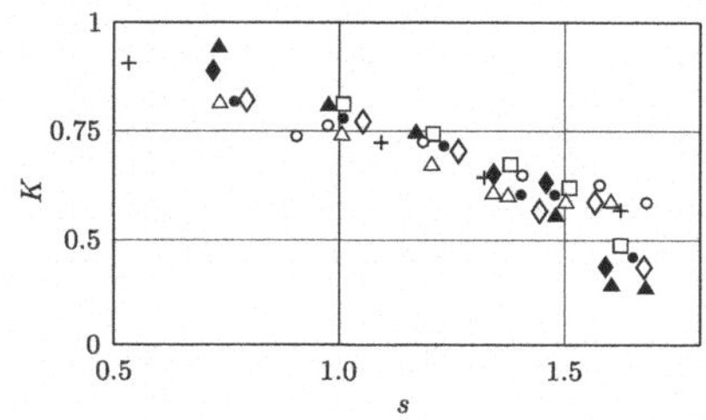

Fig. 9.16 Transition coefficient. Energy of the primary electron: 2(○), 4(+), 6(◇), 8(●), 10(□), 12(△), 14(▲), 15(◆)GeV

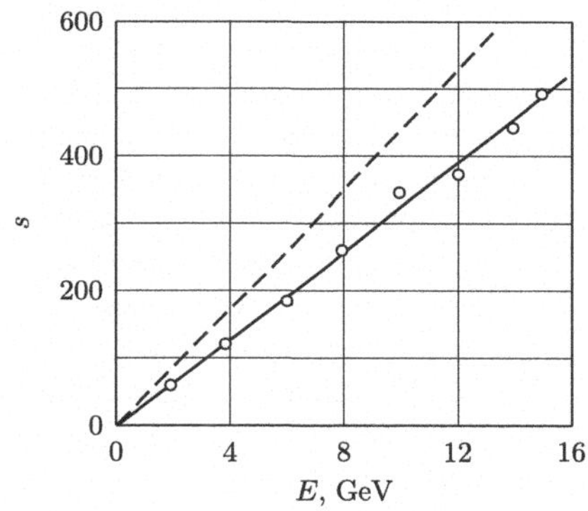

Fig. 9.17 The dependence of the average readings of the calorimeter on energy, taking into account (1) and without taking into account (2) the transition effect. Dots—experiment

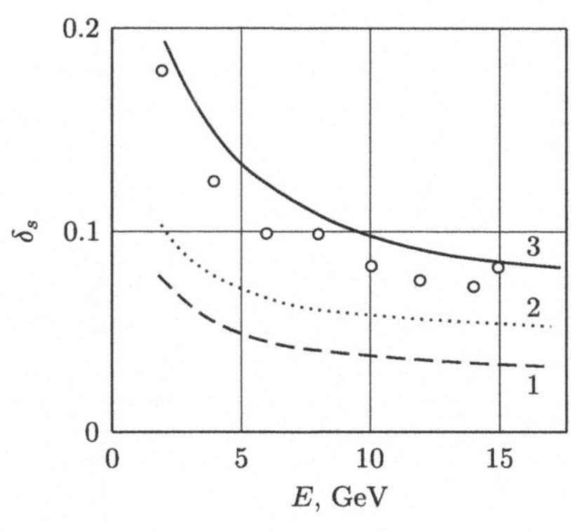

Fig. 9.18 Relative fluctuations of calorimeter readings. 1—calculation in approximation "straight-ahead", 2—calculation with scattering and transient effect, 3—calculation with taking into account fluctuations of the signal of an electron crossing the sensitive layer, dots—experiment [21]

on the shower age s, and not on E_0 and t separately. The dependence $K(s)$ for the setup under consideration is shown in Figs. 9.16, and 9.17 which show the difference between the sensitivity of the calorimeter calculated with (9.38) and without (9.37) transition effect (Fig. 9.18).

Accounting for the transition effect in the calculations of the energy resolution of the calorimeter, due to statistical fluctuations in the development of the cascade, is at first glance significantly more difficult task since the effect is largely determined by the scattering of particles in an inhomogeneous medium, while the calculation methods described in Chap. 7 are applicable only to one-dimensional problems ("straight-ahead" approximation). In reality, however, two circumstances simplify the situation if we can neglect the energy transfer through side surface of the calorimeter: the transition effect is due to low-energy electrons and fluctuations in the development of the EFC are determined by the high-energy part cascade, the scattering and transition effect which are practically missing (see Fig. 9.1). Such a separation of energy regions, affecting fluctuations and scattering, makes it possible to carry out a simple method for calculating the energy resolution of the calorimeter.

In particle physics,the term *calorimeters* relates to devices absorbing a substantial fraction of the incident particle energy and producing a signal proportional to the energy.

Calorimeters designed for high-energy physics experiments may be divided in two groups: *homogeneous calorimeters* and *sampling calorimeters*. Up to boundary effects, the energy-signal conversion mechanism in a homogeneous calorimeter finds an adequate representation in the energy degradation theory outlined above.

Sampling calorimeters consist of layers of a dense passive absorber(such as lead or copper), interleaved with active detector layers(such as silicon,plastic scintillator or liquid argon). A notable example of this type of construction is the electromagnetic calorimeter of the ATLAS experiment at the LHC, consisting of lead converter foils immersed in liquid argon, whereas, in a calorimeter of homogeneous type, a single medium(such as liquefied xenon or a crystal scintillator) serves as both the absorbing medium and the signal generator. The largest crystal electromagnetic calorimeter ever constructed is the *ECAL* of the CMS experiment at the LHC, which employs lead tungstate crystals. The ATLAS and CMS calorimeters are discussed in more detail in [22].

The energy dependence of the energy resolution is parameterized as the quadratic sum of three terms [23]:

$$\frac{\sigma_E}{E_0} = \frac{a}{\sqrt{E_0}} \circ \frac{b}{E_0} \circ c. \tag{9.39}$$

The first term, with coefficient a, is the *stochastic*, arising from fluctuations in the number of signal-generating processes (together with possible additional effects such as fluctuations in the measurement of that signal). The second term, with coefficient b, is usually referred to as the *noise term*. It receives contributions not only from noise in the readout electronics but also from effects such as *pile-up* (simultaneous energy deposition by uncorrelated particles). The third term is the *constant term* with coefficient c. It arises from several effects including:

- imperfections in calorimeter construction (dimensional tolerances, etc.);
- non-uniformities in signal collection;
- channel-to-channel calibration errors;
- fluctuations in longitudinal energy containment; and
- fluctuations in energy lost in an inert material, before or within the detection volume.

An important factor governing the energy-signal conversion in such calorimeters is the transition effect arising near the layer boundaries.

9.10 Determination of the π-Meson Beam Composition

When dealing with a multilayer calorimeter with thin, say, gas or plastic-sensitive gaps, it is natural to pass from a continuous representation of cascade curves to a discrete one, considering the sensitive region as a set of parallel planes intersecting the shower axis at equidistant points $t_k,\ k = 1, 2, \ldots, n$. In this case, the longitudinal moments of the cascade curves are approximated by integral sums. Denoting the random and ensemble averaged values of the cascade curve by y_k and $\bar{y}_k$, respectively, we get zero-moments in the form $N^{(0)} = \sum_k y_k,\ \bar{N}^{(0)} = \sum_k \bar{y}_k$.

Following the scheme suggested by Borkowsky [19] who referred in its turn to [24], we introduce the critical constant μ, so that the observed cascade belongs to EPC ensemble, if

$$R^2 = \sum_{k=1}^{n} (y_k - \bar{y}_k)^2 / A_{kk} < \mu. \tag{9.40}$$

Here A_{kk} are the diagonal elements of the matrix

$$A_{kl=} y_k \bar{y}_l - y_k \bar{y}_l \tag{9.41}$$

that is found only in electron-induced cascades.

For this method, the cascade is assumed to be an electron–photon one if the orthogonal coordinates y_i describing it are located inside the n-dimensional ellipsoid with the half-axes equal to $\sqrt{\mu A_{ii}}$. The axes of this ellipsoid are directed along the initial coordinate axes (i.e., y_i).

The presence of correlations between quantities y_i (matrix non-diagonal elements being not equal to zero) means that the distribution of points around the mean value given by the coordinates of the point $y_i = \bar{Y}_i, i = 1, 2, \ldots n$, has no spherical symmetry and is not stretched (or flattened out) along the coordinate axes. Therefore, it is reasonable to select the cascade as the electron–photon one if its coordinate point $(y_i, i = 1, 2, \ldots, n)$ gets into another ellipsoid rotated relative to the initial one given by expression (9.40). For rejection of events getting into this rotated ellipsoid which has half-axes equal to $\sqrt{\mu A'_{ii}}$, the quantity R^2 has to be calculated for every event in the following manner:

$$R^2 = \sum_{k,l=1}^{n} C_{kl}(y_k - \bar{y}_k)(y_k - \bar{y}_k) \tag{9.42}$$

where

$$C_{kl} = \sum_{i=1}^{n} a_{ik}a_{il}/A'_{ii} \tag{9.43}$$

and a_{ik} are the elements of the matrix a connecting the coordinates y_i with the new coordinates

$$x_k = \sum_{i=1}^{n} a_{kl}(y_l - \bar{y}_l) \tag{9.44}$$

which are directed along the rotated ellipsoid axes. The matrix A' is connected with the initial matrix A by the expression

$$A' = aAa^T. \tag{9.45}$$

One can easily demonstrate (averaging x_i^2 that the variance of x_i is equal to the diagonal element (A'_{ii}) of this matrix). In the system where individual x_i are not correlated

$$A_{ii} = x_i x_k = 0, \quad i, k = 1, 2, \ldots n; \quad i \neq k. \tag{9.46}$$

It is clear that, in this system, we have to take the matrix C' (i.e., matrix C in the new coordinate system) the diagonal matrix inverse to A' (i.e., $C'_{ii} = 1/A'_{ii}$). The matrix C transforms in the same way as A if the coordinate system is rotated (see (9.43) and (9.45)) and matrix a is orthogonal (that is, $a^T a = 1$). Hence, C is inverse to A in any system including the initial one

$$\sum_{l=1}^{n} C_{kl}A_{lm} = \delta_{km}. \tag{9.47}$$

In order to simplify the calculations illustrating the value of the improvement of the electron rejection from π-meson beam by the method taking into account the correlations (see Eqs. (9.42) and (9.47)) as compared with a similar method not taking the correlations into account, some assumptions are needed. First, since events caused by high-energy π-mesons are distributed in n-dimensional space more uniformly than those caused by electrons, we can therefore assume that within the ranges of y_i for events caused by electrons the hadron-induced cascades are merely distributed uniformly. The second assumption is that for the electron-induced cascade the probability of obtaining the point with coordinates y_i inside the initial ellipsoid with half-axes equal to $\sqrt{\mu A_{kk}}$ is equal to the probability of obtaining this point inside the rotated ellipsoid with half-axes equal to $\sqrt{\mu A'_{ii}}$. This assumption is also quite natural since both A_{kk} and A'_{ii} are equal to the variances of coordinates y_k and x_i for the

Table 9.1 Quantity p found from expression

E	1	3	10	40	70	100	N
$t_1 = 2, \Delta t = 2$	1.48	4.18	20.3	158	354	701	10
$t_1 = 4, \Delta t = 4$	1.14	1.50	2.65	3.65	3.82	4.23	5
n	211	130	193	47	69	43	

electron-induced events. Under these two assumptions, the increase in the efficiency of the electron rejection at the expense of taking into account the correlations between y_i is equal to the relative decrease in the ellipsoid volume with the directions of the axes found from conditions (9.46) as compared to the initial (not rotated) ellipsoid volume. The ratio of the initial ellipsoid volume to that of the desired (rotated) one is equal to

$$p = \sqrt{\prod_{k=1}^{k=n}(A_{kk})/\mathrm{Det}(A)}. \tag{9.48}$$

Here it is taken into account that in the coordinate system where the matrix A is diagonal (condition (9.46) is satisfied) the product of the diagonal elements is equal to the determinant of A and that the matrix determinant is conserved when the coordinate system is rotates.

In the paper cited, the number of charged particles with total energy greater than 2.044 MeV on the depth t_i for electron–photon cascades in lead is taken as y_i. The matrix A is obtained using the Monte Carlo program described in [25].

The calculation correctly takes into account all the processes essential for the development of the electron–photon cascades. Results a represented in Table 9.1.

References

1. Kolchuzhkin AM, Ryzhov VV, Uchaikin VV (1970) Perturbation method in the theory of the transition effect. Proc Acad Sci USSR. Ser Phys 34(10):2019 (in Russian)
2. Kolchuzhkin AM, Shevtsova II (1969) On the calculation of the transition effect for an electron-photon shower. Izv Vyssh Uchebn Zaved Fizika 2:137 (in Russian)
3. Pinkau K (1905) Errors in electromagnetic cascade measurements due to the transition effect. Phys Rev 139:1548
4. Belenky SZ (1948) Shower processes in cosmic rays. Gostekhizdat, Moscow (in Russian)
5. Ivanenko IP, Samosudov BE (1968) The angular distribution functions of cascade electrons in iron and lead. Canad J Phys 46:204
6. Nagel HH (1965) Electron-Phonon-Cascaden in Blei. Zeitschrift fur Physik Bd 186:319–346
7. Vernov SN, Vavilov ON (1946) Transition effects of the soft components of cosmic rays. Phys Rev 70:769
8. Kolchuzhkin AM, Ryzhov VV, Uchaikin VV (1970) Perturbation method in the theory of the transition effect. Proc Acad Sci USSR. Ser Phys 34(10):2019 (in Russian)

9. Crannell CJ, Crannell H, Gillespie CB et al (1969) Experimental determination of the transition effect in electromagnetic cascade showers. Phys Rev 182:1485
10. Kolchuzhkin AM, Ryzhov VV, Uchaikin VV (1971) Transition effect in electron-photon showers. Proc Acad Sci USSR. Ser Phys 35(10):2171 (in Russian)
11. Yuda T (1969) Electron-induced cascade showers in inhomogeneous media. Nucl Inst Meth Phys Res 37:301–303
12. Kolchuzhkin AM, Ryzhov VV, Shoshin EB et al (1973) On the shape of cascade curves in inhomogeneous absorbers. Izv Vyssh Uchebn Zaved Fizika 7:146–147 (in Russian)
13. Kolchuzhkin AM, Shoshin EB, Ryzhov VV (1976) Experimental studies of electron-photon cascades in inhomogeneous absorbers. Proc Acad Sci USSR. Ser Phys 40(5):947–977 (in Russian)
14. Logachev VP, Sinitsyna VG (1968) Showers formed by 600, 1000 and 1300 MeV electrons in lead. Proc Acad Sci USSR. Ser Phys 32:508 (in Russian)
15. Dmitriev VA (1958) Transition effect for electrons in the walls of an ionization chamber. Zh Eksp Teor Fiz 35:553 (in Russian)
16. Fetisov IN (1968) Measurements of the transition effect in electromagnetic cascades. Canad J Phys 46:1145–1147
17. Shoshin EV, Ryzhov VV, Kolchuzhkin AM (1975) Integral transition effect in an ionization calorimeter. Izv Vyssh Uchebn Zaved Fizika 6:138–140 (in Russian)
18. Kolchuzhkin AM, Bespalov VI (1979) Investigation of electron-photon showers by Monte-Carlo method. In: Proceedings of 16-th ICRC, vol 9. Kyoto, pp 222–227
19. Borkowsky MJ (1981) On the possibility of using the correlation existing in electron-photon cascades for determination of the π-meson beam composition. Nucl Instrum Methods 189:405–407
20. Uchaikin VV, Ryzhov VV (1988) The stochastic theory of high energy particle transport. Nauka, Siberian Branch, Novosibirsk (in Russian)
21. Van Ginneken A (1978) AEGIS, A program to calculate the average behavior of electromagnetic showers, Fermilab, FN-309
22. Dobrzynski L (2014) Instrumentation in HEP calorimetry. CMS Experiment at the LHC CERN, ESHEP2014-Cairo
23. Brown RM, Cockerill DJA (2012) Electromagnetic calorimetry. Nucl Inst Methods, Phys Res A 666:47–79
24. Borg VV, Kirillov-Ugrjumov VG, Petrukhin AA, Cherniatin VK (1972) Investigation of nuclear muono interaction. Izv AN SSSR, Ser Fiz 36:1761 (in Russian)
25. Borkowiski MJ (1979) Leningrad Inst. of Nuclear Physics Report LNPI-462 (in Russian)
26. Adams T, Adzic P, Ahuja S et al (2016) Beam test evaluation of electromagnetic calorimeter modules made from proton-damaged $PbWO_4$ crystals. J Instrum 11(04):P04012
27. Flauger W (1985) Simulation of the transition effect in liquid argon calorimeters. Nucl Instr Meth Phys Res Section A: Accelerators, Spectrometers, Detectors and Associated Equipment 241(1):72–75
28. Vernov SN, Vavilov ON (1946) Transition effects of the soft components of cosmic rays. Phys Rev 70(9–10):769
29. Christy RF, Kusaka S (1941) Burst production by mesotrons. Phys Rev 59(5):414
30. Pinkau K (1965) Errors in electromagnetic cascade measurements due to the transition effect. Phys Rev 139(6B):B1548

For Further Reading

31. Sefkow F, White A, Kawagoe K, Pöschl R, Repond J (2016) Experimental tests of particle flow calorimetry. Rev Modern Phys 88(1):015003
32. Amaldi U (1981) Fluctuationin in the calorimetry measurements. Phys Scripta 23:409–424
33. Brown RM, Cockerill DJA (2011) Electromagnetic calorimetry. Nucl Instrum Methods Phys Res A 666:47–79

Chapter 10
Monte Carlo Modeling

10.1 Introduction

Monte Carlo is one of the most versatile and widely used numerical methods. Its result is obtained by averaging over a large number of N statistically independent realizations of a random process modeled by using a random number generator. Its convergence rate $N^{-1/2}$ is dimensionally independent, so that the Monte Carlo method is very reliable, but at the same time slow [1]. This chapter provides a brief introduction to this method with some simple algorithm examples and a discussion of time reduction when calculating correlations in cascade processes.

There are two main methods for reducing the Monte Carlo calculation time: replacing standard random numbers with *quasi-random numbers* [2], and replacing the ensemble of by trajectories with modified probabilities of elementary events (*weighting methods*).

An important area of the Monte Carlo method is the particle transport with multiplication: electrons with energies above the ionization potential, neutrons in a multiplying medium, the development of electromagnetic cascades, extensive air showers, etc. The mathematical expectation of the readings of a detector that registers this process can usually be expressed in terms of a differential flow that satisfies the Boltzmann kinetic equation, and the Monte Carlo method itself can be considered as a numerical method for solving this equation, which has a characteristic statistical error. This approach, in particular, makes it possible to use various modifications of the method, which have a lower statistical error compared to direct modeling (imitation) of the process. In many physical problems (for example, in problems of microdosimetry, cosmic ray physics, etc.), the random nature of the transfer process plays an important role, and the higher moments of the detector readings, in particular, the magnitude of the fluctuation, are of interest. Conventional modifications of the Monte Carlo method are unsuitable for solving this problem, since they lead to a shift in the second and higher moments, and transfer simulation is not always effective. The latter is especially important in connection with the multiplication of particles: the direct simulation of a single cascade, which includes a large number

© Higher Education Press Limited Company 2024

V. V. Uchaikin, *Electron-Photon Cascades*, Nonlinear Physical Science,
https://doi.org/10.1007/978-981-99-7524-2_10

of secondary particles, requires many times more computer time than the simulation of a single non-branching trajectory. To solve such problems, special modifications of the Monte Carlo method are required, which make it possible to obtain unbiased estimates of higher moments, and, first of all, an estimate of the second moment, which determines the fluctuations of the detector readings. Postponing the details of EPC modeling until Chapter 12, we will touch here on the problem of increasing the efficiency of the MC method by artificially changing the probabilities, accompanied by the use of statistical weights to ensure that the final result is unbiased.

When considering the corpuscular transport problems, like neutron or gamma quantum transport, we deal with the sample statistical ensemble consisting of piecewise-linear broken lines simulating multiple scattering of particles. With a decrease in the energy of moving particles, the irregularities of the trajectories become more and more small-scale, and the trajectories themselves take on the character of Brownian motion and, due to these properties, become a convenient mathematical tool for modeling solutions to partial differential equations [3]. With an increase in energy, however, a different picture is observed: the processes of particle multiplication are switched on and begin to play an increasingly important role, similar to the multiplication of neutrons in nuclear reactors. This is especially evident in high-energy astrophysics, where cascades—sets of particles formed by a single particle and correlated with each other by conservation laws—become the subject of study. But first let's say a few words about the basic concepts of the method.

10.2 Random Numbers

The high-energy particle passage through a matter is a random process. Each element of the particle trajectory, like a free path, scattering angle, loss of energy in collision, and so on is characterized by a random variable having a known distribution. Using existing algorithms for obtaining random numbers with an arbitrary distribution can get random values of the elements of the trajectory and consequently random realizations of the trajectories in matter, being equivalent (in the statistical sense) trajectories of real particles. Because of this equivalence the process under consideration can be considered a simulation of a random phenomena.

A set of these realizations can be used for approximate calculation (estimation). The simplest (in relation to their properties) random numbers (r.n.'s) are *uniformly distributed standard* r.n.'s ξ, whose probability distribution density $p_\xi(x) = 1$ on the segment [0, 1] and the cumulative distribution function

$$F_\xi(x) \equiv \mathsf{P}(\xi < x) = x, \quad x \in (0, 1]. \tag{10.1}$$

For solving a specific problem using the Monte Carlo method, the sequence of mutually independent random numbers with more complex distributions, say η_i, may be required. Let $F_\eta(y)$ be the correspondent cumulative function. It is easy to prove, that the sequence

$$\eta_i = F^{-1}(\xi_i) \tag{10.2}$$

possess the required distribution. This algorithm is called the *inverse function method*.

For example, to get the random numbers η_i distributed exponentially over the positive semiaxis,

$$\mathsf{P}(\eta > y) = e^{-\sigma x}, \quad \sigma > 0, \tag{10.3}$$

one can use the transformation

$$\eta_i = -\frac{1}{\sigma}\ln(\xi_i) \tag{10.4}$$

(note, that ξ and $1 - \xi$ have the same uniform on (0, 1) distributions).

Let $0 < a < b$, then, according to (10.1), r.n.

$$\eta = a + (b - a)\xi \tag{10.5}$$

will be distributed uniformly on (a, b).

Let ζ be a random variable whose probability is distributed on (a, b) with probability density $q(x)$, having the highest value h. Drawing a point from the uniform distribution on the square $(a, b) \times h$ and taking its abscissa as ζ each time as its ordinate occurs under graph of $q(x)$, we obtain the desired sequence. This is the *von Neumann's accept–reject method*.

The efficiency of this method can be made higher by replacing the high side of the square $y = h =$ const by a positive function $h(x)$ with graph lying below $f(x)$, so that the probability of acceptance becomes larger. The second way to reach more high efficiency is the composition of the methods. Having the probability density $f_{2,1}(y_2|y_1)$ for the r.v. η_2 conditioned by $\eta_1 = y_1$), one can write the joint density as

$$f(y_2, y_1) = f(y_2|y_1)f(y_1),$$

and each density on the right-hand side is easily sampled, then a draw from the joint distribution may be obtained by drawing η_1 from $f(y_1)$ and then drawing η_2 from $f(y_2|\eta_1)$. Because vector (η_1, η_2) is a draw from the joint distribution it follows that the second component of the simulated vector is a draw from the marginal distribution of $f(y_2)$:

$$f(y_2) = \int f(y_2|y_1)f_1(y_1)dy_1.$$

Thus, to obtain a draw η_2 from $f(y_2)$, it is sufficient to produce a sample from the joint distribution and retain the second component. This method is quite important and arises frequently in the setting of Monte applications.

Here, we show a few algorithms of such combined kind relating to one of the most important processes in EPC: photon–electron Compton scattering.

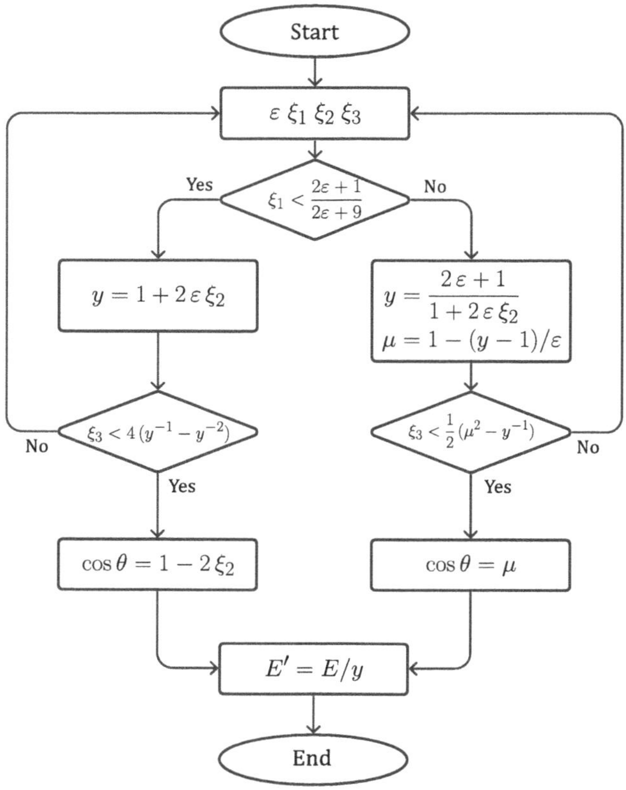

Fig. 10.1 Flowchart of the Kahn algorithm of the Compton scattering process [5]

Figure 10.1 demonstrates the flowchart of Kahn's algorithm [5] for simulation of the Compton scattering process. Here $E = \varepsilon mc^2$ and E' are energies of an incident and scattered photons, θ stands for the scattering angle, and ξ_1, ξ_2, ξ_3 are pairwise independent uniformly distributed in [0, 1] random numbers.

There are other algorithms for this process simulation. In [6], the differential with respect to the outgoing photon energy cross section

$$d\sigma/d\nu = \pi r_0^2 (mc^2) \nu_0^2$$
$$\times [\nu/\nu_0 + (m^2 + 2m\nu_0)/\nu_0 + (\nu_0^2 - 2m\nu_0 - 2m^2)/\nu_0\nu + m^2/\nu^2], \qquad (10.6)$$

where ν varies from $\nu_0 mc^2/(mc^2 + 2\nu_0)$ to ν_0, the incident photon energy. Each of the four terms in the square brackets is integrated between the above limits and their sum normalized to unity, for each incident energy E_i. The results for the first three terms are stored in a special table. The event selection proceeds by choosing a term from among the four terms with probabilities proportional to the relative magnitudes of the integrals.

Next a specific ν is selected from a distribution proportional to that of the chosen term of Eq. (10.6). The total cross sections corresponding to the integral of (10.6) over v are readily obtained in closed form. They are computed for each E_i and stored in an array.

One more version of the Compton simulator can be found in [7].

Ignoring constants being unnecessary for sampling, the author write the differential with respect to $\mu = \cos\theta$ cross section as

$$\frac{d\sigma}{d\mu} \sim f(\varepsilon, \mu) = \frac{(1+\mu^2)[1+\epsilon(1-\mu)] + \varepsilon^2(1-\mu)^2}{[1+\epsilon(1-\mu)]^3}.$$

Noting, that the direct (inversion) method here is not applicable because of the complicated dependence on μ, the author turns to the acceptance–rejection method. However, the simplest (rectangular) version of rejection function is an inefficient boundary for sampling μ, especially for high E values. Because in $E \to \infty$ (i.e., $\varepsilon \to \infty$) $f(\varepsilon, \mu) \sim 1/[\varepsilon(1-\mu)]$, one may try to choose the envelop function of type

$$h(\varepsilon, \mu) = a(\varepsilon)/[b(\varepsilon) - \mu]. \tag{10.7}$$

Imposing the conditions

$$f(\varepsilon, 1) = h(\varepsilon, 1) = 2,$$

$$f(\varepsilon, -1) = h(\varepsilon, -1) = c = 2(2\varepsilon^2 + 2\varepsilon + 10/(2\varepsilon+1)^3.$$

From these conditions, one can find:

$$b(\varepsilon) = (1 + c/2)/(1 - c/2), \quad a(\varepsilon) = 2[b(\varepsilon) - 1].$$

Sampling from $h(\varepsilon, \mu)$ in Eq. (10.7) can be achieved by the relation

$$\mu = b(\varepsilon) - [b(\varepsilon) + 1](c/2)^{\xi}$$

and energy of the scattered photon can be calculated from

$$\varepsilon' = (1 - \mu + 1/\varepsilon)^{-1}$$

in units of electron rest mass energy.

Thus, the algorithm can be summarized as follows:

1. calculate $\varepsilon,\ c(\varepsilon),\ b(\varepsilon),\ a(\varepsilon)$;
2. generate $\mu = b(\varepsilon) - [b(\varepsilon) + 1)(c/2)^{\xi_1}]$;
3. if $f(\varepsilon, \mu)/h(\varepsilon, \mu) < \xi_2$, go to (2);

4. calculate ε'.

Comparing the efficiency of this algorithm with Kahn's algorithm, Koblinger's method [9], and the algorithm used in EGS4 code system [10] showed that, up to 1 MeV, the efficiencies of the compared algorithms differ little from each other, but with increasing energy, the new algorithm quickly begins to dominate over the others (at 100 MeV it exceeds them by five times), and for this reason it should be preferred when modeling high-energy cascades (see, also, the comparison of EGS and ETRAN in [11]). Some other details of EPC simulation will be referred in Sect. 10.10 and Chap. 12.

10.3 Non-branching Trajectory in a Homogeneous Medium

Modeling a trajectory begins with its start by simulation of its birth point $\mathbf{r}_0$, direction of its motion $\boldsymbol{\Omega}_0$ and its kinetic energy E_0. Assuming that the source normalized to one emitted particle, $\int d\mathbf{r} \int d\boldsymbol{\Omega} \int dES(\mathbf{r}, \boldsymbol{\Omega}, E) = 1$, write the probability of $\mathbf{r}_0$ in volume element $\mathbf{r}$:

$$\mathsf{P}\{\mathbf{r}_0 \in d\mathbf{r}\} = d\mathbf{r} \iint S(\mathbf{r}, \boldsymbol{\Omega}, E)d\boldsymbol{\Omega}dE = S(\mathbf{r})d\mathbf{r}, \tag{10.8}$$

as well as the conditional probabilities of $\boldsymbol{\Omega}_0$ falling into $d\boldsymbol{\Omega}$ and E_0 into dE:

$$\mathsf{P}\{\boldsymbol{\Omega}_0 \in d\boldsymbol{\Omega}|\mathbf{r}_0 = \mathbf{r}\} = d\boldsymbol{\Omega}\frac{\int S(\mathbf{r}, \boldsymbol{\Omega}, E)dE}{\iint S(\mathbf{r}, \boldsymbol{\Omega}, E)d\boldsymbol{\Omega}dE} = \frac{S(\mathbf{r}, \boldsymbol{\Omega})}{S(\mathbf{r})}d\boldsymbol{\Omega}; \tag{10.9}$$

$$\mathsf{P}\{E_0 \in dE|\mathbf{r}_0 = \mathbf{r}, \boldsymbol{\Omega}_0 = \boldsymbol{\Omega}\} = dE\frac{S(\mathbf{r}, \boldsymbol{\Omega}, E)}{\int S(\mathbf{r}, \boldsymbol{\Omega}, E)dE} = \frac{S(\mathbf{r}, \boldsymbol{\Omega}, E)}{S(\mathbf{r}, \boldsymbol{\Omega})}dE. \tag{10.10}$$

Multiplying (10.8)–(10.10), we get

$$\mathsf{P}\{\mathbf{r}_0 \in d\mathbf{r}\}\mathsf{P}\{\boldsymbol{\Omega}_0 \in \boldsymbol{\Omega}|\mathbf{r}_0 = \mathbf{r}\}\mathsf{P}\{E_0 \in dE|\mathbf{r}_0 = \mathbf{r}, \boldsymbol{\Omega}_0 = \boldsymbol{\Omega}\} = S(\mathbf{r}, \boldsymbol{\Omega}, E)d\mathbf{r}d\boldsymbol{\Omega}dE,$$

i.e., having chosen $\mathbf{r}_0$, $\boldsymbol{\Omega}_0$, and E_0 in succession from the distributions

$$p_{\mathbf{r}_0}(\mathbf{r}) = S(\mathbf{r}); \; p_{\boldsymbol{\Omega}_0} = (\boldsymbol{\Omega}|\mathbf{r}_0) = S(\mathbf{r}_0, \boldsymbol{\Omega})/S(\mathbf{r}_0),$$

$$p_{E_0}(E|\mathbf{r}_0, \boldsymbol{\Omega}_0) = S(\mathbf{r}_0, \boldsymbol{\Omega}_0, E)/S(\mathbf{r}_0, \boldsymbol{\Omega}_0), \tag{10.11}$$

find a set of random variables $(\mathbf{r}_0, \boldsymbol{\Omega}_0, E_0)$ with distribution $S(\mathbf{r}, \boldsymbol{\Omega}, E)$. If the variables in the source density function are separated:

$$S(\mathbf{r}, \boldsymbol{\Omega}, E) = S_{\mathbf{r}}(\mathbf{r})S_{\boldsymbol{\Omega}}(\boldsymbol{\Omega})S_E(E), \tag{10.12}$$

then the corresponding random variables are modeled independently from each other. Indeed, substituting (10.12) into (10.11), we get

$$p_{\mathbf{r}_0}(\mathbf{r}) = S_{\mathbf{r}}(\mathbf{r}); \quad p_{\boldsymbol{\Omega}_0}(\boldsymbol{\Omega}) = S_{\boldsymbol{\Omega}}(\boldsymbol{\Omega}); \quad p_{E_0}(E) = S_E(E). \tag{10.13}$$

In turn, the coordinates and direction of particle motion are also a set of random quantities $\mathbf{r} = \{x, y, z\}$ and $\boldsymbol{\Omega} = \{\cos\vartheta, \varphi\}$, and for sampling from distributions $p_{\mathbf{r}_0}(\mathbf{r})$ and $p_{\boldsymbol{\Omega}_0}(\boldsymbol{\Omega})$ one can again use relations similar to (10.11):

$$p_{x_0}(x) = S(x); \quad p_{y_0}(y) = S(x_0, y)/S(x_0);$$

$$p_{z_0}(z) = S(x_0, y_0, z)/S(x_0, y_0); \tag{10.14}$$

$$S(x) = \iint S(x, y, z)dydz; \quad S(x, y) = \int S(x, y, z)dz \quad \ldots$$

Let, for example, an isotropic particle source with energy $E_0 = \text{const}$ uniformly distributed in a cube with edge a and center at the beginning coordinates. Then, in the source density function, the variables are divided

$$S(\mathbf{r}, \boldsymbol{\Omega}, E)d\mathbf{r}d\boldsymbol{\Omega}dE = (d\mathbf{r}/a^3)(d\boldsymbol{\Omega}/4\pi)\delta(E - E_0)dE, \tag{10.15}$$

moreover, in the Cartesian system, spatial variables $S(\mathbf{r})d\mathbf{r} = (dx/a)(dy/a)(dz/a)$; $-a/2 < x < a/2$; $-a/2 < y < a/2$; $-a/2 < z < a/2$. According to (10.5), the choice of coordinates from this distribution can be produced by the formulas:

$$x_0 = a\xi - a/2; \quad y_0 = a\xi - a/2; \quad z_0 = a\xi - a/2, \tag{10.16}$$

where all ξ are independent random numbers with uniform distribution on (0,1).

To choose a direction, we note that

$$\mathsf{P}\{\boldsymbol{\Omega}_0 \in d\boldsymbol{\Omega}\} = d\boldsymbol{\Omega}/4\pi = (1/2)d\cos\vartheta \cdot (1/2\pi)d\varphi,$$

i.e., $\cos\vartheta_0$ and φ_0 are distributed uniformly and independently on intervals $(-1, +1)$ and $(0, 2\pi)$, respectively, so one can use again the formula (10.5):

$$\cos\vartheta_0 = 2\xi - 1; \quad \varphi_0 = 2\pi\xi. \tag{10.17}$$

The energy E_0 in this case is not a random variable and therefore is simply assigned.

If the source is uniformly distributed in a ball of radius a, then Cartesian variables in the $S(\mathbf{r})$ function are no longer separated. In that case, it is convenient to pass to spherical coordinates (r, α, β), in which

$$S(\mathbf{r})d\mathbf{r} = \frac{d\mathbf{r}}{(4/3)\pi a^3} = \frac{3r^2 dr}{a^3}\,\frac{d\cos\alpha}{2}\,\frac{d\beta}{2\pi}. \tag{10.18}$$

The radial distribution $\mathsf{P}\{r_0 < r\} = \int_0^r 3r'^2 dr'/a^3 = (r/a)^3$, and according to the inverse function method

$$r_0 = a\sqrt[3]{\xi}. \tag{10.19}$$

Angles α_0 and β_0 are found from formulas similar to (10.17).

Free path length. It is simulated by Eq. (2.2).

Knowing the coordinates of the birth point of the particle, the direction of its movement, and free path length, it is easy to find the point of the first interaction particles with the medium

$$\mathbf{r}_1 = \mathbf{r}_0 + l_1 \boldsymbol{\Omega}_0, \tag{10.20}$$

or, in Cartesian projections,

$$\begin{aligned} x_1 &= x_0 + l_1 \sin\vartheta_0 \cos\varphi_0; \\ y_1 &= y_0 + l_1 \sin\vartheta_0 \sin\varphi_0; \\ z_1 &= z_0 + l_1 \cos\vartheta_0. \end{aligned} \tag{10.21}$$

Interaction type. Since in the process of free movement particle direction of motion $\boldsymbol{\Omega}_0$ and energy E_0 do not change, the state of the particle at the time of the first collision is characterized by parameters $\mathbf{r}_1 = \mathbf{r}_0 + l_1\boldsymbol{\Omega}_0$, $\boldsymbol{\Omega}_1 = \boldsymbol{\Omega}_0$, $E_1 = E_0$. As a result of the collision, the particle is scattered (survives) with probability $p_s = \Sigma_s(E_1)/\Sigma(E_1)$ or absorbed (die) with probability $p_c = 1 - p_s = \Sigma_c(E_1)/\Sigma(E_1)$ (if there are no other processes). According to (10.1),

$$\mathsf{P}\{\xi < \Sigma_s(E_1)/\Sigma(E_1)\} = \Sigma_s(E_1)/\Sigma(E_1), \tag{10.22}$$

therefore, to choose the collision outcome, it suffices to get random number ξ and compare it with the survival probability :

$$\left.\begin{array}{l} \text{if } \xi < \Sigma_s(E_1)/\Sigma(E_1), \text{ scattering;} \\ \text{if } \xi > \Sigma_s(E_1)/\Sigma(E_1), \text{ absorption.} \end{array}\right\} \tag{10.23}$$

In the case of absorption, the trajectory breaks off and begins simulation of the next trajectory. In case of scattering, go to drawing the direction and energy of the particle after scattering.

Scattering. In the process of scattering, the particle randomly changes direction of motion $\boldsymbol{\Omega}_1 \to \boldsymbol{\Omega}_2$ and energy $E_1 \to E_2$ according to differential scattering cross section:

$$\mathsf{P}\{\boldsymbol{\Omega}_2 \in d\boldsymbol{\Omega}, E_2 \in dE|\boldsymbol{\Omega}, E_1\} = \Sigma_s^{-1}(E_1)E_s(\boldsymbol{\Omega}_1 \to \boldsymbol{\Omega}_1, E_1 \to E)d\boldsymbol{\Omega}dE. \tag{10.24}$$

According to the above, drawing from this distribution can be be carried out by first choosing the direction $\boldsymbol{\Omega}_2$ from the distribution

$$p_{\boldsymbol{\Omega}_2}(\boldsymbol{\Omega}) = \Sigma_s^{-1}(E_1)\int \Sigma_s(\boldsymbol{\Omega}_1 \to \boldsymbol{\Omega}, E_1 \to E)dE = \Sigma_s^{-1}(E_1)\Sigma_s(\boldsymbol{\Omega}_1 \to \boldsymbol{\Omega}, E_1), \tag{10.25}$$

and then the energy E_2 from the distribution

$$p_{E_2}(E|\boldsymbol{\Omega}_2) = \Sigma_s(\boldsymbol{\Omega}_1 \to \boldsymbol{\Omega}_2, E_1 \to E)/\Sigma_s(\boldsymbol{\Omega}_1 \to \boldsymbol{\Omega}_2, E_1). \tag{10.26}$$

When choosing $\boldsymbol{\Omega}_2$, one should take into account that the section $\Sigma_s(\boldsymbol{\Omega}_1 \to \boldsymbol{\Omega}, E_1)$ depends on $\cos\theta = \boldsymbol{\Omega}_1\boldsymbol{\Omega}$ and uniform in the azimuthal angle. Therefore, one should first find the scattering angles θ_1 and χ_1, and then pass to angles ϑ_2, φ_2, determining the direction of the particle after scattering, according to the known formulas of spherical trigonometry:

$$\cos\vartheta_2 = \cos\vartheta_1\cos\theta_1 + \sin\vartheta_1\sin\theta_1\cos\chi_1;$$

$$\varphi_2 = \varphi_1 + \Delta\varphi_1;$$

$$\sin\Delta\varphi_1 = \sin\theta_1\sin\chi_1/\sin\vartheta_2;$$

$$\cos\Delta\varphi_1 = (\cos\theta_1 - \cos\vartheta_1\cos\vartheta_2)/\sin\vartheta_1\sin\vartheta_2. \tag{10.27}$$

If the energy of the scattered particle is uniquely related to the angle scattering (as is the case in elastic collisions), the magnitude E_2 is not played, but found from the known energy E_1 and the angle scattering θ_1.

Sometimes, it is more convenient to play energy first and then play (or calculate) scattering angle. This is how they do it when modeling Compton scattering of γ-quanta [7].

Trajectory continuation. Simulating the path length of the particle after first collision,

$$l_2 = -\ln\xi/\Sigma(E_2), \tag{10.28}$$

they find point

$$\mathbf{r}_2 = \mathbf{r}_1 + l_2\boldsymbol{\Omega}_2 \tag{10.29}$$

and so on. Thus, formulas (10.23)–(10.29) allow us to sample the point $\mathbf{x}_2$ in the phase space knowing point $\mathbf{x}_1$ and probabilistic characteristics of elementary processes. Replacing the index 1 in these formulas by the index k, and index 2 by $k+1$, we arrive at the algorithm modeling arbitrary transition $\mathbf{x}_k \to \mathbf{x}_{k+1}$, the sequential use of which allows us to continue modeling the trajectory until its absorption or leaving the scattering medium.

The simplest example of a matter inhomogeneity is its boundary with vacuum. Construction of random trajectories in such problems ends not only with the absorption of a particle but also in the event that the next path length is greater than distance from the point where the last collision occurred to borders (in the travel direction).

Consider an arbitrary inhomogeneous medium. When deriving the formula (10.28) the expression used to play the free path length was $\exp(-\Sigma_s l)$, which describes the attenuation of unscattered radiation at a distance s in a homogeneous medium. Similarly, one can obtain the corresponding formula for an inhomogeneous medium, based on a more general expression $\exp[-\tau(\mathbf{r}', \mathbf{r}, E)]$, where $\mathbf{r}' = \mathbf{r} + s\boldsymbol{\Omega}$; $\tau(\mathbf{r}, \mathbf{r}', E) = \int_0^s \Sigma(\mathbf{r} + s'\boldsymbol{\Omega}, E)ds'$ is the *optical distance along the direction particle motion*. Thus, the probability that the free path length falls into ds is equal to

$$\mathsf{P}\{s < l < s + ds\} = \Sigma(\mathbf{r} + s\boldsymbol{\Omega}, E)\exp[-\tau(\mathbf{r}, \mathbf{r} + s\boldsymbol{\Omega}, E)]ds. \tag{10.30}$$

Algorithm for sampling from distribution (10.30) can be obtained with using the distribution function method, which leads to the following formula for the free path length l:

$$-\log\xi = \int_0^l \Sigma(\mathbf{r} + s'\boldsymbol{\Omega}, E)ds' = \tau(\mathbf{r}, \mathbf{r} + l\boldsymbol{\Omega}, E). \tag{10.31}$$

If a medium consists of several homogeneous regions, then the function $\Sigma(\mathbf{r}, E)$ is piecewise constant and the interval in (10.31) becomes the sum

$$\tau(\mathbf{r}, \mathbf{r} + l\boldsymbol{\Omega}, E) = \sum_{k=1}^{m} l^{(k)}\Sigma^{(k)}(E) + l'\Sigma^{(m+1)}(E). \tag{10.32}$$

Here $l^{(k)}$ are segments on which the boundaries cut the ray drawn from the last point of collision to the particle movement direction; $\Sigma^{(k)}$—corresponding to these region attenuation coefficient; m is the number of borders that the particle crossed on the path between two collisions (it is defined as the maximum index value k for which $\sum_k l^{(k)}\Sigma^{(k)}(E) < -\log(\xi)$); l'—length particle range in the region with number $(m + 1)$, which is interaction. Thus, the formula for drawing the length of the free path has the form

$$l = \sum_{k=1}^{m} l^{(k)} - [1/\Sigma^{(m+1)}(E)]\left[\sum_{k=1}^{m} l^{(k)}\Sigma^{(k)}(E) + \log\xi\right]. \tag{10.33}$$

When determining the length of free path, the number of the region $m + 1$ is also found, in which the collision occurred. Interaction cross sections in each areas are assumed to be known, so sampling the interactions, direction of motion and energy of the particle after collisions are carried out using the ordinary Monte Carlo methods.

There is another method for constructing trajectories in an inhomogeneous medium [8]. Let us write down the kinetic equation in the form

$$\boldsymbol{\Omega}\nabla\Phi(\mathbf{r},\boldsymbol{\Omega},E)+\Sigma(\mathbf{r},E)\Phi(\mathbf{r},\boldsymbol{\Omega},E)-\hat{\mathrm{K}}_s\Phi(\mathbf{r},\boldsymbol{\Omega},E)=S(\mathbf{r},\boldsymbol{\Omega},E)$$

and make on the left-hand side the following transformation:

$$\Sigma(\mathbf{r},E)=\Sigma(E)-[\Sigma(E)-\Sigma(\mathbf{r},E)]$$

$$=\Sigma(E)-[\Sigma(E)-\Sigma(\mathbf{r},E)]\int d\boldsymbol{\Omega}'\int dE'\delta(\boldsymbol{\Omega}'-\boldsymbol{\Omega})\delta(E'-E),$$

where $\Sigma(E)$ is a positive function, satisfying the condition

$$\Sigma(E)\leqslant\Sigma(\mathbf{r},E). \tag{10.34}$$

The quantity

$$\Sigma_\delta(\mathbf{r},\boldsymbol{\Omega}'\to\boldsymbol{\Omega},E'\to E)=[\Sigma(E)-\Sigma(\mathbf{r},E)]\delta(\boldsymbol{\Omega}'-\boldsymbol{\Omega})\delta(E'-E)$$

can be considered as a differential cross section of the modified process of scattering and rewrite the initial equation in the form

$$\boldsymbol{\Omega}\nabla\Phi(\mathbf{r},\boldsymbol{\Omega},E)+\Sigma(E)\Phi(\mathbf{r},\boldsymbol{\Omega},E)-(\hat{\mathrm{K}}_s+\hat{\mathrm{K}}_\delta)\Phi(\mathbf{r},\boldsymbol{\Omega},E)=S(\mathbf{r},\boldsymbol{\Omega},E), \tag{10.35}$$

where $\hat{\mathbf{K}}_\delta$ is the integral operator of δ-scattering with kernel Σ_δ.

Eq. (10.35) describes the transport of particles in a medium with a linear attenuation coefficient $\Sigma(E)$ independent of coordinates, therefore to draw the length of free path, one can use the usual formula (10.4). However, when we follow delta-formalism, we should write the cross-section $\Sigma(E)$ as the sum of the sections: real interactions and δ-scattering (virtual interaction): $\Sigma(E)=\Sigma(\mathbf{r},E)+[\Sigma(E)-\Sigma(\mathbf{r},E)]$ and transform this formula to the form $\Sigma(\mathbf{r},E)/\Sigma(E)+[\Sigma(E)-\Sigma(\mathbf{r},E)]/\Sigma(E)=1$, which defines the relative probabilities such processes. Then the sequence of calculations for constructing random trajectories will be the following:

1. Play the free path l in a homogeneous medium with attenuation coefficient $\Sigma(E)$.
2. Calculate the coordinates of the collision point $\mathbf{r}_1=\mathbf{r}+l\boldsymbol{\Omega}$.
3. Using the coordinates $\mathbf{r}_1$, we determine in which area the interaction occurs.
4. Calculate for this region the probability of δ-scattering

$$p_\delta=[\Sigma(E)-\Sigma(\mathbf{r},E)]/\Sigma(E).$$

5. Determine the type of interaction: if $\xi < p_\delta$, then δ-scatter, otherwise one of the usual interactions.

Observe, that δ-scattering does not change the energy and direction of particle motion, and calculations for ordinary interactions are organized as described above. As a function $\Sigma(E)$ it is convenient to choose $\Sigma_{\max}(E)$—the maximum of attenuation coefficients in an inhomogeneous absorber. In this case, condition (10.34) is executed automatically. Using functions $\Sigma(E) > \Sigma_{\max}(E)$ is obviously inexpedient, since this will lead to decrease in path length and increase in the number of collisions per history.

10.4 The Measured Characteristics of Particle Transport

If, when modeling random trajectories, we use natural probabilistic laws, corresponding to real passage of particles through a matter, then the constructed trajectories are statistically equivalent to real ones, and can be directly used for evaluation of the desired characteristics of the process. So, it is easy to determine how much energy is lost by particles in some volume, how often a particle collides with atomic nucleus, and how many particles cross a given surface. These quantities are, of course, random variables. The Monte Carlo process gives approximate values of these quantities represented by sample means, obtained by averaging over sample ensemble. Results of such calculations should be accompanied by a quantitative assessment of their accuracy according to all the rules of statistical analysis.

Some characteristics of the transport process may have several equivalent definitions. So, the differential flux density of particles is expressed in terms of the density of collisions, in terms of the number of intersections of a unit area and through the path of particles in a unit the volume of the phase space. It is very convenient to formalize the process measurement of characteristics by introducing the concept of an additive functional (additive detector) $J = J[\Phi(\cdot)]$, where

$$\Phi(\mathbf{x}) \approx (1/\Delta\mathbf{x})(1/N)\sum_i q_\Phi(\mathbf{x}_i), \tag{10.36}$$

where N is the number of simulated trajectories; summation runs over all collisions of a given set of trajectories; $\mathbf{x}_i$—phase coordinates of collision points; $q_\Phi(\mathbf{x}_i)$—contribution to the flux density from one collision:

$$q_\Phi(\mathbf{x}_i) = \begin{cases} 1/\Sigma(\mathbf{x}_i), & \text{if } \mathbf{x}_i \in \Delta\mathbf{x} \\ 0, & \text{otherwise.} \end{cases} \tag{10.37}$$

Consider now a small area Δs centered at the point $\mathbf{r}$ and normal $\mathbf{n}$. The approximate value of the density flow can be obtained by counting the number of intersections in this area:

$$\Phi(\mathbf{r}, \boldsymbol{\Omega}, E) \approx (1/\Delta s\Delta\boldsymbol{\Omega}\Delta E)(1/N)\sum_i q'_\Phi(\mathbf{x}_i), \tag{10.38}$$

where

$$q'_{\Phi}(\mathbf{x}_i) = \begin{cases} \dfrac{1}{|\boldsymbol{\Omega}_i \mathbf{n}|}, & \text{if the segment } \mathbf{r}_i - \mathbf{r}_{i-1} \text{ crosses} \\ & \text{the area } \Delta s, \text{a } \boldsymbol{\Omega}_i \in \Delta\boldsymbol{\Omega},\ E_i \in \Delta E; \\ 0, & \text{otherwise.} \end{cases} \tag{10.39}$$

Formulas (10.38), (10.39) give estimation of flux density by crossing events.

$$\Phi(\mathbf{x}) \approx (1/\Delta x)(1/N) \sum_i q''_{\Phi}(\mathbf{x}_i), \tag{10.40}$$

where $q''_{\Phi}(\mathbf{x}_i)$ is that part of the length of the segment $\mathbf{r}i - \mathbf{r}_{i-1}$ which lies in the volume ΔV provided that $\boldsymbol{\Omega}_i \in \Delta\boldsymbol{\Omega}$, $E_i \in \Delta E$. Formula (10.40) determines the score flux density along run length.

It is easy to see that the formulas (10.36)–(10.40) give the average value of the flux density in the region $\Delta V \Delta\boldsymbol{\Omega} \Delta E$ or $\Delta s \Delta\boldsymbol{\Omega} \Delta E$. To make it the value differed less from the computed value $\Phi(\mathbf{r}, \boldsymbol{\Omega}, E)$, dimensions areas must be small enough. However, this reduces the probability of a particle hitting a given area, which increases statistical estimation error (10.7).

The signal of an additive detector having response function $D(\mathbf{x})$ can be written as a linear functionality from $\Phi(\mathbf{x})$:

$$J = \int D(\mathbf{x})\Phi(\mathbf{x})d\mathbf{x}. \tag{10.41}$$

To obtain an algorithm for calculating such a field characteristic radiation, we divide the entire phase space into elementary volumes $\Delta\mathbf{x}_\alpha$, write formula (10.36) for any of them: $\Phi(\mathbf{x}_\alpha) = (1/\Delta\mathbf{x}_\alpha)(1/N)\sum_i q_\Phi \mathbf{x}_i$, multiply both sides of this equality by $D(\mathbf{x}_\alpha)\Delta\mathbf{x}_\alpha$ and sum over α. Then, taking into account that under $\Delta\mathbf{x}_\alpha \to 0$ $\sum_\alpha D(\mathbf{x}_\alpha)\Phi(\mathbf{x}_\alpha)\Delta\mathbf{x}_\alpha \to \int D(\mathbf{x})\Phi(\mathbf{x})d\mathbf{x}$, we obtain

$$J \approx J_N \sum_i q_D(\mathbf{x}_i), \tag{10.42}$$

where

$$q_D(\mathbf{x}_i) = [1/\Sigma(\mathbf{x}_i)]D(\mathbf{x}_i). \tag{10.43}$$

This means that an estimate of the additive detector readings can be found by averaging along the trajectories the sum of individual collision contributions. Formula (10.43) determines the value of the contribution. It is easy to show that this formula agrees with the physical meaning of the sensitivity function. Rewrite this (10.41) as

$$J = \int \bar{q}_D(\mathbf{x}) F(\mathbf{x}) d\mathbf{x}, \tag{10.44}$$

where $F(\mathbf{x})$ is the collision density and $\bar{q}_D(\mathbf{x}) = [1/\Sigma(\mathbf{x})]D(\mathbf{x})$ is a mean contribution of a collision at the phase point $\mathbf{x}$.

Calculations in which trajectory modeling is carried out with using real transition probabilities, and the characteristics radial fields are calculated according to their physical sense, are called the *analog modeling*.

10.5 Non-analog Simulations of Detector Response

In analog simulation of the readings of the $*$ detector, its response to each interaction of a particle depends on the input and output parameters of the collision and, for this reason, is a random variable.

$$q_D(\mathbf{x}) = \begin{cases} q_c(\mathbf{x}), & \text{if the particle is absorbed;} \\ q_s(\boldsymbol{\Omega} \to \boldsymbol{\Omega}', E \to E'), & \text{if the particle} \\ & \text{scattered } (\boldsymbol{\Omega} \to \boldsymbol{\Omega}', E \to E'). \end{cases}$$

Meanwhile, formulas

$$J = \int D(\mathbf{x}) \Phi(\mathbf{x}) d\mathbf{x} = \int \bar{q}_D(\mathbf{x}) F(\mathbf{x}) d\mathbf{x} \tag{10.45}$$

and kinetic equations for the flux density $\Phi(\mathbf{x})$ and the collision density $F(\mathbf{x}) = \Sigma(\mathbf{x})\Phi(\mathbf{x})$ contain only the average values of this contribution. Therefore, taking as a node contribution of the synthetic trajectory instead of $q_D(\mathbf{x})$ another expression $\tilde{q}_D(\mathbf{x})$ such that

$$[\Sigma_c(E)/\Sigma(E)]\tilde{q}_c(\mathbf{r}, \boldsymbol{\Omega}, E) + [1/\Sigma(E)] \int d\boldsymbol{\Omega}' \int dE' \Sigma_s(\boldsymbol{\Omega} \to \boldsymbol{\Omega}', E \to E')$$

remains the same as before, then J will not change. In many cases, it is advisable to put

$$\tilde{q}_c(\mathbf{r}, \boldsymbol{\Omega}, E) = \tilde{q}_s(\mathbf{r}, \boldsymbol{\Omega} \to \boldsymbol{\Omega}', E \to E') = \bar{q}_D(\mathbf{r}, \boldsymbol{\Omega}, E),$$

i.e., to replace each random collision contribution by its mean value (a sort of *analytical averaging*). Such a replacement may reduce the variance of results.

Let, for example, it is required to find the probability of absorption of a particle in some volume ΔV. In the analog method, this requires to construct N trajectories and compute the ratio of the number of "absorbed" here trajectories to the total number of constructed ones. In other words, at such simulation $q_s = 0$, while $q_c = 1$ and $\bar{q}_D = \Sigma_c/\Sigma$. Assuming that, in each collision $\tilde{q}_D = \Sigma_c/\Sigma$, we obtain a more efficient way of estimating the number acquisitions. The increase in efficiency is due to the fact that the contribution to the calculated value will now be given by all collisions, occurred in the volume of ΔV, and not just takeovers, as was the case in analog rating.

In most problems of transport theory, the calculation of the contribution of non-scattered particles is not difficult and the main attention is paid to scattered radiation. Having written the integral transport equation in operator form:

$$\Phi = \Phi_0 + \hat{\mathrm{K}}\Phi, \tag{10.46}$$

we will see that the flux density of scattered particles Φ_s is related to the flux density of all particles Φ by the relation $\Phi_s = \hat{\mathrm{K}}\Phi$, and its contribution into the functional has the form $J_s = (D, \hat{\mathrm{K}}\Phi)$. Passing to the adjoint operator, rewrite this formula as

$$J_s = (\Phi, \hat{\mathrm{K}} + D) = (F, H_s), \tag{10.47}$$

where F stands for the collision density, and

$$H_s(\mathbf{x}) = [1/\Sigma(\mathbf{x})]\hat{\mathrm{K}} + D(\mathbf{x}) \tag{10.48}$$

is the mean value of the next collision contribution conditioned by collision in point $\mathbf{x} = (\mathbf{r}, \boldsymbol{\Omega}, E)$. Comparing (10.47) with (10.45), we see that J_s can be calculated, modeling random trajectories and counting the contribution of each collision equal to $H_s(\mathbf{x})$. By analogy with the above, this method can be be considered as an averaging of contributions not only over outcomes collisions at the point $\mathbf{x}$ but also over all possible points of the next collisions.

The formulas (10.47), (10.48) are used to evaluate local characteristics. In this case,

$$D(\mathbf{r}, \boldsymbol{\Omega}, E) = D'(\boldsymbol{\Omega}, E)\delta(\mathbf{r} - \mathbf{r}_0)$$

and

$$\left.\begin{aligned} h_s(\mathbf{x}) &= \int dE' \frac{\Sigma_s(\mathbf{r}, \boldsymbol{\Omega} \to \boldsymbol{\Omega}_D, E \to E')}{\Sigma(\mathbf{r}, E)} \frac{\exp[-\tau(\mathbf{r}, \mathbf{r}_0, E')]D'(\boldsymbol{\Omega}_D, E')}{|\mathbf{r} - \mathbf{r}_0|^2}; \\ \boldsymbol{\Omega}_D &= \frac{\mathbf{r}_0 - \mathbf{r}}{|\mathbf{r}_0 - \mathbf{r}|}. \end{aligned}\right\} \tag{10.49}$$

In contrast to the methods discussed above for estimating local characteristics, the latter method has some advantages: each collision contributes to the desired value, and the mathematical expectation of the contribution of the trajectory is exactly the same as desired value. The disadvantage of this method is relatively a slow decrease in the statistical error with an increase in N, due to the infinite value of the variance.

Expressing Φ_0 and $\hat{\mathrm{K}}$ in terms of the operators $\hat{\mathrm{G}}_0$ and $\hat{\mathrm{K}}_s$ and substituting them into the formulas (10.45), (10.46), we obtain

$$J = (D, \hat{\mathrm{G}}_0 S) + (D, \hat{\mathrm{G}}_0\hat{\mathrm{K}}_s\Phi).$$

Using the definition of adjoint operators, we transform this expression:

$$J = (\hat{\mathrm{G}}_0^+ D, S + \hat{\mathrm{K}}_s \Phi) = (h, \tilde{S}), \tag{10.50}$$

where

$$\tilde{S}(\mathbf{r}, \boldsymbol{\Omega}, E) = S(\mathbf{r}, \boldsymbol{\Omega}, E) + \int d\boldsymbol{\Omega}' \int dE' \Sigma_s(\mathbf{r}, \boldsymbol{\Omega}' \to \boldsymbol{\Omega}, E' \to E)\Phi(\mathbf{r}, \boldsymbol{\Omega}', E')$$

is the effective source density, and

$$H(\mathbf{r}, \boldsymbol{\Omega}, E) = \Phi_0^+(\mathbf{r}, \boldsymbol{\Omega}, E) = \int_0^\infty \exp[-\tau(\mathbf{r}, \mathbf{r}', E')]D(\mathbf{r}', \boldsymbol{\Omega}, E)d\xi, \tag{10.51}$$

$$\mathbf{r}' = \mathbf{r} + \xi\boldsymbol{\Omega}$$

is the contribution of a particle emitted from the point $\mathbf{r}$ in the direction $\boldsymbol{\Omega}$ with energy E averaged over the next collision in the detector. This result is a base for one more estimator of J: immediately after the birth of a particle in the source and after each scattering the average contributions $H(\mathbf{x})$ are calculated and then are summed. In particular when estimating the total flux density in plane-parallel geometry behind a barrier of thickness t, the contribution of the particle $H(z, \boldsymbol{\Omega}, E)$ after scattering by depth z is equal to $(1/|\cos\vartheta|)\exp\{-\Sigma(E)(t-z)/\cos\vartheta\}$ if the angle ϑ between the z-axis and direction of particle motion is less than $\pi/2$, and is equal to zero otherwise.

10.6 Non-analog Simulation of Trajectories

It was shown above that the average value of an additive detector readings is represented as a linear functional from the solution of some integral equation. At the same time, it can be calculated statistical modeling of the particle trajectory, so the method Monte Carlo can be considered as a numerical method for solving integral equations. Let's dwell on this in more detail.

Let us consider a random process of particles wandering in some phase space, whose points we denote by $\mathbf{x}$. Let $p_1(\mathbf{x})$ be the distribution density for the first collisions of the particle; $p_0(\mathbf{x})$ be the probability absorption of a particle upon collision at the point $\mathbf{x}$; $p(\mathbf{x} \to \mathbf{x}')d\mathbf{x}'$ be the probability of particle scattering at the point $\mathbf{x}$ followed by a collision at element $d\mathbf{x}'$ of the phase space. Integral

$$p_s(\mathbf{x}) = \int p(\mathbf{x} \to \mathbf{x}')d\mathbf{x}' \tag{10.52}$$

represents the probability that after a collision in point $\mathbf{x}$ the particle will experience at least one more collision. When of an infinite medium, this quantity is the probability of particle survival upon collision at the point $\mathbf{x}$. It is easy to see that

$$p_0(\mathbf{x}) + p_s(\mathbf{x}) = 1. \tag{10.53}$$

These definitions imply that with probability $p_1(\mathbf{x}_1)d\mathbf{x}_1 p_0(\mathbf{x}_1)$ the first particle collision will occur at the element $d\mathbf{x}_1$ and this collision will be absorption, i.e., there will be only one on the trajectory collision. Similarly $p_1(\mathbf{x}_1)d\mathbf{x}_1 p(\mathbf{x}_1 \to \mathbf{x}_2)d\mathbf{x}_2 p_0(\mathbf{x}_2)$ there is a probability that on the trajectory will have two collisions—scattering in the volume element $d\mathbf{x}_1$ and absorption in the element $d\mathbf{x}_2$. In the same way is the probability of experiencing any number of collisions n:

$$p_1(\mathbf{x}_1)d\mathbf{x}_1 p(\mathbf{x}_1 \to \mathbf{x}_2)d\mathbf{x}_2 p(\mathbf{x}_2 \to \mathbf{x}_3)d\mathbf{x}_3 \ldots$$

$$p(\mathbf{x}_{n-1} \to \mathbf{x}_n)d\mathbf{x}_n p_0(\mathbf{x}_n). \tag{10.54}$$

The contribution of the trajectory to the detector readings is determined by the sequence of collision points:

$$\mathbf{x}_1, \mathbf{x}_2, \ldots, \mathbf{x}_n \tag{10.55}$$

($\mathbf{x}_n$—phase coordinates of the collision points where particle absorption).

We will say that on the set of trajectories some functional if each trajectory (10.55) is placed in matching a certain number Q:

$$Q = \Phi_n(\mathbf{x}_1, \mathbf{x}_2, \ldots, \mathbf{x}_n), \tag{10.56}$$

where $\{\Phi_n\}$ is a given set of functions. The quantity Q is random, given the number of collisions in the trajectory and phase coordinates of collision points. Given (10.54), (10.56), the expectation of the functional Q can be written as as

$$\bar{Q} = \sum_{n=1}^{\infty} \int d\mathbf{x}_1 \int d\mathbf{x}_2 \ldots \int d\mathbf{x}_n p_1(\mathbf{x}_1) p(\mathbf{x}_1 \to \mathbf{x}_2)$$

$$p(\mathbf{x}_2 \to \mathbf{x}_3) \ldots p(\mathbf{x}_{n-1} \to \mathbf{x}_n) p_0(\mathbf{x}_n) \Phi_n(\mathbf{x}_1, \mathbf{x}_2, \ldots, \mathbf{x}_n). \tag{10.57}$$

Each term of the sum (10.57) contains the factor $p_1(\mathbf{x}_1)$ under the sign integral over the variable $\mathbf{x}_1$. Take out this common factor and rewrite (10.57) as follows:

$$\bar{Q} = \int d\mathbf{x}_1 p_1(\mathbf{x}_1) \bar{Q}_1(\mathbf{x}_1), \tag{10.58}$$

where

$$\bar{Q}_1(\mathbf{x}_1) = p_0(\mathbf{x}_1)\Phi_1(\mathbf{x}_1) + \sum_{n=2}^{\infty} \int d\mathbf{x}_2 \int d\mathbf{x}_3 \ldots \int d\mathbf{x}_n\, p(\mathbf{x}_1 \to \mathbf{x}_2)p(\mathbf{x}_2 \to \mathbf{x}_3)\ldots$$

$$p(\mathbf{x}_{n-1} \to \mathbf{x}_n)p_0(\mathbf{x}_n)\Phi_n(\mathbf{x}_1, \mathbf{x}_2, \ldots, \mathbf{x}_n). \tag{10.59}$$

This quantity is obviously the mean value of the functional Q on trajectories with a fixed value of the coordinate of the first collisions. Since the function $p_1(\mathbf{x})$ is assumed to be known, for calculation of $\bar{Q}$ by the formula (10.58), you need to know the conditional the mean $\bar{Q}_1(\mathbf{x})$.

The sum in the formula (10.59) can be converted in a similar way, taking out brackets the common factor $p(\mathbf{x}_1 \to \mathbf{x}_2)$ and writing this sum as

$$\int d\mathbf{x}_2\, p(\mathbf{x}_1 \to \mathbf{x}_2)\bar{Q}_n(\mathbf{x}_1, \mathbf{x}_2), \tag{10.60}$$

where

$$\bar{Q}_2(\mathbf{x}_1, \mathbf{x}_2) = p_0(\mathbf{x}_2)\Phi_2(\mathbf{x}_1, \mathbf{x}_2)$$

$$+ \sum_{n=3}^{\infty} \int d\mathbf{x}_3 \int d\mathbf{x}_4 \ldots \int d\mathbf{x}_n\, p(\mathbf{x}_2 \to \mathbf{x}_3)$$

$$\times\, p(\mathbf{x}_3 \to \mathbf{x}_4)\ldots p(\mathbf{x}_{n-1} \to \mathbf{x}_n)p_0(\mathbf{x}_n)\Phi_n(\mathbf{x}_1, \mathbf{x}_2, \ldots, \mathbf{x}_n)$$

is the mean value of the functional Q on trajectories conditioned two first collision points.

Substituting (10.60) into (10.59), we get the formula

$$\bar{Q}(\mathbf{x}_1) = p_0(\mathbf{x}_1)\Phi_1(\mathbf{x}_1) + \int d\mathbf{x}_2\, p(\mathbf{x}_1 \to \mathbf{x}_2)\bar{Q}_2(\mathbf{x}_1, \mathbf{x}_2), \tag{10.61}$$

which expresses the quantity $\bar{Q}(\mathbf{x}_1)$ needed to calculate expectation $\bar{Q}$, through the conditional mean $\bar{Q}(\mathbf{x}_1, \mathbf{x}_2)$. Similarly, $\bar{Q}(\mathbf{x}_1, \mathbf{x}_2)$ can be expressed in terms of $\bar{Q}(\mathbf{x}_1, \mathbf{x}_2, \mathbf{x}_3)$, etc. These calculations lead to the recurrence relation

$$\bar{Q}(\mathbf{x}_1, \mathbf{x}_2, \ldots, \mathbf{x}_n) = p_0(\mathbf{x}_n)\Phi_n(\mathbf{x}_1, \mathbf{x}_2, \ldots, \mathbf{x}_n)$$

$$+ \int d\mathbf{x}_{n+1}\, p(\mathbf{x}_n \to \mathbf{x}_{n+1})\bar{Q}_{n+1}(\mathbf{x}_1, \mathbf{x}_2, \ldots, \mathbf{x}_{n+1}), \tag{10.62}$$

which together with the formula (10.58) allows us to write mathematical expectation $\bar{Q}$ as an infinite series, equivalent to (10.57). However, under certain restrictions, imposed on functions Φ, conditional mean $\bar{Q}_{n+1}(\mathbf{x}_1, \mathbf{x}_2, \ldots, \mathbf{x}_{n+1})$ for some n is expressed in terms of $\bar{Q}(\mathbf{x}_1, \mathbf{x}_2, \ldots, \mathbf{x}_k)$, where $k \leqslant n$, and recurrence relation (10.62) becomes a closed system of equations for $\bar{Q}(\mathbf{x}_1, \mathbf{x}_2, \ldots, \mathbf{x}_k)$, $k \leqslant n$. In this case, the calculation of $\bar{Q}$ reduces to solving the system equations and integral calculation (10.58). As an example, consider functionals of recurrent type, for which

$$\Phi_n(\mathbf{x}_1, \mathbf{x}_2, \ldots, \mathbf{x}_n) = w_1(\mathbf{x}_1)\varphi_n(\mathbf{x}_1, \mathbf{x}_2, \ldots, \mathbf{x}_n); \tag{10.63}$$

$$\varphi_n(\mathbf{x}_1, \mathbf{x}_2, \ldots, \mathbf{x}_n) = \varphi_1(\mathbf{x}_1) + w_2(\mathbf{x}_1, \mathbf{x}_2)\varphi_{n-1}(\mathbf{x}_2, \mathbf{x}_3, \ldots, \mathbf{x}_n), \tag{10.64}$$

where $w_1(\mathbf{x})$, $w_2(\mathbf{x}, \mathbf{x}')$, and $\varphi_1(\mathbf{x})$ are non-negative functions. Using (10.63), (10.64), one can see that

$$\left.\begin{aligned} \varphi_2(\mathbf{x}_1, \mathbf{x}_2) &= \varphi_1(\mathbf{x}_1) + w_2(\mathbf{x}_1, \mathbf{x}_2)\varphi_1(\mathbf{x}_2); \\ \varphi_3(\mathbf{x}_1, \mathbf{x}_2, \mathbf{x}_3) &= \varphi_1(\mathbf{x}_1) + w_2(\mathbf{x}_1, \mathbf{x}_2)\varphi_1(\mathbf{x}_2) \\ &\quad + w_2(\mathbf{x}_1, \mathbf{x}_2)2(\mathbf{x}_2, \mathbf{x}_3)\varphi_1(\mathbf{x}_3); \\ \varphi_3(\mathbf{x}_1, \mathbf{x}_2, \ldots, \mathbf{x}_n) &= \sum_{i=1}^{n} \varphi_1(\mathbf{x}_i) \prod_{j=1}^{i-1} w_2(\mathbf{x}_j, \mathbf{x}_{j+1}). \end{aligned}\right\} \tag{10.65}$$

From the formulas (10.63) to (10.65), it follows that such a functional, computed on random trajectories is the sum of contributions $\varphi_1(\mathbf{x}_i)$ individual collisions, each of which is multiplied by the value

$$W_i(\mathbf{x}_1, \mathbf{x}_2, \ldots, \mathbf{x}_i) = w_1(\mathbf{x}_1) \prod_{j=1}^{i-1} w_2(\mathbf{x}_j, \mathbf{x}_{j+1})$$

called the *statistical weight*. This formula can be conveniently rewritten as in the form of a recurrence relation

$$W_1(\mathbf{x}_1) = w_1(\mathbf{x}_1);$$

$$W_i(\mathbf{x}_1, \mathbf{x}_2, \ldots, \mathbf{x}_i) = w_2(\mathbf{x}_{i-1}, \mathbf{x}_i) W_{i-1}(\mathbf{x}_1, \mathbf{x}_2, \ldots, \mathbf{x}_{i-1}). \tag{10.66}$$

To derive the equation for conditional moments, we rewrite (10.63), (10.64) as

$$\Phi_n(\mathbf{x}_1, \mathbf{x}_2, \ldots, \mathbf{x}_n) = w_1(\mathbf{x}_1)\{\varphi_1(\mathbf{x}_1) + [w_2(\mathbf{x}_1, \mathbf{x}_2)/w_2(\mathbf{x}_2)]\Phi_{n-1}(\mathbf{x}_2, \mathbf{x}_3, \ldots, \mathbf{x}_n)\}, \tag{10.67}$$

fix the values $\mathbf{x}_1$, $\mathbf{x}_2$ and average this equality over the rest of the variables. Taking into account that with such averaging the values $\omega(\mathbf{x}_1)$, $\varphi_1(\mathbf{x}_1)$ and $w_2(\mathbf{x}_1, \mathbf{x}_2)$ are no longer random, we get

$$\bar{Q}(\mathbf{x}_1, \mathbf{x}_2) = w_1(\mathbf{x}_1)\{\varphi_1(\mathbf{x}_1) + [w_2(\mathbf{x}_1, \mathbf{x}_2)/w_1(\mathbf{x}_2)]\bar{Q}(\mathbf{x}_2)\}. \tag{10.68}$$

Inserting (10.68) into (10.61) and taking into account (10.52), (10.53), we arrive at the linear integral equation

$$\bar{q}(\mathbf{x}) = \varphi_1(\mathbf{x}) + \int d\mathbf{x}' p(\mathbf{x} \to \mathbf{x}') w_2(\mathbf{x}, \mathbf{x}') \bar{q}(\mathbf{x}'), \quad (10.69)$$

where

$$\bar{q}(\mathbf{x}) = \bar{Q}(\mathbf{x}) / w_1(\mathbf{x}). \quad (10.70)$$

Thus, calculating on random trajectories the average value functional (10.63), (10.64), we thus find the integral (10.58):

$$\bar{Q} = \int d\mathbf{x} p_1(\mathbf{x}) w_1(\mathbf{x}) \bar{q}(\mathbf{x}), \quad (10.71)$$

where $\bar{q}(\mathbf{x})$ is the integral equation (10.69) solution.

This functional can be interpreted as a detector reading:

$$J = \int F_1(\mathbf{x}) F^+(\mathbf{x}) d\mathbf{x}, \quad (10.72)$$

where $F_1(\mathbf{x})$ stands for the first collision density and F^+ denotes its importance (the future contribution into the measured functional) of the particle, collided at point $\mathbf{x}$. Comparing Eq. (10.69) with Eq. (10.32) for the value of $F^+(\mathbf{x})$, we write

$$F^+(\mathbf{x}) = \bar{q}_D(\mathbf{x}) + \int d\mathbf{x}' K(\mathbf{x} \to \mathbf{x}') F^+(\mathbf{x}'). \quad (10.73)$$

The comparison shows that $\bar{Q} = J$ under conditions

$$p_1(\mathbf{x}) w_1(\mathbf{x}) = F_1(\mathbf{x}), \quad (10.74)$$

$$\varphi_1(\mathbf{x}) = \bar{q}_D(\mathbf{x}) \quad (10.75)$$

and

$$p(\mathbf{x} \to \mathbf{x}') w_2(\mathbf{x}, \mathbf{x}') = K(\mathbf{x} \to \mathbf{x}'). \quad (10.76)$$

The analysis performed shows that in order to calculate the readings additive detector in a non-breeding environment can be built trajectories using probabilities $p_1(\mathbf{x})$ and $p(\mathbf{x} \to \mathbf{x}')$ being different from $F_1(\mathbf{x})\mathbf{x}$ and $K(\mathbf{x} \to \mathbf{x}')$ describing the actual transfer process. At the same time, the contribution of each collision is multiplied by the "weight" factor (10.66), where $w_1(\mathbf{x})$ and $w_2(\mathbf{x}, \mathbf{x}')$ are defined by expressions

$$w_1(\mathbf{x}) = F_1(\mathbf{x}) / p_1(\mathbf{x}); \quad w_2(\mathbf{x}, \mathbf{x}') = K(\mathbf{x} \to \mathbf{x}') / p(\mathbf{x} \to \mathbf{x}'). \quad (10.77)$$

The formulas (10.77) are called *unbiasedness conditions*, and the described method *non-analogu simulation*. Similar results were obtained for the environment with propagation. Thus, to solve the same problem, you can choose different probabilities $p_1(\mathbf{x})$ and $p(\mathbf{x} \to \mathbf{x}')$, limited only to the following natural condition: being defined in the same phase space where the functions $F_1(\mathbf{x})$ and $K(\mathbf{x} \to \mathbf{x}')$ are given, they must not vanish where the given functions are non-zero [only in this case the expressions (10.77) make sense]. Various ways of non-analog simulations, called *Monte Carlo modifications*, give estimates with the same mathematical expectation, but different variances. One of the tasks of the theory of method Monte Carlo is to develop certain recommendations for application of these methods to solving specific problems.

10.7 The Variance of Estimators

Any characteristic of a transport process, obtained by the Monte Carlo method, is represented by the sample mean of some functional of trajectory (10.56) $J_N = (1/N)\sum_{i=1}^{N} Q_i$, $Q_i = \Phi_{n_i}(\mathbf{x}_1, \ldots, \mathbf{x}_{n_i})$, and, for this reason, is a random variable. Repeating calculations of this quantity with the same number of trajectories N, we obtain different values of J_N, since phase coordinates of collision points in each series of trajectories are different. The difference between the required average value J and sample mean J_N is the *statistical error*. If the variance of an individual trajectory contribution $\mathsf{V}Q = \overline{Q^2} - \bar{Q}^2$ is finite, then for N independent trajectories

$$\sigma_{J_N}^2 = (1/N)\sigma_Q^2, \tag{10.78}$$

and the relative statistical error

$$\delta J_N = \sigma_{J_N}/\bar{J}_N = \delta_Q/\sqrt{N} \tag{10.79}$$

where $\delta_Q = \sigma_Q/\bar{Q}$ relates to a single trajectory.

The distribution of the trajectory contribution and its moments that determine statistical error of the estimate can be investigated by methods. As in Sect. 10.6, we will assume that the point the first collision is random with probability density $p_1(\mathbf{x})$, and the transition probability $\mathbf{x} \to \mathbf{x}'$ is described by the function $p(\mathbf{x} \to \mathbf{x}')$, and

$$p_0(\mathbf{x}) = 1 - \int p(\mathbf{x} \to \mathbf{x}')d\mathbf{x}'. \tag{10.80}$$

Probability distribution density of the functional value from a single trajectory $P(Q)$ is expressed through its encounter conditioned by first collision point $P(Q|\mathbf{x}_1)$ by the total probability equation:

$$P(Q) = \int P(Q|\mathbf{x}_1)p_1(\mathbf{x}_1)d\mathbf{x}_1, \tag{10.81}$$

based on (10.63), (10.64).

On trajectories with fixed $\mathbf{x}_1$ the quantity $Q = w_1(\mathbf{x}_1)\varphi_n(\mathbf{x}_1, \mathbf{x}_2, \dots, \mathbf{x}_n)$ is random because the value $q = \varphi_n(\mathbf{x}_1, \mathbf{x}_2, \dots, \mathbf{x}_n)$ is random. The distribution densities of random variables φ and $w_1\varphi$ are linked by the relation

$$P_{w_1\varphi}(Q|\mathrm{x}_1) = \frac{1}{w_1(\mathbf{x}_1)} P_\varphi \left(\frac{Q}{w_1(\mathbf{x}_1)} \middle| \mathbf{x}_1 \right). \tag{10.82}$$

Denoting for brevity

$$q = Q/w_1(\mathbf{x}_1), \tag{10.83}$$

rewrite the total probability formula for $P_\varphi(q|\mathbf{x}_1)$, taking into account that with probability $p_0(\mathbf{x}_1)$, the trajectory brakes if collision occurs at $\mathbf{x}_1$, and with probability $p(\mathbf{x}_1 \to \mathbf{x}_2)$ continues to the point $\mathbf{x}_2$. According to (10.64), in the first case $q = \varphi_1(\mathbf{x}_1)$, in the second one $\varphi_n(\mathbf{x}_1, \mathbf{x}_2, \dots, \mathbf{x}_n) = q$, if $w_2(\mathbf{x}_1, \mathbf{x}_2)\varphi_{n-1}(\mathbf{x}_2, \mathbf{x}_3, \dots, \mathbf{x}_n) = q - \varphi_1(\mathbf{x}_1)$. Therefore

$$P_\varphi(q|\mathbf{x}_1) = \delta[q - \varphi_1(\mathbf{x}_1)]p_0(\mathbf{x}_1) + \int d\mathbf{x}_2 P_{w_2(\mathbf{x}_1,\mathbf{x}_2)}(q - \varphi_1(\mathbf{x}_1)|\mathbf{x}_2)p(\mathbf{x}_1 \to \mathbf{x}_2).$$

By analogy with (10.82),

$$P_{w_2(\mathbf{x}_1,\mathbf{x}_2)}(q - \varphi_1(\mathbf{x}_1)|\mathbf{x}_2) = \frac{1}{w_2(\mathbf{x}_1, \mathbf{x}_2)} P_\varphi \left(\frac{q - \varphi_1(\mathbf{x}_1)}{w_2(\mathbf{x}_1, \mathbf{x}_2)} \middle| \mathbf{x}_2 \right),$$

and for $P_\varphi(q/\mathbf{x}_1)$ follows the integral equation

$$P_\varphi(q|\mathbf{x}_1) = p_0(\mathbf{x}_1)\delta[q - \varphi_1(\mathbf{x}_1)] +$$

$$\int d\mathbf{x}_2 \frac{p(\mathbf{x}_1 \to \mathbf{x}_2)}{w_2(\mathbf{x}_1, \mathbf{x}_2)} P_\varphi \left(\frac{q - \varphi_1(\mathbf{x}_1)}{w_2(\mathbf{x}_1, \mathbf{x}_2)} \middle| \mathbf{x}_2 \right). \tag{10.84}$$

The random variable Q moments can be found by inserting (10.82) in (10.81), multiplying the product by Q^n and integrating over Q:

$$\overline{Q^n} = \int p_1(\mathbf{x})w_2^n(\mathbf{x})\overline{q^n}(\mathbf{x})d\mathbf{x}, \tag{10.85}$$

where $\overline{q^n}(\mathbf{x})$ are moments of probaility distribution $P_\varphi(q|\mathbf{x})$. The latter, in turn, can be obtained by multiplying all terms of the Eq. (10.84) by $q^n dq$ and integrating with respect to q. In particular, for the first and second conditioned moments, we have

$$\bar{q}(\mathbf{x}) = \varphi_1(\mathbf{x}) + \int d\mathbf{x}' p(\mathbf{x} \to \mathbf{x}')w_2(\mathbf{x}, \mathbf{x}')\bar{q}(\mathbf{x}'); \tag{10.86}$$

$$\overline{q^2}(\mathbf{x}) = \varphi_1(\mathbf{x})[2\bar{q}(\mathbf{x}) - \varphi_1(\mathbf{x})] + \int p(\mathbf{x} \to \mathbf{x}')w_2^2(\mathbf{x}, \mathbf{x}')\overline{q^2}(\mathbf{x}')d\mathbf{x}'. \quad (10.87)$$

Equation (10.85) for $n = 1$ and (10.86), of course, match (10.71), (10.69). When the conditions are met unbiased estimate (10.77) $\bar{q}(\mathbf{x}) = F^+(\mathbf{x})$; $\bar{Q} = J$, and the expressions for the second moment have the form:

$$\overline{Q^2} = \int F_1(\mathbf{x})[F_1(\mathbf{x})/p_1(\mathbf{x})]\overline{q^2}(\mathbf{x})d\mathbf{x}; \quad (10.88)$$

$$\overline{q^2}(\mathbf{x}) = \bar{q}_D(\mathbf{x})[2F^+(\mathbf{x}) - \bar{q}_D(\mathbf{x})]$$
$$+ \int K(\mathbf{x} \to \mathbf{x}')[K(\mathbf{x} \to \mathbf{x}')/p(\mathbf{x} \to \mathbf{x}')]\overline{q^2}(\mathbf{x}')d\mathbf{x}'. \quad (10.89)$$

Choosing different probabilities $p_1(\mathbf{x})$ and $p(\mathbf{x} \to \mathbf{x}')$, it is possible to achieve a significant reduction in the variance of the estimates

$$\sigma_Q^2 = \int F_1(\mathbf{x})[F_1(\mathbf{x})/p_1(\mathbf{x})]\overline{q^2}(\mathbf{x})d\mathbf{x} - J^2. \quad (10.90)$$

Let us show that by an appropriate choice of $p_1(\mathbf{x})$ and $p(\mathbf{x} \to \mathbf{x}')$, the variance of the estimate can be reduced in principle to zero. Let's choose a transitional probability in the form

$$p(\mathbf{x} \to \mathbf{x}') = \frac{K(\mathbf{x} \to \mathbf{x}')F^+(\mathbf{x}')}{\int K(\mathbf{x} \to \mathbf{x}')F^+(\mathbf{x}')d\mathbf{x}'}. \quad (10.91)$$

Inserting (10.91) in (10.89) and taking into account (10.77), (10.86), we obtain

$$\overline{q^2}(\mathbf{x}) = \bar{q}_D(\mathbf{x})F^+(\mathbf{x})$$
$$+ [F^+(\mathbf{x}) - \bar{q}_D(\mathbf{x})]\left[\bar{q}_D(\mathbf{x}) + \int K(\mathbf{x} \to \mathbf{x}')[\overline{q^2}(\mathbf{x}')/F^+(\mathbf{x}')]d\mathbf{x}'\right]. \quad (10.92)$$

It can directly be said that solution to this equation is the square of the mean:

$$\overline{q^2}(\mathbf{x}) = [\bar{q}(\mathbf{x})]^2 = F^{+2}(\mathbf{x}). \quad (10.93)$$

Choosing the density of first collisions in the form

$$p_1(\mathbf{x}) = \frac{F_1(\mathbf{x})F^+(\mathbf{x})}{\int F_1(\mathbf{x})F^+(\mathbf{x})d\mathbf{x}} = \frac{F_1(x)F^+(\mathbf{x})}{J} \quad (10.94)$$

and inserting (10.93), (10.94) into (10.90) yields $\sigma_Q^2 = 0$.

10.8 Examples of Weight MC Modifications

Consider the simplest weight modifications of the Monte Carlo method, often used in transport theory.

Let a single unit point source isotropically emitting photons with energy E_0 $\mathbf{r}_0$ be placed at point $\mathbf{r}_0$ source of particles with energy E_0. The phase density of the first collision is

$$F_1(\mathbf{r}, \boldsymbol{\Omega}, E) = \Sigma(\mathbf{r}, E)\frac{\exp\{-\tau(\mathbf{r}, \mathbf{r}_0, E_0)\}}{|\mathbf{r}-\mathbf{r}_0|^2}\delta\left(\boldsymbol{\Omega} - \frac{\mathbf{r}-\mathbf{r}_0}{|\mathbf{r}-\mathbf{r}_0|}\right)\delta(E - E_0) \tag{10.95}$$

and transition density,

$$K(\mathbf{x} \to \mathbf{x}') = \frac{\Sigma_s(\mathbf{r}, \boldsymbol{\Omega} \to \boldsymbol{\Omega}', E \to E')}{\Sigma(\mathbf{r}, E)} \times$$

$$\frac{\exp\{-\tau(\mathbf{r}, \mathbf{r}_0, E_0)\}}{|\mathbf{r}-\mathbf{r}_0|^2}\Sigma(\mathbf{r}', E')\delta\left(\boldsymbol{\Omega}' - \frac{\mathbf{r}'-\mathbf{r}}{|\mathbf{r}'-\mathbf{r}|}\right). \tag{10.96}$$

Dividing and multiplying the transition probability density by $\Sigma_s(\mathbf{r}, E)$, we get

$$K(\mathbf{x} \to \mathbf{x}') = [\Sigma_s(\mathbf{r}, E)/\Sigma(\mathbf{r}, E)]K_0(\mathbf{x} \to \mathbf{x}'),$$

where $K_0(\mathbf{x} \to \mathbf{x}')$ unlike $K(\mathbf{x} \to \mathbf{x}')$ is normalized to 1: $\int K_0(\mathbf{x} \to \mathbf{x}')d\mathbf{x}' = 1$. This means, that if we take densities $p_1(\mathbf{x}) = F_1(\mathbf{x})$, $p(\mathbf{x} \to \mathbf{x}') = K_0(\mathbf{x} \to \mathbf{x}')$ for non-analog trajectories, then absorption probability $p_0 = 1 - \int K_0(\mathbf{x} \to \mathbf{x}')d\mathbf{x}'$ becomes zero and particles continue to move until they exit the scattering region. This increases the probability of particle hitting the detector and accordingly decrease the variance. As follows from (10.66), (10.77), statistical weight in this case

$$W_k(\mathbf{x}_1, \mathbf{x}_2, \ldots, \mathbf{x}_k) = \frac{\Sigma_s(\mathbf{x}_1)}{\Sigma(\mathbf{x}_1)}\frac{\Sigma_s(\mathbf{x}_2)}{\Sigma(\mathbf{x}_2)}\cdots\frac{\Sigma_s(\mathbf{x}_{k-1})}{\Sigma(\mathbf{x}_{k-1})}$$

equals to product of probabilities of the particle survival in all previous collisions. Sampling other random elements (path length, direction, and energy after scattering) is carried out as in analog simulation. Now we note that in (10.95), (10.96) Σ is $\tilde{\Sigma}$ and τ equals to $\tilde{\tau}$, respectively, leaving unchanged differential section Σ_s. The expressions thus obtained will be used as new densities for non-analog modeling. Then the weight factors

$$w_1(\mathbf{x}) = \frac{\Sigma(\mathbf{x})}{\tilde{\Sigma}(\mathbf{x})}\exp[-\Delta\tau(\mathbf{x}_0, \mathbf{x})];$$

$$w_2(\mathbf{x}, \mathbf{x}') = \frac{\tilde{\Sigma}(\mathbf{x})\Sigma(\mathbf{x}')}{\Sigma(\mathbf{x})\tilde{\Sigma}(\mathbf{x}')} \exp[-\Delta\tau(\mathbf{x}, \mathbf{x}')],$$

$$\Delta\tau = \tilde{\tau} - \tau,$$

and the statistical weight

$$W_k(\mathbf{x}_1, \ldots, \mathbf{x}_k) = \exp[-\Delta\tau(\mathbf{x}_0, \ldots, \mathbf{x}_k)]\Sigma(\mathbf{x}_k)/\tilde{\Sigma}(\mathbf{x}_k);$$

$$\Delta\tau(\mathbf{x}_0, \ldots, \mathbf{x}_k) = \sum_{i=1}^{k} \Delta\tau(\mathbf{x}_{i-1}, \mathbf{x}_i).$$

In this case, the expression for estimating the flux density from collisions in $\Delta\mathbf{x}$

$$Q = \sum_{k=1}^{n} W_k(\mathbf{x}_1, \ldots, \mathbf{x}_k)/\Sigma(\mathbf{x}_k) = \sum_{k=1}^{n} \exp[-\Delta\tau(\mathbf{x}_0, \ldots, \mathbf{x}_k)]\tilde{\Sigma}(\mathbf{x}_k) \quad (10.97)$$

will be different from the usual expression for evaluation in an environment with section $\tilde{\Sigma}(\mathbf{x})$ only by the factor

$$\tilde{W}_k = \exp[-\Delta\tau(\mathbf{x}_0, \ldots, \mathbf{x}_k)]. \quad (10.98)$$

Setting $\tilde{\Sigma} = \Sigma_s$ in the formulas (10.97), (10.98), we obtain a new modification of analytical averaging, where the draw of the path length is made using only the scattering cross section, and the absorption probability is taken into account by introducing exponential weights $\tilde{W}_k = \exp[-\tau_c(\mathbf{x}_0, \ldots, \mathbf{x}_k)]$. In problems where the radiation field is of interest at large distances from the source, the *exponential transformation* is applied. Let, for example, the detector be at far away distance from the source along z-axis. We choose the section $\tilde{\Sigma}$ depending on the angle ϑ between the axis $0z$ and the direction of motion particles so that for $\theta = 0$ it is the smallest, and for $\vartheta = \pi$—the largest. In this case, the trajectories are more likely to reach great depths and contribute to the detector readings. AT in particular, if $\tilde{\Sigma} = \Sigma - c\cos\vartheta$, where $0 < c < \Sigma_{\min}$ is a constant, then $\Delta\Sigma = c\cos\vartheta$; $\Delta\tau = c\sum_{i=1}^{k} l_i \cos\vartheta_i = c(z_k - z_0)$ and the statistical weight is determined only by the coordinate z of the particle in present moment, regardless of its background. If the origin coincides with the center of the source, $r_0 = 0$, then

$$W_k(\mathbf{x}_1, \mathbf{x}_2, \ldots, \mathbf{x}_{k-1}, \mathbf{x}) = \exp(-cz). \quad (10.99)$$

If the detector measures the radiation field at a fixed depth z, statistical weights (10.99) can be entered after multiplying the results by $\exp(-cz)$.

Another way to improve the accuracy of particle flux density estimation, moving in some direction is a modification differential scattering cross section. Replacing in (10.96) Σ_s value $\tilde{\Sigma}_s$ satisfying the condition

$$\iint \tilde{\Sigma}_s(\mathbf{r}, \boldsymbol{\Omega} \to \boldsymbol{\Omega}', E \to E')d\boldsymbol{\Omega}'dE' = \iint \Sigma_s(\mathbf{r}, \boldsymbol{\Omega} \to \boldsymbol{\Omega}', E \to E')d\boldsymbol{\Omega}'dE',$$

and using the new section to play the direction of movement and energy of the particle after the collision, it is possible to achieve a significant increasing probability of a particle hitting the detector. The statistical weight is

$$W_k(\mathbf{x}_1, \mathbf{x}_2, \ldots, \mathbf{x}_k) = \prod_{i=1}^{k-1} \frac{\Sigma_s(\mathbf{r}_i, \boldsymbol{\Omega}_i \to \boldsymbol{\Omega}_{i+1}, E_i \to E_{i+1})}{\tilde{\Sigma}_s(\mathbf{r}_i, \boldsymbol{\Omega}_i \to \boldsymbol{\Omega}_{i+1}, E_i \to E_{i+1})}.$$

In particular, when calculating backscattering from a flat layer transition to an isotropic scattering law using appropriate weights reduces the statistical error.

The general idea behind the above methods is a change in the modeling procedure, which leads to an increase in the number of the most "valuable" trajectories that contribute the most to the calculated value. To do this, during the simulation, it is necessary to use different information about the relative trajectory values. For example, in some cases, it is possible to reduce the variance by substituting into the formulas (10.91), (10.94) in as F^+ asymptotic solutions of adjoint equations. Such modifications of the Monte Carlo method are called *simulation by importance*.

The presence of statistical fluctuations in the results obtained by the Monte Carlo method, makes us pay special attention to those tasks in which it is required to calculate the difference of values, characterizing the distribution of particles in two similar size and properties of absorbers. This may be a change in the radiation field, caused by the replacement of part of the substance of the absorber with a similar composition substance, correction to the calculation results due to refinement interaction cross sections, etc. In many cases, independent solving both problems and then subtracting $\Delta J = \tilde{J} - J$ may turn out to be unacceptable if the statistical fluctuations of $\tilde{J}$ and J are commensurate with the desired value ΔJ. To solve such problems, the *dependent sampling method* is developed.

Consider the difference between η_1 and η_2 random variables. According to elementary rules of probability theory, the average value of this the difference is $\overline{\eta_1 - \eta_2} = \overline{\eta_1} - \overline{\eta_2}$, and the variance $\mathsf{Var}(\eta_2 - \eta_1) = \mathsf{Var}(\eta_1) + \mathsf{Var}(\eta_2) - 2\mathsf{R}(\eta_1, \eta_2)$, where R is the correlation moment of random variables η_1 and η_2. For independent quantities $\mathsf{R}(\eta_1, \eta_2) = 0$ and $\mathsf{Var}(\eta_2 - \eta_1) = \mathsf{Var}(\eta_1) + \mathsf{Var}(\eta_2)$. In the case of a positive correlation between them, $\mathsf{R} > 0$ and the variance decreases.

Let the particle distribution in one medium obey the equation

$$F^+(\mathbf{x}) = \bar{q}_D(\mathbf{x}) + \int K(\mathbf{x} \to \mathbf{x}')F^+(\mathbf{x}')d\mathbf{x}',$$

and in another equation

$$\tilde{F}^{+}(\mathbf{x}) = \bar{q}_D(\mathbf{x}) + \int \tilde{K}(\mathbf{x} \to \mathbf{x}')\tilde{F}^{+}(\mathbf{x}')d\mathbf{x}'.$$

To solve the first of the equations, you can use, for example, analog method, simulating particle trajectories with probabilities $F_1(\mathbf{x})$ and $K(\mathbf{x} \to \mathbf{x}')$ and calculating the trajectory contribution Q, whose average value is equal to J. To calculate $\tilde{J}$ in the dependent test method the same trajectories are used. Considering them as displaced trajectory with respect to the second medium, we write an expression for the contribution of these trajectories to $\tilde{J}$:

$$\tilde{q} = \sum_{k=1}^{n-1} \bar{q}_D(\mathbf{x}_k)W_k; \quad W_1 = \frac{\hat{F}_1(\mathbf{x}_1)}{F_1(\mathbf{x}_1)}; \quad W_{k+1} = \frac{\tilde{(\mathbf{x}_k \to \mathbf{x}_{k+1})}}{K(\mathbf{x}_k \to \mathbf{x}_{k+1})}W_k. \tag{10.100}$$

In case of a point source $[S(\mathbf{r}, \boldsymbol{\Omega}, E) = \delta(\mathbf{r} - \mathbf{r}_0)S'(\boldsymbol{\Omega}, E)]$

$$\frac{\hat{F}(\mathbf{x}_1)}{F_1(\mathbf{x}_1)} = \frac{\tilde{\Sigma}(\mathbf{r}_1, E_1)}{\Sigma(\mathbf{r}_1, E_1)} \exp[-\Delta\tau(\mathbf{r}_0, \mathbf{r}_1, E_1)]; \tag{10.101}$$

$$\frac{\tilde{K}(\mathbf{x}_{k+1} \to \mathbf{x}_{k+1})}{K(\mathbf{x}_{k+1} \to \mathbf{x}_{k+1})} = \frac{\tilde{\Sigma}(\mathbf{r}_{k+1}, E_{k+1})\tilde{\Sigma}_s(\mathbf{r}_k, \boldsymbol{\Omega}_k \to \boldsymbol{\Omega}_{k+1}, E_k \to E_{k+1})\Sigma(\mathbf{r}_k, E_k)}{\Sigma(\mathbf{r}_{k+1}, E_{k+1})\Sigma_s(\mathbf{r}_k, \boldsymbol{\Omega}_k \to \boldsymbol{\Omega}_{k+1}, E_k \to E_{k+1})\tilde{\Sigma}(\mathbf{r}_k, E_k)}$$
$$\times \exp[-\Delta\tau(\mathbf{r}_k, \mathbf{r}_{k+1}, E_{k+1})].$$

Computing difference of the contributions for each trajectory yields

$$\Delta J = \overline{\tilde{q} - q}; \quad \tilde{q} - q = \sum_{k=1}^{n}(W_k - 1)\bar{q}_D(\mathbf{x}_k).$$

Let us consider the case when the media differ only in a certain volume ΔV. Let the i-th particle scattering occur outside this volume:

$$\Sigma(\mathbf{x}_i) = \tilde{\Sigma}(\mathbf{x}_i),$$

$$\Sigma_s(\mathbf{r}_i, \boldsymbol{\Omega}_i \to \boldsymbol{\Omega}_{i+1}, E_i \to E_{i+1}) = \tilde{\Sigma}_s(\mathbf{r}_i, \boldsymbol{\Omega}_i \to \boldsymbol{\Omega}_{i+1}, E_i \to E_{i+1}).$$

When determining the coordinates of the next collision, it may turn out that:

(a) $(i+1)$-th collision also occurred outside the volume ΔV and the particle has not crossed the volume, then $W_{i+1} = W_1$;

(b) $(i+1)$-th collision occurred outside the volume ΔV, but between the i-th and $(i+1)$-th collisions the particle has crossed the volume, then $W_{i+1} = \exp[-\Delta\tau(\mathbf{r}_i, \mathbf{r}_{i+1}, E_{i+1})]W_i$;

(c) $(i+1)$-th collision happened in the volume ΔV, then

$$W_{i+1} = \frac{\tilde{\Sigma}(\mathbf{x}_{i+1})}{\Sigma(\mathbf{x}_{i+1})} \exp[-\Delta\tau(\mathbf{r}_i, \mathbf{r}_{i+1}, E_{i+1})] W_i.$$

Another possible variant can be analyzed in a similar way: when the i-th collision occurred in the volume ΔV. From the conducted analysis, in particular, it follows that if the particle never crossed the volume, $W_k = 1$ and its contribution to the computed difference $\Delta q = \tilde{q} - q = 0$, even if the particle hits the observed area ("detector"). A decrease in volume reduces the probability of its intersection by a particle, most of the particles entering the detector volume give zero contribution and the effectiveness of the dependent test method falls.

10.9 Weighting Estimation of Correlations in Branching Processes

10.9.1 Equations for the Distribution Function and Its Moments

This section demonstrates the introduction of weights that provide an unbiased estimate in the case of branching (cascade) processes.

Let $(\Omega, \mathscr{A})$ be a measurable one-particle state space. The state ω of the initial particle generating the cascade is random: it belongs to $A \in \mathscr{A}$ with probability $p(A)$. With probability $p_0(\omega)$ a particle in the ω state can be absorbed, and with probability $1 - p_0(\omega)$, it can experience decay or scattering (the latter will be considered a special case of decay). In the process of decay, n arises, $n = 1, 2, \ldots$, secondary particles in the states $\omega_1, \ldots, \omega_n \in \Omega$. We introduce the notation

$$(\omega_1, \ldots, \omega_n) = \omega^n$$

and the product of measurable spaces

$$\left(\prod_{\nu=1}^{n} \Omega, \bigotimes_{\nu=1}^{n} \mathscr{A}\right) = (\Omega_n, \mathscr{A}_n),$$

so the outcome of the decay is $\omega^n \in \Omega_n$. Assuming a Markov process, we assume that the probability of $P_n(A_n|\omega)$ decaying into n particles such that $\omega^n \in A_n$, $A_n \in \mathscr{A}_n$ is determined by only the state of a decaying particle.

The absorption of particles in the ω state or the decay of $\omega \to \omega^n$ causes a random signal in the detector, represented by the vector $x = (x_1, \ldots, x_m)$ of the m-dimensional Euclidean space $\mathbb{R}^m$. Denote by $(X, \mathscr{B})$ the measurable space of signals, by $\pi_0(B|\omega)$ and $\pi_n(B|\omega \to \omega^n)$ the probabilities of the event $x \in B$ upon absorption in the ω state or decay of $\omega \to \omega^n$). Signals from individual events of interaction of particles with the detector are added to the overall signal from the entire cascade of particles. The probability that the total signal of a cascade generated by a particle in the ω state is equal to $x \in B$ will be denoted by $\psi(B|\omega)$, and the corresponding probability for a cascade with a random initial state by $\Psi(B)$:

$$\Psi(B) = \int_{\Omega} \psi(B|\omega) p(d\omega).$$

Conditional probability $\psi(B|\omega)$ obeys the equation

$$\psi(B|\omega) = p_0(\omega)\pi_0(B|\omega)$$

$$+\sum_{n=1}^{\infty} \int_{\Omega_n} P_n(d\omega^n|\omega)\psi(B|\omega_1) * \ldots * \psi(B|\omega_n) * \pi_n(B|\omega \to \omega^n), \tag{10.102}$$

where sign $*$ means the convolution in x:

$$\psi(B|\omega_1) * \psi(B|\omega_2) = \int_X \psi(B - x|\omega_1)\psi(dx|\omega_2),$$

$$B - x \equiv \{y : y \in X, x + y \in B\},$$

etc. Assuming that the first and second moments

$$\mu_i(\omega) = \int_X x_i \psi(dx|\omega), \quad M_i = \int_{\Omega} \mu_i(\omega) p(d\omega),$$

$$\mu_{ik}(\omega) = \int_X x_i x_k \psi(dx|\omega), \quad M_{ik} = \int_{\Omega} \mu_{ik}(\omega) p(d\omega)$$

exist, we write for them equations, following from (10.102):

$$\mu_i(\omega) = m_i(\omega) + \sum_{n=1}^{\infty} \sum_{j=1}^{n} \int_{\Omega_n} P_n(d\omega^n|\omega)\mu_i(\omega_j), \tag{10.103}$$

$$\mu_{ik}(\omega) = m_{ik}(\omega) + \sum_{n=1}^{\infty} \int_{\Omega_n} P_n(d\omega^n|\omega)$$

$$\times \Big[m_i^{(n)}(\omega \to \omega^n) \sum_{l=1}^{n} \mu_k(\omega_l) + m_k^{(n)}(\omega \to \omega^n) \sum_{j=1}^{n} \mu_i(\omega_j)$$

$$+ \sum_{j \neq i}^{n} \sum^{n} \mu_i(\omega_j)\mu_k(\omega_l) + \sum_{j=1}^{n} \mu_{ik}(\omega_j) \Big], \tag{10.104}$$

where

$$m_i(\omega) = p_0(\omega) m_i^{(0)}(\omega) + \sum_{n=1} \int_{\Omega_n} P_n(d\omega^n|\omega) m_i^{(n)}(\omega \to \omega^n),$$

$$m_{ik}(\omega) = p_0(\omega) m_{ik}^{(0)}(\omega) + \sum_{n=1}^{\infty} \int_{\Omega_n} P_n(d\omega^n|\omega) m_{ik}^{(n)}(\omega \to \omega^n),$$

$$m_i^{(n)}(\omega \to \omega^n) = \int_X x_i \pi_n(dx|\omega \to \omega^n),$$

$$m_{ik}^{(n)}(\omega \to \omega^n) = \int_X x_i x_k \pi_n(dx|\omega \to \omega^n).$$

Because of particle identity, the integrals

$$\int_{\Omega_n} 1_A(\omega_j) P_n(d\omega^n|\omega) \equiv P_n(A|\omega),$$

$$\int_{\Omega_n} 1_{A'}(\omega_j) 1_{A''}(\omega_i) P_n(d\omega^n|\omega) \equiv P_n(A' \times A''|\omega)$$

and other, where

$$1_A(\omega) = \begin{cases} 1, & \omega \in A, \\ 0, & \omega \overline{\in} A \end{cases}$$

is a set A indicator, does not depend on $j, l, \ldots$. Equations (10.103), (10.104) become

$$\mu_i(\omega) = m_i(\omega) + \int_{\Omega} K(d\omega'|\omega)\mu_i(\omega'), \tag{10.105}$$

$$\mu_{ik}(\omega) = m_{ik}(\omega) + \int_{\Omega} L_i(d\omega'|\omega)\mu_k(\omega') + \int_{\Omega} L_k(d\omega'|\omega)\mu_i(\omega')$$

$$+\int_{\Omega_2} M(d\omega' d\omega''|\omega)\mu_i(\omega')\mu_k(\omega'') + \int_{\Omega} K(d\omega'|\omega)\mu_{ik}(\omega'), \tag{10.106}$$

where

$$K(A|\omega) = \sum_{n=1}^{\infty} n \int_{\Omega_n} P_n(d\omega^n|\omega)1_A(\omega_j),$$

$$L_\alpha(A|\omega) = \sum_{n=1}^{\infty} n \int_{\Omega_n} P_n(d\omega^n|\omega)m_\alpha^{(n)}(\omega \to \omega^n)1_A(\omega_j),$$

$$\alpha = i, k,$$

$$M(A' \times A''|\omega) = \sum_{n=2}^{\infty} n(n-1)\int_{\Omega_n} P_n(d\omega^n|\omega)1_{A'}(\omega_j)1_{A''}(\omega_l).$$

10.9.2 The First Moment Modified Estimation

Consider now a random multiplication process with the same state space $(\Omega, \mathscr{A})$, but with different probabilities $p'(A)$, $P_n'(A_n|\omega)$, $\pi_n'(B|\omega \to \omega^n)$. In addition, we introduce weight factors according to the following scheme: in the initial state ω, the particle has a weight $w(\omega)$, by which the signal from its first interaction is multiplied; in the case of $\omega \to \omega^n$ decay, each of the n secondary particles has a weight $w(\omega)w(\omega \to \omega_\nu)$, $1 \leqslant \nu \leqslant n$, by which the signal from the next interactions is multiplied, and so on. The total signal of the cascade is found as the sum of these products of elementary signals and the corresponding weights. We call such a process modified one. Denoting by $\psi'(B|\omega)$ the conditional distribution of the cascade signal generated by a particle with unit initial weight, and by $\mu_i'(\omega)$, $\mu_{ik}'(\omega)$ the corresponding moments, we arrived at the following relations for the probabilities

$$\Psi(B) = \int_{\Omega} \psi'\Big(\frac{B}{w(\omega)}\Big|\omega\Big)p'(d\omega),$$

$$\frac{B}{w(\omega)} \equiv \{x : x \in X,\ xw(\omega) \in B\},$$

$$\psi'(B|\omega) = p_0'(\omega)\pi_0'(B|\omega) + \sum_{n=1}^{\infty}\int_{\Omega_n} P_n'(d\omega^n\Big|\omega)\int_X \psi'\Big(\frac{dx^{(1)}}{w(\omega \to \omega_1)}\Big|\omega_1\Big)$$

$$\times \cdots \times \int_X \psi'\Big(\frac{dx^{(n)}}{w(\omega \to \omega_n)}\Big|\omega_n\Big)\pi'_n(B - \sum_{\nu=1}^{n} x^{(\nu)}|\omega \to \omega^n) \tag{10.107}$$

and for moments

$$\mu'_i(\omega) = m'_i(\omega) + \int_\Omega K'(d\omega'|\omega)\mu'_i(\omega), \tag{10.108}$$

$$\mu'_{ik}(\omega) = m'_{ik}(\omega) + \int_\Omega L'_i(d\omega'|\omega)\mu'_k(\omega') + \int_\Omega L'_k(d\omega'|\omega)\mu'_i(\omega')$$

$$+ \iint_{\Omega_2} M'(d\omega'\, d\omega''|\omega)\mu'_i(\omega')\mu'_k(\omega'') + \int_\Omega K''(d\omega'|\omega)\mu'_{ik}(\omega'), \tag{10.109}$$

where

$$K'(A|\omega) = \sum_{n=1}^{\infty} n \int_{\Omega_n} P'_n(d\omega^n|\omega)w(\omega \to \omega_j)1_A(\omega_j),$$

$$L'_\alpha(A|\omega) = \sum_{n=1}^{\infty} n \int_{\Omega_n} P'_i(d\omega^n|\omega)m'_\alpha(\omega \to \omega_j)w(\omega \to \omega^n)1_A(\omega_j),$$

$$M'(A' \times A''|\omega) = \sum_{n=2}^{\infty} n(n-1) \int_{\Omega_n} P'_n(d\omega^n|\omega)w(\omega \to \omega_j)w(\omega \to \omega_l)1_{A'}(\omega_j)1_{A''}(\omega_l),$$

$$K''(A|\omega) = \sum_{n=1}^{\infty} n \int_{\Omega_n} P'_n(d\omega^n|\omega)w^2(\omega \to \omega_j)1_a(\omega_j).$$

Comparison of (10.105) and (10.108) implies sufficient conditions for the unbiasedness of the first moment $\mu(\omega) = \mu'(\omega)$:

$$p_0(\omega)m_i^{(0)}(\omega) + \sum_{n=1}^{\infty} \int_{\Omega_n} P_n(d\omega^n|\omega)m_i^{(n)}(\omega \to \omega^n)$$

$$= p'_0(\omega)m_i^{'(0)}(\omega) + \sum_{n=1}^{\infty} \int_{\Omega_n} P'_n(d\omega^n|\omega)m_i^{'(n)}(\omega \to \omega^n), \tag{10.110}$$

$$\sum_{n=1}^{\infty} n \int_{\Omega_n} P_n(d\omega^n|\omega)1_A(\omega_j) = \sum_{n=1}^{\infty} n \int_{\Omega_n} P'_n(d\omega^n|\omega)w(\omega \to \omega_j)1_A(\omega_j), \tag{10.111}$$

linking probabilities $\pi_n'(B|\omega \to \omega^n)$ and statistical weights $w(\omega \to \omega_\nu)$. The relations (10.110), (10.111) together with the Eq. (10.109) are a convenient starting point for the analysis of various modifications of the Monte Carlo method.

10.9.3 Some Examples

Let us consider several examples of applying the general expressions obtained above to the analysis of known modifications of the Monte Carlo method. Let both random processes—real and modified—do not branch. Then the condition (10.111) for the unbiasedness of the first moment takes the form

$$P_1(A|\omega) = \int_A P_1'(d\omega'|\omega)w(\omega \to \omega'),$$

whence, given the absolute continuity of the probability measure P_1 with respect to P_1', it follows that the statistical weight is equal to the Radon–Nikodim derivative

$$w(\omega \to \omega') = \frac{dP_1}{dP_1'}(\omega'|\omega). \tag{10.112}$$

If, in addition, the signal due to the interaction does not depend on the type of interaction and the state of the particles after it, but is a function of only the state of the particle before the collision $h_i(\omega)$ (as is the case in collision estimates), then

$$\begin{aligned} m_i^{(0)}(\omega) &= m_i^{(1)}(\omega \to \omega') = h_i(\omega), \\ m_{ii}^{(0)}(\omega) &= m_{ii}^{(1)}(\omega \to \omega') = h_i^2(\omega). \end{aligned} \tag{10.113}$$

Equation (10.113) for the mean square $\overline{x_i^2} = \mu_{ii}(\omega)$ is reduced to the form

$$\mu_{ii}(\omega) = h_i(\omega)[2\mu_i(\omega) - h_i(\omega)] + \int_\Omega P_1(d\omega'|\omega)\mu_{ii}(\omega'). \tag{10.114}$$

If the modified process also has this property (10.113), then (10.110) implies the equalities

$$m_i^{'(0)}(\omega) = m_i^{'(1)}(\omega \to \omega') = h_i(\omega). \tag{10.115}$$

Inserting (10.113) and (10.115) in (10.109) yields the equation for the equation for the second moment of the modified process,

$$\mu_{ii}'(\omega) = h_i(\omega)[2\mu_i(\omega) - h_i(\omega)] + \int_\Omega P_1(d\omega'|\omega)w(\omega \to \omega')\mu_{ii}'(\omega').$$

From general expressions, it is easy to obtain similar equations for mileage estimates when

$$m_i^{(0)}(\omega) = m_{ii}^{(0)}(\omega) = 0, \quad m_i^{(1)}(\omega \to \omega') = l_i(\omega \to \omega'),$$

$$m_{ii}^{(1)}(\omega \to \omega') = l_i^2(\omega \to \omega'), \tag{10.116}$$

where $l_i(\omega \to \omega')$ is the free path length between points ω and ω', belonging the detector sensitivity region. Inserting (10.116) into (10.110) and supposing $m_i^{'(0)}(\omega) = m_{ii}^{'(0)}(\omega) = 0$, we arrive at equation

$$\int\limits_{\Omega} P_1'(d\omega'|\omega) m_i^{'(1)}(\omega \to \omega') = \int\limits_{\Omega} P_1(d\omega'|\omega) l_i(\omega \to \omega'),$$

which is fulfilled under the condition

$$m_i^{'(1)}(\omega \to \omega') = l_i(\omega \to \omega') w(\omega \to \omega'), \tag{10.117}$$

with $w(\omega \to \omega')$ determined by (10.112). Evidently,

$$m_{ii}^{'(1)}(\omega \to \omega') = l_i^2(\omega \to \omega') w^2(\omega \to \omega'). \tag{10.118}$$

On substituting (10.116)–(10.118) into (10.106) and (10.109), for $i = k$ we get equations for second moments of basic and modified trace estimations :

$$\mu_{ii}(\omega) = \int\limits_{\Omega} P_1(d\omega'|\omega) l_i(\omega \to \omega')[l_i(\omega \to \omega') + 2\mu_i(\omega')] \int\limits_{\Omega} P_1(d\omega'|\omega) \mu_{ii}(\omega'),$$

$$\mu_{ii}'(\omega) = \int\limits_{\Omega} P_1(d\omega'|\omega) l_i(\omega \to \omega') w(\omega \to \omega')[l_i(\omega \to \omega') + 2\mu_i(\omega')]$$

$$+ \int\limits_{\Omega} P_1(d\omega'|\omega) w(\omega \to \omega') \mu_{ii}(\omega').$$

Now let the real process branch, but the modified one does not. From the conditions of unbiasedness of the first moment it follows

$$\sum_{n=1}^{\infty} n P_n(A|\omega) = \int\limits_{A} P_1'(d\omega'|\omega) w(\omega \to \omega'),$$

whence (if P_n is absolutely continuous with respect to P_1')

$$w(\omega \to \omega') = \sum_{n=1}^{\infty} n \frac{dP_n}{dP'}(\omega'|\omega).$$

If, in particular,

$$P_n(A|\omega) = \frac{p_n(\omega)}{p'_n(\omega)} P'_1(A|\omega),$$

where $p_n(\omega)$ is a probability of decay into n daughter particles, then the statistical weight

$$w(\omega \to \omega') = \sum_{n=1}^{\infty} n \frac{p_n(\omega)}{p'_n(\omega)} = \frac{\overline{n}(\omega)}{\overline{n}'(\omega)}$$

is equal to the ratio of the average numbers of particles formed during the collision in the real and modified processes. In this latter case, the equations for the second moments of the collision estimates are

$$\mu_{ii}(\omega) = h_i(\omega)[2\mu_i(\omega) - h_i(\omega)]$$

$$+ \int\limits_{\Omega_2} M(d\omega' d\omega''|\omega)\mu_i(\omega')\mu_i(\omega'') + \int\limits_{\Omega} K(d\omega'|\omega)\mu_{ii}(\omega'),$$

$$\mu'_{ii}(\omega) = h_i(\omega)[2\mu_i(\omega) - h_i(\omega)] + w(\omega) \int\limits_{\Omega} K(d\omega'|\omega)\mu'_{ii}(\omega').$$

Let us finally consider the case when the real process is not branching, but the modified one branches as follows: with the probability p'_0 the particle dies in the collision, and with the probability $p'_\nu = 1 - p'_0$ it decays into a fixed number of ν independent particles (modification of the "splitting" type). The equation for the second moment of this estimate has the form

$$\mu'_{ii}(\omega) = h_i(\omega)[2\mu_i(\omega) - h_i(\omega)] + \frac{\nu - 1}{\nu p'_\nu}[\mu_i(\omega) - h_i(\omega)]^2 + \frac{p_1}{\nu p'_\nu} \int\limits_{\Omega} K(d\omega'|\omega)\mu'_{ii}(\omega').$$

As $\nu = 1$ and $p'_\nu = p_1$, it coincides with Eq. (10.114).

10.9.4 The Modified Estimation of the Second Moment

Equation (10.107) shows that a similar method for estimating the second moment of a branching process using a non-branching process is impossible. However, it does not follow from this that the second moment of the branching process can be estimated only with the help of direct simulation. Let us show how an unbiased estimate of the moment $\mu_{ik}(\omega)$ can be made using a random process branching less intensively than the real one.

Consider the following process. A particle that starts wandering from the state ω can die in this state with probability $p_0(\omega)$ or go to the state $\omega' \in A$ with probability $P(A|\omega)$. The sequence of states of this particle before absorption is denoted by $\omega_1, \ldots, \omega_n$. In each collision of this primary particle, two secondary particles are born, marked with indices i and k, with probability $P_2(A_i \times A_k|\omega_\nu)$, experiencing their first collisions at the points $\omega^{(i)} \in A_i$ and $\omega^{(k)} \in A_k$. The probabilities of further transitions are the same for them as for the primary particle, and the trajectories of these secondary particles no longer branch. We denote the states of secondary particles by $\omega_{\nu 1}^{(i)}, \ldots, \omega_{\nu n_i}^{(i)}$ and $\omega_{\nu 1}^{(k)}, \ldots, \omega_{\nu n_k}^{(k)}$ respectively. Each such cascade with the initial state ω is assigned a number $\xi[\omega]$ ("detector readings"), calculated according to the following scheme:

$$\xi[\omega_\nu] = h[\omega_\nu] + v_i(\omega_\nu \to \omega_{\nu 1}^{(i)})\xi^{(i)}[\omega_{\nu 1}^{(i)}] + v_k(\omega_\nu \to \omega_{\nu 1}^{(k)})\xi^{(k)}[\omega_{\nu 1}^{(k)}]$$

$$+ w_{ik}(\omega_\nu \to \omega_{\nu 1}^{(i)}, \omega_{\nu 1}^{(k)})\xi^{(i)}[\omega_{\nu 1}^{(i)}]\xi^{(k)}[\omega_{\nu 1}^{(k)}] + \begin{cases} w(\omega_\nu \to \omega_{\nu+1})\xi[\omega_{\nu+1}], & \nu < n, \\ 0, & \nu = n. \end{cases}$$

Here

$$\xi^\alpha[\omega_{\nu\mu}^\alpha] = h^\alpha[\omega_{\nu\mu}^\alpha] + \begin{cases} w(\omega_{\nu\mu}^\alpha \to \omega_{\nu,\mu+1}^\alpha)\xi^\alpha[\omega_{\nu,\mu+1}^\alpha], & \mu < n_\alpha, \\ 0, & \mu = n_\alpha. \end{cases}$$

Writing $F(x|\omega)$ for cumulative distribution of the random variable $\xi[\omega]$, and $F_\alpha(x|\omega)$ for distribution of $\xi^{(\ alpha)}[\omega]$, we obtain the following equation following from the above assumptions:

$$F(x|\omega) = \iint P_2(d\omega' d\omega''|\omega) \iint dF_i(x'|\omega')dF_k(x''|\omega'')$$
$$\times\left[p_0(\omega)u(x - x_{ik}) + \int P(d\omega^*|\omega)F\left(\frac{x - x_{ik}}{w(\omega \to \omega^*)}\middle|\omega^*\right)\right], \tag{10.119}$$

where

$$x_{ik} = h(\omega) + v_i(\omega \to \omega')x' + v_k(\omega \to \omega'')x'' + w_{ik}(\omega \to \omega', \omega'')x'x'',$$

$$u(x) = \begin{cases} 0, & x < 0, \\ 1, & x \geqslant 0. \end{cases}$$

The distribution functions $F_\alpha(x|\omega)$, in turn, satisfy the equations following from (10.107).

Passing from the Eq. (10.119) to the equation for the expectation of $\xi[\omega]$

$$\overline{x}(\omega) = h(\omega) + \int P_2(d\omega'|\omega)v_i(\omega \to \omega')\overline{x}_i(\omega') + \int P_2(d\omega''|\omega)v_k(\omega \to \omega'')\overline{x}_k(\omega'') +$$

$$+\iint P_2(d\omega' d\omega''|\omega) w_{ik}(\omega \to \omega', \omega'') \overline{x}_i(\omega') \overline{x}_k(\omega'') + \int P(d\omega^*|\omega) w(\omega \to \omega^*) \overline{x}(\omega^*),$$

we obtain the unbiasedness conditions for the described estimate of the mixed moment $\mu_{ik}(\omega)$:

$$h(\omega) = \overline{m}_{ik}(\omega),$$

$$v_i(\omega \to \omega') = \frac{dL_k}{dP_2}(\omega'|\omega),$$

$$v_k(\omega \to \omega') = \frac{dL_i}{dP_2}(\omega'|\omega),$$

$$w_{ik}(\omega \to \omega', \omega'') = \frac{dM}{dP_2}(\omega', \omega''|\omega),$$

$$w(\omega \to \omega^*) = \frac{dK}{dP}(\omega^*|\omega).$$

Similarly, Eq. (10.119) can be converted into equation for the second moment of the estimate under consideration.

10.10 AEGIS Code

On the base of principles described in Sect's 10.1–10.8, there were developed MC-algorithms for simulation of the main elementary events generally determining the EPC phenomenon in terms of the Markov branching process theory. Combinations of these algorithms with organizations logical links between them led to creating few effective MC programs (codes), one of the most known among them, AEGIS, is elaborated in Fermi-Lab (see [19]).

The author characterizes this code as a simple EMS program which estimates average energy deposited as a function of position in a target of arbitrary (three-dimensional) geometry and composed of any number of materials. AEGIS is specifically written for EMS with high initial energy. The approximate expressions of the differential cross sections used for bremsstrahlung and pair production assume the energy of the incoming particle to be much larger than the electron rest energy. Nonetheless, AEGIS considers photons as low as 10^{-4} GeV because the Compton process dominates at sufficiently low energies and is accurately described by the Klein–Nishina cross-sectional formula. Pair production is neglected below 1.25 MeV and is described by the Bethe and Beitler incomplete screening formulae above this energy. When an electron is produced or is slowed down to below 5 MeV, its residual range is calculated and its energy is deposited in a randomly selected location along the produced trajectory. (For Compton electrons only the kinetic energy is deposited.) Since sub-5 MeV electrons lose almost all their energy by collision and since their range does not exceed a few g cm^{-2} this approximation is justified. Collision losses of electrons are deposited locally at uniformly spaced steps along their trajectories.

They are derived from a stopping power vs. momentum table based on the formulae of Rohrlich and Carlson. Multiple Coulomb scattering of electrons is calculated at each step using a two component angular distribution: a Gaussian distribution (in projected angle) plus a single scattering tail. The particles emerging from a bremsstrahlung or pair production vertex assume the direction of the incident particle.

The number of tracks can be simply identified with the weight assigned to electrons at each location. The basic MC scheme is identical to that for computing energy deposition. To ensure fast convergence of the calculation use of "recording particles" is made here. To estimate electron track density electron emission is simulated at each vertex with the electrons kinematical parameters selected from a distribution proportional to the differential cross section (rather than differential inelasticity).

Aiming to increase the efficiency of the code (reduce time costs and reduce the statistical error of the results), the authors apply Monte Carlo weight modifications. By introducing an additional phase variable (generation number), the authors replace the contribution of each generation with the contribution of a single particle forming a kind of electoral institution with all its shortcomings. Although it is believed that this improves accuracy by speeding up the modeling of individual bare (without branches) trajectories, this opinion is too subjective and not supported by statistical analysis. The main objection here may be a violation of the representativeness of various groups of the statistical ensemble. Another more obvious shortcoming of the generation averaging method is the impossibility of obtaining information about real fluctuations and correlations in the EPC, which play an important role in the measurement practice of astrophysicists.

Below we will return to a more detailed discussion of this issue, and now we will show algorithms for modeling elementary processes.

Sampling Procedures

In this problem there are always exactly two outgoing particles per event. Hence, sampling energy and direction of one particle determines the event completely. This leads to two alternate selection schemes: (a) choose the particle's energy, T, with a probability (roughly) proportional to the differential cress section, $d\sigma/dT$, and then decide whether to follow either this particle or its outgoing partner with a probability proportional to each particle's energy, (b) choose equally between the two outgoing particle types and use a selection function (roughly) proportional to $Td\sigma/dT$. Both schemes are used, as will be seen below.

Bremsstrahlung

The incomplete screening cross sections of Bethe and Heitler are assumed:

$$d\sigma/d\mathscr{E} = Z(Z+1)\alpha r_e^2 \mathscr{E}^{-1}\{(1 + E^2/E_0^2)[\Phi_1(\gamma) - 4\ln Z/3]$$

$$- (2E/3E_0)[\Phi_2(\gamma) - 4\ln Z/3]\}, \qquad (\gamma < 2) \tag{10.120}$$

$$d\sigma/d\mathscr{E} = Z(Z+1)\alpha r_e^2 \mathscr{E}^{-1} 4\{(1 + E^2/E_0^2 - 2E/3E_0)$$

$$\times [\ln(2E_0E/mc^2\bar{\omega}) - 1/2 - c(\gamma)]\}, \qquad (\gamma > 2) \tag{10.121}$$

where Z is the atomic number, α the fine structure constant, r_e the classical electron radius,$\mathscr{E}$ the radiated photon energy, mc^2 the electron rest energy, E_0 and E the incident and outgoing electron energies and $\gamma = (100mc^2\mathscr{E}/E_0EZ^{1/3})$. The functions $\Phi_1(\gamma)$, $\Phi_2(\gamma)$ and $c(\gamma)$ are represented in tabular form in the program. For $\gamma > 15$ $c(\gamma)$ vanishes and (10.121) then corresponds to the no screening case.

The sampling function employed is

$$S(\mathscr{E}) = N(E_0)\mathscr{E}^{-1}[1 - \mathscr{E}/E_0 - 3\mathscr{E}^2/4E_0^2]f(\gamma), \tag{10.122}$$

where $N(E_0)$ is a normalizing factor and $f(\gamma)$ is defined by

$$f(\gamma) = B - \gamma[B - \ln(200/0.8Z^{1/3}) + 1/2]/0.8 \qquad (\gamma \leq 0.8) \tag{10.123}$$

$$= \ln(2E_0E/mc^2\mathscr{E}) - 1/2 \qquad (\gamma > 0.8) \tag{10.124}$$

with $B = 5.75 - (\ln Z)/3$. $S(\mathscr{E})$ is a rough approximation to $d\sigma/d\mathscr{E}$.

The sampling proceeds in four stages: (1) $\mathscr{E}$ is selected from a distribution proportional to $\mathscr{E}^{-1}$, (2) a uniform random number is compared with the expression in square brackets in (10.122) and if larger a new $\mathscr{E}$ is chosen, (3) a uniform random number is compared with $f(\gamma)/f(0)$ and if larger a new $\mathscr{E}$ is selected. Finally (4) a uniform random number ξ is generated to decide whether to follow the photon ($\xi < \mathscr{E}/E$) or the electron. The particle is then assigned a weight

$$W_a = S^{-1}(\mathscr{E})(d\sigma/d\mathscr{E})/T_a,$$

where $T_\gamma = \mathscr{E}/E_0$ and $T_e = 1 - T_\gamma$ are the probabilities to choose a photon and electron, respectively.

The total cross sections are evaluated by integrating (10.120) over $\mathscr{E}$ between the limits $\mathscr{E} = 10^{-4}$ GeV and $(E_0 - mc^2)$. The interaction lengths are evaluated and stored for each of 71 logarithmically spaced incident energies E_i, which span the energy range of $10^{-4} \div 10^3$ GeV. The factor $N(E_0)$ of (10.122) is stored similarly in tabular form.

Compton Scattering

The Klein–Nishina formula is assumed. The differential cross sections with respect to the outgoing photon energy are

$$d\sigma/d\mathscr{E} = \pi r_e^2(mc^2)\mathscr{E}_0^2[\mathscr{E}/\mathscr{E}_0 + (m^2 + 2m\mathscr{E}_0)/\mathscr{E}_0 +$$

$$+ (\mathscr{E}_0^2 - 2m\mathscr{E}_0 - 2m^2)/\mathscr{E}_0\mathscr{E} + m^2/\mathscr{E}^2], \qquad (10.125)$$

where $\mathscr{E}$ varies from $\mathscr{E}_0 mc^2/(mc^2 + 2m\mathscr{E}_0)$ to $\mathscr{E}_0$, the incident photon energy. Each of the four terms in the square brackets (10.125) is integrated between the above limits and their sum normalized to unity, for each incident energy E_i of the 71 member set referred to above. The results for the first three terms are stored in a table. The event selection proceeds by choosing a term from among the four terms of (10.125) with probabilities proportional to the relative magnitudes of the integrals. Next a specific $\mathscr{E}$ is selected from a distribution proportional to that of the chosen term of (10.125). Note that thus far the sampling is unbiased. Finally, the decision is made to follow the photon or electron in the same way as for bremsstrahlung. Here a weighting factor is assigned: $W_\gamma = \mathscr{E}_0/\mathscr{E}$ or $W_e = \mathscr{E}_0/(\mathscr{E}_0 - \mathscr{E})$.

The total cross sections corresponding to the integral of (10.125) over $\mathscr{E}$ are readily obtained in closed form. They are computed for each E_i and stored in an array.

Pair Production

Again the incomplete screening formulae are assumed:

$$d\sigma/dE_+ = Z(Z+1)\alpha r_e^2\mathscr{E}_0^{-3}\{(E_+^2 + E_-^2)[\Phi(\gamma') - (4/3)\ln Z]$$

$$+ (2/3)E_+E_-[\Phi(\gamma') - (4/3)\ln Z]\} \qquad (\gamma' \leq 2), \qquad (10.126)$$

$$d\sigma/dE_+ = Z(Z+1)r_e^2\mathscr{E}_0^{-3}\{[E_+^2 + E_-^2 + (2/3)E_+E_-]$$

$$\times [\ln(2E_+E_-/mc^2\mathscr{E}) - 1/2 - c(\gamma')]\},$$

where $\gamma' = 100\mathscr{E}_0 mc^2/(E_+E_-Z^{1/3})$ and $E_, E_-$ are the positron and electron energies which appear symmetrically in (10.126). The selection function used is

$$S(E) = M(\mathscr{E}_0)[(4/3)(E/\mathscr{E}_0)^3 + E/\mathscr{E}_0 - (4/3)(E/\mathscr{E}_0)^2]g(\gamma')$$

with $M(\mathscr{E}_0)$ a normalizing factor and $g(\gamma')$ defined as

$$g(\gamma') = B - \gamma'[B - \ln(200/0.8Z^{1/3}) + 1/2]/0.8 \qquad (\gamma' \leq 0.8), \qquad (10.127)$$

$$g(\gamma') = \ln[2E(\mathscr{E}_0 - E)/\mathscr{E}_0 mc^2] - 1/2 \qquad (\gamma' > 0.8). \tag{10.128}$$

Note that $S(E)$ is roughly proportional to $Ed\sigma/d\mathscr{E}$. The sampling again proceeds in stages: (1) E is selected uniformly between the kinematical limits, (2) a uniform random number is compared with the expression in square brackets in (10.125) and if larger a new E is selected, and (3) a uniform random number is compared with $g(\gamma')/g(0)$ and if larger a new E is selected. The particle is taken to be an electron or positron with equal probability. The weight in either case is

$$W = 2S^{-1}d\sigma/dE. \tag{10.129}$$

The total cross section is obtained by integrating (10.126) over the kinematical limits. These are calculated and stored for each of the E_i. They are also combined with the Compton total cross sections to obtain the photon interaction lengths.

Multiple Coulomb Scattering

A rather simple straightforward procedure is employed assuming all energies are much larger than the electron rest energy. The angular distribution of the electrons traversing a thickness t is assumed to be composed of an exponential distribution in the variable θ^2 plus a single (Rutherford) scattering tail for $\theta > \theta_0 = 0.0175t^{1/2}pX_0^{1/2}$, where X_0 is the radiation length of the medium and p is the momentum in GeV/c). The exponential distribution in θ^2 is equivalent to a Gaussian distribution in projected angle. The mean square angle is assumed to be

$$\theta_m = \theta_0[1 + 0.125\log_{10}(t/0.1X_0)].$$

The total probability for single scattering $\theta > \theta_0$ as measured by the integral of the single scattering distribution (using the small angle approximation and ∞ as upper limit) is independent of thickness or momentum (it is of the order of 0.02). This simplifies the algorithm.

References

1. Haghighat A (2020) Monte Carlo methods for particle transport. CRC Press
2. Spanier J (1996) Quasi-Monte Carlo methods for particle transport problems. In: Monte Carlo and Quasi-Monte Carlo methods in scientific computing, Springer New York, pp 121–148
3. Mikhailov GA, Sabelfeld KK (1992) Optimization of weighted Monte Carlo methods. Springer-Verlag, Berlin p 225
4. Uchaikin VV (1976) Monte Carlo simulation of particle transport with multiplication. Zh Vychisl Mat Mat Fiz 16(3):758–766
5. Kahn H (1954) Applications of Monte Carlo. Rep RM-1237-AEC, Rand Corp

6. Van Ginneken A (1978) AEGIS: a program to calculate the average behavior of electromagnetic showers, *Fermilab*, FN-309, 1111.0000
7. OZmutlu EN (1992) Sampling of angular distribution in Compton scattering. AppL Radiat Isot 43(6):713–715
8. Blomquist RN, Gelbard EM (1983) An assessment of existing Klein-Nishina Monte Carlo sampling methods. Nucl Sci Eng 83:380–384
9. Koblinger L (1975) Direct sampling from the Klein-Nishina distribution for photon energies above 1.4 MeV. Nucl Sci Eng 56(2):218–219
10. Hirayama H, Namito Y, Nelson WR et al (2005) The EGS5 code system. No. SLAC-R-730, United States Department of Energy (Stanford, CA: SLAC)
11. Rogers DWO, Bielajew AF (1988) A comparison of EGS and ETRAN. In: Jenkins TM, Nelson WR, Rindi A (eds) Monte Carlo transport of electrons and photons, Plenum, New York, pp 323–342

For Further Reading

12. Spanier J (2010) Monte Carlo methods. Nuclear Computational Science, p 117
13. Vassiliev ON (2017) Monte Carlo methods for radiation transport. Fundamentals and Advanced Topics, Springer
14. Gentle JE (2003) Random number generation and Monte Carlo methods, 2nd edn. Springer, New York
15. Gilks WR, Richardson S, Spiegelhalter DJ (1996) Markov Chain Monte-Carlo in practice. Chapman & Hall, New York

Chapter 11
Stochastic Phenomenology of EAS

11.1 EAS Structure

More complex than electron–photon cascades are *extensive air showers* (EAS)—cascades initiated by primary ultrahigh-energy cosmic radiation in the atmosphere including hadron reach component [1]. They contain almost all currently known elementary particles and a search for new particles is underway.

Cosmic rays were first registered in 1912, but later it was discovered that they have a wide energy spectrum, falling so rapidly with energy ($\propto E^{-\gamma}$ with $\gamma \sim 2.7$), that around 10^{14} eV it becomes impractical to make measurements of high precision directly: the number of events falling on a detector of a size that can be accommodated on a balloon or a space-craft is simply too small. However, at this energy, sufficient particles are produced in the atmosphere as secondaries to the incoming primary cosmic rays for some to reach mountain altitudes. At $E_0 = 10^{15}$ eV, they counted around 10^6 particles covering approximately 10^4 m^2, while at 10^{20} eV, some 10^{11} particles are spread over about 10 km^2.

A visual picture of the shower development was presented in Fretter's photography [2]. Except for scale, one can see here extremely similar to those present when a high-energy particle enters the Earth's atmosphere and creates an EAS.

A power law distribution of cosmic rays was first described by E. Fermi in 1949 [3] but, until 1966, there were no predictions as to the power law index or to further structures in the energy spectrum. Observations in 1959 had indicated a steepening at around $3 \cdot 10^{15}$ eV (the "knee"), while, in 1963, it was assumed from observations made with the first large shower array, that the spectrum flattens just above 10^{18} eV. It took about 40 years to establish that there is indeed a steepening in the cosmic ray spectrum at about this energy but whether this is a cosmological effect or a consequence of a limit to which sources can accelerate particles is unclear [4].

Multiple processes play a leading role in the development of EAS [5–9]. A proton or other nucleus of primary radiation in the process of inelastic interaction with the nucleus of an air atom gives rise to a large number of secondary particles—mainly pions (neutral and charged). Charged pions, in turn, continue to generate secondary

© Higher Education Press Limited Company 2024

V. V. Uchaikin, *Electron-Photon Cascades*, Nonlinear Physical Science,
https://doi.org/10.1007/978-981-99-7524-2_11

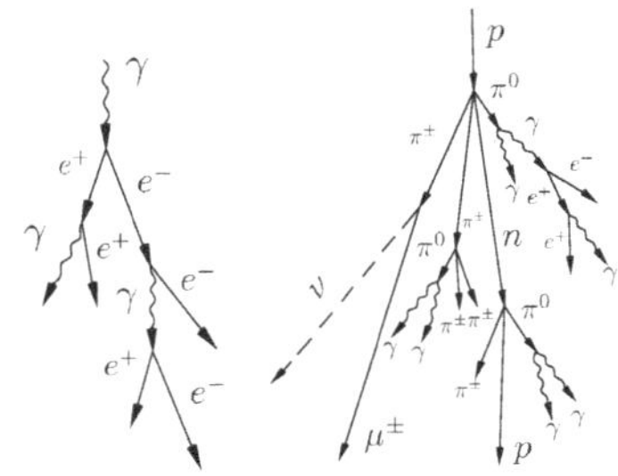

Fig. 11.1 Sketches of shower generated by photon and proton

particles, resulting in an intense energy exchange accompanied by an increase in the number of particles. Another essential process in EAS is the decay of pions, in which neutral pions give rise to EPC, and charged pions form muons. The stage initial of EAS development is shown in Fig. 11.1.

Let the detector in the plane $z = z^*$ measure the number of particles d with energy above E^*. The average value of its readings in a shower formed by an a-type particle born at a point z with energy E satisfies the equation

$$\left[-\partial_z + \sigma_a\right] \bar{N}_{ad} = \sum_b \int dE' w_{a\to b}(E \to E')\bar{N}_{bd}(z, E') \qquad (11.1)$$

with the boundary condition $\bar{N}_{ad}(z^*, E) = \varepsilon(E - E^*)\delta_{ad}$. We will assume that the decay processes are taken into account by the coefficients of this equation. In accordance with the EAS development scheme, the equation for $\bar{N}_{NN}$ is solved independently of the others, the equation for $\bar{N}_{N\pi}$ (hereinafter, the index π will be used to designate charged pions) is solved together with the equation for $\bar{N}_{\pi\pi}$, the equation for $\bar{N}_{N\mu}$ is solved together with the equations for $\bar{N}_{\pi\mu}$ and $\bar{N}_{\mu\mu}$, equation for $\bar{N}_{Ne}$—with equations for $\bar{N}_{\pi^0 e}$ and $\bar{N}_{\pi^0 e}$ (the latter includes the $\bar{N}_{\gamma e}$ solution of the electromagnetic cascade theory) in the following sequence:

$$\bar{N}_{\pi\pi} \to \bar{N}_{N\pi}, \quad \bar{N}_{\mu\mu} \to \bar{N}_{\pi\mu} \to \bar{N}_{N\mu},$$

$$\bar{N}_{\gamma e} \to \bar{N}_{\pi^0 e} \to \bar{N}_{\pi e} \to \bar{N}_{Ne}.$$

We neglect the effect of other particles on the EAS components under consideration. If we pass to mass units of length $t = \int\limits_z^{z^*} \rho(z')dz'$, then the decay cross sections in an inhomogeneous atmosphere take the form $\sigma_a^{\mathrm{dec}} = \dfrac{B_a}{\rho(t)E}$, where $B_a = m_a c/\tau_{0a}$ is a particle-type constant, and $\rho(t)$ is the density of the atmosphere at a height t (g/cm^2) above the observation level.

Neglecting the lifetime of the π^0-meson, we can write

$$\bar{N}_{\pi^0 e}(t, E) = 2 \int_{E^*}^{E} f_{\pi^0\gamma}(E \to E')\bar{N}_{\gamma e}(t, E')dE',$$

where $f_{\pi^0\gamma}$ is the spectrum of γ decay quanta of the π^0-meson normalized to unity.

Neglecting all processes of interactions (and decay) of muons, but taking into account the energy losses for ionization, we obtain

$$\bar{N}_{\mu\mu}(t, E) = \varepsilon(E - E^* - \beta_\mu t),$$

so that the contribution from the π-meson decay integral in the equation for $\bar{N}_{\pi\mu}$ takes the form

$$\int_{E^*}^{E} w_{\pi\to\mu}(E \to E')\bar{N}_{\mu\mu}(t, E')dE' = \frac{B_\pi}{\rho(t)E} \int_{E^*}^{E} f_{\pi\mu}(E - E')\varepsilon(E' - E^* - \beta_\mu t)dE'.$$

In our calculations, we assumed $f_{\pi^0\gamma}(E \to E') = 1/E$, $F_{\pi\mu}(E \to E') = 1/(1 - \theta)E$, $-\theta E < E' < E, \theta = (m_\mu/m_\pi)^2$.

The spatial distribution of EAS electrons was calculated taking into account only the Coulomb scattering of EPC electrons; the change in the direction of particle motion in other processes was neglected.

11.2 Multiple Process Models

In contrast to electromagnetic processes, multiple processes do not have a complete theory and are described in terms of various models that offer various ways of extrapolating accelerator data to the ultrahigh-energy region.

The first calculations of the passage of cosmic rays through the atmosphere and, in particular, EAS was performed under the assumption of scale invariance (scaling) of inclusive cross sections of multiple processes: $w^{sc}_{a\to b}(E \to E_1)dE_1 = w'_{a\to b}(E_1/E)dE_1/E$. Further analysis of experiments on accelerators and cosmic rays forced us to abandon this assumption [9–12]. Models with scaling violation began to be used in EAS calculations in [9, 11, 17]. Subsequently, the methods (including SBCE), developed during our research of EPC, were successfully applied in the field of the extensive air showers [19, 20].

The purpose of our EAS calculations is to study the possibilities and further improve the developed numerical method for solving inverse equations in relation to a wider range of problems, to study in detail EAS characteristics (including their fluctuations and correlations) in models with scaling violation, and to compare the results with experimental data.

The EAS characteristics were calculated by us within the framework of three non-scaling models.

Model I is essentially a variant of the empirical model [9], according to which the logarithmic growth of the cross section is provided by the processes of particle production in the pionization region ($x \lesssim 0.1$):

$$\sigma_a^{inel}(E) = \sigma_a^0[1 + \alpha_a \ln(E/E_0)], \qquad a = N, \pi,$$

where $E_0 = 20\text{GeV}, \alpha_N = 0.04, \sigma_N^0 = 250\,\text{mb}, \alpha_\pi = 1.3\,\alpha_N, \sigma_\pi^0 = \sigma_N^0/1.3$. Inclusive distributions are approximated by the following formulas:

$$w_{a\to b}(E \to E')dE' = \sigma_a^0 \xi_a(E) f_{ab}(E \to E'/E)dE'/E,$$

$$f_{NN}(E \to x) = 1 + \alpha_N 1(x_0 - x)1(x - x_0 E_0/E)/x,$$

$$f_{N\pi^0}(E \to x) = \frac{1,2(1-x)}{(x^2+a^2)^{1/2}} e^{-5.56x} + \frac{\eta_N 1(x_0 - x)}{x^2+a^2} = \frac{1}{2} f_{N\pi}(E \to x),$$

$$f_{\pi\pi^0}(E \to x) = \frac{1,3(1-x)}{(x^2+a^2)^{1/2}} e^{-4.76x} + \frac{\eta_\pi 1(x_0 - x)}{x^2+a^2},$$

$$f_{\pi\pi}(E \to x) = 0.6 + 2 f_{\pi\pi^0}(E \to x).$$

Here $x_0 = 0.1$, $a^2 = 0.128/E$, $\eta_a = (\alpha_a/3)\ln(E/E_0)/\ln \dfrac{(x_0 + \sqrt{x_0^2 + a^2})}{a}$. The values of the coefficients $\xi_a(E)$ are determined from the condition of the energy conservation law

$$\frac{1}{\sigma_a^{inel}} \sum_b \int_0^E w_{a\to b}(E \to E')E'dE' = E.$$

The above distributions provide (in the specified energy range) an increase in the average multiplicity according to the law $\sim E^{0.44}$ and a slight change in the average value of the inelasticity coefficient (from 0.53 at 10^2GeV to 0.67 at 10^7GeV).

Model II is based on the theory of supercritical number using an additive quark model [14].

Model III is a model of quark-gluon strings with parameters refined from the latest ISR and SPS accelerator data [18].

Table 11.1 gives a general idea of the ratio of the quantitative characteristics of the models under consideration.

The results of calculations for these models are given in [19, 20].

It is instructive to compare two kinds of showers, generated by photon and proton, as it is done in the article [2].

Taking from this article Fig. 11.2 illustrates typical showers generated by (a) a 0.32 TeV gamma ray and (b) a 1 TeV proton, showing the much greater width of the latter, due to transverse momentum given to pions in strong interactions. To make

Table 11.1 Quantitative characteristics of models

Character-ristika	Model	Primary hadron energy, GeV			
		10^2	10^4	10^6	10^8
σ_p, cm^2/g	I	0.278-1	0.1265-1	0.1451-1	0.1638-1
	II	0.1074-1	0.1321-1	0.1568-1	0.1815-1
	III	0.113-1	0.1279-1	0.1523-1	0.1758-1
σ_π, cm^2/g	I	0.844-1	0.103-1	0.122-1	0.140-1
	II	0.818-1	0.104-1	0.127 -1	0.150-1
	III	0.881-1	0.102-1	0.1296-1	0.159-1
$\langle K \rangle$	I	0.53	0.60	0.65	0.69
	II	0.50	0.52	0.55	0.57
	III	0.59	0.69	0.75	0.80
$\langle K_{ch} \rangle_\pi$	I	0.72	0.77	0.81	0.83
	II	0.82	0.86	0.88	0.89
$\langle \eta_{ch} \rangle_p$	I	4.66	21.9	151.7	1323
	II	8.68	28.5	59.1	96
	III	4.65	14.9	44.7	128.6
$\left\langle \frac{\Sigma E_{\pi^0}}{E} \right\rangle_p$	I	0.238	0.268	0.268	0.279
	II	0.270	0.286	0.293	0.297
	III	0.198	0.230	0.250	0.266

this clearer, the lateral scale has been expanded by a factor of 5. Only a fraction of the particle track length radiating Cherenkov light has been shown, to avoid saturation. The more collimated gamma ray shower results in a pool of Cherenkov light on the ground with a radius about 120 m (at typical mountain altitude), beyond which the light intensity falls rapidly.

A simplification which is very useful in interpreting Cherenkov signals from showers is to regard the light reaching the ground as made up of three parts. The main bulk of particles, near shower maximum, produce the light seen at 100–130 m from the shower axis: here the intensity of the light is a good measure of the calorimetric energy deposition in the atmosphere. Close to the shower axis (within 40 m, say) one sees the light from local particles—the tail of the shower—so the intensity here fluctuates very much from shower to shower, and should be closely correlated with the signal in particle detectors on the ground. And particles actually reaching the ground are surrounded by a $1/r$ distribution of light intensity, so there are localized peaks due to penetrating muons, in hadronic showers, and due to electrons reaching the ground within about 30m of the axis even in the more energetic gamma showers.

Muons can also be detected over a large area with Cherenkov imaging: a muon passing within about 40m of a 10m telescope will produce a short arc on the image (part of a circle of 1.3° radius, becoming a complete arc if the muon strikes the

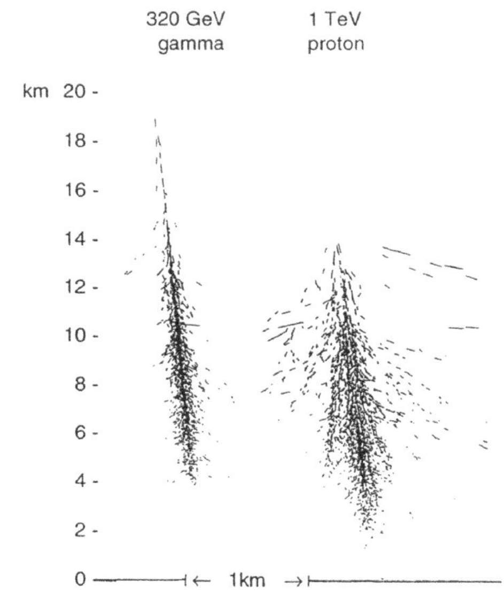

Fig. 11.2 Real forms of showers generated by a 320 GeV photon and a 1 TeV proton. The lateral scale is exaggerated by a factor of 5. The picture is thinned down by showing each track as a sparsely dashed line [2]

light collecting mirror). These separate arcs may typically appear about a degree away from the image of the main cascade, and add to the "width" of the image, causing it to be rejected as "not gamma-like". However, there is on average less than one muon arc fragment per image at the energies most commonly studied, somewhat below 1 TeV, so again, muons do not play the main part in hadron rejection. Although hadronic showers should get narrower at high energies, and presumably less different from gamma showers, except for an increasingly reliable contribution from muons, it is found in practice that (using pixels 15 arc minutes apart, with unfilled spaces between the phototube-sensitive areas) if showers are characterized by several image parameters (typically width, length, distance from source, and angular deviation of image axis from line pointing to source), discrimination against hadrons improves up to at least 10 TeV.

11.3 Parametric Sensitivity Analysis

One of the most important tasks of the theoretical analysis of EASs is to determine the sensitivity of the observed characteristics to the characteristics of hadronic interaction models. It is difficult to solve this problem by a simple comparison of calculations performed by different authors in different models, since usually several interaction parameters differ, other physical constants may differ, and different calculation methods give different errors, which are often not given. The method proposed in [23, 24] seems to be more promising. It is based on the parametrization of a multiple process and the calculation of the change in EAS characteristics when changing each of the α_i parameters, separately within the same method. At the same time, the EAS cascade curves (nuclear active and electronic components) were also parameterized using four components Y_j, which determine the magnitude and position of the maximum, as well as the longitudinal dimensions of the increasing and decreasing parts of the cascade curve. Based on the results of calculations, the sensitivity of the EAS parameters to the parameters of hadronic interactions was estimated, which is determined by the formula

$$\Gamma_j^i = \frac{\alpha_i}{Y_j}\frac{\partial Y_j}{\partial \alpha_i}. \tag{11.2}$$

Calculations have shown that the inelasticity coefficients and the index of the degree of growth of the average multiplicity have the greatest influence on the shape of the cascade curve. But for the practical application of these results, information on the longitudinal development of EAS is needed, which creates certain difficulties. In addition, the calculation model used by the authors (independence of the cross sections of inelastic interactions on energy and the initial values of the parameters) does not seem to be relevant at present.

We have made an attempt to further develop this approach. In contrast to [21, 22], the observed characteristics Φ_j^+ are chosen as EAS parameters, and a more realistic model according to modern concepts is used as a reference one. The sensitivity was found by numerically solving the Eq. (4.52) written directly for the derivatives themselves (Sect. 4.5).

The interaction model was parameterized as follows:
Cross section of inelastic interactions

$$\sigma_a(E) = \sigma_{0a}\{1 + \beta_a[ln(E/E_0)]^\gamma\},$$

Inclusive distribution of secondary particles

$$W(E \to E') \equiv \frac{1}{\sigma_a} w(E \to E') = \frac{A/E}{(E'/E)^2 + a^2},$$

Average multiplicity

$$\bar{n}(E) = n_0 E^a,$$

Average coefficient of inelasticity

$$\bar{K}(E) = K_2 E^{\varkappa}.$$

Note that the parameters A and a are related to the inelasticity coefficient and multiplicity by the normalization relations:

$$\bar{n}(E) = \int_0^E W(E \to E')dE', \tag{11.3}$$

$$\bar{K}(E) = \int_0^E \left(\frac{E'}{E}\right) W(E \to E')dE'.$$

For a certain choice of these parameters, the inclusive distribution in the pionization region coincides with that given in [24].

Representing the inclusive distribution in an analytical form that allows the calculation of integrals (11.3) in terms of elementary functions allows one to obtain an analytical expression for the operator $L_i^+ = \partial L^+/\partial\alpha_i$:

$$(L_i^+)_{ac} = \frac{\partial\sigma^{tot}}{\partial\alpha_i}\delta_{ac} - \frac{\partial\sigma^{in}(E,\alpha)}{\partial\alpha_i}\int_0^E dE' W_{ac}(E, E'),$$

if the parameter α_i enters the cross section of inelastic interactions and

$$(L_i^+)_{ac} = \sigma^{in}(E)\int_0^E dE' \frac{\partial W_{ac}(E, E', \alpha)}{\partial\alpha_i},$$

$$\frac{\partial W(E, E', \alpha)}{\partial\alpha_i} = \frac{\partial W}{\partial A}\cdot\frac{\partial A}{\partial\alpha_i} + \frac{\partial W}{\partial a}\cdot\frac{\partial a}{\partial\alpha_i}$$

in the case where α_i characterizes an inclusive distribution.

Of particular interest are the variations in the value function caused by the change in the partial coefficients of inelasticity at a constant value of the average particle energy. In this case,

$$\left(\frac{\partial L^+}{\partial K_\gamma}\right) = -\sigma_\pi^{in}(E)\left[\frac{1}{K_{\gamma\pi}}\int_0^E dE' W_{\pi\pi^0}(E, E') + \frac{1}{K_{\gamma\pi}-1}\int_0^E dE' W_{\pi\pi}(E, E')\right].$$

Equations (4.52) for derivatives of N_e, N_μ and N_h with respect to σ_{0p}, $\sigma_{0\pi}$, β_p, β_π, α, K_{t0t} and $K_{\gamma\pi}$ were solved by the numerical method mentioned above with the following model parameters: $\sigma_{0p} = 0.011$ cm ^{2}/ r, $\sigma_{0\pi} = 0.00926$ cm^2 r, $\beta_p = \beta_\pi = 0.05$, $n_{0p} = 11.7$, $n_{0\pi} = 12.8$, $\alpha = 0.25$, $K_{0p} = 0.5$, $\varkappa = 0$, $K_{\gamma\pi}$ =0.27.

The use of derivatives to study value variations suggests that in the considered range of α_i parameters, the dependence of Φ_j^+ on them is close to linear. To determine the area of this linearity, we calculated the variations of the value functions for a given change in the model parameters in two ways. In Tables 11.2 and 11.3 the values of Φ_j^+ are given, obtained by directly solving adjoint equations for the values of parameters $\alpha_i + \delta\alpha_i$ (1) and by using derivatives on the value by parameters (2). As can be seen from these tables, the dependence of the total number of EAS electrons and muons on the parameter β, which characterizes the logarithmic increase in the cross section

Table 11.2 The number of EAS electrons from the primary proton of energy 10^{16} eV

t, g/cm^2	$\delta\beta_\pi = -0.02$		$\delta\beta_\pi = -0.02, \delta\beta_p = -0.02$		$\delta\alpha_p = 0.25, \delta\alpha_\pi = 0.25$	
	1	2	1	2	1	2
130	7.32+4	7.30+4	6.42+4	7.42+4	2.13+5	1.68+5
370	3.57+6	3.58+6	3.26+6	3.28+6	4.72+6	4.89+6
610	7.37+6	7.37+6	7.13+6	7.20+6	6.71+6	6.88+6
720	6.55+6	6.56+6	6.50+6	6.59+6	5.44+6	5.40+6
880	4.09+6	4.09+6	4.30+6	4.31+6	3.05+6	2.85+6
1026	2.08+6	2.08+6	2.32+6	2.29+6	1.43+6	1.31+6

Table 11.3 Average number of muons in a shower from a proton of energy E_0

E_0, TeV	$\delta\alpha_p = 0.25, \delta\alpha_\pi = 0.25$		$\delta\beta_\pi = -0.02, \delta\beta_p = -0.02$		$\delta K_{\gamma\pi} = 0.06$	
	1	2	1	2	1	2
100	1.66+3	1.65+3	1.55+3	1.56+3	1.30+3	1.23+3
1 000	1.41+4	1.42+4	1.24+4	1.25+4	9.83+3	9.06+3
10,000	1.22+5	1.23+5	9.99+4	1.01+5	7.59+4	6.76+4

of inelastic interactions, is close to linear. At the same time, a change in the partial inelasticity coefficient $K_{\gamma\pi}$ in pion–nucleus interactions by 0.06 leads to significant changes in the value function and does not belong to the region of its linear changes. The dependence of the total number of EAS electrons and muons on the degree of increase in the average multiplicity is weaker than the dependence on the inelasticity coefficients.

Linearized representation (2) describes the change in the total number of EAS particles somewhat worse when a large number of parameters are varied. In Table 11.4, the values of the value function obtained with the following change in the model parameters are given: $\delta\alpha = 0.1$; $\delta\sigma_{0p} = 0.00055$, $\delta\sigma_{0\pi} = -0.00093$, $\delta\beta_p = \delta\beta_\pi = -0.02$. As can be seen from Table 11.4, the number of EAS muons with energies above 1 and 10 GeV, as well as the total number of electrons at depths of 250–600 g/cm^2, is reconstructed most accurately. The same EAS characteristics as the number of high-energy hadrons and muons are more sensitive to the elementary interaction parameters under consideration, and the used variations of the parameters lie outside the range of linear variation of the value functions. The difference becomes especially noticeable at energies of the primary particle on the order of 10^{18}eV. Note that to reduce the errors introduced by the nonlinear behavior of the value functions, you can use the method of iterative refinement of the parameters using Eq. (4.51). Figures 11.3, 11.4, 11.5 and 11.6 show the sensitivity of value functions to the above parameters of elementary interaction. As can be seen from Fig. 11.3, at a fixed measurement accuracy, the higher the muon sensitivity to the model parameters, the higher the energy threshold of the detector. In this case, the sensitivity to the parameters of the cross section of pion–nucleus interactions, other things being equal, is

Table 11.4 EAS characteristics calculated by solving conjugate equations (1) and by the method of differentiation with respect to parameters (2)

Charactertics t, g/cm^2 EAS	Method	$E_0 = 3 \cdot 10^{14}$eV	10^{15}	10^{165}	10^{17}	10^{18}
Electrons $t = 250$	1	6.96+4	1.77+5	1.12+6	7.03+6	4.41+7
	2	6.88+4	1.75+5	1.11+6	6.85+6	4.26+7
Electrons $t = 670$	1	1.69+5	5.91+5	6.70+6	7.02+7	6.87+8
	2	1.66+5	5.76+5	6.50+6	6.85+7	6.75+8
Electrons $t = 1026$	1	3.65+4	1.46+5	2.15+6	2.95+7	3.80+8
	2	3.52+4	1.39+5	2.05+6	2.77+7	3.53+8
Muons $t = 1026$, $E > 10$ GeV	1	1.29+3	3.62+3	2.86+4	2.28+5	1.85+6
	2	1.27+3	3.57+3	2.81+4	2.23+5	1.82+6
Muons $t = 1026$, $E > 1$ GeV	1	4.61+3	1.33+4	1.10+5	9.22+5	7.76+6
	2	4.72+3	1.36+4	1.13+5	9.49+5	7.98+6
Muons $t = 1026$, $E > 100$ GeV	1	8.15+1	2.02+2	1.43+3	1.12+4	8.83+4
	2	7.20+1	1.76+2	1.28+3	1.00+4	7.80+4
Hadrons $t = 690$, $E > 1$ TeV	1	2.55+0	9.36+0	1.06+2	1.07+3	1.04+4
	2	2.23+0	8.33+0	9.47+1	9.58+2	9.35+3
Hadrons $t = 690$, $E > 3$ TeV	1	5.06-1	2.09+0	2.71+1	2.94+2	2.93+3
	2	4.24-1	1.79+0	2.37+1	2.55+2	2.55+3
Hadrons $t =690$, $E > 10$ TeV	1	7.66-2	4.07-1	6.33+0	7.58+1	7.98+2
	2	5.93-2	3.30-1	5.38+0	6.46+1	6.75+2

noticeably higher than the parameters of nucleon–nucleus interactions. The fact of the high sensitivity of the total number of electrops and hadrons to partial inelasticity coefficients in nucleon– and pion–nucleus interactions also agrees with the above results. A characteristic property of EASs is a significantly higher sensitivity of the total number of hadrons to the parameter $\sigma_{0\pi}$ of the inelastic interaction cross section than to σ_{0p}. At the same time, the total number of hadrons is a less sensitive character-

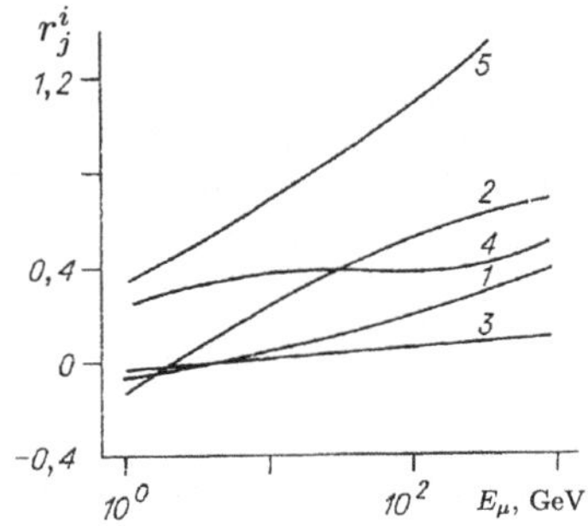

Fig. 11.3 Sensitivity of the number of muons with higher energy E_μ to parameters. 1—σ_{0p}, 2—$\sigma_{0\pi}$, 3—β, 4—α, 5—K_{tot}

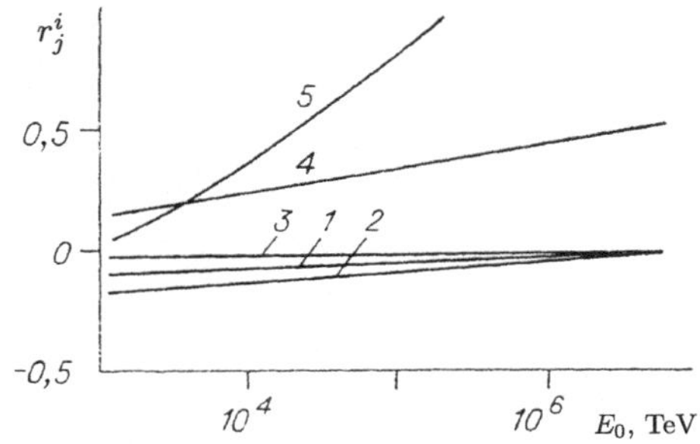

Fig. 11.4 Sensitivity of the number of muons with energies above 1 GeV from a proton of energy E_0 to parameters. 1—σ_{0p}, 2—$\sigma_{0\pi}$, 3—β, 4—α, 5—K_{tot}

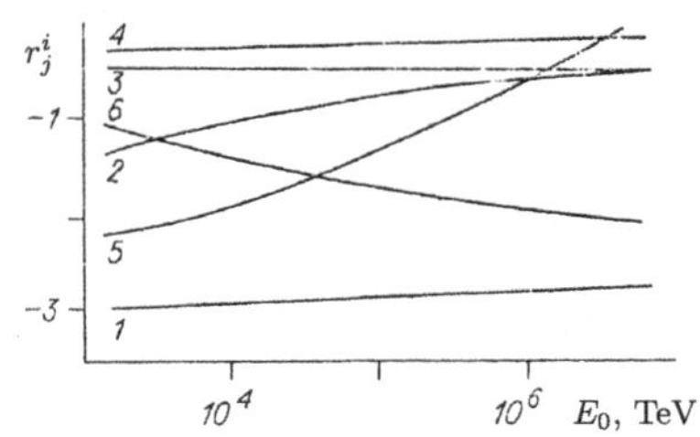

Fig. 11.5 Sensitivity of the number of hadrons with energies above 1 GeV from a proton of energy E_0 to parameters. 1—$\sigma_{0\pi}$, 2—σ_{0p}, 3—β, 4—α, 5—K_{tot}, 6—$K_{\gamma\pi}$

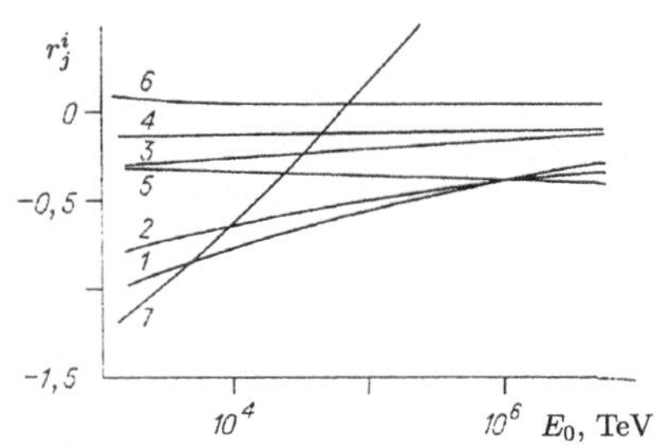

Fig. 11.6 Sensitivity of the number of EAS electrons from a proton of energy E_0 to parameters. 1—σ_{0p}, 2—$\sigma_{0\pi}$, 3—β, 4—α, 5—K_{tot}, 6—$K_{\gamma\pi}$

istic to the degree of multiplicity growth than to the cross section and pion–nucleus interactions.

At energies of the primary particle of the order of 10^{18}eV, the sensitivity of hadrons to the total inelasticity coefficient in proton–nucleus interactions is weak. To demon-

Fig. 11.7 Cascade curves of EAS electrons. **a**—$E = 10^4$ GeV, **b**—$E = 10^7$ GeV. I—interaction model [28], II—interaction model [29]. In the oval the contributions of the σ_p, n_{ch} and K variations are shown

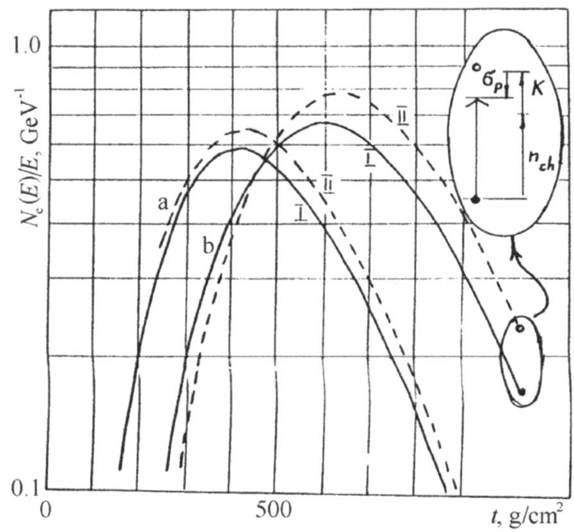

Fig. 11.8 EAS muon number at sea level versus primary proton energy. Muon threshold energy $E_{th} = 1$ GeV. I—interaction model [28], II—interaction model [29]. In the oval the contributions of the σ_p, n_{ch} and K variations are shown

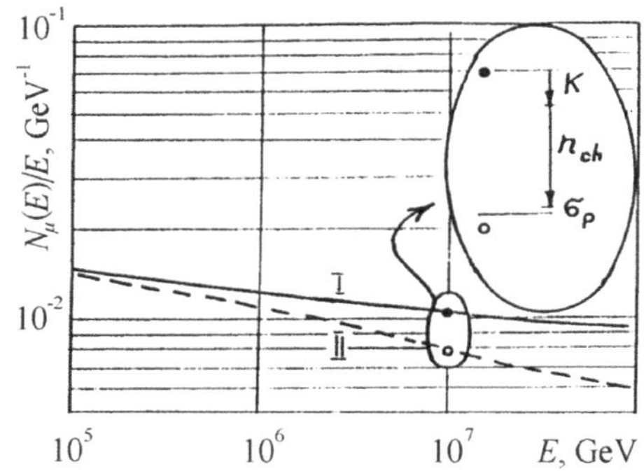

strate the application of this approach to the similar problem in the highest energy EAS simulation, the variations of the EAS characteristics due to the change in the interaction model were investigated [25]. The results obtained within the framework of the two different interaction models [26, 27] are presented in Figs. 11.7, 11.8, and 11.9.

From these figures, it seems that EAS characteristics in model II [29] obtained from EAS data in model I [29] do not contradict the results of the direct calculations. Inside ovals, one can see the comparative contributions of the cross section (σ_p), multiplicity (n_{ch}) and inelasticity (K) variations to the EAS characteristics.

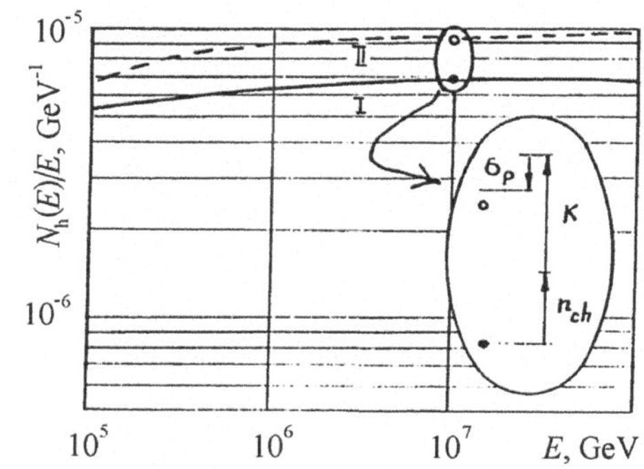

Fig. 11.9 EAS hadron number at mountain level (t_{obs} = 700 g/cm^2) versus primary proton energy. Hadron threshold energy $E_{th} = 10^3$ GeV. I—interaction model [28], II—interaction model [29]. In the oval the contributions of the σ_p, n_{ch} and k variations are shown

11.4 Functional Sensitivity Analysis

A more detailed analysis of the sensitivity of the EAS characteristics to the interaction model, including the study of the influence of the model characteristics in different energy regions, can be carried out using functional derivatives. Let us represent the set of measured EAS characteristics in the form of functionals of the set of functions $u_1(x_1), u_2(x_2), \ldots, u_k(x_k)$ describing the properties of the source, the medium in which the particles propagate, and the detector:

$$I_m = F_m(u_1(\cdot), \ldots, u_k(\cdot)). \tag{11.4}$$

Such functions are usually the energy spectra of primary particles $S_a(E)$, the cross sections of nucleon– and pion–nucleus inelastic interactions $\sigma_p(E)$ and $\sigma_\pi(E)$, the inclusive spectra of particle collisions $W_{a\to b}(E \to E')$, etc., x_i are the corresponding sets of arguments.

We call the u_i-value (generalized value) of the point x_i with respect to the functional I_m the derivative

$$\lambda_{mi}(x_i) = \frac{\delta F_m\,(u_1(\cdot), \ldots, u_k(\cdot))}{\delta u_i(x_i)}. \tag{11.5}$$

In this terminology, the usual value of the particle $\Phi_a^{+(m)}(x)$ (here we also use the notation of Chap. 9) is the S_a-value of the point x with respect to the linear functional

$$I_m = \sum_a (\Phi_a^{+(m)}, S_a) \equiv (\Phi^{+(m)}, S). \tag{11.6}$$

If we denote by $D_a^{(m)}(x)$ the sensitivity function of the additive detector, then the differential particle flux density $\Phi_a(x)$ can be written as $D_a^{(m)}$-value of the point x with respect to the functional

$$I_m = \sum_a (\Phi_a, D_a^{(m)}) \equiv (\Phi, D^{(m)}). \tag{11.7}$$

Finally, if u_i characterizes the process of interaction of particles with matter, the transfer of which is described by the linear equation

$$L\Phi = S, \tag{11.8}$$

then, for linear functionals (11.6), (11.7), we arrive at the usual perturbation theory (Sect. 9.2):

$$\lambda_{im}(x_i) = -\left(\Phi^{+(m)}, \frac{\delta L}{\delta u_i(x_i)}\Phi\right),$$

$$\delta I_m = \sum_i \int \lambda_{mi}(x_i)\delta u_i(x_i)dx_i.$$

Given that the ordinary value satisfies the equation

$$L^+\Phi^+ = D, \tag{11.9}$$

we obtain for the sensitivity of the linear functionals (11.6) and (11.7) to the characteristics of the interaction of particles with the medium three representations

$$\lambda_{mi}(x_i) = \left(\frac{\delta\Phi^{+(m)}}{\delta u_i(x_i)}, S\right) = \left(D^{(m)}, \frac{\delta\Phi}{\delta u_i(x_i)}\right) = -\left(\Phi^{+(m)}, \frac{\delta L}{\delta u_i(x_i)}\Phi\right).$$

Since the EAS is formed by a single particle $S(x) = \delta(x - x_0)$, the following form is most convenient for their analysis:

$$\lambda_{mi}(x_i; x_0) = \left(\frac{\delta\Phi^{+(m)}}{\delta u_i(x_i)}, S\right) = \frac{\delta\Phi^{+(m)}(x_0)}{\delta u_i(x_i)}. \tag{11.10}$$

Differentiating (11.9) with respect to $u_i(x_i)$, we get an equation for the considered value as a function of the argument $x_0 \equiv x$:

$$L^+\lambda_{mi}(x_i; x) = -\frac{\delta L^+}{\delta u_i(x_i)}\Phi^{+(m)}(x) \equiv \widetilde{D}_i^{(m)}. \tag{11.11}$$

Equations (11.9) and (11.11) can be solved by the same scheme, described, for example, in paragraph 7.1. Based on the results of these calculations, it is possible to study the influence of the perturbations under consideration on the observed EAS characteristics in the vicinity of a given model

$$\delta I_m = \sum_i \int \lambda_{mi}(x_i)\delta u_i(x_i)dx_i. \tag{11.12}$$

Generalized value calculations give us information not only about the most sensitive EAS characteristics but also allow us to calculate their variations when the model of nuclear interactions changes. Let us first consider the qualitative characteristics of the differential sensitivity.

It is quite obvious that a change in the inclusive spectrum by unity in the region of high energies of secondary particles leads to greater variations in the EAS components than in the region of low energies. This property of differential sensitivity follows from the monotonic dependence of the value functions Φ^+ on energy. At the same time, when developing new and refining existing multiple production models using sensitivity functions, it is necessary to take into account the restrictions imposed on the $\delta W(E_1, E_2)$ variations of inclusive distributions, which follow from the energy conservation law

$$\sum_c \int_0^{E_0} \delta W_{ac}(E_0, E) E dE = 0.$$

In the simplest case, when the inclusive distribution changes in two identical small energy intervals around E_1 and E_2, the magnitudes of these changes should be related by the relation $\Delta W_1 = -\dfrac{E_2}{E_1} \Delta W_2$. Thus, it is of interest to compare sensitivity to variations that take into account the law of conservation of energy, i.e.,

$$G(E_1, E_2) \equiv \lambda(E_1, E_2) E_1/(E - 2\Phi^+). \tag{11.13}$$

The values of the G functions for the average EAS characteristics from the primary proton of energy 10^{17}eV are shown in Fig. 11.10. The calculations were carried out for seven[1] of various EAS characteristics:

(1) muons with energies above 5 GeV at a depth of 690 g/cm^2;
(2) hadrons with energies above 1 TeV at a depth of 690 g/cm^2;
(3) hadrons with energies above 10 TeV at a depth of 690 g/cm^2;
(4) electrons at a depth of 690 g/cm^2;
(5) muons with energies above 10 GeV at sea level;
(6) number of electrons at sea level;
(7) electron number density at a distance of 500 m from the shower axis.

Only the curves corresponding to the muon sensitivity decrease monotonically in the fragmentation region. The different slopes of the muon curves, on the one hand,

[1] The numbering given below is used to indicate the corresponding EAS characteristics in Figs. 11.10 and 11.11 and in Tables 11.5 and 11.6.

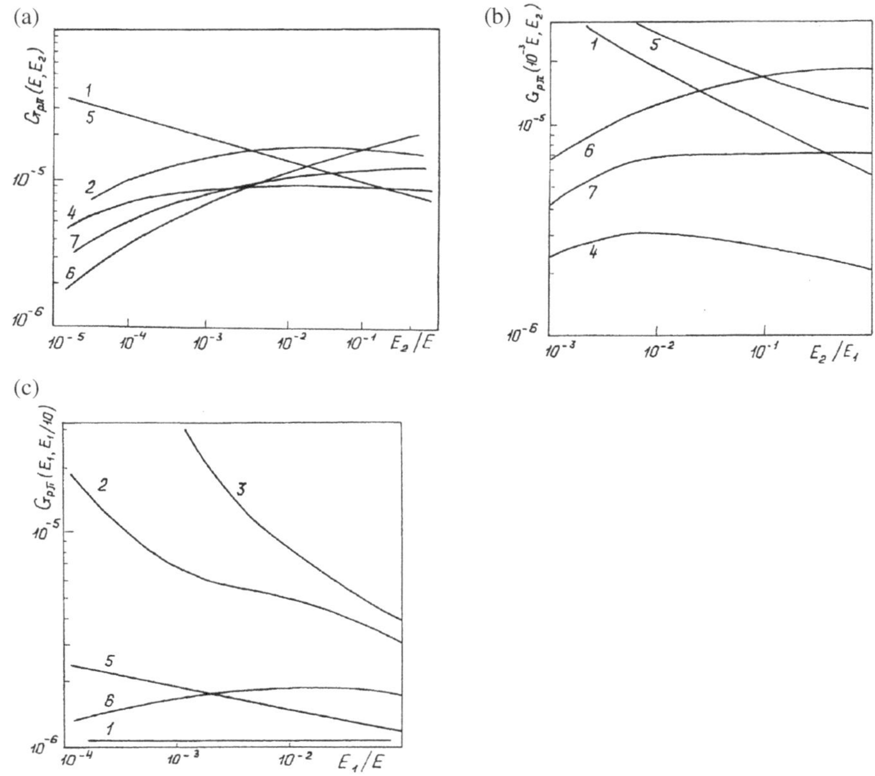

Fig. 11.10 The sensitivity of the average EAS characteristics from primary energy proton $E_0 = 10^{17}$eV. **a**, **b**—dependence on the energy of secondary particles; **c**—on the energy of the interacting nucleon

and the electron and hadronic curves, on the other hand, indicate the possibility of significant variations in the N_μ/N_e ratio for small changes in the dependence N_h/N_e. Attention is drawn to the weak dependence of the sensitivity of the EAS components on the energy of the interacting nucleon. The exception is high-energy hadrons.

If we neglect the production of nucleons in pion–nucleus interactions, then the average EAS characteristics will depend linearly on the inclusive distributions $W_{p\pi}(E_1, E_2)$. This means that for any EAS component, the equality

Table 11.5 Function comparison $\Phi_p^+(E_0, t_0)$

E_0, eV	Interaction	Process	EAS characteristics						
			1	2	3	4	5	6	7
10^{17}	First	$p \to \pi\pm$	0.48	0.26	0.21	0.29	0.47	0.23	0.26
		$p \to \pi^0$	0	0	0	0.21	0	0.13	0.16
	Other	$p \to \pi\pm$	0.50	0.69	0.76	0.27	0.50	0.41	0.34
		$p \to \pi^0$	0	0	0	0.18	0	0.24	0.19
	All	$p \to \pi \pm 0$	0.98	0.95	0.97	0.95	0.97	1.01	0.95
	First	$p \to \pi\pm$	0.48	0.22	0.17	0.28	0.47	0.22	0.25
		$p \to \pi^0$	0	0	0	0.20	0	0.10	0.15
10^{18}	Other	$p \to \pi\pm$	0.49	0.69	0.79	0.28	0.50	0.45	0.36
		$p \to \pi^0$	0	0	0	0.21	0	0.25	0.23
	All	$p \to \pi \pm 0$	0.97	0.91	0.96	0.97	0.97	1.02	0.99

$$\Phi_p^+(E_0, t_0) = \sum_c \int_0^{E_0} dE_1 \int_0^{E_1} dE_2 \lambda_{pc}(E_1, E_2) W_{pc}(E_1, E_2). \tag{11.14}$$

In order to check the equations and the correctness of the calculations, we compared the $\Phi_p^+(E_0, t_0)$ functions calculated by solving Eq. (11.9) and using the equality (11.14). The ratios of these values are given in Table 11.5. Small deviations of these ratios from unity can be explained by errors in the approximation of the two-dimensional functions λ_{pc} used in the calculation of the integral (11.14). In Table 11.5, the values of the relative contributions of the first interaction of the proton and the neutral component are also given. Comparison of characteristics 4 and 6 shows that the pions produced in the first interaction of the nucleon make 50% of the contribution to the electromagnetic component at mountain level and 35% at sea level. About 50% remains the contribution of the first interaction to the number of muons. This information is useful for interpreting EAS fluctuations. It should also be noted that the dominant contribution to hadrons is the secondary interactions of the nucleon.

In contrast to the functions $G_{p\pi}(E_1, E_2)$, the sensitivity of the average EAS characteristics to inclusive distributions $W_{\pi\pi}(E_1, E_2)$ strongly depends on the energy E_1. The energy dependence E_2 is similar to the functions $G_{p\pi}(E_1, E_2)$. The production of pions by pions leads to a nonlinear dependence of EAS characteristics on $W_{\pi\pi}(E_1, E_2)$ and narrows the range of possible $\delta W_{\pi\pi}(E_1, E_2)$ variations, in which the changes in the detector readings can only be described using the λ-functions of (E_1, E_2). Nevertheless, this area remains quite wide. This statement can be illustrated by comparing EAS characteristics calculated with inclusive distributions corresponding to the two models directly and using the relation

Table 11.6 Approximation coefficients $G_i(E_1, E_2)$

Process	*	E^*,GeV	α_1	α_2	α_3	α_4	α_5	α_6
	1	5	4.68	0.913	–0.090	–0.488	0.033	–0.030
	2	10^3	4.45	1.119	0.062	–0.294	0.084	0.019
$p \to \pi\pm$	4	1	8.387	0.736	0.109	3.613	0.065	–0.054
	5	10	4.883	0.905	–0.012	–0.287	0.029	0.095
	6	1	8.381	0.869	0.313	4.072	0.074	–0.066
	7	1	8.861	0.754	0.293	4.346	0.069	–0.087
	4	1	5.891	0.909	–0.052	0.995	0.057	–0.085
$p \to \pi^0$	6	1	5.192	1.195	0.018	0.848	0.086	–0.015
	7	1	5.593	1.023	0.027	0.928	0.060	–0.071
	1	3	4.883	1.082	0.825	–1.515	0.082	0.398
	2	10^3	5.313	1.232	1.065	–0.527	0.088	0.107
	3	10^4	9.824	0.265	2.614	5.593	0.558	–0.912
$\pi \to \pi\pm$	4	1	9.305	0.808	1.126	3.413	0.089	0.305
	5	10	4.884	1.116	0.791	–1.568	0.103	0.540
	6	1	9.692	0.893	1.267	4.333	0.079	0.150
	7	1	10.300	0.768	1.304	4.717	0.085	0.182
	4	1	5.942	1.080	0.852	–0.031	0.078	0.319
$\pi \to \pi^0$	6	1	5.520	1.328	0.841	0.115	0.083	0.232
	7	1	6.040	1.148	0.899	0.335	0.076	0.250

Note *–Characteristic

$$\Phi_2^+(E_0) = \Phi_1^+(E_0) + \sum_i \int_0^{E_0} dE_1$$
$$\times \int_0^{E_1} dE_2 \lambda_i(E_1, E_2)(W_{2i}(E_1, E_2) - W_{1i}(E_1, E_2)).$$

For the convenience of calculations, the functions $G(E_1, E_2)$ were approximated by analytical expressions

$$G(E_1, E_2) = 10^{-\alpha_1} x^{\alpha_2 - \alpha_3 y} [\ln E_2/E^*]^{\alpha_4 + \alpha_5 y} 10^{-\alpha_6 y} x,$$

where $x = E_1/E_1$; $y = \ln(E_1/E^*)$, E^* is the threshold energy. Approximation coefficients are given in Table 11.6. The values of EAS characteristics calculated in two ways are given in Table 11.7. A comparison of these data points to the possibility of describing significant variations in EAS characteristics using sensitivity functions.

The sensitivity of the average readings of the seven EAS characteristics described above hadrons is shown in Fig. 11.11. The $\lambda_{\sigma\pi}$ function tends to zero at $E = E_0$, which is related to the absence of E_0 pions in the showers. The most sensitive to variations in the cross section of inelastic interactions of protons are hadrons and the total number of electrons at sea level. The number of electrons at sea level is more

Table 11.7 Comparison of EAS characteristics

Model	*	$E_h > 1\text{TeV}$	$E_h > 10\text{TeV}$	$E_\mu > 1\text{GeV}$	$N_e, t = 1030$
$W \sim \dfrac{A}{x^2 + a^2}$	1	$6.5 \cdot 10^2$	$3.8 \cdot 10^1$	$1.96 \cdot 10^5$	$2.7 \cdot 10^7$
$W \sim \dfrac{A(E)}{(x + a)^3/2}$	1	$7.9 \cdot 10^2$	$6.4 \cdot 10^1$	$1.50 \cdot 10^5$	$4.1 \cdot 10^7$
	2	$8.0 \cdot 10^2$	$6.0 \cdot 10^1$	$1.40 \cdot 10^5$	$4.5 \cdot 10^7$

*Note**–Method

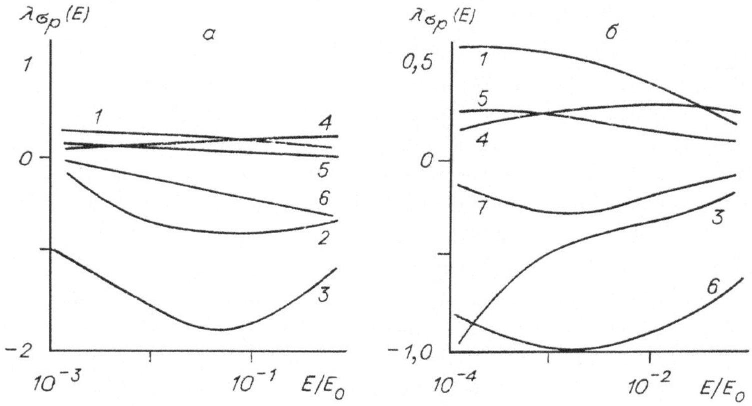

Fig. 11.11 Sensitivity of average EAS characteristics to cross sections for inelastic interactions of hadrons. **a**—to $\sigma_p(E)$, **b**—to $\sigma_\pi(E)$

sensitive to variations in the pion cross section than the total number of hadrons or the electron density at a distance of 500 m from the shower axis.

As mentioned above, differential sensitivity can be used to calculate parametric derivatives. Control calculations carried out using the data in Table 11.6 agree with the results obtained in (11.3), where Eq. (11.2) were used to calculate the parametric sensitivity, the solution of which is more economical.

Thus, the functional method makes it possible to determine the region of inclusive distributions responsible for the formation of one EAS component or another, and to conduct a qualitative and quantitative comparison of various models. A comparison of data on the sensitivity of the electromagnetic component at sea and mountain levels shows a significant increase in the role of the pionization region with a decrease in the depth of shower registration.

Calculations show a slight difference between the functions $\Phi^+(E)$ for the primary pion and proton, which, in turn, leads to the same sensitivity of the EAS components to variations in the distributions $W_{p\pi}(E_1, E_2)$ and $W_{pp}(E_1, E_2)$. Using the coefficients given in Table 11.7, one can calculate the variations in the number of electrons by replacing the uniform distribution of the inelasticity coefficient with the delta function

$$\delta N_e = N_e \int_0^1 (3.3 + 0.8 \lg x) x (\delta(x - 0.5) - 1) dx \approx 0.04 N_e,$$

which agrees well with the results of calculations in Ref [30]. Since the function $(3.3 + 0.8 \lg x)$ changes slowly, any variations of $W_{pp}(x)$ that satisfy the condition $\int_0^1 x \delta W_{pp}(x) dx = 0$, will lead to a slight change in the average number of EAS particles.

In the study of EASs, the determination of the relative role of the inclusive distributions $W_{p\pi}$ and $W_{\pi\pi}$ is of great importance. In [31], the sensitivity of the average EAS characteristics was studied by the Monte Carlo method. For each hadron, information was collected about the ratio of the energy of the born particle to the energy of its parent. Comparing the numbers of muons and hadrons that came to the observation level and those produced directly by a proton or pions, the authors of [31] conclude that pion interactions play a decisive role. This conclusion does not contradict the higher sensitivity of the EAS components to variations in the $W_{\pi\pi}(E_1, E_2)$ cross section obtained in this work, but the criterion used here for comparing the roles of pion and proton interactions is not entirely correct. Even in the case of $W_{\pi\pi}(E_1, E_2) \equiv 0$, the EAS electronic component will decrease only by 60%. The influence of the inclusive distribution of particles produced in proton interactions manifests itself indirectly through the value, i.e., their average contribution to the component under study. Note that the calculations [31] were carried out for the primary energy of 10^{15} eV, and estimates of the role of the pionization region can be made more accurately based on the data in Table 11.7. The x distribution [31] is expressed directly in terms of sensitivity functions and inclusive distributions

$$f(x) = \sum_i \int_0^{E_0} \lambda_i(E_1, x E_1) W_i(E_1, x E_1) dE_1.$$

11.5 Solving Equations for Second Moments

To compare the calculated characteristics of EAS with the experimentally measured characteristics of showers, selected according to some attribute (the value of some characteristic at the observation level), it is necessary to set the chemical composition and energy spectrum of the primary radiation, find a way to recalculate the EAS characteristics from the primary proton to more complex nuclei, and know the statistical fluctuations and correlations of the characteristics under consideration.

The equations for the second moments $\overline{N^2_{ad}}$, according to Sect. 4.6, have the form

$$[\partial_t + \sigma_a]\overline{N^2_{ad}}(t, E) = \sum_b \int dE_1 w_{a\to b}(E \to E_1)\overline{N^2_{bd}}(t, E_1)$$

$$+ \sum_{bc} \int dE_1 \int dE_2 w_{a\to bc}(E \to E_1, E_2)\bar{N}_{bd}(t, E_1)\bar{N}_{cd}(t, E_2) \qquad (11.15)$$

with boundary conditions $\overline{N^2_{ad}}(0, E) = 1(E - E^*)\delta_{ad}$. Equations have a similar form for mixed moments (4.59). They can be resolved in the same order and by the same numerical method as the equations for the first moments if two-particle inclusive distributions are known $w_{a\to bc}$.

The first calculations of EAS fluctuations by the method described above were carried out with a rough approximation of the two-particle function, constructed from the following considerations. If the particles were born independently of each other, the two-particle function would be equal to the product of the single-particle ones:

$$\sigma_a^{-1} w^{\text{ind}}_{a\to bc}(E \to E_1, E_2) = \sigma_a^{-2} w_{a\to b}(E \to E_1) w_{a\to b}(E \to E_2). \qquad (11.16)$$

The presence of kinematic correlations limits the range of variables in which it is nonzero:

$$w_{a\to bc}(E \to E_1, E_2) = w_{a\to bc}(E \to E_1, E - 2)1(E - E_1 - E_2). \qquad (11.17)$$

Substituting (11.16) into the right side (11.17), we get a function of two variables defined in the correct range of their change. It is clear, however, that the conditions (11.16) (independence) and (11.17) (a consequence of kinematic correlations) are physically incompatible. Therefore, the function defined above will not satisfy the normalization conditions for the two-particle distribution. To correct this situation, we introduce the normalization factor $\varkappa_a(E)$:

$$w_{a\to bc}(E \to E_1, E_2) = \varkappa_a(E)\sigma_a^{-1} w_{a\to b}(E \to E_1)$$

$$\times\ w_{a\to c}(E \to E_2)1(E - E_1 - E_2). \qquad (11.18)$$

The values of this factor, determined from the law of conservation of energy

$$\sigma_a^{-1}\left\{\sum_b \int dE_1 w_{a\to b}(E \to E_1)E_1^2 + \sum_{bc} \int dE_1 \int dE_2 w_{a\to bc}(E \to E_1, E_2)E_1 E_2\right\} = E^2.$$

To speed up the calculation of double integrals on the right side of Eq. (11.15), they were represented as

$$F_{abc}(t, E) = \zeta_{abc}(t, E)F^{\text{ind}}_{abc}(t, E), \qquad (11.19)$$

where

$$F_{abc}^{\text{ind}}(t,E) = \sigma_a^{-1}\left[\int dE_1 w_{a\to b}(E \to E_1)\bar{N}_{bd}(t,E_1)\right.$$
$$\left. + \int dE_2 w_{a\to c}(E \to E_2)\bar{N}_{cd}(t,E_2)\right].$$

The calculation speedup was achieved by the fact that the coefficient (11.19) was previously calculated for a thinned set of variables, in the process of solving Eq. (11.15), its necessary values were found by interpolation.

Fluctuations of various EAS components from a proton with a fixed primary energy up to 10^8 GeV are calculated in this way. It was found that in the range $10^5 - 10^8$ GeV, the relative fluctuations ($\sqrt{D}/\bar{N}$) of the number of electrons at sea level decrease from 110 to 53%, and the fluctuations of the number of muons ($E^* > 1$ GeV) at the same level, from 22 to 13%. These results approximately agree with the EAS fluctuation calculations (45 ± 7 and $7.7 \pm 1.2\%$ at $E = 10^8$ GeV) performed by the Monte Carlo method in similar models [32, 33]. Under the assumption of a log-normal distribution of the EAS components [19], a recalculation to the complex chemical composition of the primary radiation [34] was carried out in the superposition model, which showed satisfactory agreement [35] with experimental data [1, 36, 46]

11.6 EAS Covariance Matrix

Of course, the two-particle function (11.18) does not properly reflect dynamic correlations in multiple processes. An approximate allowance for dynamic correlations can be made on the basis of data on fluctuations of eventual characteristics (inelasticity coefficients, multiplicities), which are easier to extrapolate to the region of superhigh energies. It is convenient to do this using Eqs. (4.66)–(4.69) for the covariance matrix of the measured EAS characteristics, which allow one to distinguish the fluctuation components due to fluctuations in the ranges, interactions with known cross sections (decays), and fluctuations in multiple processes. Consider the calculation of the last component.

Let $q, r, \ldots$ be linear functionals of EAS at a fixed level of observation. The function introduced in Sect. 4.6 $C(t,x) = C_a(t,x) = \text{cov}(q,r)$ is essentially the covariance matrix of the considered system of functionals, whose diagonal elements are equal to the variances of the corresponding functionals.

According to (4.6), the contribution of multiple processes to the EAS covariance matrix is given by the equation

$$[\partial_t + \sigma_a]C_a^{mn}(t,E) - \sum_b \int dE' w_{a\to b}(E \to E')C_b^{pl}(t,E') = B_a^{mn}(t,E), \tag{11.20}$$

where the right side

$$B_a^{\text{mhp}}(t, E) = \sigma_a^{\text{mhp}} \text{cov}^{\text{mhp}}(q_a'', r_a'') = \sigma_a^{\text{mhp}} C_a''(t, E), \tag{11.21}$$

$$q_a'' = \sum_i \bar{q}_{b_i}(E - i), \qquad r_a'' = \sum_i \bar{r}_{b_i}(E - i), \tag{11.22}$$

characterizes fluctuations in multiple hadron production (mhp).

We represent q_a'' as the sum of three terms corresponding to the leading (surviving) particle (for definiteness, we will consider the primary particle to be a proton), π^0-mesons generating electron–photon showers, and π^+-mesons that form the hadronic component of the EAS:

$$q_N'' = \bar{q}_N(E_N) + \sum_{\pi^0} \bar{q}_0(E_i) + \sum_{\pi} \bar{q}_\pi(E_i). \tag{11.23}$$

Let $\xi = (K_0, K_{ch}, n_0, n_{ch})$ be the set of eventual characteristics of the multiple process: partial inelasticity coefficients $K_0 = \sum_{\pi^0} E_i/E$, $K_{ch} = \sum_\pi E_i/E$ and the corresponding multiplicities n_0, n_{ch}. Sum

$$K_0 + K_{ch} = K = 1 - E_n/E$$

is equal to the total inelasticity coefficient. We fix the parameters ξ. In this case, only the last two terms in the sum (11.23) remain random:

$$q_N''|_\xi = \bar{q}_N((1 - K)E) + \sum_{\pi^0} \bar{q}_0(E_i)|_\xi + \sum_{\pi} \bar{q}_\pi(E_i)|_\xi. \tag{11.24}$$

Their fluctuations are due to various possible ways of distributing a fixed energy between a fixed number of particles in each component. Neglecting at high energies the magnitude of these fluctuations in comparison with the contribution of fluctuations of the parameters ξ to the result, we rewrite (11.24) as

$$q_N''(t, E, \xi) = \bar{q}_N(t, (1 - K)E) + \int_0^E W_{N\to\pi^0}(E \to E'|\xi)\bar{q}_0(t, E')dE'$$

$$+ \int_0^E W_{N\to\pi}(E \to E'|\xi)\bar{q}_\pi(t, E')dE',$$

where $W_{a\to b}(E \to E'|\xi)$ is the inclusive distribution of secondary particles of type b under the condition that the parameters ξ are fixed. We will assume that this distribution depends only on those components that belong to this type of particles:

$$W_{a\to b}(E \to E'|\xi) = W_{a\to b}(E \to E'|\xi_b). \tag{11.25}$$

Conditional inclusive distributions (spectra) (11.25) satisfy the obvious normalization conditions:

$$\int_0^{K_b E} W_{a\to b}(E \to E'|\xi_b)dE' = n_b, \tag{11.26}$$

$$\int_0^{K_b E} W_{a\to b}(E \to E'|\xi_b)(E'/E)dE' = K_b. \tag{11.27}$$

Moreover, after averaging over ξ_b, they should lead to unconditional inclusive distributions:

$$\langle W_{a\to b}(E \to E'|\xi_b)\rangle = W_{a\to b}(E \to E'). \tag{11.28}$$

Let us introduce the notation

$$\Theta_{N\to N}(t, E) = \Big\langle \bar{q}_N(t, (1-K)E)\Big\rangle \Big/ \bar{q}'_N(t, E), \tag{11.29}$$

$$\Theta_{N\to\pi^0}(t, E) = \Big\langle \int W_{N\to\pi^0}(E \to E'|\xi)\bar{q}_0(t, E')dE'\Big\rangle \Big/ \bar{q}'_N(t, E), \tag{11.30}$$

$$\Theta_{N\to\pi}(t, E) = \Big\langle \int W_{N\to\pi}(E \to E'|\xi)\bar{q}_\pi(t, E')dE'\Big\rangle \Big/ \bar{q}'_N(t, E). \tag{11.31}$$

For the mean values of the relative contributions of various groups of secondary particles to the functional q''_a, we obtain

$$q''_N(t, E, \xi) = \bar{q}'_N(t, E)\Big\{X_{N\to N}(E, \xi)\Theta_{N\to N}(t, E)$$

$$+ X_{N\to\pi^0}(E, \xi)\Theta_{N\to\pi^0}(t, E) + X_{N\to\pi}(E, \xi)\Theta_{N\to\pi}(t, E) + 1\Big\}, \tag{11.32}$$

where

$$X_{N\to N}(E, \xi_N) = \frac{\bar{q}_N(t, (1-K)E)}{\langle q_n(t, (1-K)E)\rangle} - 1, \tag{11.33}$$

$$X_{N\to b}(E, \xi_b) = \frac{\int W_{N\to b}(E \to E'|\xi_b)\bar{q}_b(E')dE'}{\langle \int W_{N\to b}(E \to E'|\xi_b)\bar{q}_b(E')dE'\rangle} - 1 \tag{11.34}$$

are functions of random values of eventual parameters. Using a similar representation for another functional

$$r''_N(t,E,\xi) = \bar{r}'_N(t,E)\Big\{Y_{N\to N}(E,\xi)\Phi_{N\to N}(t,E)$$

$$+\ Y_{N\to\pi^0}(E,\xi)\Phi_{N\to\pi^0}(t,E) + Y_{N\to\pi}(E,\xi)\Phi_{N\to\pi}(t,E) + 1\Big\}$$

and inserting (11.32)–(11.34) into (11.21) taking account (11.28), arrive at the following expression:

$$C''_N(t,E) = \left[\sum_{bc}\langle X_{N\to b}Y_{N\to c}\rangle \Theta_{N\to b}\Phi_{N\to c}\right]\bar{q}'_N\bar{r}'_N. \tag{11.35}$$

Note that from the definition of the quantities introduced above it follows:

$$\sum_b \Theta_{N\to b} = \sum_b \Phi_{N\to b} = 1,\ \langle X_{N\to b}\rangle = \langle Y_{N\to b}\rangle = 0.$$

It is known [20] that the dependence of many functionals on the energy of the primary particle can be represented as

$$\bar{q}(t,E) = A(t)E^{\alpha(t)}, \tag{11.36}$$

where the coefficient A and the exponent α do not depend on the energy. Our calculations confirmed the applicability of this approximation in a fairly wide range of changes in the energy of the primary particle. Thus, at sea level for the average number of muons with energies above 1 GeV $\alpha_\mu = 0.932$, $A_\mu = 0.313\cdot 10^{-1}$ ($10^5 \leqslant E \leqslant 10^8$GeV), and for the average number of electrons $\alpha_e = 1.169$, $A_e = 0.110\cdot 10^{-1}$ ($10^5 \leqslant E \leqslant 10^8$GeV), etc. As the calculations showed, it is sufficient that such an approximation satisfactorily describes $\bar{q}(E)$ in the interval $(0.01E, E)$, since it makes the main contribution to fluctuations (an analogous result also took place in the case of EPC (Sect. 7.4)). Approximation (11.36) simplifies the expressions (11.33), (11.34), leaving only moments of conditional inclusive spectra in them

$$\widetilde{W}_{N\to b}(\alpha_b|\xi_b) = \int_0^{K_bE} W_{N\to b}(E\to E'|\xi_b)(E'/E)dE',$$

the order of which coincides with the exponent α_b of the considered functional of the secondary particle of type b:

$$X_{N\to N}(E,\xi) = \frac{(1-K)^{\alpha_N}}{\langle(1-K)^{\alpha_N}\rangle} - 1, \tag{11.37}$$

$$X_{N\to b}(E,\xi) = \frac{\widetilde{W}_{N\to b}(\alpha_b|\xi_b)}{\langle\widetilde{W}_{N\to b}(\alpha_b|\xi_b)\rangle} - 1. \tag{11.38}$$

The quantities (11.37) and (11.38) are the centered and normalized moments of the conditional inclusive spectra. To approximate the dependence of the moments on α in the region of interest to us ($0 < \alpha \lesssim 1.5$), we need to use additional information or additional assumptions. The simplest class of functions satisfying the conditions (11.26), (11.27) are functions of the form

$$W_{a\to b}(E \to E'|\xi_b) = n_b U_{a\to b}(E'/\bar{E})/\bar{E} \tag{11.39}$$

where $\bar{E} = K_b E/n_b$. Taking into account that the inclusive spectra of nonleading particles rapidly decrease with increasing E', and the multiplicity increases with increasing primary energy, we extend the integration domain in the normalization integrals to a semi-infinite interval:

$$\int_0^\infty U_{a\to b}(x)dx = \int_0^\infty U_{a\to b}(x)dx = 1.$$

In this approximation

$$\widetilde{W}_{a\to b}(\alpha_b|\xi_b) = \widetilde{U}_{a\to b}(\alpha_b) n_b^{1-\alpha_b} K_b^{\alpha_b}, \tag{11.40}$$

and random inclusive moments do not depend on the specific form of the function $U_{a\to b}$:

$$X_{a\to b}(\alpha_b|\xi_b) = \frac{n_b^{1-\alpha_b} K_b}{\langle n_b^{1-\alpha_b} K_b\rangle} - 1. \tag{11.41}$$

The formulas (11.39)–(11.41) can be obtained on the basis of an elementary statistical approach assuming an equiprobable energy distribution $K_b E$ between n_b produced particles. In this case

$$W_{a\to b}(E \to E_1|\xi_b) = n_b(n_b - 1)[1 - E_1/K_b E]^{n_b-2} \Big/ (K_b E), \tag{11.42}$$

$$W_{a\to bb}(E \to E_1, E_2|\xi_b) = n_b(n_b - 1)^2(n_b - 2)$$
$$\times\ [1 - (E_1 + E_2)/K_b E]^{n_b-3} \Big/ (K_b E)^2. \tag{11.43}$$

In the region $E \ll n_b \bar{E}$, the expression (11.42) coincides in form with the spectrum of the Raman model

$$W_{a\to b}(E \to E_1|\xi_b) \approx n_b e^{-E_1/\bar{E}},$$

the calculation of the moments of which leads to the formula (11.40). The calculation of the fluctuations of $\sum_i (E_i/E)^\alpha$ using the two-particle spectrum (11.43) leads to the conclusion that for $n \to \infty$ they decrease according to the Poisson law, and at high energies, as was done above, they can be neglected.

11.7 Inelasticity and Multiplicity Fluctuations

We now turn to a discussion of the distributions of eventual parameters. The most important of them for the developed EAS are the coefficients of inelasticity characterizing the distribution of energy between the various components of multiple processes. In particular, it was noted [37] that the coefficient of inelasticity is one of the main characteristics of an elementary act, which determines the behavior of others. Accordingly, we begin our discussion with the inelasticity coefficients, representing their joint distribution as

$$\psi(K_0, K_{ch}, K) = \psi(K_0, K_{ch}|K)W(K).$$

Putting aside for now the question of the distribution of the total inelasticity coefficients $\psi(K_0, K_{ch}|K)$.

It follows from considerations of isotopic invariance that the distribution of energy between the various isotopic components π^0, π^+, and π^- of a multiple process should be uniform on average [38]. If we assume that this distribution is not only uniform on average but also equiprobable (in the sense of the microcanonical distribution) for a fixed value of the total inelasticity coefficient, we can obtain distributions of partial coefficients. In this case, we confine ourselves to considering n-mesons, and we will neglect the other components of multiple processes, having a weak effect on EAS development.

In this approximation, we get that

$$\psi(K_0, K_+, K_-|K) = (2/K^2)\delta(K_0 + K_+ + K_- - K),$$

where the coefficient in front of the δ-function is determined from the normalization condition:

$$\int_0^K dK_0 \int_0^{K-K_0} dK_+ \int_0^{K-K_0-K_+} dK_- \psi(K_0, K_+, K_-|K) = 1.$$

The joint distribution density of the coefficients K_o and $K_{ch} = K_+ + K_-$ has the form

$$\psi(K_0, K_{ch}|K) = \int_0^{K_{ch}} \psi(K_0, K_{ch} - K_-, K_-|K)dK_- = (2K_{ch}/K^2)\delta(K_0 + K_{ch} - K).$$

By integrating over one of the variables, we find the conditional distributions of each of the partial coefficients:

$$\psi(K_0|K) = (2/K^2)(K - K_0), \qquad K_0 < K,$$

$$\psi(K_{ch}|K) = (2/K^2)K_{ch}, \qquad K_{ch} < K.$$

Averaging these distributions over the total inelasticity coefficient, we obtain

$$\psi(K_0, K_{ch}) = \frac{2K_{ch}}{(K_0 + K_{ch})^2} W(K_0 + K_{ch}), \tag{11.44}$$

$$\psi(K_0) = 2\int_{K_0}^{1} \frac{(K - K_0)W(K)}{K^2} dK \simeq 2[K_0 - 1 - \ln K_0], \tag{11.45}$$

$$\psi(K_{ch}) = 2K_{ch} \int_{K_{ch}}^{1} \frac{W(K)}{K^2} dK. \tag{11.46}$$

From these distributions, the following formulas for mean values, variances, and correlation coefficient follow:

$$\langle K_0 \rangle = \frac{1}{3}\langle K \rangle, \tag{11.47}$$

$$\langle K_{ch} \rangle = \frac{2}{3}\langle K \rangle, \tag{11.48}$$

$$\mathsf{D}K_0 = \frac{1}{6}\mathsf{D}K + \frac{1}{18}\langle K \rangle^2, \tag{11.49}$$

$$\mathsf{D}K_{ch} = \frac{1}{2}\mathsf{D}K + \frac{1}{18}\langle K \rangle^2, \tag{11.50}$$

$$R(K_0, K_{ch}) = -\frac{3\delta^2 K - 1}{\sqrt{(1 + 3\delta^2 K)(1 + 9\delta^2 K)}}, \tag{11.51}$$

where

$$\delta^2 K = \mathsf{D}K/\langle K \rangle^2. \tag{11.52}$$

Let us compare the obtained distributions with the experimental data. To this end, we use the following approximation of the accelerator data on the distribution of the total inelasticity coefficient [39]:

$$W(K) = K^{p-1}(1-K)^{q-1}/B(p,q);$$

$$p = 2.7; \qquad q = 2.2. \tag{11.53}$$

In accordance with formulas (11.47)–(11.52), we obtain $\langle K \rangle = 0.55$, $\delta^2 K = 0.14$; $\langle K_0 \rangle = 0.18$, $\delta^2 K_0 = 0.71$; $\langle K_{ch} \rangle = 0.37$, $\delta^2 K_{ch} = 0.28$; $R(K_0, K_{ch}) = -0.33$. These characteristics of the partial inelasticity coefficients turn out to be close to the experimental ones. Moreover, the distributions themselves, given by the formulas (11.45), (11.46) with the approximation (11.53) are close to the experimental ones (Fig. 11.12 and 11.13). In the ultrahigh-energy region, for EAS calculations, instead of the beta distribution (11.53), we will use the distribution of the inelasticity coefficient, which follows from this model (p. 11.1).

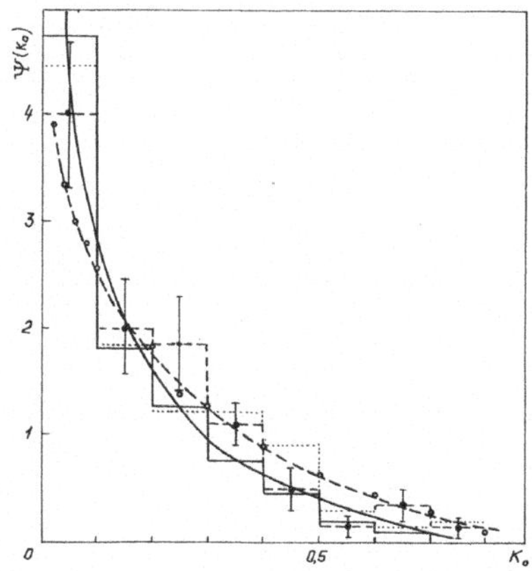

Fig. 11.12 Distribution of the partial coefficient K_0. 1) Full histogram: experim. pN-interaction 21 GeV [37]; 2) full dots: experim. πN-interaction, 60 GeV [41]; 3) dotted histogram: experim. pPb interaction $2 \div 32$ TeV [42], full curve: calculation $\psi(K_0) = K_0 - 1 - \ln K_0$, dashed curve: calculation when $W(K)$ is taken from the quark-gluon string model (E = 10 TeV)

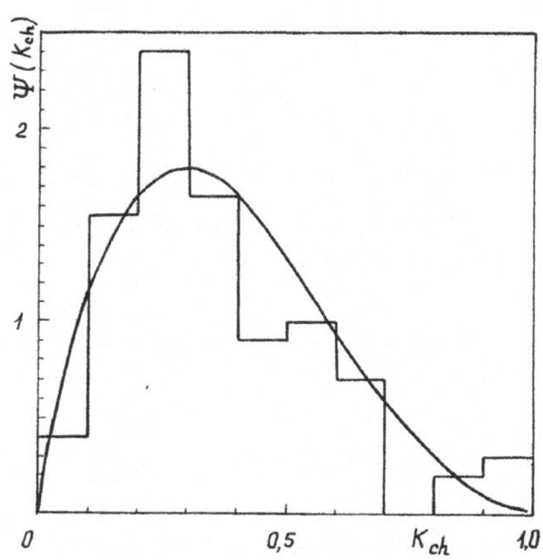

Fig. 11.13 Distribution of K_{ch} 200 GeV [37, 40]) of Histogram. Histogram: experim. pN, 200 GeV. Curve is calculated by (11.46)

We write the joint distribution of eventual parameters in the following form:

$$\psi(\xi) = \psi(n_0, n_{ch}|K_0, K_{ch})\psi(K_o, K_{ch}).$$

Let us assume that the multiplicity distribution of one component (charged or neutral) is determined by the energy transferred to it and does not depend on the energy of the other component:

$$\psi(n_0, n_{ch}|K_0, K_{ch}) = \psi'(n_0|K_0 E)\psi(n_{ch}|K_{ch} E),$$

commonly used multiplicity distributions, averaged over inelasticity coefficients and depending only on the initial energy:

$$\psi(n_{ch}, E) = \int_0^1 \psi'(n_{ch}|K_{ch})\psi(K_{ch})dK_{ch}.$$

From this, in particular, similar relations for the moments follow:

$$\langle n_{ch}^{\gamma}\rangle = \int_0^1 \overline{n'^{\gamma}_{ch}}(K_{ch} E)\psi(K_{ch})dK_{ch}. \tag{11.54}$$

Let the average multiplicity depend on the energy according to the power law:

$$\langle n_{ch}\rangle = cE^{\delta}. \tag{11.55}$$

Putting $\gamma = 1$ in (11.54) and using (11.55), we get

$$c = \int_0^1 [\bar{n}'_{ch}(K_{ch} E)/(K_{ch} E)^{\delta}]K_{ch}^{\delta}\psi(K_{ch})dK_{ch}.$$

For this integral to be independent of E, it is sufficient that the relation $\langle\bar{n}'_{ch}\rangle(K_{ch} E)/(K_{ch} E)^{\delta}$ is independent of it, i.e.,

$$\bar{n}'_{ch}(K_{ch} E) = c\frac{K_{ch}^{\gamma}}{\langle K_{ch}^{\gamma}\rangle}E^{\delta} = \langle n_{ch}\rangle K_{ch}^{\delta}/\langle K_{ch}^{\delta}\rangle. \tag{11.56}$$

It has been experimentally established that in a wide range of energies the distribution of the ratio $z = n_{ch}/\langle n_{ch}\rangle$ does not depend on the energy and types of colliding particles (KNO-scaling). In particular, the moments do not depend on the energy

$$\langle z^{\gamma}\rangle = \langle n_{ch}^{\gamma}\rangle/\langle n_{ch}\rangle^{\gamma} = \text{const}, \qquad \gamma < 3. \tag{11.57}$$

Substituting (11.54) and (11.56) into (11.57), we see that for the latter condition to be satisfied, it is sufficient to be independent of the $K_\pi E$ moments

$$\bar{z}'^\gamma = \overline{n'^\gamma_{ch}}(K_{ch}E)/[\bar{n}'_{ch}(K_{ch}E)]^\gamma,$$

which is equivalent to the independence of the distribution of the ratio of the conditional ("associated") multiplicity to its mean value:

$$z' = n'_{ch}/\bar{n}'_{ch}(K_{ch}E). \tag{11.58}$$

This property of multiple processes, observed in experiments, is defined as "particular" KNO-scaling [39]. The following relation between the moments of two scaling variables is obvious:

$$\overline{z'^\gamma} = [\langle K^\delta_{ch}\rangle^\gamma/\langle K^{\delta\gamma}_{ch}\rangle]\langle z^\gamma\rangle. \tag{11.59}$$

This formula can be used to find "particular" moments. Taking the distribution of the multiplicity of charged particles in the form
$\psi(z) == \frac{\pi z}{2}e^{-\pi z^2/4}$, we get

$$\langle z^\gamma\rangle = (2/\sqrt{\pi})^\gamma \Gamma(1+\gamma/2). \tag{11.60}$$

Substituting (11.63) into (11.62), we find the moments of partial multiplicity. We will assume that they are the same for the charged and neutral components. Then, using a formula similar to (11.61), one can find the mean value and fluctuations of the multiplicity of the neutral component. The results of calculations using the distribution (11.53) show satisfactory agreement with experiment (Figs. 11.14 and 11.15).

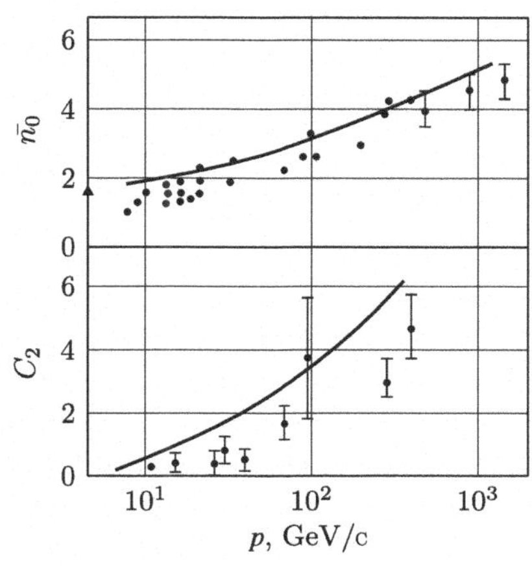

Fig. 11.14 Multiplicity of neutral particles depending on momentum (in the laboratory coordinate system) for $K^\pm p$, $\pi^\pm p$ and $p^\pm p$ interactions, $C_2 = Dn_0 - \bar{n}_0$ [8]. Curve—our calculation

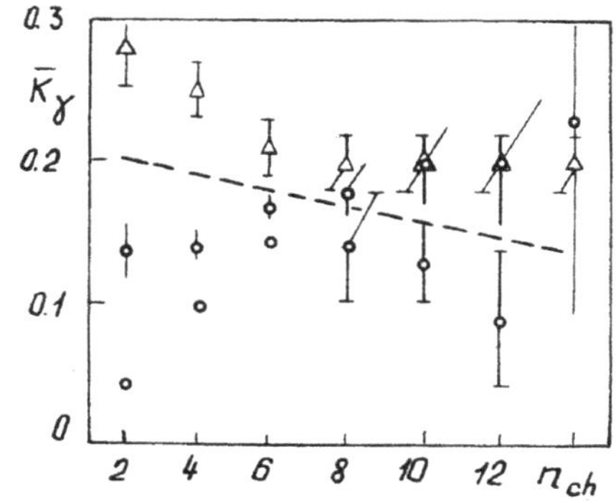

Fig. 11.15 Dependency $\bar{K}_0 \approx K_\gamma(n_{ch})$ for pp (•), pp^- (∘) and $\pi^- p$ (△) interactions [43]. Curve—our calculation

$$\langle X_{N\to b} Y_{N\to b}\rangle = \frac{\overline{z'^{2-\alpha_b-\beta_b}}}{\overline{z'^{1-\alpha_b}}\ \overline{z'^{1-\beta_b}}} \frac{\langle K_b^{\delta(2-\alpha_b-\beta_b)+\alpha_b+\beta_b}\rangle}{\langle K_b^{\delta(1-\alpha_b)+\alpha_b}\rangle\langle K_b^{\delta(1-\beta_b)+\beta_b}\rangle} - 1, \qquad b = c. \tag{11.61}$$

Recall that the moments z' are related to the moments of the scaling variable z of charged particles by the relation (11.62). In addition, in calculating the above relations, it is convenient to use the following formulas, which follow from Sect. 11.7:

$$\langle (1-K)^\alpha K_0^\beta\rangle = \frac{2}{(\beta+1)(\beta+2)}\langle (1-K)^\alpha K_\beta\rangle,$$

$$\langle (1-K)^\alpha K_\pi^\beta\rangle = \frac{2}{\beta+2}\langle (1-K)^\alpha K_\beta\rangle,$$

$$\langle K_0^\alpha K_\pi^\beta\rangle = 2B(\alpha+1, \beta+1)\langle (K^{\alpha+\beta})\rangle.$$

Substituting the expressions found above into (11.35) and calculating $\Theta_{N\to b}$, $\Phi_{N\to b}$ using formulas (11.29)–(11.31) considering (11.28) and (4.61)

$$\Theta_{N\to b}(t, E) = \int W_{N\to b}(E \to E')\bar{q}_b(t, E')dE' \Big/ \left[\bar{q}_b(t, E) + \frac{1}{\sigma_b}\partial_t \bar{q}_b\right],$$

we will complete the preparation of Eq. (11.20) for solution.

The numerical solution of the equations for the first moments and dispersions of EAS characteristics was performed using the data of the quark-gluon string model for hadron–nucleus interactions [18]. According to this model, $\sigma_{prod}(E)$ grows according to a logarithmic law, inclusive spectra are characterized by strong scaling violation in the pionization region and weak violation in the fragmentation region. From the same model, the distributions of energetically separated particles were obtained, necessary for calculating the matrix elements (11.61) for nucleon– and pion–nucleus

Table 11.8 Square of relative fluctuations of the number of electrons in EAS at sea level as a function of proton energy

δ^2	Energy (GeV)					
	10^5	10^6	10^7	10^8	10^9	10^{10}
$\delta_1^2 \times 10^1$	5.37	3.03	1.78	1.04	0.057	0.262
$\delta_2^2 \times 10^2$	16.20	6.67	3.08	1.16	0.482	0.202
$\delta_3^2 \times 10^2$	13.30	4.39	1.87	0.69	0.249	0.102
$\delta_4^2 \times 10^2$	5.15	3.24	2.56	2.25	2.40	3.24
$(\delta_4^2)' \times 10^1$	0.376	0.493	0.635	0.87	1.16	1.46
$\delta_5^2 \times 10^2$	1.51	0.68	0.504	0.641	0.663	1.17
$(\delta_5^2)' \times 10^2$	2.05	1.78	1.55	1.48	1.36	1.24
$\delta_6^2 \times 10^2$	3.71	2.38	1.69	1.11	1.14	1.35
$\sum_{i=1}^{6} \delta_i^2 \times 10$	9.36	4.77	2.75	1.63	1.05	0.868

interactions. In accordance with the above analysis of the equations, the square of the relative fluctuations of the components EAS can be represented as $\delta^2 = \delta_{\text{prob}}^2 + \delta_{\text{vz}}^2$ where $\delta_{\text{prob}}^2 = \sum_{i=1}^{3} \delta_i^2$; $\delta_{\text{vz}}^2 = \sum_{i=4}^{6} \delta_i^2$.

Different values of the index i correspond to different fluctuation sources: 1—the first proton run, 2—the remaining runs of the leading nucleon, 3—the runs of all hadrons, 4—the first interaction of the proton, 5—the remaining interactions of the leading nucleon, 6—interactions of other hadrons. Calculations of δ_4^2 and δ_5^2 were also performed using matrix elements (11.61) calculated on the basis of the uniform distribution of K, which, as shown above, leads to good agreement with the experimental data on K_0 in hadron–nucleus interactions (the corresponding results are marked with a prime: $(\delta_4^2)'$ and $(\delta_5^2)'$).

The results of calculations of fluctuations of the electron and muon components of the EAS from the primary proton with an energy in the interval $10^5 - 10^{10}$GeV are presented in Tables 11.8, 11.9 and 11.10 [19]. A comparison of our results with the data of other calculations is shown in Figs. 11.16 and 11.17.

The following conclusions follow from the theoretical analysis of EAS statistics.

1. The main role in the formation of fluctuations of the EAS electronic component at sea level in the region of energies lower than 10^8GeV, particle ranges play; fluctuations of the first range make a significant contribution, the influence of fluctuations of the remaining ranges of the leading nucleon is comparable with the influence of the first interaction, the role of the ranges of other particles, and other interactions is small. In the region of $10^9 - 10^{10}$GeV, the role of the first interaction becomes very significant. The first interaction plays an even greater role in the case of observing the electron component at the level of mountains, when we approach the maximum of the cascade curve, where the influence of ranges on fluctuations is relatively small. This conclusion is in qualitative (and in cases where the influence of interactions is small, also in quantitative) agreement with the calculations of other authors [18, 19, 44–46].

Table 11.9 Square of relative fluctuations of the number of electrons in EAS at the level of mountains ($z^* = 700\text{g/cm}^2$) as a function of proton energy

δ^2	Energy (GeV)					
	10^5	10^6	10^7	10^8	10^9	10^{10}
$\delta_1^2 \times 10^2$	5.86	1.54	0.654	0.433	0.803	1.42
$\delta_2^2 \times 10^3$	9.52	8.63	1.68	1.19	1.31	1.18
$\delta_3^2 \times 10^3$	8.61	2.54	1.35	0.808	0.657	0.753
$\delta_4^2 \times 10^3$	6.56	4.08	4.27	4.36	6.40	9.16
$(\delta_4^2)' \times 10^2$	0.336	0.506	0.66	0.858	1.08	1.38
$\delta_5^2 \times 10^3$	1.68	0.698	0.581	1.27	1.70	2.94
$(\delta_5^2)' \times 10^3$	9.46	9.43	9.41	9.22	9.03	8.88
$\delta_6^2 \times 10^3$	7.89	6.67	5.55	6.47	7.28	10.40
$\sum_{i=1}^{6} \delta_i^2 \times 10^2$	9.29	3.81	2.01	1.84	2.54	3.86

Table 11.10 The square of relative fluctuations in the number of EAS muons ($E_\mu > 1\text{GeV}$) at sea level in dependence on proton energy

δ^2	Energy (GeV)					
	10^5	10^6	10^7	10^8	10^9	10^{10}
$\delta_1^2 \times 10^3$	6.31	3.17	1.64	0.859	0.435	0.219
$\delta_2^2 \times 10^4$	9.95	4.64	2.25	0.95	0.498	0.291
$\delta_3^2 \times 10^4$	1.580	4.64	1.930	0.781	0.377	0.206
$\delta_4^2 \times 10^2$	5.71	5.95	6.45	6.76	7.29	7.84
$(\delta_4^2)' \times 10^2$	6.01	8.01	9.73	12.00	14.60	17.60
$\delta_5^2 \times 10^2$	1.37	1.23	1.11	1.25	1.38	1.58
$(\delta_5^2)' \times 10^2$	1.50	1.54	1.63	1.54	1.56	1.50
$\delta_6^2 \times 10^2$	1.75	1.40	1.32	1.20	1.11	0.905
$\sum_{i=1}^{6} \delta_i^2 \times 10^2$	9.72	8.92	9.09	9.31	9.83	10.40

2. Fluctuations of the muon component (in contrast to the electron component) are determined mainly by fluctuations in the elementary act. In this case, the first interaction plays the main role. It is important to note that fluctuations in the partial inelasticity coefficients are an essential factor reflecting the influence of the elementary interaction process. When simulating EASs by the Monte Carlo method, information about the distribution of the total inelasticity coefficient, the multiplicity distribution, total cross sections, and single-particle inclusive spectra is usually used. Fluctuations of the partial inelasticity coefficients are obtained as a result of the multiple process simulation procedure and, as a rule, are not controlled [45].

As a result, fluctuations, for example, k_0 are much lower than the experimental data [41]. Because of this, the fluctuations of the muon component can be underestimated by a factor of two or more (Table 11.11).

Since the data presented here on the fluctuations of the muon component are in marked contradiction with most other calculations (an exception, however, is the

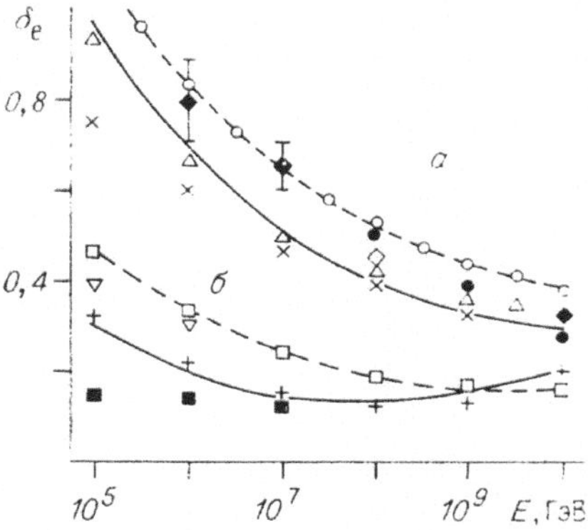

Fig. 11.16 Dependence of relative fluctuations of the total number of electrons EAS from the energy of the primary proton. **a**—sea level(♦—[33], •—[32], ×—[44]); **b**—the level of the mountains, z^*—690g/cm^2 (■—[45], ▽—[46]). Curves: dashed—[19], solid—calculation in the model of quark-gluon strings

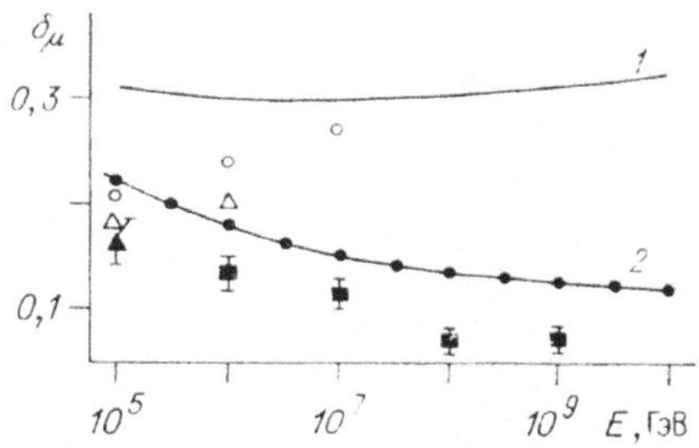

Fig. 11.17 The dependence of the relative fluctuations of the total number EAS electrons on the energy of the primary proton. Mountain level, z^*—690g/cm^2, $E_\mu > 5$GeV (○—[45], △—[46]); sea level, $E_\mu > 1$GeV (■—[33], ▲—[45]). 1—calculation in the quark-gluon string model, 2—[19]

Table 11.11 Relative fluctuations of the total number of particles δ_N and $\bar{\delta}(r)$ in EAS from the primary proton

E_0,GeV	δ^N	r, m					
		30	100	200	400	600	1000
10^5	1.04	0.9	0.74	0.64	0.55	0.50	0.44
	0.15	0.52	0.23	0.16	0.16	0.19	0.24
10^7	0.59	0.48	0.39	0.33	0.28	0.26	0.23
	0.16	0.33	0.18	0.17	0.19	0.22	0.27
10^9	0.39	0.31	0.27	0.24	0.22	0.22	0.21
	0.21	0.26	0.21	0.22	0.24	0.26	0.31

Note The upper number is for electrons, the lower one is for photons of Cherenkov radiation

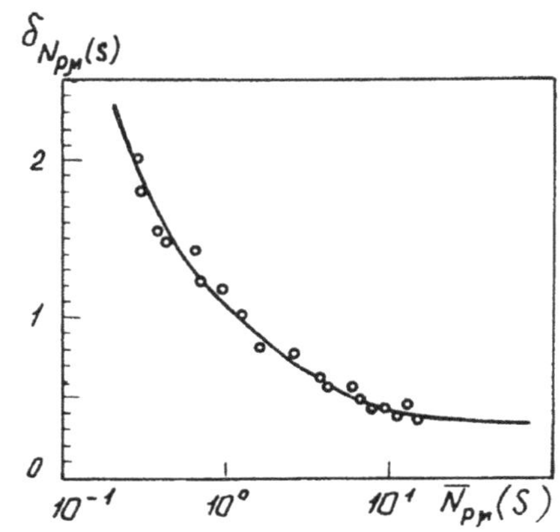

Fig. 11.18 Dependence of fluctuations of the number of muons hitting the site on their average values in the shower from the primary proton of energy 10^6GeV. Points—Monte Carlo calculation, solid line—according to the formula (11.62)

work [47], where due to variations in the simulation of the interaction fluctuations in the number of muons varied from 10 to 40%), we note once again that it follows from our calculations that smaller values of N_μ fluctuations do not agree with experimental data on fluctuations of partial inelasticity coefficients for hadron–nucleus interactions in the region energies 2–30 GeV.

Another effect of the increased influence of fluctuations in the elementary act manifested itself in the spatial distribution of muons. Due to the fact that the influence of particle range fluctuations, which deform the spatial distribution, the muon transverse coordinates are statistically independent. Hence, it follows that the square of the relative fluctuations of the number of muons arriving at the area S is related to the square of the fluctuation of the total number of muons by the relation

$$[\delta N_{p\mu}(S)]^2 = [\delta N_{p\mu}]^2 + \frac{1}{\bar{N}_{p\mu}(S)} - \frac{1}{\overline{N_{p\mu}}}, \tag{11.62}$$

where $\bar{N}_{p\mu}$ is the average number of muons arriving at this area in the shower from the primary proton, and $\overline{N_{p\mu}}$ is the average number of all muons at the observation level (with energy above the threshold). Monte Carlo calculations [48] confirmed the validity of this assumption (Fig. 11.18). Note that the above approximation significantly refines the commonly used estimate based on the Poisson distribution.

As shown in Chap. 10, a similar situation (independence of the transverse coordinates of peripheral particles) takes place in electron–photon cascades. In this approximation, fluctuations in the number of EAS electrons arriving at an area S at a distance r from the shower axis are described by the formula

$$[\delta N_{pe}(r, S)]^2 = [\overline{\delta\rho}_{pe}(r)]^2 + \frac{1}{S\bar{\rho}_{pe}(r)}$$

where the first term is the density fluctuations calculated neglecting the fluctuations in the EPC (and therefore independent of S), and the second term reflects the influence of area S on the fluctuation value. We emphasize that this formula follows from the analysis of conjugate equations for the moments of the measured number of particles in an EAS.

11.8 Two-Component Model of Fluctuations

A phenomenological analysis of fluctuations in the spatial distribution of electrons in EASs measured by a detector of finite dimensions has been carried out. The simplest assumptions about the statistical characteristics of the random distribution of particles in the observation plane make it possible to use average cascade curves, relative fluctuations of the total number of particles, and average spatial distributions to solve the problem. It is shown that the square of relative fluctuations of the number of particles arriving at a fixed area σ_k is represented as a sum of two terms, the first of which tends to a finite limit, and the second behaves like σ_k^{-1} at $\sigma_k \to 0$. The results of the analysis are confirmed by numerical calculations.

The peculiarity of determining statistical (from shower to shower) fluctuations in number of particles arriving at a detector of finite size is that the size cannot tend to zero, as is done when determining the density of the average number of particles, but must remain a non-vanishing parameter of the problem. Direct simulation of a high-energy shower up to the area under consideration of small sizes requires a long computer time. The partial shower averaging used in a number of algorithms [33, 45] leads to fluctuation distortion. In particular, stopping the simulation of the particle trajectory with the replacement of the stochastic value (i.e., the random contribution of its descendants to the detector reading) by its average value eliminates the dependence of fluctuations on the size of a small detector, that obviously contradicts experiment. Therefore, the term "fluctuations in the spatial distribution density" used here is not quite incorrect: the random density of the number of point particles is described by a generalized function, whose mean square, and therefore, variance does not exist. The Poisson ensemble, sometimes used for fitting the random point number, assumes its independency, which contradicts both theoretical scheme and observation data.

In [48], we presented a phenomenological approach to this problem based on the polynomial form of a distribution of points over the observation plane, conditioned by its total number. This model ensures a continuous transition from Poisson fluctuations in the small-size case to the fluctuations of the total number of particles, being in agreement with the numerical calculations in the intermediate region. It has turned out that such a model can easily be constructed on the simplest assumption about the nature of the distribution of particles without involving free (not related to the known characteristics of showers) parameters.

Let us divide the whole plane perpendicular to the shower axis into elements σ_k ($k = 1, \ldots, m$) and denote by N_k the random number of particles incident on σ_k

and by $N = \sum_{k=1}^{m} N_k$ the total number of particles in the observation plane, and by $P(n_1, \ldots, n_m|n)$ the probability of the event $\{N_1 = n_1, \ldots, N_m = n_m\}$ provided that $N = n$. The main assumption of the model is that the distribution of $P(n_1, \ldots, n_m|n)$ is described by the polynomial law

$$P(n_1, \ldots, n_m|n) = \frac{n!}{n_1! \ldots n_m!} q_1^{n_1} \ldots q_m^{n_m}, \tag{11.63}$$

where $q_k > 0$, $\sum_{k=1}^{m} q_k = 1$, $\sum_{k=1}^{m} n_k = n$. Physically, this means that positions of various particles in the observation plane do not depend (in the statistical sense) on the other particle positions.

To calculate the fluctuations of the spatial distribution of particles, we pass from the probability distribution (11.63) to the generating function

$$F(u) \equiv \langle u_1^{N_1} \ldots u_m^{N_m} \rangle = \langle (q_1 u_1 + \ldots + q_m u_m)^N \rangle. \tag{11.64}$$

The first and second factorial moments of N_k are obtained by differentiation of (11.64) in u_k:

$$\bar{N}_k = (\partial F/\partial u_k)_{u=1} = \langle N q_k \rangle, \tag{11.65}$$

$$N_k^{[2]} = (\partial^2 F/\partial u_k^2)_{u=1} = \langle N(N-1) q_k^2 \rangle.$$

The relative fluctuations square

$$\delta_{N_k}^2 = \langle N(N-1) q_k^2 \rangle \langle N q_k \rangle^{-2} - 1 + \langle N q_k \rangle^{-1} \tag{11.66}$$

consists of two parts, different in asymptotic behavior as $\sigma_k \to 0$. The first one

$$(\delta_{N_k}^2)_{\text{PC}} = \langle N(N-1) q_k^2 \rangle \langle N q_k \rangle^{-2} - 1, \tag{11.67}$$

having a finite limit, we call the *regular component* (RC) of the fluctuation, the second one

$$(\delta_{N_k}^2)_{\text{PC}} = \langle N q_k \rangle^{-1}, \tag{11.68}$$

infinitely increasing at $q_k(\sigma_k) \sim \sigma_k \to 0$, will be called the *Poisson component* (PC).

If assume q_k to be independent of N, then from (11.65) to (11.68) follow

$$q_k = \bar{N}_k/\bar{N}, \tag{11.69}$$

$$(\delta_{N_k}^2)_{\text{PC}} = \delta_N^2 - 1/\bar{N}, \tag{11.70}$$

$$\delta_{N_k}^2 = \delta_N^2 - 1/\bar{N} + 1/\bar{N}_k. \tag{11.71}$$

It is easy to see that this result satisfies the above requirements in the limiting cases of large ($\bar{N}_k \to \bar{N}$) and small ($\bar{N}_k \ll \bar{N}$) detectors. To analyze the situation in the intermediate region, we used the results of the Monte Carlo simulation of the distributions of muons with energies above 5 GeV at the level of mountains in an EAS from a primary proton with an energy of 10^6 GeV [48]. The study of fluctuations in the number of muons arriving at areas of different sizes placed at different distances from the shower axis confirmed the applicability of formula (11.71) for the muon component (Fig. 11.19).

A similar comparison for the electron component of the shower cannot be carried out due to the lack of results of a complete (analog) simulation of EAS development up to the drawing of electron trajectories crossing a small area. Therefore, we will consider another way of testing the model, based on comparison with the results of incomplete EAS simulations. Let us assume, for definiteness, that only the nuclear-active component is modeled, and the random contributions of π^0-decay photons to the spatial distribution of electrons are replaced by their average values (this is done in Monte Carlo calculations [33, 45], the same interpretation can also be given to the results of our numerical calculations [19]). Let $x_1, x_2, \ldots$ be the phase points of photon production in the random realization of the shower, and $N(x_i \to \sigma_k)$ be the random number of electrons coming to the area σ_k of the electron–photon cascade formed by the photon from x_i. The total number of electrons arriving at σ_k in this shower implementation is

$$N_k = \sum_i N(x_i \to \sigma_k). \tag{11.72}$$

With incomplete modeling, instead of (11.72) we have another random variable

$$N_k' = \sum_i \bar{N}(x_i \to \sigma_k). \tag{11.73}$$

It is easy to verify that the average values of these quantities coincide, and the fluctuations are related by the relation

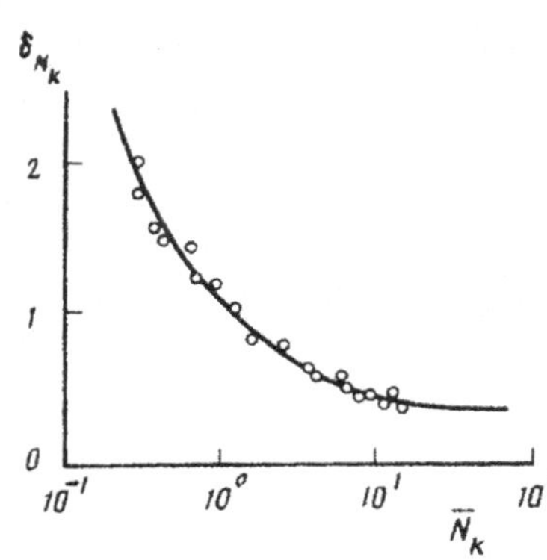

Fig. 11.19 Dependence of relative fluctuations of the number of muons ($E_\mu > 5$GeV, mountain level) arriving at the σ_k area, on their average value in a shower from a primary proton with an energy of 10^6GeV. The curve presents calculations by formula (11.68), points—calculation by the Monte Carlo method [48]

$$\delta^2_{N_k} = \delta^2_{N'_k} + Q/\bar{N}_k^2, \tag{11.74}$$

$$Q = \int S(x)[\overline{N^2}(x \to \sigma_k) - \bar{N}^2(x \to \sigma_k)]dx,$$

and $S(x)$ is the mean density of photons in a shower. When $\sigma_k \to 0$ (the area shrinks to a point whose position on the plane is given by the two-dimensional vector $\mathbf{r}$)

$$\bar{N}'_k \approx f(\mathbf{r})\sigma_k,$$

$$\overline{N^2}(x \to \sigma_k) - \bar{N}^2(x \to \sigma_k) \sim \bar{N}(x \to \sigma_k),$$

$$Q \sim \int S(x)\bar{N}(x \to \sigma_k)dx = \bar{N}_k$$

and instead of (11.74), we get

$$\delta^2_{N_k} = \delta^2_{f(\mathbf{r})} + 1/\bar{N}_k. \tag{11.75}$$

This result is consistent with the decomposition of fluctuations into two components, and the role of PC fluctuations is played by fluctuations of quantity (11.73) or (which is the same) fluctuations of quantity

$$f(\mathbf{r}) = \sum_i \bar{f}(x_i \to \mathbf{r}),$$

$$f(x_i \to \mathbf{r}) = \lim_{\sigma_k \to 0}[\bar{N}(x_i \to \sigma_k)/\sigma_k],$$

which is appropriate to call the RC (*regular component*) of the random density of the spatial distribution of shower particles. Let's represent it in the form

$$f(\mathbf{r}) = N\rho(\mathbf{r}), \tag{11.76}$$

where $\rho(\mathbf{r})$ satisfies $\int \rho(\mathbf{r})d\sigma = 1$.

It is easy to see that formula (11.70) (without the second term, which is negligible compared to the first)

$$(\delta^2_{N_k})_{\text{PC}} \approx \delta^2_{f(\mathbf{r})} = \delta^2_N \tag{11.77}$$

corresponds to representation (11.76) when the function $\rho(\mathbf{r})$, which describes the shape of the spatial distribution, is deterministic and does not depend on N. Wherein

$$q_k = \rho(\mathbf{r})\sigma_k, \tag{11.78}$$

and PC fluctuations are due only to fluctuations of the total number N.

Table 11.12 Electron number density $f(r)$ (upper numbers) and PC fluctuations (lower numbers) in EASs from protons for sea level

r, M	E_p, GeV			
	10^5	10^6	10^7	10^8
10	1.86	$3.40 \cdot 10^1$	$5.61 \cdot 10^2$	$8.76 \cdot 10^3$
	1.03	0.75	0.56	0.44
20	$3.14 \cdot 10^{-1}$	5.67	$9.07 \cdot 10^1$	$1.36 \cdot 10^3$
	0.91	0.65	0.48	0.38
100	$1.82 \cdot 10^{-2}$	$3.12 \cdot 10^{-1}$	4.70	$6.55 \cdot 10^1$
	0.74	0.52	0.39	0.31
200	$2.47 \cdot 10^{-3}$	$4.03 \cdot 10^{-2}$	$5.74 \cdot 10^{-1}$	7.53
	0.64	0.45	0.33	0.27
400	$2.92 \cdot 10^{-4}$	$4.46 \cdot 10^{-3}$	$5.92 \cdot 10^{-2}$	$7.22 \cdot 10^{-1}$
	0.55	0.38	0.28	0.23
600	$8.11 \cdot 10^{-5}$	$1.19 \cdot 10^{-3}$	$1.51 \cdot 10^{-2}$	$1.75 \cdot 10^{-1}$
	0.50	0.34	0.26	0.22
1000	$1.60 \cdot 10^{-5}$	$2.21 \cdot 10^{-4}$	$2.64 \cdot 10^{-3}$	$2.89 \cdot 10^{-2}$
	0.44	0.30	0.23	0.21
$\bar{N}$	$6.15 \cdot 10^3$	$1.07 \cdot 10^5$	$1.67 \cdot 10^6$	$2.43 \cdot 10^7$
δ_N	0.97	0.66	0.51	0.39

Note The last two lines are averages and relative
Fluctuations of the total number of particles at the observation level

Table 11.13 The same EAS characteristics as in Table 11.12, for mountain level (690g/cm^2)

r, M	E_p, GeV			
	10^5	10^6	10^7	10^8
10	$1.33 \cdot 10^{-1}$	$1.99 \cdot 10^2$	$2.54 \cdot 10^3$	$2.89 \cdot 10^4$
	0.35	0.24	0.19	0.19
20	1.95	$2.74 \cdot 10^1$	$3.30 \cdot 10^2$	$3.53 \cdot 10^3$
	0.28	0.20	0.18	0.20
100	$8.96 \cdot 10^{-2}$	1.14	$1.25 \cdot 10^1$	$1.23 \cdot 10^2$
	0.21	0.17	0.19	0.22
200	$1.02 \cdot 10^{-2}$	$1.20 \cdot 10^{-1}$	1.23	$1.13 \cdot 10^1$
	0.18	0.17	0.21	0.24
400	$9.79 \cdot 10^{-4}$	$1.05 \cdot 10^{-2}$	$9.92 \cdot 10^{-2}$	$8.51 \cdot 10^{-1}$
	0.17	0.19	0.23	0.27
600	$2.40 \cdot 10^{-4}$	$2.43 \cdot 10^{-3}$	$2.18 \cdot 10^{-2}$	$1.79 \cdot 10^{-1}$
	0.17	0.21	0.25	0.29
1000	$4.03 \cdot 10^{-5}$	$3.78 \cdot 10^{-4}$	$3.18 \cdot 10^{-3}$	$2.47 \cdot 10^{-2}$
	0.19	0.23	0.27	0.31
$\bar{N}$	$3.80 \cdot 10^4$	$5.33 \cdot 10^5$	$6.44 \cdot 10^6$	$6.99 \cdot 10^7$
δ_N	0.31	0.20	0.16	0.12

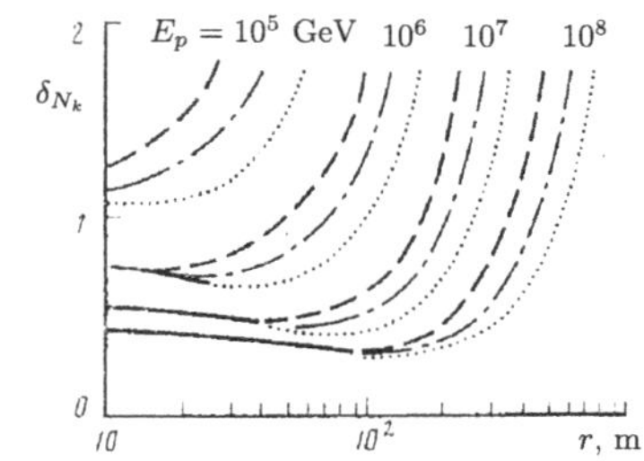

Fig. 11.20 Relative fluctuations of the number of electrons arriving at areas 1 (dashed lines), 2 (dash-dotted lines) and 5 m^2 (dotted) at different distances from the axis EAS from a proton with energy E_p

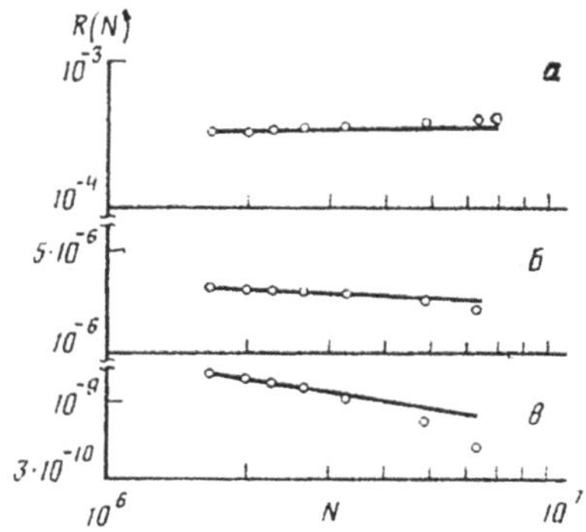

Fig. 11.21 Exact values of $R(r, N)$ (points) and approximation by formula (11.80) (curves) for the electron density at distances 10 (a), 100 (b) and 1000 m (c) in a shower from a primary proton with an energy of 10^7GeV (sea level)

Calculations of fluctuations in the spatial distribution of the electron EAS components have shown that relation (11.77) is not satisfied for it, where $\delta^2_{f(\mathbf{r})}$ depending on the distance r can be as more or less than δ^2_N. Tables 11.12 and 11.13 show the results of our numerical calculations of EASs from protons with energies $10^5 - 10^8$ GeV [9]. For description of the process of multiple hadron production, model of quark-gluon strings is used [18]. It can be seen from the tables that for the level of the sea in the entire indicated energy interval, a total regularity: at small distances $\delta_f > \delta_N$, at large distances ($r \geqslant 30$m) $\delta_f < \delta_N$. A similar situation for the level of mountains is observed only up to energy 10^6 GeV, at higher energies in the entire range of considered distances (10–1000 m), δ_f exceeds δ_N. The influence of the PC, which depends on the detector area, consists in the rapid increase in fluctuations at large distances from the shower axis (Fig. 11.20). The features noted above point to the significant role of fluctuations in the shape of the $\rho(\mathbf{r})$ distribution and to the change in the sign of the correlations $\rho(\mathbf{r})$ and N by changing the distance from the shower axis. The simplest way to obtain this effect is to replace the random variables $\rho(\mathbf{r})$ and N with deterministic functions $\rho_0(r, s)$ and $N_0(s)$ of the random age parameter s. This representation makes it easy to explain the behavior δ_f with a change in r: a random increase in N after the shower maximum means a decrease in s, leading to a narrowing of the distribution $\rho_0(r, s)$ and vice versa, so that the product $N_k = N\rho(\mathbf{r})\sigma_k$ fluctuates at small distances within larger limits than N, and at larger distances it fluctuates to a smaller extent.

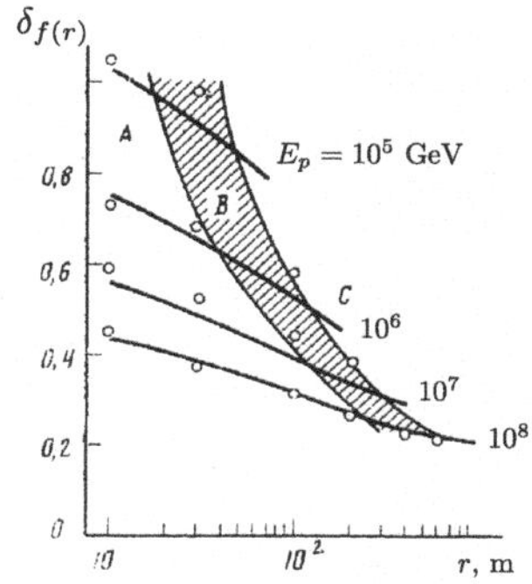

Fig. 11.22 Fluctuation in the number of EAS electrons arriving at the detector $\sigma_k = 5\text{m}^2$. A is the region where the main contribution $\geqslant 90\%$ is from PC, C is the PC region, the shaded transition region is B. Curves—numerical calculations of RC, points—approximation (11.85)

Outside the region of maximum, the relationship between s and $N_0(s)$ is mutual (one-to-one and $\rho_0(r, s)$ can be expressed in terms of N):

$$\rho_0(r, s(N)) = R(r, N). \tag{11.79}$$

This dependence can be conveniently approximated by a power function

$$R(r, N) = \alpha(r)\nu^{\gamma(r)}, \qquad \nu = N/\bar{N}. \tag{11.80}$$

The applicability of such an approximation is confirmed, in particular, by Fig. 11.21.

Using (11.78)–(11.80), instead of (11.65) and (11.67), we get

$$\bar{N}_k = \langle \nu^{\gamma+1} \rangle \alpha(r) \sigma_k \bar{N} \tag{11.81}$$

and

$$(\delta^2_{N_k})_{\text{PC}} = \delta^2_{f(r)} \approx \langle \nu^{2\gamma+2} \rangle \langle \nu^{\gamma+1} \rangle^{-2} - 1. \tag{11.82}$$

The moments $\langle \nu^\alpha \rangle$ under the assumption of a log-normal distribution of the total number of electrons [19] have the form (Fig. 11.22)

$$\langle \nu^\alpha \rangle = (1 + \delta^2_N)^{\alpha(\alpha-1)/2}. \tag{11.83}$$

Table 11.14 shows that this approximation satisfactorily agrees with the results of direct numerical integration of the distributions over N obtained by us using the Monte Carlo method with high statistics (10^6 showers). From (11.81) to (11.83), we find simple formulas for the density of the average number of particles and PC fluctuations of the electronic component:

Table 11.14 The reduced (formula (11.83)) moments of the distribution over the number of EAS electrons from a 10^7 GeV proton at sea level

α	$\langle\nu^\alpha\rangle^{1/\alpha}$		α	$\langle\nu^\alpha\rangle^{1/\alpha}$	
	Numeric integral	Approxim (11.83)		Numeric integral	Approxim (11.83)
0.2	0.89	0.89	2.0	1.13	1.15
0.6	0.95	0.95	2.6	1.22	1.25
1.2	1.03	1.03	3.0	1.28	1.32
1.6	1.08	1.09	3.6	1.45	1.53

Table 11.15 Values of the index $\gamma_k(r)$ in approximation (11.80) for EAS electrons (primary proton, sea level)

r, m	E_p, GeV			
	10^5	10^6	10^7	10^8
10	$4.43{\cdot}10^{-2}$	$8.36{\cdot}10^{-2}$	$8.62{\cdot}10^{-2}$	$1.18{\cdot}10^{-1}$
30	$-4.62{\cdot}10^{-2}$	$-3.71{\cdot}10^{-2}$	$-5.29{\cdot}10^{-2}$	$-2.39{\cdot}10^{-2}$
100	$-1.88{\cdot}10^{-1}$	$-2.06{\cdot}10^{-1}$	$-2.17{\cdot}10^{-1}$	$-1.95{\cdot}10^{-1}$
200	$-2.80{\cdot}10^{-1}$	$-3.03{\cdot}10^{-1}$	$-3.31{\cdot}10^{-1}$	$-2.95{\cdot}10^{-1}$
400	$-3.69{\cdot}10^{-1}$	$-4.04{\cdot}10^{-1}$	$-4.29{\cdot}10^{-1}$	$-3.97{\cdot}10^{-1}$
600	$-4.20{\cdot}10^{-1}$	$-4.64{\cdot}10^{-1}$	$-4.68{\cdot}10^{-1}$	$-4.22{\cdot}10^{-1}$
1000	$-4.83{\cdot}10^{-1}$	$-5.24{\cdot}10^{-1}$	$-5.28{\cdot}10^{-1}$	$-4.48{\cdot}10^{-1}$

$$\bar{f}(r) = \alpha(r)\bar{N}(1+\delta_N^2)^{(\gamma+1)\gamma/2}, \tag{11.84}$$

$$\delta_{f(r)}^2 = (1+\delta_N^2)^{(1+\gamma)^2} - 1. \tag{11.85}$$

The coefficients α and γ can be determined by comparison with the results of numerical calculations (Tables 11.12 and 11.13). In this case, formula (11.85) is simply a more or less convenient approximation of the results of numerical calculations. But formulas (11.84), (11.85) can also be used for other purposes: making sure that $\alpha(r)$ is close $\bar{\rho}(r) = R(r, \bar{N})$, the approximate values of the coefficient y can be determined from formula (11.80) (Table 11.15):

$$\tilde{\gamma} = \lg[R(r, N)/R(r, \bar{N})]/\lg[N/\bar{N}],$$

$$N/\bar{N} \in (1-\delta_N, 1+\delta_N). \tag{11.86}$$

In this case, formula (11.85) makes it possible to obtain approximate $\tilde{\delta}_{f(r)}$ fluctuations even without numerical calculations, relying only on the known average shower characteristics and fluctuations of the total number of particles shows satisfactory agreement between $\tilde{\delta}_{f(r)}$ and $\delta_{f(r)}$ in the region where the PC contribution is not dominant. This is an important argument in favor of the phenomenological approach developed in this paper, as being simpler and less related to the features of one or another model of an elementary act of a multiple process. As the observation level increases, we approach the maxima of the cascade curves, in the vicinity of which the relationship between s and N ceases to be one-to-one and $\rho_0(r, s)$ cannot be approximated by formula (11.80). For this reason, formula (11.81) does not give correct results for the level of mountains. The calculation of fluctuations in this case can be made using Table 11.13 and formulas (11.75). In any case, the fluctuation is decomposed into two non-correlated components, RC and PC, in connection with which it seems appropriate to call the model described above a two-component model of fluctuations in the spatial distribution of shower particles.

References

1. Grieder PK (2010) Extensive air showers: high energy phenomena and astrophysical aspects-a tutorial, reference manual and data book. Springer-Verlag, Berlin, Heidelberg
2. Hillas AM (1996) Differences between gamma-ray and hadronic showers. Space Sci Rev 75:17–30
3. Fermi E (1949) On the origin of the cosmic radiation. Phys Rev 75:1169
4. Kampert KH, Watson AA (2012) Extensive air showers and ultra high-energy cosmic rays: a historical review (2012). arXiv:1207.4827v1. [physics.hist-ph]. (19 Jul 2012)
5. Nikitin YuP, Rozental IL (1980) High energy nuclear physics. Atomizdat, Moscow (in Russian)
6. Murzin VS, Sarycheva LI (1974) Multiple processes at high energies. Atomizdat, Moscow (in Russian)
7. Nikitin YuP, Rozental IL (1976) Theory of multiple processes. Atomizdat, Moscow (in Russian)
8. Grishin VG (1982) Inclusive processes in hadron interactions at high energies. Atomenergoizdat, Moscow (in Russian)
9. Ivanenko IP, Kanevsky BL, Roganova TM (1979) On the violation of scale invariance in the pionization region during the transition from accelerator to superhigh energies. Yadernaya Fizika 29:694–707 (in Russian)
10. Kalmykov NN, Khristiansen GB (1976) Violation of scale invariance in hadron interactions at superhigh energies. Pisma v Zhurn Exp Theoret Phys 23:595–599 (in Russian)
11. Erlykin AD, Kuzina NP (1981) Analysis of the passage of cosmic rays through the atmosphere in the light of the problem of scaling breaking. Vopr Atom Sci Technol Ser Techn Phys Exp, Kharkov 2(8):34–50 (in Russian)
12. Adamov DS, Erlykin AD (1983) Violation of scaling in the fragmentation region of the spectra of pp and pp interaction at an energy of 1011–1014 eV. Brief Commun Phys 12:32
13. Baiburina SG, Borisov AS, Guseva ZM et al (1984) Investigation of nuclear interactions in the energy range $10^{14} - 10^{17}$ eV by the method of X-ray emulsion chambers in cosmic rays (experiment "Pamir"). Proc of Phys Inst Acad Sci USSR 154:3–141 (in Russian)
14. Kalmykov NN, Khristiansen GB (1983) On a possible explanation of scaling violation in hadron interactions at energies above 103 TeV. Pisma v Zhurn Exp Theor Phys 37(5):247–249 (in Russian)
15. Kasahara K (1978) On the apparent break of scaling features in high-energy cosmic-ray components in the atmosphere. Nuovo Cimento 46A:333–364
16. Dunaevskii AM, Pashkov SV, Mukhamedshin RA et al (1984) Calculations of high-energy nuclear-electromagnetic cascades. Proc Phys Inst Acad Sci USSR 154:142–217 (in Russian)
17. Ivanenko IP, Roganova TM (1983) Cascade showers caused by ultrahigh energy particles. Nauka, Moscow (in Russian)

18. Shabelsky YuM (1986) Inclusive spectra of secondary particles in p and collisions in the quark-gluon string model. Yadernaya Fizika 11(7):186–196 (in Russian)
19. Kalmykov NN, Uchaikin VV, Chernyaev GV (1985) Calculation of EAS characteristics in the framework of the theory of supercritical polaron taking into account the additive quark model. Proc Acad Sci USSR. Ser Phys 49(7):1299–1302 (in Russian)
20. Lagutin AA, Uchaikin VV, Chernyaev GV (1987) To the theory of Extensive Air Shower fluctuations. J Nucl Phy 45(3):757–769 (in Russian)
21. Uchaikin VV, Chernyaev GV (1989) Phenomenological analysis of fluctuations of lateral distribution of shower particles. J Nucl Phy 50(9):736–742 (in Russian)
22. Oda H, Nakatsuka T, Nakai H et al (1975) Theoretical studies of the longitudinal development of extensive air showers. 1 Prog Theor Phys 54(5):1472–1486
23. Oda H, Nakatsuka T, Nakai H et al (1976) Theoretical studies of the longitudinal development of extensive air showers. 2 Prog Theor Phys 55(6):1844–1856
24. Dedenko LG, Nikolsky SI, Stamenov IN (1976) On the spatial distribution of the electron flux in extensive air showers. Brief Commun Phys 1:30
25. Hinkley SW, Tsokos CP (1975) Stochastic processes in particle-number fluctuations in an electronphoton shower. Acta Biotheor 24:58–74
26. Linde IA, Kanevsky BL (1981) Method of solution of cascade theory non-linear estimation for function of distribution over the number of particles in nuclear-electromagnetic shower. Proc. of 17-th ICRC Paris 5:268–271
27. Varfolomeev AA, Drabkin LB (1970) Calculation of electromagnetic showers in lead with an initial energy of 6 GeV. Cosmic Rays 12:32–59 (in Russian)
28. Ivanenko IP, Kanevsky BL, Roganova TM (1979) Yadernaya Fizika 29:694 (in Russian)
29. Kalmykov NN, Khristiansen GB (1983) Pis'ma. Zh Eksp Teor Fiz 37:247 (in Russian)
30. Litvinov VA (1985) Independent use of multiplicity and inelasticity coefficients in an empirical model of nuclear interactions, Tomsk. Dep. in VINITI 17.06.85, No. 5597-85, p 18 (in Russian)
31. Kalmykov NN, Khristiansen GB (1978) On the role of fragmentation and pionization regions of inclusive spectra in the development of EAS. In: Proceedings of an SSSR. Series physics, vol 42, no 7, pp 1497–1500 (in Russian)
32. Dedenko LG, Dymova IA (1974) Fluctuations of prospatial distribution of electrons in EAS in the energy range $10^{17} - 10^{18}$ eV. In: Proceeding of an SSSR. Series physic, vol 38, pp 1033–1036 (in Russian)
33. Dixon HE, Turver KE (1974) Computer simulation of cosmic-ray air showers. III. Fluctuations in shower development. Proc Roy Soc A 339:171–195
34. Nikolskii SI (1981) Hadron interactions in cosmic rays at energies above the accelerator range. Sov Phys Usp 24:925–948
35. Chernyaev GV, Lagutin AA, Uchaikin VV (1983) The model of hadron-nuclear interaction in the superhigh-energy range. In: Proceeding of 18-th ICRC, vol 9. Bombay, pp 2–18
36. Ashton F, Nejabat H (1981) The hadron component of extensive air showers of size in the region of 3105 particles at sea level. In: Proceeding of 17-th ICRC, vol 6. Paris, pp 203–205
37. Murzin VS, Sarycheva LI (1968) Cosmic rays and their interaction. Atomizdat, Moscow (in Russian)
38. Watson KN (1952) The hypothesis of charge independence for nuclear phenomena. Phys Rev 85:850–852
39. Anoshin AI, Solovyov MI, Suleymanov M et al (1979) Particular KNO-scalepng in hadron on hadron and hadron-nucleus interactions. Yadernaya Fizika 30:389–396 (in Russian)
40. Hayakawa S (1969) Cosmic ray physics. Wiley-Interscience, New York
41. Tretyakova MI (1984) Inelastic interactions protons and pions with nucleons and photoemulsions in the energy range 10-400 GeV and composition quark structure of hadrons, Inelastic interactions of hadrons and nuclei at high energies. Alma-Ata, pp 95–133 (in Russian)
42. Nam RA, Nikolsky SI, Pavlyuchenko VP et al (1977) Verification of scale invariance in the energy range 1-40 TeV using ionization calorimeter. (Preprint/Phys. Institute of the Academy of Sciences of the USSR; No. 104) (in Russian)

43. Uchaikin VV, Lagutin AA, Plyasheshnikov AV (1979) On fluctuations in the number of particles in the magnetic cascade. Yadernaya Fizika 30:429–436
44. Vdovchik E (1975) Determination of the mass of primary braidsmic rays from measurements of wide atmospheric showers. In: Proceeding of academy of sciences of the USSR. Series physical, vol 39, no 6, pp 1147–1159
45. Popova L (1984) Phenomenological model of high energy hadron interactions. Acta Universitatis Lodziensis, Lodz
46. Dedenko LG (1970) Electron and muon components at the level of mountains. In: Proceeding of academy of sciences of the USSR. Series physical, vol 34, no 19, pp 1974–1976
47. Atrashkevich VB, Fomin YA, Khristiansen GB (1965) Calculation of fluctuations in the development of atmospheric showers using the Monte Carlo method. In: Proceeding of academy of sciences of the USSR. Series physical, vol 29, no 9, pp 1696–1701
48. Konopelko AK, Litvinov VA, Sheshnikov AV et al (1986) On fluctuations in a wide air showers. In: Proceeding of academy of sciences of the USSR. Series physical, vol 50, no 11, pp 2211–2213

For Further Reading

49. Grieder PKF (2010) Extensive air showers. Springer, Berlin Heidelberg
50. Gaisser TK (1992) Cosmic rays and particle physics, 2nd edn. Cambridge University Press, Cambridge
51. Capdevielle JN et al (1992) The Karlsruhe EAS simulation code CORSIKA. KFK Report 4998
52. Rao MSV, Sreekantan BV (1998) Extensive air showers. World Scientific, Singapore

Chapter 12
Cherenkov Radiation of EPC

12.1 Radiation in a Homogeneous Medium (B-Approximation)

The modern astrophysics deals with high-energy neutrino telescopes, using large volume of natural optically transparent media, where EPC charged particles moving with velocities exceeding the speed of light in that media c/n emit the Cherenkov radiation. Experimental installations on deep-water detection (see, for example, Russian project NT - 200, at the depth at 1 km in Like Baikal in Siberia; AMANDA in deep polar ice at the South Pole; NESTOR 3500 m deep in the Mediterranean near Pylos, Greece) are usually spatial lattices of recording underwater elements (Cherenkov light detectors). Distance between recording elements should not be much more the length of transparency of water. The layer of water above installation provides protection from cosmic rays muons. Water between recording elements serves as a detector. Neutrinos interact with nuclei. Interactions give rise to secondary particles of cascade. Electrons of EPC generate the Cherenkov light serving as an information carrier [1–3].

In the paper [4], we analyzed the fluctuations of Cherenkov radiation and its correlations with the number of particles in B-approximation of EPC theory.

Let us denote by $\Psi_{e,\gamma}(N, q|t, E)$ the joint distribution of the particle number N and the Cherenkov radiation amount q emitted to t r.u. units from the point of the primary electron (e) or photon (γ) with energy E appearance. In view of the large number of Cherenkov photons emission, the radiation process is considered as continuous:

$$\Delta q = a\Delta t,$$

where Δq is the Cherenkov radiation emitted by an electron with an energy above the threshold E_1 on the path Δt. The coefficient a is assumed to be independent of the energy E and of the depth t. Thus, the variable N is discrete, the variable q is continuous, and the normalization of the function has the form

© Higher Education Press Limited Company 2024

V. V. Uchaikin, *Electron-Photon Cascades*, Nonlinear Physical Science,
https://doi.org/10.1007/978-981-99-7524-2_12

$$\sum_{N=0}^{\infty} \int_{0}^{\infty} dq \Psi(N, q|t, E) = 1.$$

Besides,

$$\sum_{N=0}^{\infty} \Psi(N, q|t, E) = \Psi(q|t, E)$$

is the distribution density of Cherenkov radiation, and

$$\int_{0}^{\infty} dq \Psi(N, q|t, E) = \Psi(N|t, E)$$

– distribution of the number of particles. For the joint distribution of N and q, the boundary conditions have the form;

$$\Psi_e(N, q|0, E) = \delta_{N1}\delta(q),$$

$$\Psi_\gamma(N, q|0, E) = \delta_{N0}\delta(q).$$

Consider an electron with energy E at a distance t from the point of observation. When passing from the point t and the point $t - \Delta t$ ($\Delta t > 0$), the electron with probability

$$W_e(E, E')dE'\Delta t$$

can emit a bremsstrahlung quantum with an energy in the interval dE' and initiate two independent showers and with probability

$$1 - \Delta t \int_{0}^{\infty} W_e(E, E')dE'$$

pass a segment of the path Δt without bremsstrahlung. Using the theorem on the distribution of the sum of independent random variables up to terms of the first order in Δt, we obtain the following relation:

$$\Psi_e(N, q|t, E) = \Psi_e(N, q - a\Delta t|t - \Delta t, E - \beta\Delta t)$$

$$\times [1 - \sigma_e \Delta t] + \Delta t \int_{0}^{E} dE' W_e(E, E')$$

$$\times \sum_{N'=0}^{N} \int_0^q dq' \Psi_\gamma(N', q'|t, E')\Psi_e(N - N', q - q'|t, E - E'),$$

where

$$\sigma_e = \int_0^E W_e(E, E')dE'.$$

Expanding Ψ_e into a series in Δt, we arrive at the following equation:

$$\partial_t \Psi + a\partial_q \Psi_e + \sigma_e \Psi_e = \int_0^E dE' W_e(E, E') \sum_{N'=0}^{N} \int_0^q dq' \Psi_\gamma(N', q'|t, E')$$

$$\times \Psi_e(N - N', q - q'|t, E - E') - \beta\partial_E \Psi_\gamma. \tag{12.1}$$

Similarly, we obtain the equation for EPC from a primary photon:

$$\partial_t \Psi_\gamma + \sigma_0 \Psi_\gamma = \int_0^E dE' W_p(E, E')$$

$$\times \sum_{N'=0}^{N} \int_0^q dq' \Psi_e(N', q'|t, E')\Psi_e(N - N', q - q'|t, E - E'). \tag{12.2}$$

The derivative with respect to q, absent in the second equation, describes the continuous emission of Cherenkov radiation by electrons measured by the detector, which does not have a noticeable effect on the development of the cascade itself.

Multiplying Eqs. (12.1), (12.2) by N, summing over N and integrating over q, we obtain

$$\partial_t \bar{N}_e + \beta\partial_E \bar{N}_e + \sigma_e \bar{N}_e - \int_0^E dE' W_e(E, E')[\bar{N}_\gamma(t, E') + \bar{N}_e(t, E - E')] = 0,$$

$$\partial_t \bar{N}_\gamma + \sigma_0 \bar{N}_\gamma - 2\int_0^E dE' W_p(E, E')\bar{N}_e(t, E') = 0. \tag{12.3}$$

Equation (12.3) for the average number of particles formed by a primary electron or photon at a distance t from the beginning of the line coincides with the adjoint equations of the usual cascade theory. Assuming for simplicity the threshold energy of secondary particles is also equal to E_1, we write down the obvious relations

$$\bar{N}_e(t, E) = \sum_{N=0}^{\infty} N \int_0^{\infty} dq \Psi_e(N, q|t, E) = \bar{N}_e^{(e)}(t, E, E_1), \tag{12.4}$$

$$\bar{N}_\gamma(t, E) = \sum_{N=0}^{\infty} N \int_0^{\infty} dq \Psi_\gamma(N, q|t, E) = \bar{N}_e^{(\gamma)}(t, E, E_1),$$

the right-hand sides of which contain the usual cascade functions for the primary energy E and the threshold one E_1.

Represent Eq. (12.2) in the matrix form

$$\hat{\mathbf{L}}\bar{\mathbf{q}} = \mathbf{a}, \qquad \mathbf{a} = \begin{pmatrix} a \\ 0 \end{pmatrix}, \tag{12.5}$$

where

$$\bar{\mathbf{q}}(t, E) = \sum_{N=0}^{\infty} \int_0^{\infty} q \boldsymbol{\Psi}(N, q|t, E) dq \tag{12.6}$$

stands for the average amount of Cherenkov radiation emitted along the path t by a shower formed respectively by a primary electron or photon. Differentiating Eq. (12.5) with respect to t and comparing with (12.3), we obtain

$$\bar{\mathbf{N}}(t, E) = \frac{1}{a} \frac{\partial \bar{\mathbf{q}}(t, E)}{\partial t}, \tag{12.7}$$

whence

$$\bar{\mathbf{q}}(t, E) = a \int_0^t \bar{\mathbf{N}}(\tau, E) d\tau = a\mathbf{S}(t, E). \tag{12.8}$$

The obtained relations have a simple physical meaning: in the model of a homogeneous atmosphere, the average amount of radiation $\bar{\mathbf{q}}(t, E)$ at a depth t is proportional to the average area $\mathbf{S}(t, E)$ under the corresponding part of the cascade curve.

Numerical calculations of the area under the cascade curve were carried out in [5]. To calculate the fluctuations of the Cherenkov radiation, it is necessary to represent this quantity in an analytical form.

Applying the saddle-point method for inverse transformations, we obtain

$$\mathbf{S}(t, E) = G(1, \varepsilon) \frac{E}{\beta} - \frac{H_1(p) D(p) \mathbf{B}(p) G(p, \varepsilon) \exp\{yp + \lambda_1(p) t\}}{p[2\pi(\lambda''_1(p)\lambda_1^2(p)t - \lambda''_1(p)\lambda_1(p) + \lambda_1'^2(p))]^{1/2}}, \tag{12.9}$$

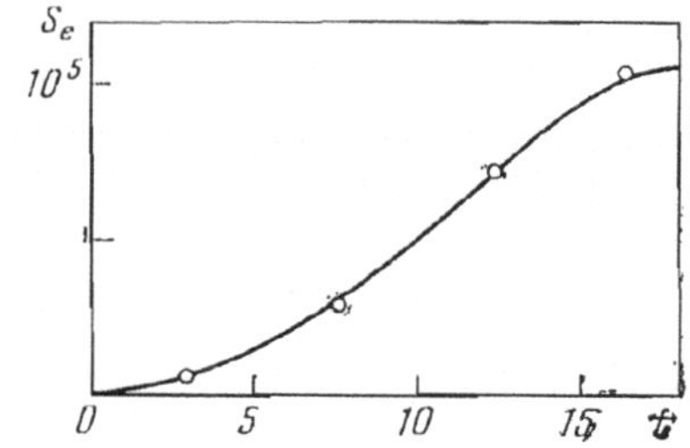

Fig. 12.1 The area under cascade curve $S_e(t, E)$ for the primary electron; $y = 11.7$, $E_1 = 0$. Curve—calculation by formula (12.9), dots—data from [5]

where

$$\mathbf{B}(s) \equiv \begin{pmatrix} B_e(s) \\ B_\gamma(s) \end{pmatrix} = \begin{pmatrix} 1 \\ B(s)/[\lambda_1(s) + \sigma_0] \end{pmatrix}$$

and the saddle point p, in contrast to the usual cascade parameter s, satisfies the equation

$$t = -\frac{y}{\lambda_1'(p)} + \frac{1}{\lambda_1(p)}, \quad p > 1. \tag{12.10}$$

The physical meaning of these expressions is obvious: the area under the cascade curve in the section from zero to t is equal to the difference between the total area and the area under the curve in the section from t to ∞. It can be seen from Fig. 12.1 that the values of $S(t, E)$ calculated using these formulas practically coincide with the results of numerical calculations published in [5].

Similarly, from (12.1), (12.2) we obtain equations for the second moments:

$$\overline{\mathbf{N}^2}(t, E) = \sum_{N=0}^{\infty} N^2 \int_0^{\infty} dq \boldsymbol{\Psi}(N, q|t, E),$$

$$\overline{\mathbf{q}^2}(t, E) = \sum_{N=0}^{\infty} \int_0^{\infty} q^2 \boldsymbol{\Psi}(N, q|t, E) dq,$$

$$\boldsymbol{\varphi}(t, E) \equiv \begin{pmatrix} \overline{N_e q_e}(t, E) \\ \overline{N_\gamma q_\gamma}(t, E) \end{pmatrix} = \sum_{N=0}^{\infty} N \int_0^{\infty} q \boldsymbol{\Psi}(N, q|t, E) dq.$$

The equations for $\overline{\mathbf{N}^2}$ were considered in [6], where the results of calculating the quantity $\overline{N^2}/\bar{N}^2 = \delta_N^2 + 1$, depend at a fixed energy only on s.

Remaining systems of equations

$$\hat{\mathbf{L}}\overline{\mathbf{q}^2} = \mathbf{U}, \tag{12.11}$$

$$\hat{\mathbf{L}}\boldsymbol{\varphi} = \mathbf{V} \tag{12.12}$$

differ from each other only in free terms:

$$U_e(t, E) = 2\int_0^E dE' W_e(E, E')\bar{q}_e(t, E - E')\bar{q}_\gamma(t, E') + 2a\bar{q}_e(t, E),$$

$$U_\gamma(t, E) = 2\int_0^E dE' W_\gamma(E, E')\bar{q}_\gamma(t, E - E')\bar{q}_e(t, E'); \tag{12.13}$$

$$V_e(t, E) = \int_0^E dE' W_e(E, E')[\bar{q}_e(t, E - E')\bar{N}_\gamma(t, E')$$

$$+ \bar{q}_\gamma(t, E')\bar{N}_e(t, E - E')] + a\bar{N}_e(t, E),$$

$$V_\gamma(t, E) = 2\int_0^E dE' W_\gamma(E, E')\bar{q}_e(t, E')\bar{N}_e(t, E - E'). \tag{12.14}$$

It is easy to see that, by virtue of equality (12.7), the free terms are related by the relation

$$\mathbf{V}(t, E) = \frac{1}{2a}\frac{\partial \mathbf{U}(t, E)}{\partial t}$$

whence the analogous relations for the moments follow

$$\boldsymbol{\varphi}(t, E) = \frac{1}{2a}\frac{\partial \overline{\mathbf{q}^2}(t, E)}{\partial t}. \tag{12.15}$$

This result can also be obtained directly from the definition of the quantities included in expression (12.15). Let us consider a random (not averaged) cascade curve $N_e(t, E)$ and the value of Cherenkov radiation associated with it

$$q_e(t, E) = a\int_0^t N_e(t', E)dt'. \tag{12.16}$$

Let's square (12.16)

$$q_e^2(t, E) = a^2 \int_0^t dt' \int_0^t dt'' N_e(t', E) N_e(t'', E)$$

and differentiate it:

$$\frac{\partial q_e^2(t, E)}{\partial t} = 2aN_e(t, E)q_e(t, E). \tag{12.17}$$

Averaging (12.17) over the ensemble of individual showers and removing the derivative operator from under the sign of the mean, we arrive at relation (12.15), which allows one to confine oneself to solving one system of equations (12.11) with bullet boundary conditions to calculate both magnitudes. Let us transform the free terms of $\mathbf{U}(t, E)$. Substituting expressions (12.8) and (12.9) into (12.13) and neglecting (as done in [6]) the dependence of p on E under the integrals, we obtain

$$\mathbf{U}(t, E) = \mathbf{P}(t, E) + \frac{a^2 \lambda_1(p) t}{\lambda_1''(p)\lambda_1^2(p)t - \lambda_1''(p)\lambda_1(p) + \lambda p_1'^2(p)} \mathbf{R}(t, E), \tag{12.18}$$

where

$$R_e(t, E) = \bar{N}_e(t, E)\bar{N}_\gamma(t, E)\gamma_1(p),$$

$$R_\gamma(t, E) = \bar{N}_e^2(t, E)\gamma_2(p),$$

$$P_e(t, E) = 2\left[aG(1, \varepsilon)\frac{E}{\beta}\right]^2 \gamma_1(1) + 2a\bar{q}_e(t, E)$$

$$- \frac{2a^2 H_1(p) D(p) G(p, \varepsilon) G(1, \varepsilon) \exp\{\lambda_1(p)t + y(p+1)\}}{p[2\pi(\lambda_1''(p)\lambda_1^2(p)t - \lambda_1''(p)\lambda_1(p) + \lambda_1'^2(p))]^{1/2}}$$

$$\times \left[A(p+1) - A(p) + \frac{B(p)[C(p) - C(p+1)]}{\lambda_1(p) + \sigma_0}\right],$$

$$P_\gamma(t, E) = 2\left[aG(1, \varepsilon)\frac{E}{\beta}\right]^2 \gamma_2(1)$$

$$- \frac{2a^2 H_1(p) D(p) G(p, \varepsilon) G(1, \varepsilon) \exp\{\lambda_1(p)t + y(p+1)\}}{p[2\pi(\lambda_1''(p)\lambda_1^2(p)t - \lambda_1''(p)\lambda_1(p) + \lambda_1'^2(p))]^{1/2}} [B(p) - B(p+1)].$$

In accordance with this, we represent the solutions of equations (12.11) as the sum of two terms

$$\overline{\mathbf{q}^2}(t, E) = \boldsymbol{\Phi}^0(t, E) + \boldsymbol{\Phi}^1(t, E), \tag{12.19}$$

satisfying the equations

$$\hat{\mathbf{L}}\boldsymbol{\Phi}^0 = \mathbf{P}, \tag{12.20}$$

$$\hat{\mathbf{L}}\boldsymbol{\Phi}^1 = \frac{a^2\lambda_1''(p)t}{\lambda_1''(p)\lambda_1^2(p)t - \lambda_1''(p)\lambda_1(p) + \lambda_1'^2(p)}\mathbf{R}. \tag{12.21}$$

We are looking for a solution to equations (12.20) and (12.21) in the form

$$\boldsymbol{\Phi}^0(t, E) = \left[aG(1, \varepsilon)\frac{E}{\beta}\right]^2 + \boldsymbol{\varphi}^0(p)\exp\{\lambda_1(p)t + y(p+1)\}, \tag{12.22}$$

$$\boldsymbol{\Phi}^1(t, E) = \frac{a^2\mathbf{f}(p, t)D^2(p)G^2(p, \varepsilon)\lambda_1''(p)t}{\lambda_1''(p)\lambda_1^2(p)t - \lambda_1''(p)\lambda_1(p) + \lambda_1'^2(p)} \times \exp\{2\lambda_1(p)t + 2yp\}. \tag{12.23}$$

Substituting (12.22) into (12.20) and solving the equations for $\boldsymbol{\varphi}^0$, we obtain

$$\boldsymbol{\Phi}^0(t, E) = \left[aG(1, \varepsilon)\frac{E}{\beta}\right]^2 - \frac{a^2H_1(p)D(p)G(1, \varepsilon)G(p, \varepsilon)\mathbf{B}'(p)\exp\{\lambda_1(p)t + y(p+1)\}}{p[2\pi(\lambda_1''(p)\lambda_1^2(p)t - \lambda_1''(p)\lambda_1(p) + \lambda_1'^2(p))]} - \frac{[2a^2G^2(1, \varepsilon)E/\beta - 2a\bar{q}_e(t, E)][\lambda_1(p) + \sigma_0]\mathbf{B}'(p)}{[\lambda_1(p) - \lambda_1(p+1)][\lambda_1(p) - \lambda_2(p+1)]}, \tag{12.24}$$

$$\mathbf{B}'(p) \equiv \begin{pmatrix} B_e'(p) \\ B_\gamma'(p) \end{pmatrix} = \begin{pmatrix} 1 \\ B(p+1)/[\lambda_1(p) + \sigma_0] \end{pmatrix}.$$

Substituting (12.24) into (12.19), we obtain the following expressions for the variance

$$\sigma_{e,\gamma}^2(t, E) = \overline{q_{e,\gamma}^2} - \bar{q}_{e,\gamma}^2 = \frac{a^2D^2(p)G^2(p, \varepsilon)\exp\{2\lambda_1(p)t + 2yp\}}{\lambda_1''(p)\lambda_1^2(p)t - \lambda_1''(p)\lambda_1(p) + \lambda_1'^2(p)} \times \left[f_{e,\gamma}(t, p)\lambda_1''(p)t - \frac{1}{2\pi p^2}H_1^2(p)B_{e,\gamma}^2(p)\right].$$

It is easy to verify that the functions $f_e(t, p)$ and $f_\gamma(t, p)$ at $t \gg 1$ satisfy Eqs. (54), (55) from [4]. Using the results of this work for $p < 1.47$, we obtain

$$(\delta_q)_{e,\gamma}^2 = \frac{a^2H_1^2(p)D^2(p)G^2(p, \varepsilon)B_{e,\gamma}^2(p)\exp\{2\lambda_1(p)t + 2yp)\}}{2\pi p^2[\lambda_1''(p)\lambda_1^2(p)t - \lambda_1''(p)\lambda_1(p) + \lambda_1'^2(p)]\bar{q}_{e,\gamma}^2(t, E)}(\delta_N(p))_{e,\gamma}^2. \tag{12.25}$$

For $p > 1.47$, the second term in Eq. (12.10) is small compared to the first one, and as p increases, its relative value decreases. Setting $p = s$ in this region, we get

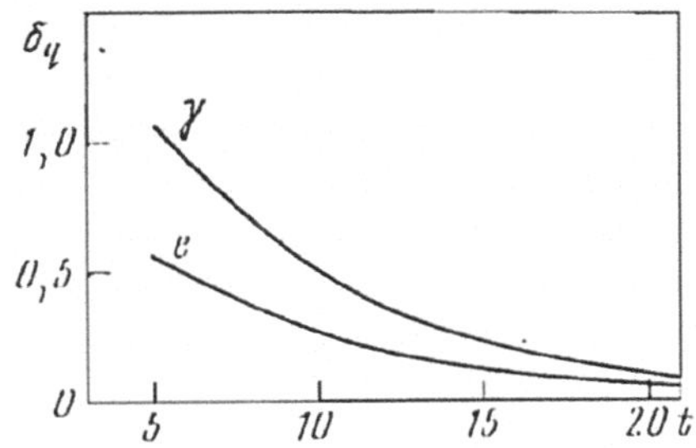

Fig. 12.2 Relative fluctuations of the Cherenkov radiation of showers produced primary electron (e) and photon (γ); $y = 11.7$, $E_1 = 0.25\beta$ [4]

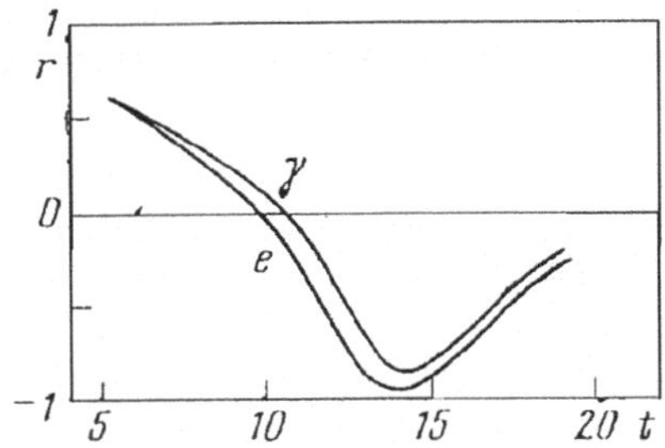

Fig. 12.3 Correlation coefficient of Cherenkov radiation and the number of shower particles for the same parameters [4]

$$(\delta_q)_{e,\gamma} = \frac{a\bar{N}_{e,\gamma}(t, E)}{\lambda_1(s)\bar{q}_{e,\gamma}(t, E)}(\delta_N(s))_{e,\gamma}. \tag{12.26}$$

Calculation example using formula (12.25) for $y = 11.7$ and $E_1/\beta = 0.25$ is shown in Fig. 12.2. As expected, the relative fluctuations rapidly decrease according to a law close to exponential, and the magnitude of the fluctuations in the case of a primary photon is much larger than for an electron; in particular, at the shower maximum they are 40 and 20%, respectively.

To calculate the correlation coefficient

$$r_{N,q} = \frac{\varphi - \bar{N}\bar{q}}{\sigma_N \sigma_q}, \tag{12.27}$$

we use formulas (12.7) and (12.15), as a result of which we obtain the following relation:

$$r_{N,q} = \frac{1}{2a\sigma_N\sigma_q}\frac{\partial\sigma_q^2}{\partial t}. \tag{12.28}$$

The corresponding curves are shown in Fig. 12.3. A characteristic feature of the behavior of the correlation coefficient with increasing t is a monotonic decrease with transition from positive values at $t \lesssim y$ to negative values at $t \gtrsim y$ with subsequent tending to zero. This behavior is related to the shape of typical cascade curves: at $t \lesssim y$, most of the Cherenkov radiation is emitted near $t = y$ (with decreasing t the number of electrons decreases rapidly), as a result of which there is a strong positive correlation between q and N. For $t \gtrsim y$, a large fraction of the Cherenkov radiation is emitted at the shower maximum. Since the integral of the cascade curve is proportional to the primary energy and does not fluctuate, a random increase in the number of particles in the region of the maximum is accompanied by a decrease in the number of particles in the region beyond the maximum. At great depths, Cherenkov radiation practically does not fluctuate and becomes independent of the number of particles. This explains the autocorrelation between q and N and its decrease as $t \to \infty$.

12.2 Cherenkov Radiation in Water (Monte Carlo)

A relativistic track ($\beta = 1$) in water or ice ($n \simeq 1.33$) produces about $N_0 \approx 250$ cm^{-1} optical photons in a wavelength interval between 300 nm and 500 nm, which is a typical sensitive region of photo-detectors, used in the aforementioned neutrino telescopes. The Cherenkov-angle for a relativistic track ($\beta = 1$) in ice is $\theta_{C0} = \arccos(1/n) \approx 41°$. A large fraction of detected Cherenkov-photons in high-energy neutrino telescopes originates from electromagnetic cascades. These are initiated by a high-energy electromagnetic particle which produces a shower of secondary particles by subsequent bremsstrahlung and pair production processes. The primary particle can originate from radiative energy losses of a high-energy muon (bremsstrahlung and pair production), or from the decay of $\pi^0 \to 2\gamma$ in hadronic cascades, or be a high-energy electron from a charged-current interaction of an electron neutrino. The physical length of a shower is typically less than 10 m. This is short, compared to the scale of neutrino telescopes and the full Cherenkov-light is created locally and expands with time as an almost spherical shell with a characteristic angular distribution of the intensity. Due to the large number of particles the full tracking of each particle in Monte- Carlo simulations of cascades in neutrino telescopes is very time consuming. However, the development of electromagnetic cascades is very regular because fluctuations are statistically suppressed by the large number of interactions and large number of involved particles. Hence, it can be well approximated by the average development (Fig. 12.4).

We give in this section some review of theoretical results in this direction and concerning application of methods described in this book. The results are mainly obtained by my colleagues from Altai State University and published in [8–10].

Dimensions of the front disk of EPC in water (tens of centimeters) are much smaller than the longitudinal size of the shower (meters) and the distance between neutrino telescope detectors (tens of meters). Therefore, when calculating the measured characteristics of the Cherenkov radiation, the EPC sizes are often neglected, and a spatially distributed source moving with high speed in water is replaced by a luminous point with a given angular distribution, which moves along the shower axis. In the framework of this *luminous point model* (LPM) all Cherenkov photons entering the detector at a fixed moment in time, were born at a certain point in space, located on the axis of the shower. On this assumption, the measured signal can easily

Fig. 12.4 A simulated electromagnetic cascade. A primary electron of 100 GeV has been injected at the left pointing towards the right. Shown are all generated charged secondary particles (red for negative and blue for positive charge) as the result of a Geant4 simulation. Neutral particles, like photons are not shown [7]

be linked with the depth-angular distribution of the radiated Cherenkov photons. The LPM usage allows investigators to significantly reduce the amount of settlement information required for planning and data processing of deep sea detection experiments.

Knowing an initial particle energy allows to define the depth-angle distribution $n(E, t, \vartheta)$ and the mean number of all Cherenkov photons $n(E, t, \vartheta)dtd\Omega$ radiated in the depth element dt into $d\Omega$ at an angle ϑ to the shower axis. Writing $n(E, t)$ for the depth distribution of the Cherenkov photon sources (i.e. the mean number of all photons emitted by cascade unit at t),

$$n(E, t) = \int_{4\pi} n(E, t, \vartheta)d\Omega \tag{12.29}$$

let us introduce the normalized to 1 conditioned by E, t angular structure function

$$f(\vartheta|E, t) = \frac{n(E, t, \vartheta)}{n(E, t)}, \tag{12.30}$$

$$\int_0^\infty dt \int_{4\pi} f(\vartheta|E, t)d\Omega = 1. \tag{12.31}$$

Integrating (12.31) and (12.30) over depth yields the corresponding degradation spectra obeying the corresponding integral equations, integration of their solution over other (angular and energy) variables gives the unnormalized depth degradation distribution of emitted Cherenkov photons. Similarly to electron cascade curves, t-dependence $n(E, t)$ with a fixed E has a local maximum, t_m whose position shifts slowly towards larger values of depth as primary energy grows. The value of t_m in case of primary photon is somewhat greater ($\sim 0.25 \div 0.30$ m) in comparison to the electron initiated shower. Calculations show that the following interrelation takes place (Table 12.1):

$$n(E, t) \simeq 300 \cdot N_e(E, t). \tag{10}$$

Table 12.1 Ratio $n(E,t)/N_e(E,t)$ in EPC from primary photon with the energy E

E	t/t_m								
GeV	0.05	0.10	0.20	0.40	0.70	1.00	1.50	2.00	3.00
1	303	303	302	301	300	299	298	296	296
10	304	303	301	300	299	298	297	296	–

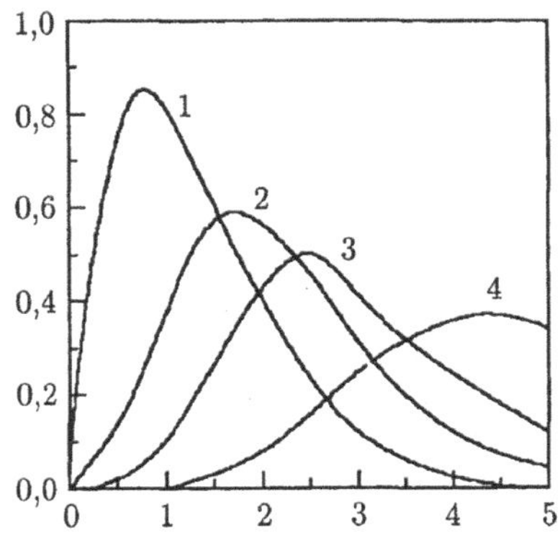

Fig. 12.5 Depth-distributions of Cherenkov photons emitted by EPC's with energies $E_\gamma = 1$ GeV (1), 10 GeV (2), 10^2 GeV (3), $E = 10^4$ GeV (4) [9]

Numerical analysis showed also, that for the primary energy region $E \geq 1$ GeV, the shape of the normalized equilibrium angular distribution $f(E, \vartheta)$ does not depend practically on the energy and type of the primary particle, i.e. $f(E, \vartheta) \simeq f(\vartheta)$. The following conclusions regarding this distribution can be made from the data presented in [9] (Figs. 12.5 and 12.6).

- Function $f(\vartheta)$ has a sharp local maximum near $\vartheta = \arccos(1/m$ ($m = 1.33$) is the refraction parameter of water). This maximum is caused by contribution from high energy electrons moving along the shower axis and emitting the Cherenkov light along a conic surface having the opening angle $\vartheta_0 \simeq \vartheta_0(E = \infty) = \arccos(1/m)$.
- The function $f(\vartheta)$ has a long flat tail at large values of ϑ. The main contribution to this tail is given by cascade electrons having large values of the multiple scattering angle.
- At a fixed energy E, the angular distribution becomes wider with increasing depth. This is explained by the increase in scattering cascade particles due to the softening of their energy spectrum.

MC-analysis showed as well, that E-dependence of the normalized angular distribution $f(E, t, \vartheta)$ can be significantly weakened, if instead of depth t to use a new variable $u = t/t_m$, where t_m is the depth of the cascade electron curve maximum.

As it was established by numerical calculations, the mean characteristics of EPC Cherenkov light can be expressed in terms of the depth-angle distribution of emitted photons, if one neglects the size of the shower particle disk and the contribution of the scattered light. An obvious advantage of this model, suggested by Prof. I. P.

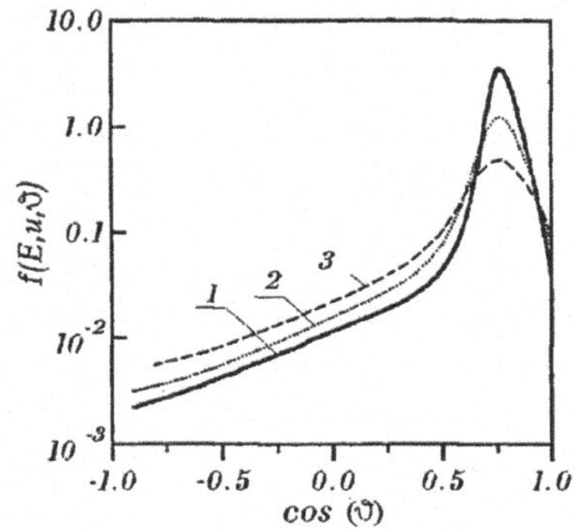

Fig. 12.6 Normalized angular distributions of Cherenkov photons by EPC from primary electron with $E = 10^3$ GeV. $u = t/t_m$. 1—$u = 0.3$, 2—$u = 1$, 3—$u = 3$ [9]

Table 12.2 The primary energies E_∞ above which the angular distribution does not depend on the energy and type of primary particle

	u			
	0.3	0.5	1.0	2.00
Primary electron	10^4	10^2	10	1
Primary photon	10^3	10^2	3	1
Type of primary particle	10^2	10	3	1

Ivanenko and called the *luminous point* model (LPM), is a possibility of an essential reduction of amount of calculational data needed for planning and data analysis of deep underwater experiments.

The data in Table 12.2 allow to estimate the applicability conditions of this model. The Cherenkov photons propagation through the water was simulated by SAMC method. The average photon free paths to absorption and scattering were considered to be independent of photon wavelength and were taken equal to 15 m. The scattering law was also considered independent of the wavelength. Details of description relate to the Cherenkov detector Quasar NT-200 collaboration. The basic characteristics of this detector is its effective registration area

$$S_{\text{eff}} = C(\theta)R^2,$$

where $R = 17.5$ cm, θ is the angle between the detector optical axis $\vec{\Omega}_0$ and photon arrival direction $\vec{\Omega}$.

12.3 Cherenkov Radiation in Atmosphere

Ground-based gamma-ray astronomy exploits the air shower produced by the interaction of a primary gamma ray in the Earth's atmosphere. For gamma-rays of approximately $10^{11} \div 10^{14}$ eV (VHE) few air-shower particles reach ground level. However, sufficiently high-energy secondary particles will emit Cherenkov radiation, resulting in illumination of $\sim 10^5$ m^2 patch of ground for a few nanoseconds. The *Imaging Atmospheric Cherenkov Technique* (IACT) is based on the observation of this emission by a number of large reflecting telescopes, placed within the light pool with a typical spacing of $\sim$100 m. The Cherenkov light is focussed onto ultra-fast, but typically rather coarsely pixelated, cameras, resulting in roughly elliptical images of the shower emission in multiple telescopes.

Traditionally, this event reconstruction is performed using the *Hillas parameters* of the camera images [11], being the moments of the camera image which, given the approximately elliptical nature of typical camera images, already capture much of the available image information.

Since 1955, when the first measurements of the Cherenkov light flashes from air showers have been performed, tremendous progress happened in developing the registration technique. Currently operating ground based imaging air Cherenkov telescopes, successfully used for the observations of very high energy γ-rays, have already reached the energy threshold of $\simeq$200 GeV.

For understanding the performance of future *low energy* (or $\simeq$100 GeV class) detectors relevant Monte Carlo simulations are needed. Here an overview is given for the space-angular and temporal characteristics of Cherenkov light emission from air showers induced by primary γ-rays of energy $10 \div 100$ GeV. For that detailed Monte Carlo simulations, a few packages for calculating electromagnetic cascades in a user-defined, complex medium have been elaborated.

The first of them was MOCCA (= MOnte Carlo CAscade) [12] which was developed to study such cascades, with the objectives of (a) following particles down to very low energies (well below 1 MeV) from an arbitrarily high primary energy, (b) avoiding the use of analytical expressions at large distances, (c) permitting changes in the calculation to be made readily—both to investigate the possible significance of features of very energetic interactions that are not yet well understood (in particular investigating to which apparently minor features the results were very sensitive and therefore uncertain), (d) in a short computing time—so that questions could be readily investigated by making several runs.

Based on the original article [13], we consider main elements of this code and later demonstrate some results.

The first its part (EMCCS) intended for the generation of a EPC in the atmosphere, involves all electron-photon interactions listed in Chap. 5: electron-positron pair production, Compton scattering, bremsstrahlung, ionization losses, Coulomb scattering, Cherenkov light emission. Most of the algorithms used in this code are slightly modified those described in Chaps. 10 and 12.

Modeling electron-positron penetration was realized by using the scheme of catastrophic collision, which lead to emission of photon/electron of a relatively high energy, $E > Q$, exceeding the intermediate energy threshold of Q. The same approach was used for the simulation of the e^+, e^- ionization processes ($Q = Q_I$) and bremsstrahlung ($Q = Q_R$). The emitted electron must have a kinetic energy above a certain intermediate energy threshold, $T > Q_I$.

Dealing with the Cherenkov radiation of showers in atmosphere, one has to take into account its inhomogeneity, arising as a result of molecular heat motion and gravitation forces. When the energy of the emitted electrons is much higher than the threshold value for atmosphere ($\beta \simeq 1$), the radiation angle at the high h km above the sea level can be calculated as

$$\Theta_{\text{Ch}}^2 = 2\eta, \quad \eta = (1 - n) = \eta_0 e^{-h/\lambda}, \quad \eta_0 = 2.93 \cdot 10^{-4}, \quad \lambda \simeq 7.1 \text{ km}.$$

The total number of Cherenkov light photons emitted (but not absorbed) by an air shower

$$Q = 2Q_0(\eta_0 t_0/\rho_0) \int N_e(t)dt = 2Q_0(\eta_0 t_0/\rho_0)(E_0/\varepsilon_0)\delta, \tag{12.32}$$

where δ denotes the part of the cascade electrons above the energy threshold of the Cherenkov light production in air (it equals about 0.25 for $E_0 = 10$ Gev). Note that Eq. (12.32) does not take into account the absorption of the Cherenkov photons, which decrease it on around 20%. The total amount of Cherenkov photons from a γ-ray-induced air shower at sea level

$$Q_{\text{ph}}^{\gamma} = (28 + 4 \ln E_0) \cdot E_0 \cdot 10^3,$$

where E_0 is the primary energy of a shower (in GeV).
The number of Cherenkov photons emitted per unit path length of a particle with the charge Ze, moving with the velocity $v > 1$ is given by the formula

$$a(x) = \frac{dN}{dx} = 2\pi\alpha Z^2 \int_{\lambda_1}^{\lambda_2} \left(1 - \frac{1}{[(v/c)n(\lambda)]^2}\right) \lambda^{-2} d\lambda.$$

Particles with $v < c/n(\lambda)$ cannot emit Cherenkov light at wavelength λ. At visible wavelengths this results in an energy threshold of more than 20 MeV (for electrons or positrons). The amount of light emitted by particles above threshold depends on the index of refraction n, which in turn depends on other atmospheric parameters, determining its transparency. Total shower light is collected from all altitudes development and the shape of the recorded pulse is related to the profile of the cascade process and the height distribution of these parameters over of the atmosphere. Different atmospheric density profiles, generally, result in different indices of refraction near shower maximum and, thus, in different amounts of Cherenkov light emitted.

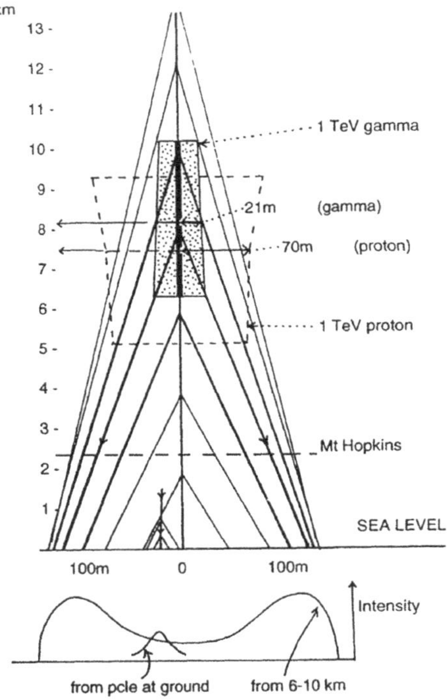

Fig. 12.7 Geometry of Cherenkov rays in air showers

Most of the light arriving in the inner region of fairly flat light density of about 120 m radius at 2000 m altitude is emitted near and after the shower maximum and is particularly affected by the longitudinal shower development as compared to the total amount of Cherenkov light.

The stippled box, pictured at Fig. 12.7, contains the main region of emission of light in gamma-ray showers: 25% is emitted above the top of the box and 25% below the bottom. The median altitude is 8.1 km. At these heights, half the emission occurs at radial distances within the box. The dashed-line box indicates the corresponding main reagion of emission in proton showers of the same energy. Cherenkov light is emitted from ultra-fast particles travelling along the axis at the angles shown (with exaggerated horizontal scale), resulting in a peak of intensity on the ground just beyond 100 m from the centre.

Muons can also be detected over a large area with Cherenkov imaging: a muon passing within about 40 m of a 10 m telescope will produce a short arc on the image (part of a circle of 1.3° radius, becoming a complete arc if the muon strikes the light collecting mirror). These separate arcs may typically appear about a degree away from the image of the main cascade, and add to the "width" of the image, causing it to be rejected as "not gamma-like". However, there is on average less than one muon arc fragment per image at the energies most commonly studied, somewhat below 1 TeV, so again, muons do not play the main part in hadron rejection. Although hadronic showers should get narrower at high energies, and presumably less different from gamma showers, except for an increasingly reliable contribution from muons, it is found in practice that (using pixels 15 arc minutes apart, with unfilled spaces between the phototube sensitive areas) if showers are characterized by several image parameters (typically width, length, distance from source and angular deviation of image axis from line pointing to source) discrimination against hadrons improves up to at least 10 TeV.

The opening angle u of the Cherenkov light c cone above threshold also depends on n,

$$\cos \Theta_{\mathrm{Ch}} = c/nv,$$

which for $(n-1) \ll 1$ corresponds to

$$\Theta_{\mathrm{Ch}} \approx \sqrt{2(n-1)}\ \mathrm{rad}.$$

For dealing with particle detectors, a pressure correction is usually sufficient to account for different atmospheric density profiles by passage to total depths in cascade units, or equivalently, in mass/area units.

Since $n(\lambda) - 1$ changes by only 5% over the wavelength range 300–600 nm, the range typically covered by photomultipliers, air-shower Cherenkov simulations are usually simplified by assuming a wavelength-independent index of refraction n, obtained at an effective wavelength. Another frequent simplification is to assume that $n(\lambda) - 1$ is proportional to air density, although it is a more complex function of pressure, temperature, and water vapour content.

The estimation of the degree of influence of these factors on the average photon flux of Cherenkov radiation at different levels of its registration, was performed by means of the analog Monte Carlo simulation.[1]

Some exciting detections and observations of TeV γ-ray emission from a number of galactic and extragalactic objects [14, 15] have shown the high performance of currently operating Imaging Atmospheric Cherenkov Technique (IACT). Several projects for future detectors have been proposed lately. The major forthcoming instruments, such as CANGAROO IV, H.E.S.S., MAGIC and VERITAS, will have significantly better sensitivity to γ-ray fluxes in the energy range from 50 GeV up to 50 TeV. All of them use as a messenger the Cherenkov radiation of EAS.

For a lack of a collimated test beam of TeV γ-ray photons, the ground-based Cherenkov detectors heavily rely on the Monte Carlo simulations of the Cherenkov light emission from air showers which are used to understand the performance of detector. Basic parameters of the instrument (detection area, angular and energy resolution, efficiency of cosmic ray rejection, etc.) can be derived from the simulations.

[1] This was one of our first Monte Carlo EPC calculations. Later, this code was heavily modified and, through the efforts of A. V. Plyasheshnikov and A. K. Konopelko, turned into the currently known ALTAI code, including simulation of electron scattering, pair production, Compton scattering and photoelectric photon absorptions. The trajectories of γ-quanta were modeled in exact accordance with the interaction cross sections, the Bethe–Heitler, and the Klein–Nishina-Tamm formulas. The detailed description of the ALTAI can be found in [13].

The crucial point is a precise measurement of the primary shower energy. For that one should include properly the simulations of all processes of Cherenkov light emission in the air shower as well as photon propagation on the way from the emitting shower particle to the telescope camera. Measurements of γ-ray fluxes, energy spectra, upper limits strongly depend on the absolute energy calibration of a telescope.

Here we give a description of ALTAI computational code developed for detailed Monte Carlo simulations of the Cherenkov light emission in TeV air showers. Among the other existing codes intended for such simulations, KASCADE [16], CORSIKA [17], this code has a distinctive advantage—it's rather high speed of shower simulations due to a specific algorithm used for processing the multiple scattering of charged shower particles. This approach does not consume a lot of CPU time and allows to perform fast and accurate simulations. Together with the additional routine recently developed for the simulations of telescope response [18] the ALTAI code was extensively used for production of a standard Monte Carlo database used in the HEGRA (High Energy Gamma Ray Astronomy) VHE γ-ray experiment (Konopelko A. K., Plyasheshnikov A. V., Aharonian F., et al., HEGRA Collaboration). The ALTAI code consists of two major programs which simulate the development of the electromagnetic (EMCCS) and proton-nuclei cascade in Earth atmosphere. Giving a brief description of ALTAI, we refer the reader for details to the original article [13].

12.4 ALTAI Code

12.4.1 Electromagnetic Cascade

Here we overview the part of the code intended for the generation of an electromagnetic cascade in the atmosphere (EMCCS). Note that at present all cross-sections for electron and photon interactions are established and are well described in detail in the relevant energy range. The electromagnetic part of the code accounts for the following interaction processes: electron-positron pair production, Compton scattering for the primary photons and bremsstrahlung, ionization losses, Coulomb scattering for the primary electrons and positrons.

Multiple Scattering

The relativistic electron suffers an enormous number of interactions along its path length in the matter. Apparently, a direct simulation of all interactions is not possible. That is why all numerical codes for the shower simulation use a specific technique for grouping electron interactions (see for review [19–21]). In this approach one can simulate only the so-called catastrophic e^-, e^+ interactions. These interactions lead to emission of photon/electron of a relatively high energy, $E > Q$, which exceeds the intermediate energy threshold of Q. The same approach was used for the sim-

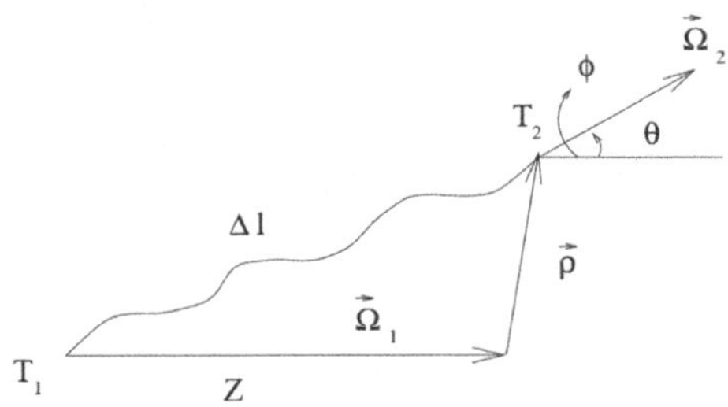

Fig. 12.8 Transformation of phase coordinates of a charged particle at the multiple scattering segment

ulation of the e^-, e^+ ionization process ($Q = Q_I$) and bremsstrahlung ($Q = Q_R$). The emitted electron must have a kinetic energy above a certain intermediate energy threshold, $T > Q_I$. A random path length between neighboring catastrophic collisions is divided into small segments of $\Delta l \leq 5\text{gr/cm}^2$. The phase coordinates of a charged particle are sampled at the end of the segment Δl according to the cumulative effect of all low energy collisions along the segment.

Energy Losses

The energy losses of a charged particle, ΔE, in the segment, Δl, were sampled according to the Landau-Vavilov formula [22]. Plyasheshnikov and Kolchuzkin [23] have tabulated this formula for a specific conditions of shower simulations.

Apparently the length of a multiple scattering segment, Δl, is one of the basic parameters of this method. By use of rather small segments one can reduce systematic error introduced by approximations of analytical solutions for the phase transformations at the multiple scattering segment. On the other hand this may slow down the procedure of shower simulation. We found that the optimum length of the multiple-scattering segment is within the range of $\sim 1-5\,\text{gr/cm}^2$. Very accurate analytical solutions of a multiple-scattering process for a e^-, e^+ transport allow us to use in calculations such segment length without introducing systematic errors in three-dimensional shower development.

In comparison to other computational codes Altai uses more accurate distributions derived analytically from the multiple scattering theory. On the contrary the standard approach used for example in EGS-IV code does not include fluctuations of the energy losses at the multiple-scattering segment. In addition, in the EGS-IV code the lateral displacement of charged particle at the multiplescattering segment is taken as $\vec{\rho} = 0$ and correspondingly the longitudinal displacement is of $z = \Delta l$. All this necessitates a small segment length and makes significantly more time consuming the shower simulations. Thus for the simulations of Cherenkov light from the TeV air shower ALTAI code is faster by as much as a few times compared with the EGS-IV.

12.4.2 Emission of Cherenkov Light

A charged particle (e, μ) in an air shower can emit Cherenkov light when its energy exceeds a certain threshold energy $E > E_{\text{th}}$. The threshold energy is determined as

$$E_{\text{th}} = \frac{m_0 c^2}{\sqrt{2(n-1)}}, \tag{12.33}$$

where n is the refraction index in air, m_0 is the particle mass. Thus at sea level the energy threshold is $E_{\text{th}} \simeq 20$ MeV for electron and $\simeq$4 GeV for muons. To describe the altitude dependence of the refraction index we used the following expression (Beliaev et al. [24])

$$n = 1 + \eta, \qquad \eta(H) = \eta_0/\rho_0 \cdot \rho(H), \tag{12.34}$$

where $\eta_0 = 2.9 \cdot 10^{-4}$, $\rho_0 = 1.22 \cdot 10^{-3}\,\text{g/cm}^3$, and $\rho(H)$ is the air density at a height H above the sea. The model of a standard atmosphere (Elterman 1968) was used in the simulations.

The mean number of Cerenkon photons emitted in a 1 cm pathlength by electron is described as

$$\frac{dQ}{dl} = 2\pi\alpha\left(\frac{1}{\lambda_1} - \frac{1}{\lambda_2}\right)\sin^2\theta_c, \qquad \sin^2\theta_c = 1 - \frac{1}{(1-(mc^2/E)^2)n^2}, \tag{12.35}$$

where α is a fine structure constant $\alpha = 1/137$ and $\theta_c \simeq 1°$ is the opening angle of the Cherenkov light cone. The spectral region of emission is defined by wavelengths λ_1 and λ_2. In simulations we included sampling of the Cherenkov light attenuation in the atmosphere due to the Raleigh scattering, ozone and aerosol absorption. The cross-sections for these processes were calculated using the data of Driscoll and Vaughan (1978).

In the shower simulation procedure (PARCHR, GSTCHR) (Konopelko 1990), the total number of Cherenkov photons n_{ph} emitted at the multiple scattering segment Δl is defined. The use of rather small multiple scattering segments allows to assume that all Cherenkov photons are emitted from the center of the multiple scattering segment. To a rather good approximation an intersection of the Cherenkov light cone with the observation plane forms a circle of a radius r_c which can be calculated as

$$r_c = (Z - Z_{\text{obs}}) \cdot \theta_c, \tag{12.36}$$

where Z and Z_{obs} are the height of Cherenkov light emission and the height of the observation level above the sea, respectively. The number of photons hitting the telescope mirror can be calculated as

$$\langle \tilde{n}_{\text{ph}} \rangle = \frac{\Delta L}{2\pi r_c}(1 - p(Z, Z_{\text{obs}}))\langle n_{\text{ph}} \rangle, \tag{12.37}$$

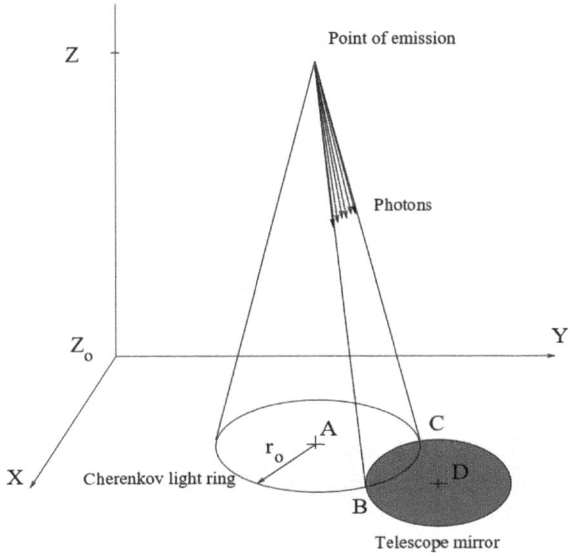

Fig. 12.9 Geometry of propagation of Cherenkov light photons

where ΔL is length of the circle arc between points B and C (see Fig. 12.9). $p(Z, Z_{\text{obs}})$ is a probability of the Cherenkov light attenuation in the atmosphere along the way of a photon propagation. For specific response functions of the telescope camera one may calculate corresponding average number of photoelectrons as $\langle n_{\text{ph.e.}} \rangle = \xi \langle n_{\text{ph}} \rangle$, where ξ is a photon-to-photoelectrons conversion efficiency. Finally one can use the Poisson distribution in order to simulate the random number of photons (photoelectrons) hitting the telescope. The photons reaching the telescope mirror were uniformly distributed over the circle arc BC (see Fig. 12.9).

Each photon has a set of 6 variables, $\vec{v}(x, y, \theta_x, \theta_y, z_0, t)$. z_0 is a height of the photon emission (in g/cm^2) and t is a time of photon arrival to the detector. Such database of the photon is used for detailed simulations of the telescope camera response (Hemberger, PhD thesis, University of Heidelberg, MPI H-V-19 (1998)).

12.4.3 Hadron-Nuclei Cascade

For simulations of the hadron-nuclei cascade ALTAI uses the phenomenological model of hadron interactions [13] which, in the most part, is based on the available accelerator data. The energy spectra of secondary hadrons generated in a pA interactions were approximated by use of the radial scaling model (Hillas 1979). In this approach the energy spectra may be presented as

$$x dN/dx = F_{(p\to q)}(x) H_q(x, E), x = E/E_0, \tag{12.38}$$

where E_0, E are the energies of the primary and secondary particles, respectively. Indeces p, q denote the type of primary and secondary particle $(p, n, \pi^+, \pi^-, \pi^0)$. The basic functions are given in [13].

For the total cross-sections of inelastic hadron interactions we used the data given by Shabelski [25] in the following form

$$\sigma_{\pi^\pm} = 1.2\sigma_{(p,n)}, \tag{12.39}$$

$$\sigma_{(p,n)} = \begin{cases} 258, & E < E_1, \\ 273(1 + 3.29 \cdot 10^{-3} y^2), & E_1 < E < E_2, \\ 258(1 + 0.07y), & E > E_2, \end{cases}$$

$$E_1 = 100\,\text{GeV}, \qquad E_2 = 10^4\,\text{GeV}, \qquad y = \ln(E/10^3)\,\text{GeV}. \tag{12.40}$$

The cross-sections in Eqns. (12.39)–(12.40) are measured in mBarn and correspond to the particle interaction with air on assumption that the mean value of atomic number for air nuclei is of 14.4.

We used a simplified geometrical representation [26] for the total cross-section of the nucleus-nucleus interactions as follows

$$\sigma_A = R^2 \sigma_N, \qquad R = \frac{A^{1/3} + A_0^{1/3} - b}{1 + A_0^{1/3} - b}, \tag{12.41}$$

where A_0 is the effective atomic mass of air ($A_0 = 14.4$) and b is the effective radius of nuclei overlapping zone ($b = 1.17$). σ_N defines the cross-section of hadron interaction with air.

The transverse momenta of secondary hadrons were calculated according to the distribution given by Ranft [27]

$$f(p_\perp) = p_\perp \frac{\exp(-Bp_\perp^2) + C\exp(-Dp_\perp)}{(2D)^{-1} + CD^{-2}} \tag{12.42}$$

where the corresponding parameters (B, C, D) were defined for the interactions with air as described in the cited article, $p_\perp$ is measured in GeV/c.

The ALTAI code was developed for the simulations in the energy range relevant for the very high energy γ-ray astronomy $E \leq 50$ TeV. In this energy range the mean number of kaons $(K^\pm, K^0)$ produced in hadron-nuclei interactions is very small compared with the number of emitted pions $(\pi^\pm, \pi^0)$. Besides this, in this energy region kaons and pions exhibit similar properties of inelastic interactions (see, e.g., Grishin 1982). For this reason, the kaon production process was excluded from simulation by introducing appropriate corrections for the probabilities of pion production in the $(p, n, \pi^\pm) \to X$ inelastic collisions.

For each inelastic hadron interaction (MULTIP) the leading particle is sampled. It is assumed that the energy is uniformly distributed within the range $(0, E_0)$

where E_0 is the energy of a primary interacting particle. Note that the type of the leading particle is sampled randomly amongst proton and neutron for interaction $(p, n) + A \rightarrow (p, n)$, and amongst charged pion or neutral pion for interaction $(\pi^{\pm}, \pi^0) + A \rightarrow (\pi^{\pm}, \pi^0)$. Thus the total inelasticity coefficient was determined as $\kappa = 1 - E/E_0$. At the next step we sampled multiple production of other secondary particles $(\pi^{\pm}, \pi^0)$. The type of secondary particle was randomized assuming that π^0 pions forms, on average, one third of all pions generated in hadron interactions. The energy of secondary particle was simulated using the spectra given by Eq. (12.38). We have completed the production of secondary particles when their total energy exceeds the energy of a primary particle. The energy of the last simulated particle was renormalized in order to conserve the total energy in each inelastic interaction.

The transverse momentum of the secondaries was simulated using Eq. (12.42). By the renormalization of the transverse momentum of leading particles we allowed to fulfill the momentum conservation for each hadron interaction without distortion of stochastic properties of a multi-particle production process. The test calculations have shown that such an approach describes very well the initial inclusive spectra of the secondary particles if the number of secondaries is relatively high ($\gg$1). Note that as almost identical algorithm was developed and tested by Barashenkov and Toneev [28].

Note that we have included in the code the propagation of a single muon generated in the hadron-nuclei cascade due to the $\pi^{\pm}$ decay (SGG, TRMUON). The emission of the Cherenkov light from the muons was included in simulations according to the scheme described above.

12.4.4 Nucleus-Nucleus Interactions

Implementing in the ALTAI code a model of independent nucleon interactions of colliding nuclei with nuclei fragmentation included (TAFRAC, TRAN, FRAG, OVERLAP), its authors assume that all nucleons of the projectile nucleus have the same energy determined as E_0/A_p. E_0 and A_p are the energy and atomic number of projectile nucleus, respectively. All nucleons of the projectile nucleus which overlap with the target nucleus interacted independently with each other. The non-overlapping part of the projectile nucleus decayed into individual nucleons and heavier fragments. The energy of fragment with the atomic number A is defined as $E_f = (E_0 \cdot A)/A_p$ and a random number of a fragments is sampled according to the probabilities of different channels summarized in Table 12.3.

12.4.5 Comparison with Other Codes and Data

An overview of Monte Carlo results on lateral, temporal, and angular characteristics of the Cherenkov light in air showers of 10 GeV–1 TeV was given by Konopelko

Table 12.3 Mean number of fragments created by different nuclei (Murzin 1988)

Primaries | Secondaries	$Z > 10$	C, N, O, F	Li, Be, B	He
$Z > 10$	0.17	0.29	0.26	1.34
C, N, O, F	–	0.11	0.24	1.00
Li, Be, B	–	–	0.15	0.51

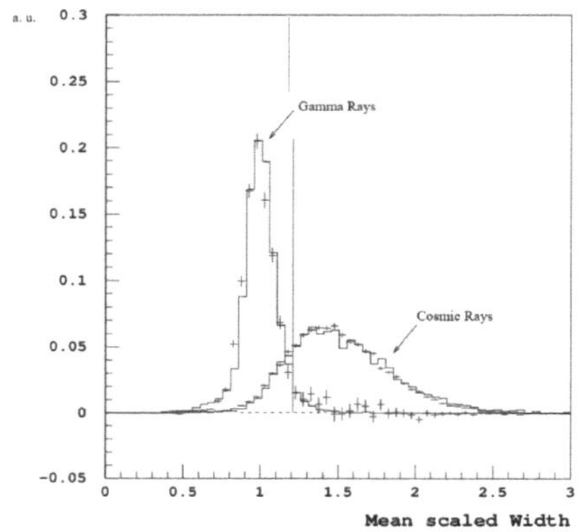

Fig. 12.10 Distribution of a shape parameter (mean scaled Width) of Cherenkov light images simulated with ALTAI code (histograms) and recorded with HEGRA system of IACTs (crosses)

[29]. Most of the Cherenkov light characteristics calculated with the ALTAI code reproduce rather well the results obtained with the MOCCA code [30]. The *stereoscopic* observations of BL Lac object Mkn 501 in 1997 with the HEGRA system of imaging air Cherenkov telescopes allowed the first measurements of the parameters of the Cherenkov light images from the γ-ray-induced air showers. The HEGRA data are in excellent agreement with the results of Monte Carlo simulations obtained with ALTAI code [31]. Using the Mkn 501 data sample comprising 38,000 γ-ray events we have tested in great detail the parameters of Cherenkov light image orientation and shape [32] (see Fig. 12.10) as predicted by the simulations.

The HEGRA stereoscopic system of 5 IACTs was used for the measurements of the lateral distribution of the Cherenkov light in the γ-ray-induced air showers. These measurements have been compared again with the Monte Carlo simulations using the ALTAI code [34]. The simulations using the ALTAI code reproduce very well the measured lateral distributions of the Cherenkov light. Note that the measured shape of the Cherenkov light lateral distributions is almost independent of the detector simulation procedure but strongly depends on the development in space of a multi-TeV γ-ray shower in the atmosphere.

We made several comparisons of the shape and size (total number of photoelectrons) of the Cherenkov light images for the proton- and nuclei induced air shower calculated with the ALTAI and CORSIKA codes [17]. The CORSIKA code was used with the HDPM model of the proton-nuclei interactions. This model is based on supercollider data and describes rather well the proton-nuclei air showers of

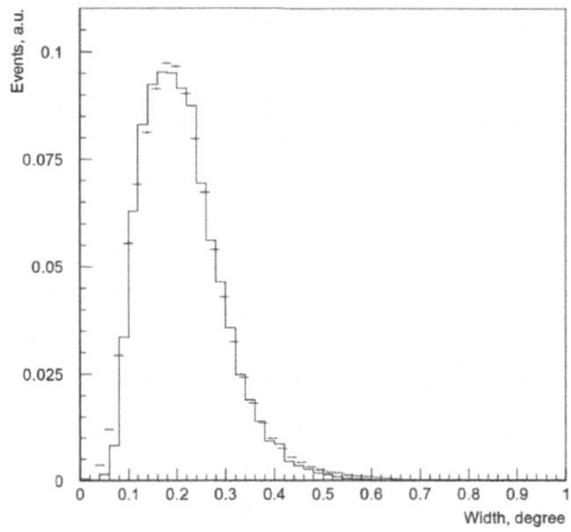

Fig. 12.11 Distribution of image shape parameter Width for the ALTAI simulated events (histogram) and cosmic ray images recorded with a 10 m Whipple IACT (crosses) [33]

energy below 10 TeV. In addition, the HDPM model of CORSIKA allows to perform relatively fast calculations with respect to other more modern shower generators, e.g. VENUS/QGSJET. Despite the different algorithms and schemes in the two codes the results appeared to be almost identical [35]. The simulations with the ALTAI and CORSIKA codes well reproduce each other. Note that simulations with the ALTAI code are essentially less time consuming. The HEGRA data for the cosmic ray air showers were compared with the ALTAI simulations. Good agreement between simulations and data provided a precise measurement of the cosmic ray proton spectrum in the energy range $\sim 1-5$ TeV.

We see in Fig. 12.10 the distribution of so called mean scaled Width parameter for primary γ-rays and cosmic rays extracted from the HEGRA data as well as from the Monte Carlo simulations. The simulations fit very well the data. Recently, we compared the relevant results of the ALTAI simulations with the cosmic ray data taken with a 10 m Whipple imaging air Cherenkov telescope during the 1995/1996 Crab Nebula observations [33]. For these simulations we have used the standard detector response functions offered by the Whipple collaboration. The simulations reproduced very well the shape of recorded Cherenkov light images (see Fig. 12.11).

As mentioned above the upper energy of the shower simulations is about 50 TeV. It is limited by the simplified phenomenological model of the hadronic cascade. Although in the case of pure electromagnetic showers one can extend simulations using the ALTAI code to much higher energies without breaking any model restrictions, such extension may need the introduction of a number of changes into the code for the correct treatment of a larger number of particles.

12.5 Semianalytical Monte Carlo Method

12.5.1 The SAMC Philosophy

The numerical method of kinetic problems, described above and called the Monte Carlo method (because of using random numbers) has become an effective means of theoretical analysis of astrophysical problems including high-energy cascade showers. Known virtues of this method are consistent accounting for space-time structure of showers and their fluctuations from shower to shower, playing the crucial fluctuations in its development, which often have a significant effect on the accuracy and reliability of observational conclusions. No hard problems arise in case of a nonhomogeneous medium, and the arrangement and properties of the measuring devices (detectors) may be easily involved into the computational process. Shortly speaking, MC-method realizes statistical modeling (simulation) of a real process. If the elementary probabilities prescribed to model events coincide with their natural counterparts, we talk on *analogous* modeling or *imitation*. For the sake of brevity, we will use for Analogous Monte Carlo Method abbreviation AMCM.

It should be noted however, that the AMCM has an obvious disadvantage limiting its use in analysis of high-energy cascade processes. This is the rapid (approximately linear) increasing in *computational time consumption* (CTC) with growing E_0 because of its branching nature.

The first group should include the approach (see, for example, [29, 36]), in which only the high-energy part of the cascade is modeled, containing particles with energy above some limiting value E^*. It is chosen so that the CTC would not be too large. Contributions of subcascades, generated by particles with energy lower than E^*, are obtained by alternative methods. These may be analytical or numerical methods, tabulated experimental ata or again AMCM.

The advantage of the second group schemes is the ability to take into account fluctuations in the development of the cascade process. The main disadvantage of this approach is that when solving problems, it requires consistent consideration of the properties of the detectors and inhomogeneity of the medium, as well as the memory volume for the storage of initial and current data.

In the schemes of the second group, the acceleration of calculations is achieved by cutting branches of the cascade process with compensation of the bias by introduction of the corresponding statistical weights. In particular, in the Van Ginneken scheme [37] such a cutoff is complete, i.e. the branching cascade process is replaced nonbranching trajectory of a single particle. In other codes (say, in the MOSSA [38], CASCADE [39] programs) partial cutting of branches is used. We will use for such hybrid modifications symbol HMCM. A successful version of this kind proved to be the *semianalytical Monte Carlo method* (SAMC) based on the combination of the analytical solution of the adjoint cascade equations describing the development of the high energy part of the cascade and the complete Monte Carlo simulations of the lower part of the shower. In the first work, represented SAMC [40], it was interpreted as an iteration method of sources, reflected in subsequent publications developing this method,

The basic idea of SAMC method is that all particles in EPC may be sorted with respect to their energy into two groups: *high energy* group and *low energy* group. In the first one, some special assumptions are declared to simplify the cascade equations making them more appropriate for standard analytical methods. The second, low energy group is simulated by using the complete Monte Carlo approach. Let us denote the energy separating these groups by E_m. Increasing this value yields a rapid reduction of the systematic errors caused by simplified treatment of the high energy part of the cascade. On the other hand, the CTC significantly increases with growth of E_m. Its optimum value depends on a nature of the characteristics under calculations and usually lies in the range of $1 \div 10$ GeV.

Each particle in the EPC is specified by the set of variables $x = (a, \mathbf{r}, \mathbf{p})$, where a stands for a type of particle(e,γ), $\mathbf{r}$, $\mathbf{p}$ designate its phase-space coordinates (radius- and momentum-3d vectors).

The EPC is such a structure, where new particles are born in collisions of moving particles with atoms (nuclei) of the medium, and in this sense, the collision points in the phase space can be considered as sources of these particles and operate with their distributions as with the phase density of off-cascade sources. The only, but very significant difference of internal sources from outer ones is that the formers (unlike the latters) are not statistically independent. Due to the Markovian nature of the cascade process, which reflects cause-and-effect relationships in an electron-photon system, it is more convenient in transfer theory to use the concept of the influence function $g(x \to x_1)$, rather than the correlation function, defining it as the phase density of those born at the point x_1 particles that are descendants of one particle—the ancestor of the entire cascade, born at the point x.

Consider now a process of measurement of some additive functional Q from a random realization of the cascade. Because of its additivity, the measurement result can be written as the sum

$$Q = Q_H + Q_L, \tag{12.43}$$

where Q_H is a contribution in it of particles with $E > E_m$, and Q_H is a total contribution of another, low-energy particles part (with $E < E_m$). For the same reason, the second term in this equation becomes the double sum

$$Q_L = \sum_b \sum_k q_b^{(k)}, \tag{12.44}$$

where the sum runs over all secondary particles born by particles with $E > E_m$ and having at that moment the energy $E < E_m$. The quantity $q_b^{(k)}$ determines the contribution into the measured functional (say, J) of a subcascade generated by b-particle with the number k.

Averaging Eqs. (12.43) and (12.44) over the statistical ensemble yields

$$\bar{Q} = \bar{Q}_H + \bar{Q}_L, \quad \bar{Q}_L = \sum_b \int\limits_{E<E_m} S_b(x)\bar{q}_b(x)dx, \tag{12.45}$$

where $S_b(x)$ is the mean number of low-energy b-particles born by high-energy particles in a unite phase volume in vicinity $x = (\mathbf{r}, \boldsymbol{\Omega}, E)$; $\bar{q}_b(x)$ is the mean contribution into J by the subcascade, generated by the low-energy b-particle born at x.

The function $S_b(x)$ in (12.45) has a meaning of the differential density of some "imaginary" (or, *equivalent*) source of low-energy b-particles. This analogy inspires advance calculating the contributions $\bar{Q}_L$ for large number of low-energy cascades, birth points and types whose primary particles are sampled in accordance with a set of differential densities $S_b(x)$.

12.5.2 Equations for Equivalent Sources

The differential densities $S_b(x)$ satisfy the special system of adjoint (backward) cascade equations. For sufficiently large values of E_m, it can be significantly simplified and solved analytically or numerically.

Let us discuss the method for deriving these equations, following the work [40]. Assuming, that the primary particle generated the cascade moves along the z-axis at the origin of the system, the primary and secondary particles energy will be measured in units $E_m : \varepsilon_0 = E/E_0, \; \varepsilon = E/E_m$. Taking into account only bremsstrahlung for electrons and pair-production for photons and ignoring the lateral displacements of the equivalent source particles, authors of [40] arrived at the following equations:

$$[\partial_t + \sigma_r(\varepsilon_0)]S_{eb}(\varepsilon_0, t, \varepsilon, \vartheta)$$

$$+ \int [S_{eb}(\varepsilon_0, t, \varepsilon, \vartheta) - S_{eb}(\varepsilon_0, t, \varepsilon, |\boldsymbol{\vartheta} - \boldsymbol{\theta}|)]\, W_c(\varepsilon_0, \boldsymbol{\theta})d\boldsymbol{\theta}$$

$$- \int\limits_1^{\varepsilon_0} W_r(\varepsilon_0, \varepsilon_0 - \varepsilon_0')S_{eb}(\varepsilon_0', t, \varepsilon, \vartheta)d\varepsilon_0' - \int\limits_1^{\varepsilon_0} W_r(\varepsilon_0, \varepsilon_0')S_{\gamma b}(\varepsilon_0', t, \varepsilon, \vartheta)d\varepsilon_0' = 0;$$

$$[\partial_t + \sigma_p(\varepsilon_0)]S_{\gamma b}(\varepsilon_0, t, \varepsilon, \vartheta) - 2\int\limits_1^{\varepsilon_0} W_p(\varepsilon_0, \varepsilon_0')S_{eb}(\varepsilon_0', t, \varepsilon, \vartheta)d\varepsilon_0' = 0. \tag{12.46}$$

In this equations, σ_p, W_p are the total and differential pair production cross section respectively, σ_r and W_r are the similar characteristics for the bremsstrahlung, W_c is the angle-differential cross section of Coulomb scattering. First subindexes relates to the particle generating the subcascade, the second one indicates a type of the low-energy secondary particle. The boundary conditions to solutions of these equations are of the form:

$$S_{ab}(\varepsilon_0, t = 0, \varepsilon, \vartheta) = G_{ab}(\varepsilon_0, \varepsilon)\delta(\theta)/2\pi\theta, \tag{12.47}$$

where

$$G_{ee}(\varepsilon_0, \varepsilon) = W_r(\varepsilon_0, \varepsilon_0 - \varepsilon), \quad G_{\gamma e}(\varepsilon_0, \varepsilon) = 2W_p(\varepsilon_0, \varepsilon),$$

$$G_{e\gamma}(\varepsilon_0, \varepsilon) = W_r(\varepsilon_0, \varepsilon), \quad G_{\gamma\gamma}(\varepsilon_0, \varepsilon) = 0.$$

In case of a homogeneous medium, Eq. (12.46) can be represented in a universal form independent of of material properties of the medium, if (1) to measure the depth t in cascade units, (2) to involve the full screening formulas for cross sections, and to use for scattering description the Landau approximation, containing the mean squared angle per c.u. $\overline{\Theta^2} \simeq (\varepsilon_s/\varepsilon_0)^2$, with $\varepsilon_s = K/E_m$ including the multiple scattering energy $K = 21$ MeV.

Introducing the universal angular variable $\theta = \vartheta/\varepsilon_s$ completes the transition to the desired universal form of the high-energy part of cascade system in dimensionless variables $t = x/X_0,\ \varepsilon = E/E_m,\ \theta = \vartheta/\varepsilon_s$:

$$\left[\partial_t + \sigma_r(\varepsilon_0) - (2\varepsilon_0)^{-2}\theta^{-1}\partial_\theta\theta(\partial_\theta)\right] S_{eb}(\varepsilon_0, t, \varepsilon, \theta)$$

$$-\int_1^{\varepsilon_0} W_r(\varepsilon_0, \varepsilon_0 - \varepsilon_0')S_{eb}(\varepsilon_0', t, \varepsilon, \theta)d\varepsilon_0' - \int_1^{\varepsilon_0} W_r(\varepsilon_0, \varepsilon_0')S_{\gamma b}(\varepsilon_0', t, \varepsilon, \theta)d\varepsilon_0' = 0;$$

$$\left[\partial_t + \sigma_p(\varepsilon_0)\right] S_{\gamma b}(\varepsilon_0, t, \varepsilon, \theta) - 2\int_1^{\varepsilon_0} W_p(\varepsilon_0, \varepsilon_0')S_{e,b}(\varepsilon_0', t, \varepsilon, \theta)d\varepsilon_0' = 0,$$

with the corresponding boundary condition

$$S_{ab}(\varepsilon_0, t = 0, \varepsilon, \theta) = G_{ab}(\varepsilon_0, \varepsilon)\delta(\theta)/2\pi\theta.$$

This system was solved by authors of [40] with the use of traditional (for EPC) technique of the integral transformations and inversion by the saddle-point method. The result was represented in the form

$$S_{ab}(\varepsilon_0, t, \varepsilon, \theta) = P_a(\theta; s, \varepsilon)S_{ab}(\varepsilon_0, t, \varepsilon).$$

Here s, as usual, denotes the cascade parameter, obeying the equation

$$\lambda_1'(s)t + \ln\varepsilon_0 = 0.$$

The functions

$$S_{eb}(\varepsilon_0, t, \varepsilon) = \frac{\exp[\lambda_1(s)t + s\ln\varepsilon_0]}{\sqrt{2\pi\lambda_1''(s)t[\lambda_1(s) - \lambda_2(s)]}}\left\{[\lambda_1(s) + \sigma_0]\hat{G}_{eb}(s, \varepsilon) + C(s)\hat{G}_{\gamma b}(s, \varepsilon)\right\}, \tag{12.48}$$

$$S_{\gamma b}(\varepsilon_0, t, \varepsilon) = \frac{B(s)}{\lambda_1(s) + \sigma_0} S_{eb}(\varepsilon_0, t, \varepsilon),$$

where $\hat{G}_{ab}(s, \varepsilon)$ are the Mellin transforms of the boundary conditions, and standard cascade functions relate now to the effective source. The multiplier

$$P_b(\theta; s, \varepsilon) = \frac{2f_1(s)}{\pi}\int_1^\infty \frac{[\lambda_1(s) + \sigma_0]\hat{G}_{eb}(\hat{\varepsilon}, \varepsilon)\tau_{s/2+1}[2\theta\hat{\varepsilon} f_1^{1/2}(s)]}{\left\{[\lambda_1(s) + \sigma_0]\hat{G}_{eb}(s, \varepsilon) + C(s)G_{\gamma b}(s, \varepsilon)\right\}\hat{\varepsilon}^{s+1}} d\hat{\varepsilon}$$

$$+ \frac{2f_1(s)}{\pi}\int_1^\infty \frac{C(s)\hat{G}_{\gamma b}(\hat{\varepsilon}, \varepsilon)\tau_{s/2}[2\theta\hat{\varepsilon} f_1^{1/2}(s)]}{\left\{[\lambda_1(s) + \sigma_0]\hat{G}_{eb}(s, \varepsilon) + C(s)\hat{G}_{\gamma b}(s, \varepsilon)\right\}\hat{\varepsilon}^{s+1}} d\hat{\varepsilon} \tag{12.49}$$

is independent of the primary particle type and energy normalized as

$$2\pi\int_0^\infty P_b(\theta; s, \varepsilon)\theta d\theta = 1.$$

Notice, that

$$\tau_\alpha(z) = \int_0^\infty \frac{J_0(zu)udu}{[1 + u^2]^\alpha} = (z/2)^{\alpha-1}\frac{K_{1-\alpha}(z)}{\Gamma(\alpha)} \tag{12.50}$$

and

$$f_1(s) = -\frac{s\lambda_1'(s)[\lambda_1(s) - \lambda_2(s)]}{\lambda_1(s) + \sigma_0}. \tag{12.51}$$

Clearly, the technique is applicable only in case of using formulas in the total screening approximation. In another case, the numerical method should be used.

12.5.3 Sampling From the Equivalent Source

After manipulating the equations for artificial sources, the latter turned out to be distributed along a beam coinciding with the axis of the main cascade, continuing the direction of the particle generating it. In this case, the differential densities S_{ab} can be represented as a function of four independent variables—depth, primary and secondary energy, as well as the angle between the shower axis and the particle emission direction. First, we discuss their sampling in the approximation ($\varepsilon_0 \to \infty$) introduced above, and then consider some corrections for a more accurate (numerical) solution of these equations.

Passing from the dept t to the cascade parameter s, let us consider the function

$$S_{a,b}(\varepsilon, s) = 2\pi |t'(s)| \int_{\varepsilon_t}^{1} d\varepsilon \int_{0}^{\infty} S_{ab}(\varepsilon_0, t(\varepsilon_0, s), \varepsilon, \theta)\theta d\theta, \tag{12.52}$$

determining the number of b-particles produced by a high-energy part of the cascade in a unit interval of the cascade parameter near the value s with an energy in $(\varepsilon_t, 1)$, where $\varepsilon_t = E_t/E_m$ is the energy tracking of low-energy particles in matter, measured in units E_m.

Obviously, the most natural way to obtain a random value of the cascade parameter of the born particle will be its sampling from a probability density proportional to the sum $S_{ae}(\varepsilon)$ assigning this particle a statistical weight

$$w_a(\varepsilon_0) = \int_{0}^{\infty} [S_{ae}(\varepsilon_0, s) + S_{a\gamma}(\varepsilon_0, s)]ds. \tag{12.53}$$

In this case, it is possible to completely avoid fluctuations of the statistical weights, because for any emitted low-energy particle, the value of weights (12.53) will be the same.

Investigations, described in the Plyasheshnikov thesis [10] have shown, that this method of drawing cascade parameter is good enough at calculating cascade characteristics near the shower maximum. At small and great depths it gives a rather large statistical error, because at $\varepsilon_0 \gg 1$ the functions $S_{ab}(\varepsilon_0, s)$ have a sharp maximum at the point $z = 1$, where particles predominantly appear. For this reason, the drawing of the cascade parameter was offered uniformly from some finite interval $(0, s_m)$, computing the weight as

$$w_a(\varepsilon_0, s) = s_m[S_{ae}(\varepsilon_0, s) + S_{a\gamma}(\varepsilon, s)]. \tag{12.54}$$

Integrating (12.52) with the use (12.48) yields

$$S_{eb}(\varepsilon_0, s) = \left[\frac{-\lambda_1''(s) \ln \varepsilon_0}{2\pi\lambda_1'(s)^3}\right]^{1/2} \frac{\varepsilon_0^{s-\lambda_1(s)/\lambda_1'(s)}}{\lambda_1(s) - \lambda_2(s)} \left[(\lambda_1(s) + \sigma_0)\hat{F}_{eb}(s, \varepsilon_t) + C(s)\hat{F}_{\gamma b}(s, \varepsilon_t)\right],$$

with

$$\hat{F}_{ab}(s, \varepsilon_t) = \int_{\varepsilon_t}^{1} \hat{G}_{ab}(s, \varepsilon)d\varepsilon,$$

and

$$S_{\gamma b}(\varepsilon_0, s) = \frac{B(s)}{\lambda_1(s) + \sigma_0} S_{eb}(\varepsilon_0, s).$$

The functions $S_{ab}(\varepsilon_0, s)$ show an evident way of sampling low-energy b-particles in accordance with probabilities

$$p_{ab}(\varepsilon_0, s) = \frac{S_{ab}(\varepsilon_0, s)}{S_{ae}(\varepsilon_0, s) + S_{a\gamma}(\varepsilon_0, s)}, \quad b = \{a, b\}.$$

In $\varepsilon_0 = \infty$—approximation, they don't depend on a primary particle energy and type:

$$p_{ae}(\varepsilon, s) = p_e(s) = 1 - p_\gamma(s),$$

$$p_\gamma(s) = \frac{[\lambda_1(s) + \sigma_0]\hat{F}_{e\gamma}(s, \varepsilon_t)}{[\lambda_1(s) + \sigma_0][\hat{F}_{ee}(s, \varepsilon_t) + \hat{F}_{e\gamma}(s, \varepsilon_t)] + C(s)\hat{F}_{\gamma e}(s, \varepsilon_t)}. \tag{12.55}$$

To draw the energy of a low-energy electron or photon, it is necessary to know the normalized energy spectrum at fixed values of the parameters ε_0 and s. It can be expressed in terms of differential density according to the formula

$$P_{ab}(\varepsilon; \varepsilon_0, s) = \frac{\int_0^\infty S_{ab}(\varepsilon_0, t, \varepsilon, \theta)\theta d\theta}{\int_{\varepsilon_t}^1 d\varepsilon \int_0^\infty S_{ab}(\varepsilon_0, t, \varepsilon, \theta)\theta d\theta},$$

which in the approximation under consideration takes the form

$$P_{ab}(\varepsilon; \varepsilon_0, s) \equiv P_b(\varepsilon; s) = \frac{[\lambda_1(s) + \sigma_0]\hat{G}_{eb}(s, \varepsilon) + C(s)\hat{G}_{\gamma b}(s, \varepsilon)}{[\lambda_1(s) + \sigma_0]\hat{F}_{eb}(s, \varepsilon_t) + C(s)\hat{F}_{\gamma b}(s, \varepsilon_t)}. \tag{12.56}$$

Note that the emission angle of a low-energy particle may be modeled by using the normalized angular distribution (12.49). Observe, that although the adjoint Eq. (12.46) and their solution are stationary, they do not contain information about the time of birth of low-energy cascading particles. Neverheless, the results can be applied for estimation of the EPC temporal characteristics under some minor simplifications.

12.5.4 Primary-Energy Correction

Algorithms for drawing the phase coordinates of particles of an "imaginary" source, described above are based on the use of asymptotical ($E_0 \to \infty$) formulas for source densities S_{ab}. To get more accurate data for them, Dr. Plyasheshnikov suggested to replace formula (12.54) by the product

$$w_{ab}(\varepsilon_0, s, \varepsilon, \theta) = s_m[S_{ae}(\varepsilon_0, s) + S_{a\gamma}(\varepsilon_0, s)]\delta u_{ab}(\varepsilon_0, s)\delta v_{ab}(\varepsilon_0, s)\delta w_{ab}(\varepsilon_0, s, \varepsilon, \theta),$$

where δu_{ab} and δv_{ab} are E_0-corrections to depth- and energy-distributions of the equivalent source respectively, obtained by applying the SBCE-method to solving the densities

$$S_{ab}(\varepsilon_0, t, \varepsilon) = 2\pi \int_0^\infty S_{ab}(\varepsilon_0, t, \varepsilon, \theta)\theta d\theta.$$

The third factor δw_{ab} is an E_0-correction to angular distribution form by using a more accurate value for $\overline{\theta^2}$, so that

$$\delta w_{ab}(\varepsilon_0, s, \varepsilon, \theta) = \frac{P_a(\mu_{ab}\theta; s, \varepsilon)}{P_a(\theta; s, \varepsilon)}\mu_{ab}^2, \tag{12.57}$$

with the function $P_a(\theta; s, \varepsilon)$ defined by Eq. (12.49) and the scaling factor μ_{ab^2} determined in terms of angles calculated with and without taking into account the finiteness of ε_0:

$$\mu_{ab}(\varepsilon_0, s, \varepsilon) = \left[\frac{\overline{\theta^2}(\infty, s, \varepsilon)}{\overline{\theta^2}(\varepsilon_0, s, \varepsilon)}\right]_{ab}^{1/2}. \tag{12.58}$$

The numerical values of these corrections are shown in Tables 12.4 and 12.5 [10].

Table 12.4 Function value $\boldsymbol{\delta u}_{\gamma e}(\varepsilon_0, s)$

	s							
ε_0	0.1	0.2	0.3	0.4	0.7	1.0	1.45	1.90
2	5.643	3.509	2.643	2.172	1.541	1.317	1.216	1.153
4	3.884	2.276	1.684	1.393	1.077	1.015	1.026	1.020
10	1.970	1.230	1.030	0.964	0.949	0.985	1.024	1.018
30	0.908	0.821	0.869	0.911	0.972	1.002	1.019	1.012
10^2	0.584	0.798	0.907	0.950	0.985	1.003	1.014	1.000
10^3	0.640	0.907	0.960	0.973	0.989	1.002	1.009	1.006
10^4	0.775	0.942	0.968	0.977	0.991	1.002	1.006	1.003
10^7	0.910	0.964	0.978	0.984	0.993	1.000	1.003	1.002

Table 12.5 Function value $\delta v_{ee}(\varepsilon_0, s, \varepsilon)$

ε_0	s								
s	4			10			10^2		
ε	0.4	1.0	1.9	0.4	1.0	1.9	0.4	1.0	1.90
0.95	0.950	1.050	1.010	0.920	1.030	1.010	0.972	1.020	1.000
0.75	1.020	1.020	0.989	0.939	1.010	0.993	0.995	1.000	0.995
0.50	1.020	0.965	0.990	1.020	0.991	0.993	1.010	0.994	0.996
0.20	0.995	0.957	1.000	1.040	0.977	1.000	1.010	0.988	1.000
0.10	0.962	0.950	1.010	1.040	0.974	1.000	1.010	0.986	1.000
0.03	0.973	0.947	1.010	1.040	0.371	1.000	1.010	0.985	1.000

12.5.5 The SAMC Errors and CTMC Estimation

Among the important problems to be addressed in the course of SAMC application, the most important one is choosing an optimal value of the separation energy $E_{\max}$. It is clear that some compromise should be found in this interplay between the error growing with $E_{\max}$ decrease, and the CTC growing with $E_{\max}$ increase. The best way to solve this problem is a numerical experiment. The trial calculations have shown, that the relative error of the longitudinal characteristics development will not exceed a few percent at depths $t < 3t_{\max}$ ($t_{\max}$ is shower maximum depth), if to choose $E_{\max} \geq 40E_{\text{cr}}$.

To study the influence of the $E_{\max}$ value on the accuracy of characteristics, related to scattering of shower particles, angular and radial distributions of electrons EPC in the atmosphere with energies above $E_{th} \in (1 \div 25)$ MeV in the range of values parameter $E_{\max} \in (0.1 \div 3t_{\max})$ GeV, primary energies $E_0 \in (10^2 \div 10^6)$ GeV and depths $t \in (0.3 \div 3)t_{\max}$, it turned out that for $E_{\max} \geq 10E_{\text{cr}}$ the angular distribution is almost independent of the separation parameter.

As to lateral distributions, the following restrictions for $E_{\max}$ values region have been established:

$$E_{\max} \geq 10E_{\text{cr}}, \quad (\overline{r^2{}_a})^{1/2}(E > E_{\max}) \leq r_{\min}/3,$$

where $r_{\min}$ is the shortest radial distance from that under consideration, and

$$(\overline{r^2{}_a})^{1/2} \approx Kt_0.$$

In addition to the systematic error mentioned above, the results of calculations according to the SAMC scheme, as well as any other calculations using random numbers, also contain a statistical error. Numerical analysis showed that the depth-dependence of the relative statistical error δ(standard error/mean value) for cascade curves, depth distributions of absorbed energy, angular and radial distributions of the number of particles and absorbed energy, root-mean-square angles and radii, etc. takes the least

Table 12.6 Ratio (AMC CTMC)/(SAMC CTMC) for Plyasheshnikov scheme (left column) and (van Ginneken)/(Plyasheshnikov) right column. Air, $E_{th} = 4$ MeV, $E_\gamma = 3$ Gev

E_0/E_{th}	$3 \cdot 10^4$	$3 \cdot 10^5$	$3 \cdot 10^3$	$3 \cdot 10^5$	$3 \cdot 10^6$	$3 \cdot 10^7$	$3 \cdot 10^8$
$t/t_m = 0.5$	10	60	250	3	5	15	40
$t/t_m = 1$	2	6	12	2	5	15	25
$t/t_m = 2$	11	80	500	4	6	10	15

its value near the shower maximum. Up to $t = 2t_{\max}$, δ exceeds its smallest (in the maximum) value no more than twice.

Unlike the analog Monte Carlo method, the SAMC statistical error δ weakly depends on the primary energy. With growth E_0 by ten times (with the same number of particles emitted by the "imaginary" source) δ increases by approximately $10 \div 15\%$. For angular and radial distributions their δ minimal values are usually $1.5 \div 2$ times greater than the cascade curve δ at the same depth and in rather wide region differ from their least values no more then $20 \div 30\%$.

Comparison of SAMC CTMC with the ordinary AMC CTMC was carried out on cascade curves by using formula $T = \delta^2\tau$ (see, for example, [10]), where τ is time spent on modeling one cascade. Some results of this comparison with data of van Ginneken [37] are represented in Table 12.6. It shows, that for $E_0/E_{th} \geq 10^4$ SAMC has a higher speed of computations, moreover, the difference in speeds increases with increasing E_0/E_{th} and at $E_0/E_{th} \sim 10^9$ (primary energy $E_\gamma \simeq 10^6$ GeV) reaches a few tens.

In the process of intensive adaptation of the SAMC to astrophysical problems and supplementing it with SBCE-code, the computational code ALTAI was created [41].

To the end of this section, we show results of the first MC calculation performed, before creating the Altai code. The region of integrating was divided into rectangles $\Delta t_i \times \Delta E_j$, and the integral itself was represented as a sum of their contributions:

$$N = \sum_{i,j} a_{ij} R_{ij},$$

where

$$a_{ij} = \frac{1}{\Delta t_i \Delta E_j} \int\limits_{\Delta t_i} dt \int\limits_{\Delta E_j} dE a(E, t),$$

and

$$R_{ij} = -\int\limits_{\Delta t_i} dt \int\limits_{\Delta E_j} dn(E, t)$$

is the path of lighting electrons in the rectangle $\Delta t_i \Delta E_j$.

The choice of interval boundaries is made taking into account the specific form of the function $a(E, t)$. Near the threshold energy, the length of the energy intervals

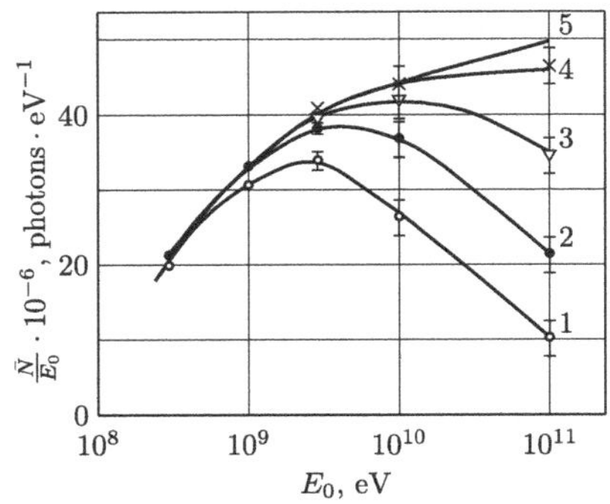

Fig. 12.12 Dependence of the average number of Cherenkov photons on the primary energy for different observation heights: 1 – 200 g · cm^{-2}, 2 – 300 g · cm^{-2}, 3 – 400 g · cm^{-2}, 4 – 616 g · cm^{-2}, 5—sea level

is 10 MeV, then it increases, and starting from $E = 10^3$MeV, the distance between the boundaries is an order of magnitude. 18 × 20 elements of the matrix a_{ij} were calculated by the Monte Carlo method according to the usual scheme for calculating the multidimensional integral. The dependences of temperature and other characteristics of the atmosphere on height, included in $a(E, t)$, are determined from the data of the ARDC model [42], and the refractive index of air is determined from the formula given in [43].

The vertical electron-photon showers in the atmosphere come from primary photons with energies $E_0 = 3 \cdot 10^8, 10^9, 3 \cdot 10^9, 10^{10}, 10^{11}$eV. For each value of E_0, a series of 15 showers were calculated. For each of the constructed showers, the number of Cherenkov photons with a wavelength 3000 Å $< \lambda <$ 6000 Å emitted from the beginning of the shower to the observation level t_k,

$$N(E_0, t_k) = \sum_{i=1}^{k} \sum_{j=1}^{20} a_{ij} R_{ij}.$$

Then, for each series, the average value of this number $\overline{N}(E_0, t_k)$ and relative fluctuations are found

$$\delta(E_0, t_k) = \frac{\sigma[N(E_0, t_k)]}{\overline{N}(E_0, t_k)}.$$

The calculation results are shown in Figs. 12.12 and 12.13. For $E_0 = 10^{11}$eV, our results for the average value of $(5.04 \pm 0.09) \cdot 10^6$ photons at sea level and $(4.67 \pm 0.17) \cdot 10^6$—at a height of 3860 m are consistent with the data of [44] ($4.8 \cdot 10^6$ and $4.75 \cdot 10^6$ photons, respectively).

It is not difficult to give a physical explanation for the obtained energy dependences of $\overline{N}/E_0$ and δ. As the energy of the primary photon increases, the range of all shower electrons increases in proportion to the energy. However, the range of "luminous" electrons increases more rapidly, since with an increase in the primary energy the shower reaches denser layers of the atmosphere, where the threshold energy of the Cherenkov radiation is lower. Relative fluctuations decrease in this case. As soon as an

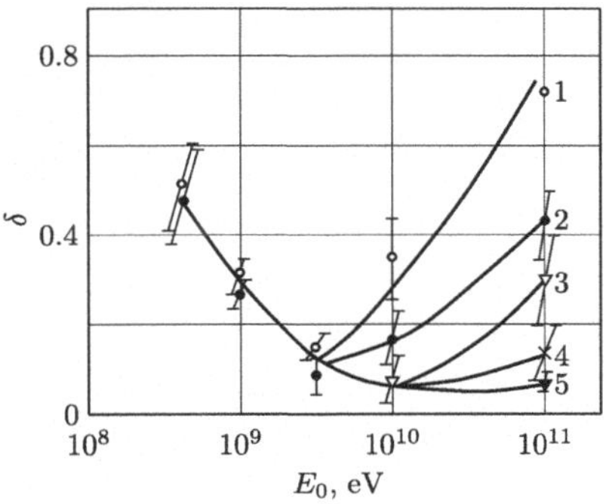

Fig. 12.13 Dependence of relative fluctuations of Cherenkov radiation on primary energy. The designations are the same as in Fig. 12.12

appreciable fraction of the range of "luminous" electrons falls below the observation level, the opposite tendency arises: as E_0 increases, the fraction of the Cherenkov shower radiation observed at this level decreases, while the fluctuations increase.

Later the codes GEANT, EGS, FLUKA, MOCCA, KASCADE, CORSIKA, ALTAI were built and involved into ground-based astrophysical experiments processing.

A Few Words of the ALTAI Code

The latter, ALTAI[2] has been elaborated by a scientific group, whose work on this project was primarily initiated and substantially advanced at the Tomsk Technological University, Tomsk, Russia, under the direction and with participation of Prof. A. M. Kolchuzhkin, to whose memory this book is dedicated. The composition of this group, as well as its location, has changed over time. The first result in the frame of this project was represented in [45, 46].

The most part the ALTAI code was developed and programmed at the Altai State University, Barnaul, Russia. Most of this project was carried out at ASU, and its completion (in the astrophysical application) was performed by A. V. Plyasheshnikov and A. K. Konopelko during their years at the Max-Planck-Institut fuer Kernphysik, Heidelberg (the end of 1990th—the beginning 2000th). In the final review of the project, they write: "The authors thank Dr. K. V. Vorobjev, Prof A. M. Lagutin, Dr. V. A. Litvinov and Prof V. V. Uchaikin for their valuable input and support of this activity. The computational code ALTAI was used and further developed at the Max-Planck-Institut fuer Kernphysik, Heidelberg. The authors would like to acknowledge the contribution and support of all members of the Heidelberg group in particular Prof F. Aharonian, Dr. M. Hemberger, Prof W. Hofmann, J. Kettler and Prof H. J. Voelk. AKK thanks Prof T. C. Weekes for support and supervision of a short-term project at the University of Arizona, Tucson, and at the Whipple Observatory, Harvard-Smithsonian Center for Astrophysics, Amado".

---

[2] It is the abbreviation for Atmospheric Light Telescope Array Image; Mountain Altai is the pristine wilderness in the south-west of Russia.

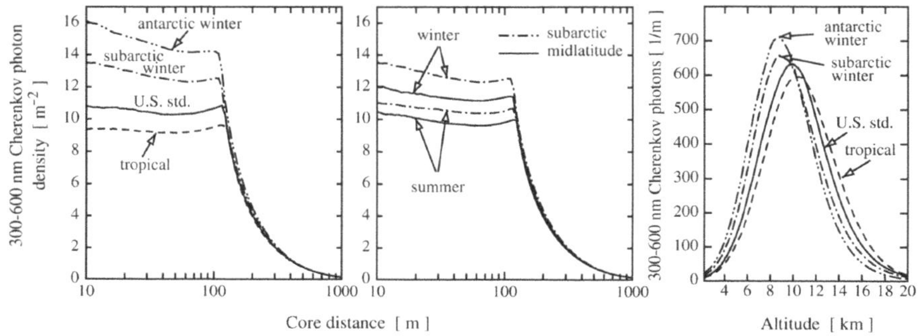

Fig. 12.14 Left and middle: Average lateral distributions of Cherenkov light photons in the wavelength range 300–600 nm for vertical 100 GeV gamma-ray showers in CORSIKA 5.71 simulations with different atmospheric profiles (2000 showers simulated for each profile). Absorption of Cherenkov light is taken into account. Observation altitude is 2200 m above sea level. Left panel: models from tropical to antarctic winter. Middle: Seasonal variations. Right: Longitudinal distribution of Cherenkov light emission [47]

Among the other existing codes intended for such simulations, MOCCA (Hillas 1979 [12]), KASCADE [16], CORSIKA [17], this code has a distinctive advantage—it's rather high speed of shower simulations due to a specific algorithm used for processing the multiple scattering of charged shower particles. This approach does not consume a lot of CPU time and allows to perform fast and accurate simulations. Together with the additional routine recently developed for the simulations of telescope response (Hemberger 1998) the ALTAI code was extensively used for production of a standard Monte Carlo database used in the HEGRA (High Energy Gamma Ray Astronomy) VHE γ-ray experiment (Konopelko et al. [13]) (Fig. 12.14).

12.6 The Lateral Light Distribution

The first simple calculations relating to Cherenkov radiation in atmosphere gave the following formula for the radius of the Cherenkov light-cone at sea level:

$$r = (5.8 \cdot 10^{-4})^{1/2} h \cdot \exp(-h/2\lambda). \qquad (12.59)$$

It has a maximum value at $r = 126$ m for the light coming from the height $h_{\max} = 2\lambda = 14$ km above the sea level. Reference [48] claimed for the first time on existence of a peculiarity (hump) in the lateral distribution of the Cherenkov light photons at the radius from the shower axis $r \sim 120 \div 140$ m and rapid decrease in the photon density beyond the distance 140 m from the shower core.

More precise calculations taking into account the Coulomb scattering of the electrons have been performed by [48]. They derived analytical formulas for the Cherenkov light contribution from different heights above sea level to the detector

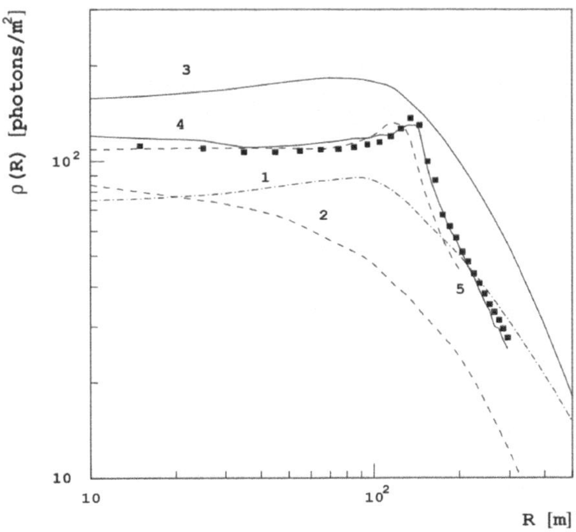

Fig. 12.15 Lateral distribution function of Cherenkov photon density, $\rho(r)$, in 1 TeV γ-ray-induced air shower at sea level (wavelength range is of $\Delta\lambda = (0.3 \div 0.6)\mu m$) calculated by [49] (curve 1), [50] (curve 2); [51] (curve 3); [30] (curve 4); [40] (curve 5). Black dots show the ALTAI results from [29]

located at different impact distances from the shower core up to 1 km. It was noted for the first time that the Cherenkov light intensity at sea level could be considered as a superposition of two components: the *direct emission* from the high energy electrons moving along the shower core, when the following relation is valid, $\Theta_{\mathrm{Ch}} > \langle\theta_s^2\rangle^{1/2}$ (the mean-squared angle of the electron scattering), and the *diffuse emission* from the low energy electrons characterized by the condition $\Theta_{\mathrm{Ch}} < \langle\theta_s^2\rangle^{1/2}$. The direct emission mostly contributes at the distances from the shower core $r \simeq 100 \div 140$ m whereas the diffuse emission generated at greater ages of the shower development in denser layers of atmosphere is distributed almost homogeneously over a large area around the shower core.

The numerical calculations of the Cherenkov light in EAS carried out by [49] gave the first systematic and rather precise theoretical predictions on the different characteristics of the Cherenkov light generated in air showers. The results were renewed later by the Monte Carlo simulations (Fig. 12.15).

Note, that calculating the space-time distribution of the Cherenkov light of showers is an essentially more complicated computing problem, having no simple analytical solutions. Taking into account multivariate nature of irregular information receiving from various lines of a measuring complex transforming the coming radiation into information flows directed to computer systems that provide continuous processing, cannot be denied, that only the Monte Carlo method can do such an amount of work (Fig. 12.16).

All these and other numerical calculations show a prominent hump at a distance of about 135 m from the core. The nature and origin of this hump are investigated analytically as well as through Monte Carlo calculations and it is shown to be due to electrons of energy higher than 1 GeV, above which the root mean square scattering angle is less than the Cherenkov emission angle. The conditions under which such

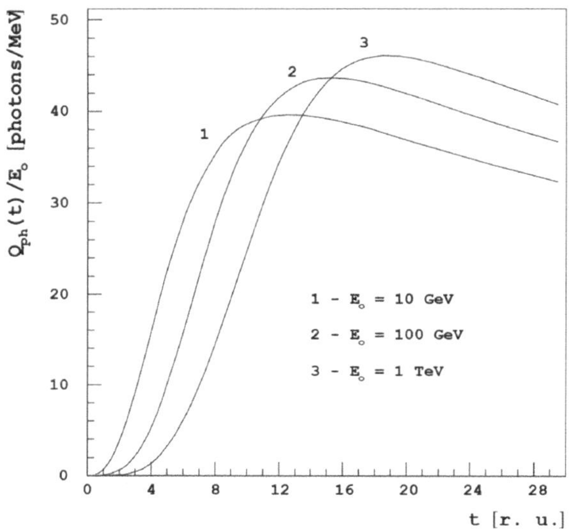

Fig. 12.16 An average number of Cherenkov light photons Q_{Ch}, in 10 GeV, 100 GeV and 1 TeV γ-ray air shower at different depths of cascade development depth (Monte Carlo simulations)

a hump can arise are discussed in [36]. Noting, that such a hump in the lateral distribution was noticed in the calculation of [37, 49], but not in that of Browning and [36, 38] paid special attention to this phenomenon.

Let's take a closer look at their argumentation. It was realised long ago, the authors write, that a hump might be present in the Cherenkov lateral distribution in gamma-ray cascades since the core distance, r, at which the Cherenkov photons arrive is equal to $h\theta_C$,, where h is the height of production of Cherenkov photons and Θ_C, the Cherenkov emission angle, and the increase in h is, to a large extent, compensated by the decrease in θ_c. Photons produced at altitudes of about 7–20 km arrive at a distance of about 130 m from the core. However, it was thought that Coulomb scattering and deflection of electrons in the Earth's magnetic field would remove the hump [39].

Rao and Sinha have reasoning as follows. High-energy electrons, once created, lose energy continuously as they travel down in the atmosphere, mainly by the bremmstrahlung process, thereby increasing their cumulative RMS scattering angle ($\propto 1/E$) and emitting Cherenkov radiation all the time until they fall below the Cherenkov threshold. The Cherenkov angle also increases as the particle travels down since the atmospheric density and hence the refractive index increase with depth, but this increase is much slower ($\propto \sqrt{x}$ compared with e^x for scattering). There is a certain energy and depth above which the cumulative RMS scattering angle is less than the Cherenkov angle at that depth and, in this region, the emitted photons arrive at a core distance nearly equal to $h\theta_c$, at the observation level, whereas the photons emitted by the lower-energy electrons spread out. If the contribution to the density of photons at that distance from these high-energy electrons is comparable to or larger than that due to lower-energy electrons, then a hump will appear at that distance. Maximum contribution from electrons of a certain energy comes from the depth where they reach maximum since the number of photons emitted per

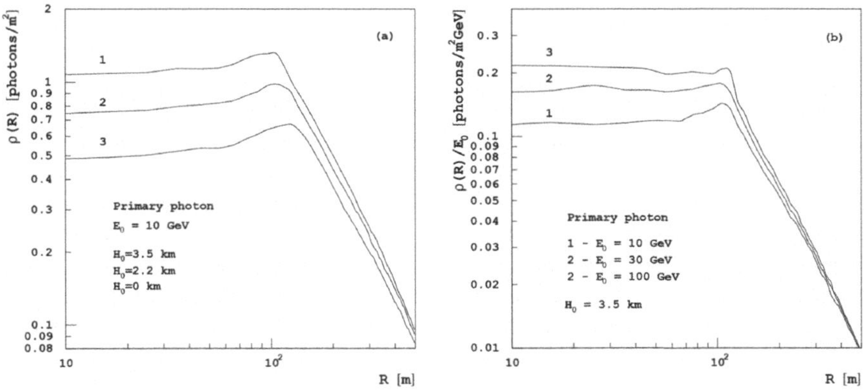

Fig. 12.17 Lateral distribution density of Cherenkov photon density $\rho(r)$ at sea level from 10 GeV-photon-induced shower calculated at three different heights above sea level (left panel). Normalized density of Cherenkov light photons, $\rho(r)/E_0$, in EAS from a photon with primary energies 10 GeV, registered at three levels (left panel). Calculated with ALTAI code

Fig. 12.18 Lateral distribution at different atmospheric depths in gamma-ray cascades of energy 100 GeV: 540 g·cm^{-2} (**a**), 700 g·cm^{-2} (**b**), 800 g·cm^{-2} (**c**), 920 g·cm^{-2} (**d**) and 1030 g·cm^{-2} (**e**) (wide-angle detector). The Rao and Sinha calculations [36]

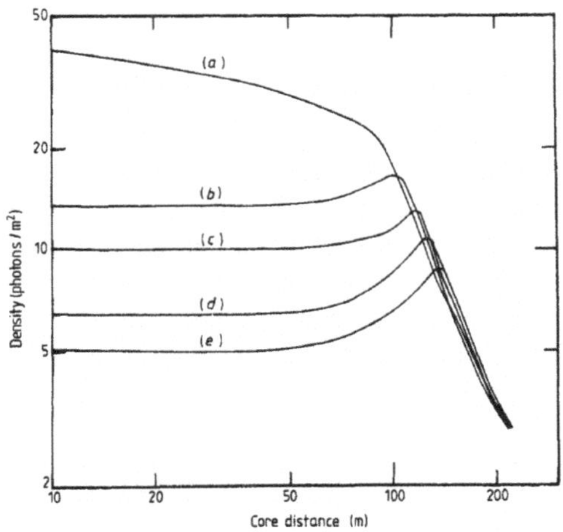

gcm^{-2} is nearly independent of the depth in the atmosphere (it varies only from 300 to 330 for atmospheric depths between 10 and 500 g cm^{-2}). Therefore the problem of identifying the depth and energy of electrons responsible for the hump reduces to finding the depth of maximum of such electrons for which the cumulative scattering angle and the Cherenkov angle are about equal (Fig. 12.17).

The hump in Fig. 12.18 is present at all depths, though the prominence is decreasing with depth, except at 540 g cm^{-2} where the shower maximum is too close. The hump completely disappears at the depth of 540 g·cm^{-2} as at this altitude the contribution from higher-energy electrons closer to the observation level becomes appreciable at near-core distances. As the primary gamma-ray energy increases, the altitude at which the hump disappears decreases slowly. Thus the hump can be seen at mountain altitudes also, except at very high altitudes and not too high primary energies.

Looking at both these figures, one can see, that as the altitude of observation increases the shower maximum for a given primary energy comes closer to the observation level and the situation is similar to the case of a fixed observation level but with increasing primary energy. Thus we expect the hump to be present but with its prominence decreasing with altitude. In addition, the core distance at which the hump appears becomes smaller with altitude as the height to the maximum decreases. Figure 12.18 shows the lateral distribution at different atmospheric depths for vertically incident gamma-ray showers of energy 100 GeV and these features can be clearly seen. The hump completely disappears at the depth of 540 g·cm^{-2} as at this altitude the contribution from higher-energy electrons closer to the observation level becomes appreciable at near-core distances. As the primary gamma-ray energy increases the altitude at which the hump disappears decreases slowly. Thus the hump can be seen at mountain altitudes also, except at very high altitudes and not too high primary energies.

12.7 Fluctuations in the Cherenkov Light Amplitude

An important issue which arises when detecting Cherenkov light from air showers is the fluctuations in a number of registered photons in individual events. These fluctuations a act temporal and angular distributions of the observed light ashes. Naively one can assume that the fluctuations in photon amplitude have a pure Poisson nature. In fact the Monte Carlo simulations [35] revealed much larger fluctuations than one can expect from Poissonian statistics. The results of Monte Carlo simulations for photon-induced EPC of energy 10, 50, 100 GeV at 2.2 km above the sea are shown in Fig. 12.19.

The amplitude fluctuations depend on primary energy, impact distance from the shower as well as on the area of the optical reflector. The following expression is relevant:

$$\delta_N^2 = \sigma_N^2/\bar{N}^2 \gg 1/\bar{N}. \tag{12.60}$$

In reality optical detectors collect the Cherenkov light photons coming form the air shower not independently but by a certain number of Cherenkov light "clusters". The photons belonging to a cluster come from the same emitting cascade electron and are highly correlated in time-angle space. Thus, the observed photon amplitude is a sum over clusters with various photon numbers:

$$N = N_1 + \cdots + N_n \simeq \bar{N}_1 n. \tag{12.61}$$

Simplifying on the assumption on the cluster independence yields

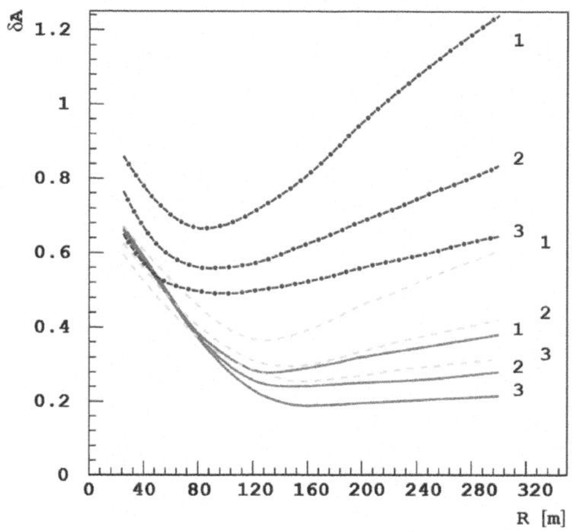

Fig. 12.19 Fluctuations in a number of detected Cherenkov photons registered by the optical detector with the mirror size of 250 m^2 (solid), 50 m^2 (dashed), 10 m^2 (dotted-dashed), from photon-induced EPC with $E_0 = 10$ GeV (1), 50 GeV (2), and 100 GeV (3)

$$\delta(N) = \bar{N}_1/N = 1/n. \tag{12.62}$$

This can roughly explain the high level of the fluctuations (compared, say, with the Poisson law) and also their dependence on the energy, impact distance and area of the optical reflector $\delta^2(N)(E, R, S)$. It means that fluctuations of photon amplitude depend in general on the number of *detected clusters*. The high energy electrons in the cascade maximum contribute in the impact energy range ~125 m that provides a large number of independent clusters whereas at small impact distances from the shower core photons come mainly from a few electrons at the cascade tail.

12.8 Time Structure of the Cherenkov Signal

The time pulse of the Cherenkov light measured at a certain distance from the shower core strongly correlates with the longitudinal development of the shower itself. The simultaneous registration of Cherenkov light time pulses by a number of detectors scattered over a large area at the observation level is usually used for accurate reconstruction of the arrival direction of multi-TeV air showers (e.g. AIROBICC, Themistocle). At the same time the structure of the measured Cherenkov light pulses could be used for the separation of γ-ray-induced air showers from cosmic ray air showers. For the first time Tümer et al. [52] applied this technique experimentally by sorting out the pulses with irregular shape (multi-hump events). The Monte Carlo studies performed by using ALTAI-Code [41] have shown that fluctuations of the pulse shape are much lower for EPC compared with EAS. This difference ultimately could be used effectively for the shower classification. Recently the data acquisition systems based on the Flash ADC readout (see for detail [53]) make it possible to measure simultaneously the Cherenkov light pulses in each channel of a multi-channel

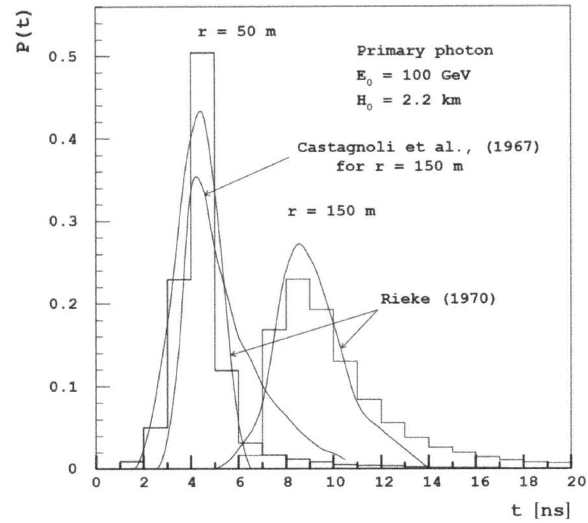

Fig. 12.20 Time pulses of Cherenkov light photons in 100 GeV γ−induced. Curves represent earlier results of authors indicated, histograms are obtained by means of ALTAI-Code

imaging camera installed at the telescopes designed to observe γ-rays of 500 GeV. This technique makes straightforward testing the efficacy of the CR rejection based on timing information. For the forthcoming low energy Cherenkov detectors the timing technique could also be used. In this concern it is interesting to look for the time pulses from 10 GeV air showers at different impact distances from the shower core in observation plane.

The exact treatment of the time-structure of the Cherenkov light in air showers using analytical methods usually fails because it needs to take into account the curvature and thickness of shower particle front as well as the Cherenkov light emission angle in respect to the track of the charged particles. The latter is actually the subject of independent studies (see for detail [51]). The Monte Carlo simulations could serve for that.

The first Monte Carlo calculations by γ-induced EPC of 100 GeV predicted by [50] demonstrated significant differences in the predicted absolute time delays and in the pulse shape (see Fig. 12.20). For comparison simulations under mostly the same conditions have been carried out using the ALTAI Code (see Figs. 12.20, 12.21). Figure 12.20 shows a good agreement of the results with calculations of [50]. Note that Cherenkov time profiles of EPC from photons with primary energies 10^{15}, 10^{16}, 10^{17} eV were calculated by [37].

For better understanding the Cherenkov pulse structure, one should notice some its details. First, if we ignore the EPC lateral widening, all electrons will move along the shower axis, and the time-delay of the Cherenkov photons emitted by some point at this axis with respect to the electron motion along the shower core is expressed as

$$t(R) = \int \frac{dl}{v(l)} - \frac{h}{c}, \tag{12.63}$$

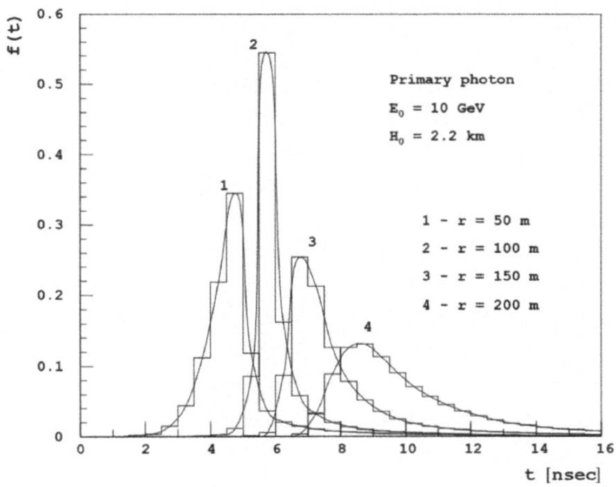

Fig. 12.21 Average Cherenkov light time pulses calculated at different distances from shower core in an air shower induced by γ-ray of 10 GeV. The observation level is at 2.2 km above the sea

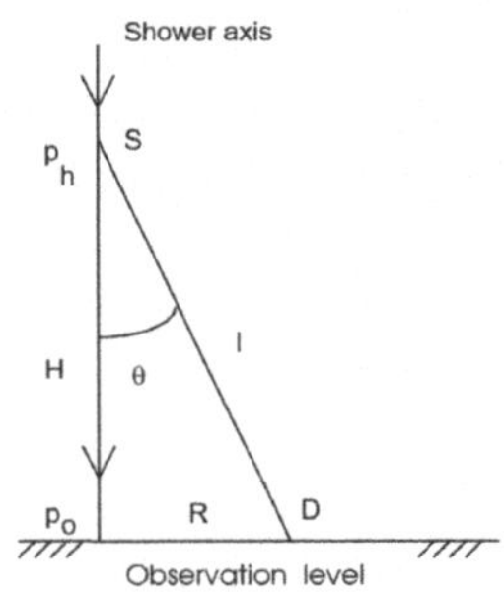

Fig. 12.22 The geometry of Cherenkov radiation-registration

where R is the impact distance between the detector and the shower core, h stands for the height above the observation level, and $v(l)$ is the speed of light on the way to the detector (see Fig. 12.22). This speed depends on the refractive index, and the latter is linked with the density of the air, which varies with height h. Thus, using the exponential approximation of the atmosphere density yields

$$v(h) = c/n = c/[1 + (\eta_0/\rho_0)\rho(h)], \tag{12.64}$$

where ρ_0 is the density at sea level. For a gamma-ray-induced EPC the time delay at the observation level becomes $t(r) \propto \sqrt{h^2 + R^2}$.

As follows from Eq. (12.64) the Cherenkov light emitted by the cascade electrons deeply penetrating in the atmosphere will be detected earlier than the photons coming from the upper layers, if the detector is not far from the shower core ($R \leq 150$ m). The situation changes to the opposite, when the detector is placed at larger impact distances.

The exact treatment of the time structure of the Cherenkov light in air showers using analytical methods usually 0.2 [54] fails because it needs to take into account the curvature and thickness of shower particle front as well as the Cherenkov light 0.1 emission angle in respect to the track of the charged particles. The latter is actually the efficiency of cosmic ray rejection based on timing information.

The time pulse of the Cherenkov light measured at a certain distance from the shower core strongly correlates with the longitudinal development of a shower. The simultaneous registration of Cherenkov light time pulses by a number of detectors scattered over a large area at the observation level is usually used for accurate reconstruction of the arrival direction of multi-TeV air showers (e.g. AIROBICC, Themistocle). At the same time the structure of the measured Cherenkov light pulses could be used for the separation of-ray-induced air showers from cosmic ray air showers. For the first time [52] applied this technique experimentally by sorting out the pulses with irregular shape (*multi-hump events*). The Monte Carlo studies of Konopelko et al. [41] have shown that the fluctuations of the pulse shape are much lower for γ-ray-induced showers compared with other showers. This difference ultimately could be used effectively for the shower classification. The data acquisition systems based on the Flash ADC readout (see for detail [53]) make it possible to measure simultaneously the Cherenkov light pulses in each channel of a multi-channel imaging camera installed at the telescopes designed to observe γ-rays of 500 GeV. This technique makes straightforward testing the efficacy of cosmic ray rejection based on timing information.

For the forthcoming low energy Cherenkov detector, the timing technique can also be used. In this concern, it is instructive to look for the time pulses from 10 GeV air showers at different impact distances from the shower core in the observational plane.

12.9 Cherenkov Image of EPC

For the first time [54] made a photograph of the Cherenkov light ash from an air shower. Since then a lot of experimental and theoretical studies have been done for understanding the two-dimensional angular distribution (image) of the Cherenkov light from air showers. The special interest in that is associated with a dramatic γ-ray astronomy based on the detection of the Cherenkov light images with the optical reflectors.

Nowadays, the arrays of imaging atmospheric Cherenkov telescopes (IACTs) are created for this invention. The effective detection area of stereoscopic multi-IACT systems is determined essentially by the total geometrical area of the array which may be covered by independent standard > 3 telescope ‘cells’ with a linear size of about 100 m. The idea of the ‘cell’ makes straight forward the study of the characteristics of IACT arrays. For an individual shower the data set consists of Cherenkov images registered by n IACTs. In the second-moment approach the images are described by several parameters which determine the position,orientation, shape and brightness of the image in the camera. For example, Width, Length parameters are the rms spread of light along the minor and major axes of the image, respectively; Distance is the

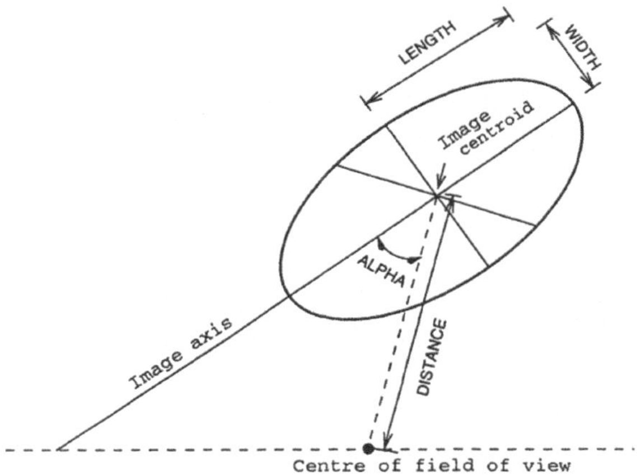

Fig. 12.23 Geometrical interpretation of the image parameters

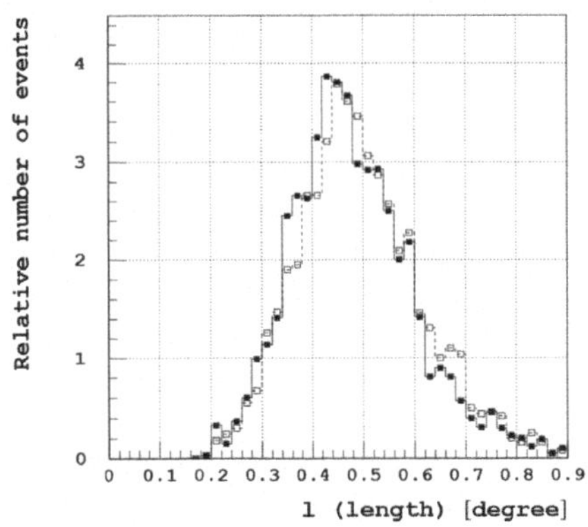

Fig. 12.24 The probability distribution for the image length, l. Primary oxygen nucleus, $E = 10$ TeV, $q_0 = 10$ ph.e. •—CORSIKA, ◊—simulation code [35]

angular distance from the image centroid to its center, Angle between the major axis of the image and the line connecting the image centroid and the center of the FoV, Size is the total number of photoelectrons in the image. There are few other useful derivative image parameters, in particular the Azwidth parameter which is a very convenient combination of the shape and orientational parameters of the image. For a detailed description of these and other (less effective) second-moment parameters of the Cherenkov light images, see [55, 56].

The Cherenkov light image is measured by the multi-channel camera installed in the focus of the optical reflector. The detected image is actually the matrix of the PMTs (channels) responses for each event. According to the procedure of [55], the shape of the image is characterized by the parameters IACT (see Fig. 12.23).

These parameters provide a strong suppression (at least in 100 times) of the cosmic ray background events saving half of the γ-rays. The parameters of the cosmic ray images (very soon the images of γ-rays also) measured in experiment could be directly compared with the Monte Carlo simulations (Figs. 12.24 and 12.25).

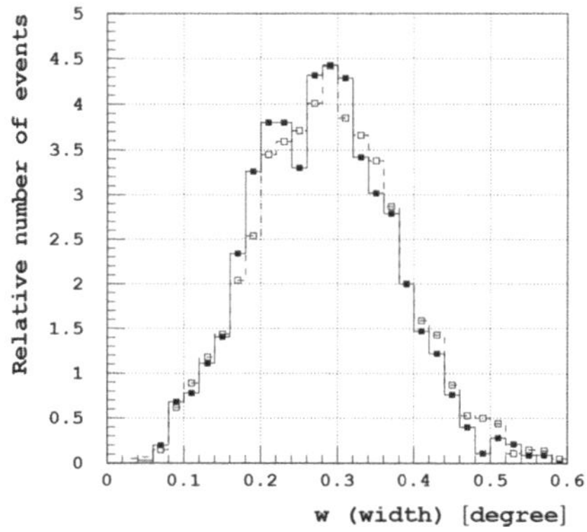

Fig. 12.25 The probability distribution for the image width, w. Primary oxygen nucleus, $E = 10$ TeV, $q_0 = 10$ ph.e. •—CORSIKA, ◊—simulation code [35]

A significant improvement of the imaging technique comes from stereoscopic observations with more than 2 telescopes operating in coincidence mode. The stereoscopic observations make it possible to reconstruct the arrival direction of individual air showers as well as to measure the distance from the telescope to the shower core in the observation plane using the layout and orientation of a several images detected for a certain event. It dramatically improves the angular and energy resolution of the technique and consequently increases the detection sensitivity for observing γ-ray source candidates (see for detail [57]).

In [35], the Monte Carlo calculations of the Cherenkov radiation from air showers produced by different groups of CR nuclei (see Figs. 12.24, 12.25) were described. The image parameters being most sensitive to the atomic number of the primary nuclei, were determined. The method of separation of showers produced by different nuclei based on the second-moment parametrization of the Cherenkov images currently used in the observation of TeV gamma-sources [6] was discussed. Due to the isotropic distribution of CR images in the focal plane of the telescope, only the shape parameters of the image can be invoked for such an analysis. It was shown, that in the case of a single IACT the width (w) parameter, characterizing the lateral development of the shower, allows us to distinguish the proton component from heavier nuclei (including α-particles) and thus to measure the spectrum of CR protons with high accuracy. The combination of width with other parameters (the intrinsic fluctuations of the image intensity, length (l), size, density, etc.) makes it possible to separate nuclei with significantly different atomic numbers A, for example protons and α-particle from the nuclei of the (CNO) group and heavier ones. Simultaneous detection of showers by > 2 IACTs, sufficiently separated from each other, gives information about the shower development in different projections and allows the determination of the distance to the core of each individual shower. This improves significantly the efficiency of nuclei identification, and allows the flux evaluation for five different groups of CR nuclei: protons, CNO, Si, Fe.

The recent observations of TeV γ-radiation from several astrophysical sources convincingly demonstrate the high performance of imaging atmospheric Cherenkov telescopes (IACTs), and set a solid basis for ground-based γ-ray astronomy (see e.g. [58]). Remarkable features of the IACT technique are the high detection rate and the ability of accurate reconstruction of air shower parameters. The differences in the development of electromagnetic and hadron showers are transferred to their Cherenkov images, which ultimately results in an effective separation of γ-ray and cosmic-ray (CR) induced showers (see e.g. [30]). The images of γ-ray showers are compact and regular compared with images of hadron showers. It is remarkable that there is also a difference between images of hadron showers produced by different groups of CRs. Because of the shower maximum position increases rapidly with A, the determination of the shower maximum, using the angular distribution [59] or the time structure [60] of the Cherenkov radiation of showers, yield methods for the measurement of the chemical composition of CRs at energies above 100 TeV amu^{-1}.

The χ^2 Parameter in the Image Analysis

In the process of IACT observations, air showers are registered that have a wide range of primary energies and core locations. The primary energy, E, and the core location, R_0, influence the magnitude, size and shape of the image. As a consequence some additional fluctuations, not connected with the nature of the primary particle, are introduced in the distribution of the image light intensity. For a successful application of intrinsic fluctuations to the separation of air showers one should diminish these 'unwanted' fluctuations. In principle, this can be achieved by taking into account the dependence of selection criteria on E and R_0, but a single IACT does not provide us with these quantities. For this purpose we use in our analysis an alternative approach aimed to reduce the dependence of the selection criteria on the image parameters (size, ω and l) correlating essentially with the quantities E and R_0. It is described as follows.

Let us locate the origin O of the XYZ coordinate system in the centre of gravity of the image and denote as $\rho(x, y)$ the density of the image light intensity in the focal plane of the IACT optical reflector. Let us also consider the l- and ω-parameters as a measure of distance along the major (X) and the minor (Y) image axes. As a result we come to a modified distribution function

$$\phi(u, v) = (lw \text{ size})^{-1}\varrho(x, y), \quad \int du \int dv \phi(u, v) = 1, \quad u = x/l, \quad v = y/w, \tag{12.65}$$

whose values for fixed u, v correlate with size, l and w (and, consequently, with E and R_0) essentially weaker than does the initial density $\rho(x, y)$.

The two-dimensional density $\phi(u, v)$ could be characterized by two one-dimensional functions $\phi_l(u)$ and $\phi_w(v)$ which are the projections of $\phi(u, v)$ to the corresponding coordinate axes.

Table 12.7 Fluctuations of quantities N_l, N_w. The selection criteria are the following: (200 ph.e. $< S <$ 1000 ph.e.), ($0.3° < l < 0.7°$), ($0.2° < w < 0.4°$); $\Delta u = \Delta v = 0.15$, $K = 10$

		k				
		1	3	5	7	9
$u_k(v_k)$		–0.75	–0.45	–0.15	0.15	0.45
$\sigma_{N_{l,k}}$	P	0.38	0.28	0.24	0.20	0.22
	O	0.30	0.21	0.18	0.16	0.17
	Fe	0.27	0.16	0.12	0.11	0.13
$\sigma_{N_{w,k}}$	P	0.26	0.22	0.21	0.23	0.27
	O	0.20	0.16	0.15	0.15	0.18
	Fe	0.14	0.12	0.11	0.11	0.13

$$\phi_l(u) = \int \phi(u, v)dv, \qquad \phi_w(v) = \int \phi(u, v)du. \tag{12.66}$$

In turn the functions $\phi_l(u)$ and $\phi_w(v)$ can be parametrized by a number of integrals over Δu and Δv bins:

$$N_{l,k} = \int_{u_k}^{u_{k+1}} \phi_l(u)du, \quad u_{k+1} = u_k + \Delta u, \quad k = 1, 2, \ldots, K, \tag{12.67}$$

$$N_{w,k} = \int_{v_k}^{v_{k+1}} \phi_w(v)du, \quad v_{k+1} = v_k + \Delta v, \quad k = 1, 2, \ldots, K. \tag{12.68}$$

The set of integrals N_w and N_l determine the image profile in the longitudinal and transverse directions. Thus, fluctuations in the image shape reflect themselves in the fluctuations of the integrals N_w and N_l.

In Table 12.7 we show the data on the fluctuations of quantities N_l, N_w calculated assuming the primary energy spectrum. One can see that the normalized fluctuations in the image shape correlate significantly with the atomic number of the primary nucleus. In particular, the difference in fluctuations corresponding to protons and iron nuclei is as large as a factor of ~2. To construct the selection criteria dealing with the image fluctuations we calculate a set of mean values and mean-squared deviations of the quantities $N_{l,k}$, $N_{w,k}$, $k = 1, 2, \ldots K$ (see (12.67) and (12.68)) corresponding to the group of nuclei assigned for selection. After that, for detected images from overall data sample, we calculate the χ^2 value according to the following expression:

$$\chi_t^2 = K^{-1} \sum_{k=1}^{K} \left\{ [(\bar{N}_{t,k}^{(j)} - \bar{N}_{t,k}^{(i)})/\sigma_{N_{t,k}^{(i)}}]^2 \right\}, \quad t = l, w. \tag{12.69}$$

The separation of the chosen nuclei (i) from the other nuclei could be performed using the simple cut on the χ^2 parameter:

$$\chi_t^2 < \chi_0^2, \qquad t = l, w. \tag{12.70}$$

Note that for a number of telescopes we apply simultaneously this cut for all detected images.

To understand the meaning of the selection parameter (12.69) let us calculate the mean value of χ^2 for an arbitrary nuclei group (j):

$$\bar{\chi_t^2} = K^{-1} \sum_{k=1}^{K} \left\{ [\sigma_{N_{t,k}^{(j)}}/\sigma_{N_{t,k}^{(i)}}]^2 + [(\bar{N}_{t,k}^{(j)} - \bar{N}_{t,k}^{(i)})/\sigma_{N_{t,k}^{(i)}}]^2 \right\}, \quad t = l, \omega. \tag{12.71}$$

Because of the renormalization of the distribution (12.66) the integrals NNt depend only weakly on the atomic number (i.e. $\bar{N}^{(i)} \cong \bar{N}^{(j)}$) and that is why the first term in the right hand side of equation (12.71) gives the main contribution to the $\bar{\chi}_t^2$ parameter. At the same time $\sigma_{N^{(j)}}$ decreases with j (see Table 12.7, the index j runs up while the nuclei atomic number increases). Consequently, $\bar{\chi}_t^2$also decreases with j, and the probability distribution of χ_t^2 becomes more narrow. Thus to select the images corresponding to small values of χ_t^2 one can reject air showers which belong to nuclei groups with numbers $j < i$, where the index i corresponds to the assigned nuclei group.

Note that in general the χ_t^2 parameter depends on values of parameters Δu, Δv, K. However the present analysis shows that the efficiency of nuclei separation is quite insensitive to the exact values of these parameters. Everywhere in analysis (except for Table 12.65) we use the following values $\Delta u = \Delta v = 0.15$, $K = 40$.

References

1. Cao Z, Chen S, Liu R, Yang R (2023) Ultra-high-energy gamma-ray astronomy. Annu Rev Nucl Part Sci 73:341–363
2. Hillas AM (2013) Evolution of ground-based gamma-ray astronomy from the early days to the Cherenkov telescope arrays. Astropart Phys 43:19–43
3. Völk HJ, Bernlöhr K (2009) Imaging very high energy gamma-ray telescopes. Exp Astron 25:173–191
4. Uchaikin VV, Lagutin AA (1975) Calculation of fluctuations of Cherenkov radiation of an electron-photon shower. Nucl Phys 21(6):1257–1264 (in Russian)
5. Ivanenko IP (1958) Energy spectrum of cascade electrons in light materials. Zh Eksp Teor Fiz 6(1):105
6. Gerasimova NM (1964) Fluctuations in the number of particles in an electron-photon shower. Proc Phys Inst Acad Sci USSR 26:192–206 (in Russian)
7. Rädel L, Wiebusch C (2013) Calculation of the Cherenkov-light yield from electromagnetic cascades in ice with Geant4. Astropart Phys 44:102–113

8. Plyasheshnikov A, Lagutin A, Makarov V, Misaki A (1997) Influence of light scattering and finite size of shower particle disk on the space-temporal structure of Cherenkov light emitted in water by high energy electromagnetic cascade. Proc 25th ICRC Durban 7:137–140
9. Plyasheshnikov AV, Lagutin AA, Misaki A, Tiumentsev A (1998) Cherenkov radiation of electron-photon cascades in water. In: Proceedings of the Altai State University, Special issue, Barnaul, pp 47–58
10. Plyasheshnikov AV (2000) Simulation of cascade processes in gamma-astronomy and cosmic-ray physics. Dissertation Dr. Phys Math Sci Barnaul
11. Plyasheshnikov A, Lagutin A, Misaki A (1998) Cherenkov radiation of electron-photon cascades in water. Bull Altai State Univ:47–58
12. Hillas AM (1997) Shower simulation: lessons from MOCCA. Nuclear Phys B (Proc Suppl) 52B:29–42
13. Konopelko AK, Plyasheshnikov AV (2000) ALTAI: computational code for the simulations of TeV air showers as observed with the ground-based imaging atmospheric Cherenkov telescopes. Nucl Instrum Methods Phys Res, Sect A 450(2–3):419–429
14. Cao Z, Chen S, Liu R, Yang R (2023) Ultra-high-energy gamma-ray astronomy. Annu Rev Nucl Part Sci 73:341–363
15. Völk HJ, Bernlöhr K (2009) Imaging very high energy gamma-ray telescopes. Exp Astron 25:173–191
16. Kertzman MP, Sembroski GH (1994) Computer simulation methods for investigating the detection characteristics of TeV air Cherenkov telescopes. Nucl Instrum Methods Phys Res Sect A: Accel, Spectrometers, Detect Assoc Equip 343(2–3):629–643
17. Heck D, Knapp J, Capdevielle JN, Schatz G, Thouw T (1998) CORSIKA: A monte carlo code to simulate extensive air showers. Forschungszentrum Karlsruhe GmbH, Karlsruhe
18. Konopelko A, Aharonian F, Hemberger M, Hofmann et al (1989) Effectiveness of TeV-ray observations at large zenith angles with a stereoscopic system of imaging atmospheric Cerenkov telescopes. J Phys G: Nucl Part Phys 25(9)
19. Berger M (1963) Monte Carlo calculation of the penetration and diffusion of fast charged particles. In: Methods of computational physics, vol I, Academic Press , N.Y., pp 135–217
20. Kolchuzhkin AM, Uchaikin VV (1978) An introduction to the theory of the particle penetration through the matter. Atomizdat (in Russian), Moscow
21. Gaisser TK, Engel R, Resconi E (2016) Cosmic rays and particle physics. Cambridge University Press
22. Vavilov PV (1957) Ionization losses of heavy charged particles with high energies. Journ Exper Theor Phys 37(46):920–923
23. Kolchuzhkin AM, Plyasheshnikov AV (1975) The radial distribution of electron flux from a point monodirectional source. AtNaia Energ (in Russian) 38(6):327
24. Beliaev AA, Ivanenko IP, Kanevsky BL et al (1980) Electromagnetic cascades in cosmic rays at very high energies. Nauka, Moscow (in Russian)
25. Shabelski YM (1987) Inelastic interactions of high energy nuclei in atmosphere. In: Proceedings of the 20th international cosmic ray conference Moscow, vol 5, p 147
26. Murzin VS (1988) Introduction to cosmic ray physics. Moscow State University
27. Ranft J (1972) Estimation of radiation problems around high-energy accelerators using calculations of the hadronic cascade in matter. Part Accel 3:129–161
28. Barashenkov VS, Toneev VD (1972) Interactions of high energy particles with atomic nuclei (in Russian). Atomizdat, Moscow
29. Konopelko AK (1997) Space-angular and temporal parameters of Cherenkov light emission in air showers of energy from 1 TeV down to 10 GeV. In: Proceedings of the Kruger Park workshop on tev gamma-ray astrophysics, towards a major atmospherics cherenkov detector-V, pp 8–11
30. Hillas AM (1996) TeV gamma-ray astrophysics. Space Sci Rev 75:17–30
31. Aharonian F (1999) The time averaged TeV energy spectrum of MKN 501 of the extraordinary 1997 outburst as measured with the stereoscopic Cherenkov telescope system of HEGRA
32. Konopelko A, Hemberger M, Aharonian F et al (1999) Performance of the stereoscopic system of the HEGRA imaging air Cerenkov telescopes: Monte Carlo simulations and observations. Astropart Phys 10(4):275–289

33. Konopelko AK (1999) Whipple collaboration electronic preprint. SAO, Tucson
34. Aharonian FA, Akhperjanian AG, Barrio JA et al (1999) Measurement of the radial distribution of Cherenkov light generated by TeV γ-ray air showers. Astropart Phys 10(1):21–29
35. Plyasheshnikov AV et al (1998) Study of mass composition and energy spectrum of primary cosmic radiation by the imaging atmospheric Cherenkov technique. J Phys G: Nucl Part Phys 24:653–672
36. Rao MVS, Sinha S (1988) The origin of the hump in the Cerenkov lateral distribution in gamma-ray showers and a possible means of separating them from proton showers. J Phys G: Nucl Phys 14:811–827
37. Patterson JR, Hillas AM (1983) The relation of Cerenkov time profile widths to the distance to maximum of air showers. J Phys G: Nucl Phys 9(3):323
38. Browning R, Turver KE (1977) Cerenkov radiation from computer simulations of the γ-ray-initiated atmospheric showers. Nuov Cim A 38:223–238
39. Porter NA (1973) Geomagnetic effects on the angular distribution of Cerenkov light from extensive air showers. Lett. Nuovo Cimento 8:481–484
40. Sinha S (1995) On the fluctuations of Cerenkov photon densities in very high-energy gamma ray showers. J Phys G: Nucl Part Phys 21(3):473
41. Konopelko AK et al (1991) About a possibility of background rejection in VHE gamma-ray astronomy on the base of differences in the temporal structure of Cherenkov radiation of air showers. In: Proceedings of the 22th ICRC, Dublin, vol 1, p 484
42. Zuev VE (1970) Propagation of visible infrared waves in the atmosphere. Soviet Radio, Moscow (in Russian)
43. Goody RM (1966) Atmospheric radiation, part 1. Fundamentals of theory, Moscow, Mir (in Russian)
44. Zatsepin VI, Chudakov AE (1962) Method of finding local sources of high-energy photons. Zh Eksp Teor Fiz 14(2):469
45. Kolchuzhkin AM, Bespalov VI (1979) The investigation of electron-photon showers by Monte Carlo method. In: Proceedings of 16-th ICRC. Kyoto vol 9, pp 222–227
46. Kolchuzhkin AM, Uchaikin VV, Bespalov VI (1973) Calculations of the Cherenkov radiation of electron-photon shower in atmosphere. Izvestia AN SSSR, Ser Phys 7:1442–1446 (in Russian)
47. Bernlöhr K (2000) Impact of atmospheric parameters on the atmospheric Cherenkov technique. Astropart Phys 12:255–268
48. Goldanskii VI, Zhdanov GB (1954) On cherenkov radiation of cosmic particles in the atmosphere. Zh Eksp Teor Fiz 26:405
49. Zatsepin VI (1964) The angular distribution of intensity of cerenkov radiation from extensive cosmic-ray air showers. Zh Eksp Teor Fiz 47(2):689
50. Castagnoli C, Dardo M, Penengo P (1967) Longitudinal distribution of cerenkov light from extensive air showers. Phys Rev 160(5):1186
51. Ivanenko IP et al (1980) Electron-photon cascades in cosmic rays of the very high energy. Moscow State University Press
52. Tümer OT et al (1990) A promising large area VHE gamma ray detector with excellent hadron rejection capability. In: Proceedings of the 21th ICRC, Adelaide, vol 2, pp 155
53. Hess M et al (1997) Search for TeV gamma-ray emission from supernova remnants. In: Proceedings of the 25th ICRC, Durban, vol 5, OG, p 133
54. Hill DA, Porter NA (1961) Space-angular and temporal parameters of Cherenkov light. Nature 191:690
55. Hillas AM (1985) Cerenkov light images of EAS produced by primary gamma rays and by nuclei. In: Proceedings of the 19th ICRC, La Jolla, vol 3, p 445
56. Fegan DJ (1996) The art and power of verenkov imaging. Space Sci Rev 75:137–151
57. Aharonian FA et al (1997) The potential of ground based arrays of imaging atmospheric Cherenkov telescopes. Astropart Phys 6(343–368):369–377
58. Cawley M, Weekes T (1995) Instrumentation for very high energy gamma-ray astronomy. Exp Astr 6:7–42

59. Boothby et al (1997) A new measurement of cosmic-ray composition at the knee. Nucl Phys B (Proc Suppl) 52:166
60. Lindner A et al (1995) Composition and energy measurement with a new method applied to the HEGRA data in the energy range of the Knee. In: Proceedings of the 24th ICRC, Rome, vol 1, p 422
61. Ivanov AA, Matarkin SV, Timofeev LV (2022) Measuring temporal characteristics of the Cherenkov radiation signal from extensive air showers of cosmic rays with a wide field-of-view telescope addendum to the Yakutsk array. Phys Rev D 105(4):042008
62. Grigoriev VM, Knurenko SP, Pravdin MI (2012) Testing calculation results of cherenkov radiation flux using CORSIKA. Phys Particles Nuclei Lett 9(6–7):567–570

Chapter 13
Relativistic EPC in Intergalactic Medium

13.1 Introduction

The development of electromagnetic cascades in photon gas and magnetic fields is a rather common phenomenon in astrophysics. In photon fields such cascades can be created on almost all astronomical scales, from compact objects like accreting black holes, fireballs in gamma-ray bursts, sub-pc jets of blazars to large-scale (up to 100 kpc) AGN jets and 1 Mpc size clusters of galaxies. Since the energy density of 2.7° K CMBR significantly exceeds it of intergalactic magnetic fields, these interactions initiate electron–photon cascades. The contributions of γ-rays from these cascades should constitute a noticeable fraction into the diffuse extragalactic background.

In the general case, the description of the cascade development represents an intricate problem. The solution of the nonstationary problem with the inclusion of the angular and spatial distribution of the particles, energy loss, and curving of the trajectories of cascade electrons in a magnetic field and of the nonlinear effects of the cascade on a medium seems to be possible by the simulation method only. High consumption of computer time does not make it possible, however, to recommend this method for making calculations in a wide range of values of variables and parameters, which are necessary because of the poor knowledge of the physical conditions of the source. Furthermore, the existence of a large number of minor effects makes it difficult to study the main regularities of the cascade process.

Recall, that the cosmic radio intensity in the microwave region is very strong and isotropic with the photon number density is as high as 10^3 cm^{-3}. Thus, the mean free paths of high energy electrons and γ-rays, respectively, for the inverse Compton and pair production collisions can be as short as 10^{22} cm. The inverse Compton collision produces a photon of high energy, and the photon creates a pair of electrons. The succession of these processes is analogous to that in matter, in which photons are radiated by the bremsstrahlung of electrons and electrons are generated by the pair creation in the nuclear Coulomb field. In different astrophysical environments γ-ray production may proceed with high efficiency through bremsstrahlung, inverse Compton scattering and synchrotron (and/or curvature) radiation, respectively. The

© Higher Education Press Limited Company 2024

V. V. Uchaikin, *Electron-Photon Cascades*, Nonlinear Physical Science,
https://doi.org/10.1007/978-981-99-7524-2_13

use of this phenomenon in astrophysical investigations is effective when the cooling time that characterizes the rate of the process does not significantly exceed (i) the source (accelerator) age, (ii) the characteristic time of non-radiative losses caused by adiabatic expansion of the source or particle escape, and (iii) the cooling time of competing radiation mechanisms that result in low-energy photons outside the γ-ray domain. As long as the charged particles are effectively confined in the γ-ray production region, at some circumstances, these conditions could be fulfilled even in environments with relatively low gas and photon densities or weak magnetic fields. These reasons inspire the further development of numerical approach to electromagnetic cascades not only in matter but also in photon gas and in magnetic fields as more common media in astrophysics.

We start a discussion of the problem by citing the paper [1]. Giving a qualitative description of the intergalactic space, its author noticed:

The cascade process in the intergalactic space may take part in γ-ray astronomy. Since the opacity of space is great for γ-rays of energies above the threshold of pair creation, it might be considered that the detection of extragalactic sources is impossible and the intensity of isotropic γ-rays is unappreciably small in this energy region. However, the reduction of the absorption coefficient is not so much as to essentially alter the conclusion thus far believed. Nevertheless, the role of the cascade process in space should be taken into account seriously in discussing the astrophysical implication of γ-ray astronomy. Since the microwave photons are isotropic and have the energy distribution represented by the Planck law with temperature corresponding to 7×10^{-4} eV, the threshold energy E_{th} is distributed over a wide range depending on θ and ε. Taking the average over angle and energy, we find that the pair creation cross section becomes appreciable at energies above 0.9×10^{14} eV, which may be called the *critical energy* (E_{cr}), above which the cascade process takes place.

Above the critical energy, the pair creation cross section increases rapidly and reaches a maximum at about 10^{15} eV, and then decreases slowly toward high energy. The cross section for the inverse Compton effect above the maximum behaves nearly in the same way. Thus, in the framework of the qualitative approach, there is no need treating electrons and photons separately in the first approximation. One has only to know the cross section for an incident particle with energy E to produce a secondary particle with energy in dE'. It is approximately given by

$$W(E \to E')dE' = 2\pi r_0^2 E_{\text{cr}} dE'/E^2$$

and yields the total cross section $2\pi r_0^2 (E_{\text{cr}}/E)$ being different from the exact expression by a factor $\ln(E/E_{\text{cr}})$. The logarithmic factor appears also in the energy loss rate by the inverse Compton effect, but the amount of energy lost in the collision mean free path is small compared with the electron energy. Hence, one can neglect the energy loss like as in the approximation A in the ordinary cascade theory.

The qualitative reasoning leads us to the conclusion that the problem solution consists of two terms, one representing the simple absorption and the other that being subject to the cascade process. The second term attenuates more slowly than the first one, and the attenuation length increases as the cascade develops. However, the attenuation length is not much different from that of the first term, because of the inverse energy dependence of the cross section. For a cascade process starting with a particle of energy E_0, the attenuation length increases approximately by a factor of $\ln E_0$. On the other hand, the number of particles becomes as great as E_0, and most of the particles at the maximum cascade development have energies as low as E_{cr}. As energy goes down below E_{cr}, the intergalactic space becomes transparent, and one can expect a greater γ-ray flux at energies below E_{cr}.

13.2 The Ivanenko–Sizov Theory

In [2], considered the stationary one-dimensional (in-depth) picture of development of an electron–photon cascade in an isotropic photon field, averaged over the spatial and angular variables. The field was supposed to have a power-law spectrum with the upper and lower cut-offs, and to a monochromatic field, the EPC is initiated by a monoenergetic beam of electrons or gamma-quanta or by the primary radiation of the power-law spectrum. Two types of interactions were taken into account: the inverse Compton effect and the photoproduction of electron–positron pairs. The one-dimensional EPC development of an electron-photon cascade in an isotropic photon field was numerically investigated. Two cases were investigated: the field with power-law spectrum limited on both sides,

$$n(\omega)d\omega = n_0 \frac{\omega_1^{\alpha}}{\omega^{\alpha+1}} d\omega (\alpha > -1, \quad \omega_1 \leq \omega \leq \omega_2), \tag{13.1}$$

and the monochromatic field $n(\omega) n_0 \delta(\omega - \omega_0)$. Two types of processes were taken into account: the inverse Compton effect and the photoproduction of electron–positron pairs. The first approximation was limited to a linear problem (without the interaction effect on the background photon field). The magnetic field deflecting the electron trajectory from the cascade axis approximately was taken into account by cutting the electron spectrum from the low-energy side. The relationship between the geometric field thickness z and the thickness in cascade units $t_0 = (4\pi r_0^2 n_0)$ in the general case of a nonuniform field is of the form

$$t = \int_0^z \frac{dz}{t_0(z)}, \quad t_0(z) = (4\pi r_0^2 n_0(z))^{-1}.$$

The differential spectra of electrons $P(t, E)$ and gamma-quanta $\Gamma(t, E)$ have been obtained by the numerical solving the cascade equations

$$\partial_t P(t, E) = L_1(P, \Gamma) + 2W_{\gamma e}(E_0 \rightarrow E) \exp(-\Sigma_\gamma(E_0)t), \tag{13.2}$$

$$\partial_t \Gamma(t, E) = L_2(P, \Gamma), \tag{13.3}$$

where L_1, L_2 are determined by equations

$$L_1(P, \Gamma) = \int_E^{E_0} dE' P(t, E') W_{ee}(E' \to E) + 2 \int_E^{E_0} dE' \Gamma(t, E') W_{\gamma e}(E' \to E) - \Sigma_e(E) P(t, E), \tag{13.4}$$

$$L_2(P, \Gamma) = \int_E^{E_0} dE' P(t, E') W_{e\gamma}(E' \to E) - \Sigma_\gamma(E) \Gamma(t, E), \tag{13.5}$$

and $\Sigma_{e,\gamma}$ stand for the total cross sections of electron or photon, respectively, whereas $W_{ab}(E' \to E)$ for inclusive elementary processes $a(E') \to b(E)$. The second term on the right-hand side of Eq. (13.2) arises in the case of excitation of the cascade by gamma-quanta with energy E_0 due to the presence of the function $\delta(E - E_0)$ in the boundary conditions (the first generation of electrons). At an energy below the pair production threshold ($E < E_{\rm Th}$, $\mu_\gamma = 0$), integrating Eq. (13.3) with respect to the depth t yields

$$\Gamma(t, E) = \int_E^{E_0} dE' W_{e\gamma}(E' \to E) \int_0^t dt' P(t', E'), \quad E < E_{\rm Th}. \tag{13.6}$$

In the monochromatic field, the total cross sections (and, consequently, the cascade characteristics) depend on the energy E of the particles, interacting with the field, only via the product ωE, where ω is the energy of the field photons. At energies much greater than the pair photoproduction threshold $E_{\rm Th} = m^2c^4/\omega$, the approximations

$$\Sigma_\gamma(E, \omega) \approx 2\Sigma_e(E, \omega) \approx \frac{1}{E\omega} \tag{13.7}$$

and

$$W_{\gamma e}(E \to E', \omega) dE' \approx \Sigma_\gamma(E, \omega) \frac{dE'}{E} \tag{13.8}$$

are used. Equation (13.7) is valid to within a factor logarithmically dependent on the energy, Eq. (13.8) indicates that, at a high gamma-quantum energy, the spectrum of secondary particles may be considered as flat. A similar relation can also be written for the cross section for the inverse Compton effect $W_{e\gamma}(E \to E', \omega) = W_{ee}(E \to E - E', \omega)$.

In case of the power-law spectrum (13.1), the total cross sections and the cascade characteristics depend on the product $\omega_1 E$ and the ratio ω_1/ω_2. In the E-dependence dependence of Σ_γ, having a characteristic "roof-like" shape for $\alpha > 0$ (Fig. 13.1), two energy values are singled out: $E_{\rm Th} = m^2c^4/\omega_2$ and $E_{\rm KN} = 2m^2c^4/\omega_1$. At $E < E_{\rm Th}$ the cross section becomes zero, and near $E_{\rm KN}$ it has a knee.

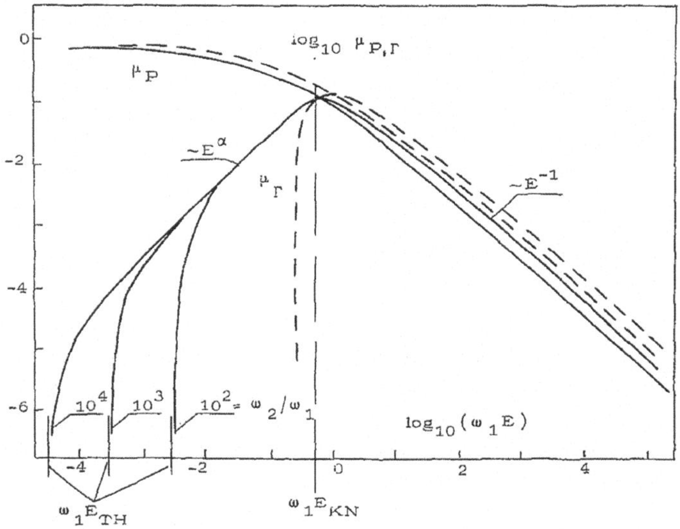

Fig. 13.1 Total cross section for the inverse Compton effect (μ_p) and for the pair photoproduction (μ_r) in a photon field of the power-law spectrum for $\alpha = 1$ (solid curves) and in a monoenergetic photon field for $\omega = \omega_1$ (dashed curves)

Solutions $P(t, E)$, $\Gamma(t, E)$ of Eq. (13.2) on the positive quarter of plane (t, E) were found by the peace-wise polynomial method: they were constructed sequentially in cells of the network in the direction of energy decrease from $E = E_0$

$$[E_1, E_0], [E_2, E_1], \dots, [E_j, E_{j-1}], \dots,$$

and of depth increase from $t = 0$ to $t_{\max}$:

$$[0, t_1], [t_1, t_2], \dots, [t_{k-1}, t_k], \dots$$

The numerical values of $(P, \Gamma)(t, E_0)$ were calculated directly from Eq. (13.2) neglecting the integral terms in Eq. (13.4), whereas the values of (P, Γ) $(0, E)$ follow from the boundary conditions. Requiring the satisfaction of Eq. (13.2) at nine points of each cell $[E_j, E_{j-1}]$, $[t_{k-1}, t_k]$ of the network determines the coefficients $(P, \Gamma)_{mn}$ of the function quadratic in E and t

$$(P, \Gamma)(t, E) = A(E)B(t) \sum_{m,n=0,1,2} (P, \Gamma)_{mn}(E - E_j)^m (t - t_k)^n,$$

where

$$A(E) = E^{-1}, \quad B(t) = \exp(-\Sigma_\gamma(E_0)t).$$

At the cell boundary, the joining with the solutions in the neighboring cells was carried out. The solutions obtained are continuous functions of E and t. The accuracy was checked by comparing the derivatives at the boundary of the cells and was established at a level of 0.1%. The step in both variables was chosen depending on the accuracy attained and was on the average 50 intervals per order of magnitude.

13.3 EPC From Monoenergetic Source

The $E - t$ distributions in EPC from γ-quanta with various energies E_0 are presented in Fig. 13.2. All cascade characteristics are given per primary gamma-quantum. Figure 13.2a shows t-dependence of Γ/ω for two values of production ωE_0: 10^2 (dashed lines) and 10^4 (continuous lines), and for several values of ωE from 10 to 0.02. Observe, that below the pair photoproduction threshold, the cascade curves increase monotonically, but above it have a maximum.

The development of the cascade over the depth is determined by the relation between the numbers of generated and absorbed gamma-quanta with a given energy (correspondingly, the first and the second terms on the right-hand side of Equation (13.5)). For $E_{\mathrm{Th}} \ll E \ll E_0$ at depths smaller than $t_{\max}$, there exists a balance between the two quantities, leading to the t-independence of Eq. (13.5). The derivative $\partial\Gamma/\partial t$ in Eq. (13.3) has a constant positive value and the Γ increases as t. In the absence of absorption, the function $P(t', E')$ in Eq. (13.6) increases linearly with the depth and $\Gamma\ t^2$. When $t > t_{\max}$ and $E > E_{\mathrm{Th}}$ the second term in Eq. (13.5) predominates and the cascade curves decrease with the depth exponentially.

Let us consider the spectrum $\Gamma(t, E)$ above E_{Th}. It is seen from Fig. 13.2a, that $t_{\max} \propto E_0$ and at a fixed E_0 does not depend on E while the Γ at $t_{\max}$ does depend on E_0 at a fixed E, which weakly (rather, logarithmically, see the explanation to Eq. (13.7)) depends on E_0. These properties of Γ in the photon field are drastically distinguished from that in matter and follow from the energy dependence of the cross sections (13.7).

The depth t is explicitly involved in Eq. (13.2) only through the complex $\Sigma_\gamma(\omega, E_0)t \approx t/\omega E_0$ which can be regarded as a new dimensionless scaling variable in expressing the depth dependence in terms of $t/\omega E_0$. The cascade curves for different E_0 at a fixed E overlap [3, 4].

The cascade model considered below is based on two elementary processes: the production of electron–positron pairs by photons and the inverse Compton scattering of electrons. For energy-differential cross sections of interactions with a homogeneous and isotropic monoenergetic photon field of energy ω and concentration n, we use the notation $W_{ab}(E \to E', \omega)$, where E and E' are the energies of the primary (a) and one of the secondary particles (b).

According to [5], the photoproduction cross sections for $W_{\gamma e}$ pairs and the inverse Compton scattering $W_{e\gamma}$ on the cascade length unit $t_0 = (4\pi r_0^2 n)^1$ under the condition that the field photon energy $\omega \ll E, E'$ are taken in the form

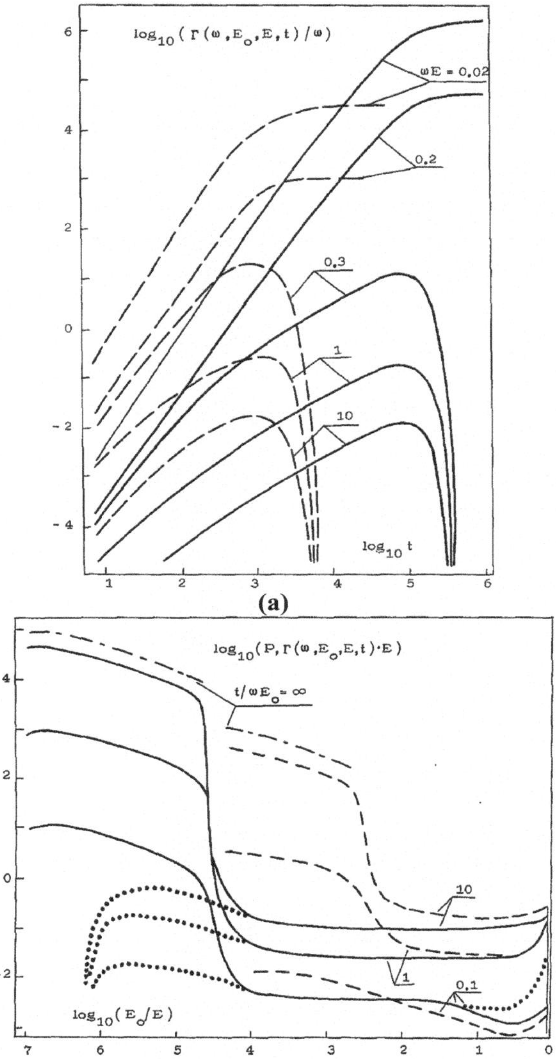

Fig. 13.2 Cascade curves **a** and cascade gamma-quantum spectra **b** for two energy values of a primary gamma-quantum, $\omega E_0 = 10^2$ (dashed curves) and $\omega E_0 = 10^4$ (solid curves). The dots in Figure **b** represent the cascade electron spectra multiplied by 0.5

$$W_{\gamma,e}(E \to E', \omega) = \frac{1}{bE}\left[\frac{1}{x(1-x)} - 2 - \frac{2}{bx(1-x)} + \frac{2}{b^2x^2(1-x)^2}\right.$$

$$\left. - \frac{1}{bx^2(1-x)^2} + \frac{4}{bx(1-x)}\ln\left(bx(1-x)\right)\right]\varepsilon(E - E_{th}(\omega)), \qquad (13.9)$$

$$W_{e,\gamma}(E \to E', \omega) = \frac{1}{bE}\left[\frac{1}{(1-x)} + 1 - x + \frac{2}{bx(1-x)} + \frac{2}{b^2x^2(1-x)^2}\right.$$

$$\left. - \frac{1}{bx^2(1-x)^2} + \frac{4}{bx(1-x)}\ln\left(bx(1-x)\right)\right]\varepsilon(E - E_p(\omega)), \qquad (13.10)$$

here $x = E'/E$, $E_{th}(\omega) = m^2c^4/\omega$ is the pair photoproduction threshold energy, $b = 4E/E_{th}(\omega)$. The kinematically allowed regions of the secondary particle energy E' are determined by the inequalities

$$E_{\gamma 1} \le E' \le E_{\gamma 2}, \quad 0 \le E' \le E_e, \qquad (13.11)$$

$$E_{\gamma 1,\gamma 2} = 0.5E(1 \pm (1 - E_{th}(\omega)/E)), \quad E_e = \frac{E}{1 + E_{th}(\omega)/4E}. \qquad (13.12)$$

The absorption cross sections for photons and electrons in the photon field are obtained by integrating the corresponding differential cross sections

$$\Sigma_\gamma(E, \omega) = \int_{E_{\gamma 2}}^{E_{\gamma 2}} dE' W_{\gamma e}(E \to E', \omega), \qquad (13.13)$$

$$\Sigma_e(E, \omega) = \int_0^{E_e} dE' W_{e\gamma}(E \to E', \omega). \qquad (13.14)$$

The differential cross sections for the interaction of particles with a photon field described by the $n(\omega)$ distribution are derived by integrating the expressions (13.9) and (13.10):

$$W_{\alpha\beta}(E \to E') = \int n(\omega) W_{\alpha\beta}(E \to E', \omega) d\omega. \qquad (13.15)$$

In the choice of a power-law distribution of field photons

$$n(\omega) = \begin{cases} n_0 \alpha \omega_1^{\alpha_x - 1}/\omega^{\alpha_x}, & \omega_1 \le \omega \le \omega_2, \\ 0, & \omega < \omega_1,\ \omega > \omega_2, \end{cases} \qquad (13.16)$$

interaction cross sections per cascade length unit $X_0 = (4\pi r_0^2 n_0)^{-1}$ for $\alpha_x > 0$ are [6]

$$W_{\alpha\beta}(E \to E') = \alpha \omega_1^{\alpha_x - 1}\left[F_{\alpha\beta}(E, E'; \omega_2) - F_{\alpha\beta}(E, E'; \omega_{\alpha\beta})\right], \qquad (13.17)$$

where

$$F_{\gamma e}(E, E'; \omega) = \frac{m^2c^4}{4E^2}\frac{1}{\omega^{\alpha_x}}\left\{-\frac{1}{\alpha_x x(1-x)} + \frac{2}{\alpha_x} + \frac{2}{(\alpha_x+1)bx(1-x)}\right.$$

$$-\frac{4}{(\alpha_x+2)b^2x^2(1-x)^2} + \frac{1}{(\alpha_x+1)bx^2(1-x^2)}$$

$$\left.-\frac{4}{(\alpha_x+1)(bx(1-x))}\left[\ln(bx(1-x)) + \frac{1}{\alpha_x+1}\right]\right\}, \quad (13.18)$$

$$F_{e\gamma}(E, E'; \omega) = \frac{m^2c^4}{4E^2}\frac{1}{\omega^{\alpha_x}}\left\{-\frac{1}{\alpha_x(1-x)} - \frac{1-x}{\alpha_x} - \frac{2x}{(\alpha_x+x)b(1-x)}\right.$$

$$+\frac{4x^2}{(\alpha_x+2)b^2(1-x)^2} + \frac{x^3}{(\alpha_x+1)b(1-x)^2}$$

$$\left.+\frac{4x}{(\alpha_x+1)b(1-x)}\left[\ln\frac{(b(1-x))}{x} + \frac{1}{\alpha_x+1}\right]\right\}, \quad (13.19)$$

$$\omega_{\gamma e} = \max\left\{\omega_1, \frac{m^2c^4}{4Ex(1-x)}\right\}, \quad \omega_{e\gamma} = \max\left\{\omega_1, \frac{xm^2c^4}{4E(1-x)}\right\}.$$

The threshold energy of pair photoproduction in this field is $E_{th}(\omega_2)$.

Absorption cross sections for photons and electrons $\Sigma_{\gamma,e}$ in the background photon field $n(\omega)$ were calculated, as in the case of a monoenergetic photon field, by integrating the corresponding differential cross sections (13.17). The integration limits are given by the expressions (13.11) and (13.12) with threshold energy $E_{th}(\omega_2)$. The results of these cross section calculations are shown in Fig. 13.3. Interaction cross sections in a monoenergetic photon field, as one can see from expressions (13.9)–(13.14), are functions of the product of the energies of interacting particles ωE, $\omega E'$, so $\Sigma_{e\gamma}(E, \omega = 10^{-2})$ (Fig. 13.3a) can be used to find sections for arbitrary ω. Cross section Σ_γ has a maximum at $E_m(\omega) = 3E_{th}(E_m = 100$ MeV for $\omega = 10^{-2}$ MeV). In high-energy region, $\Sigma_\gamma(\omega E) \simeq 2\Sigma_e(\omega E) \simeq (2E/E_{th}) \ln(4E/E_{th})$. Cross section Σ_e at $E \to 0$ tends to Tomson's limit 2/3.

Cross sections of interaction in a power-kind photon field, as follows from substitution (13.17)–(13.19) into (13.15), depend on ω_1, $\omega_1 E$, ω_2/ω_1 and α_x. However, in the case of a sufficiently wide distribution $\omega_1 \ll \omega_2$ and incident spectra $\alpha_x > 0$, the first the term in expression (13.17) is much less than the second term at energies $E \gg E_{th}$, and the cross sections are practically independent of ω_2. Wherein, in the region of large energies $E > 2m^2c^4/\omega_1$ we have $\Sigma_{e\gamma} \approx \Sigma_{e\gamma}(E; \omega_1)$. In the energy

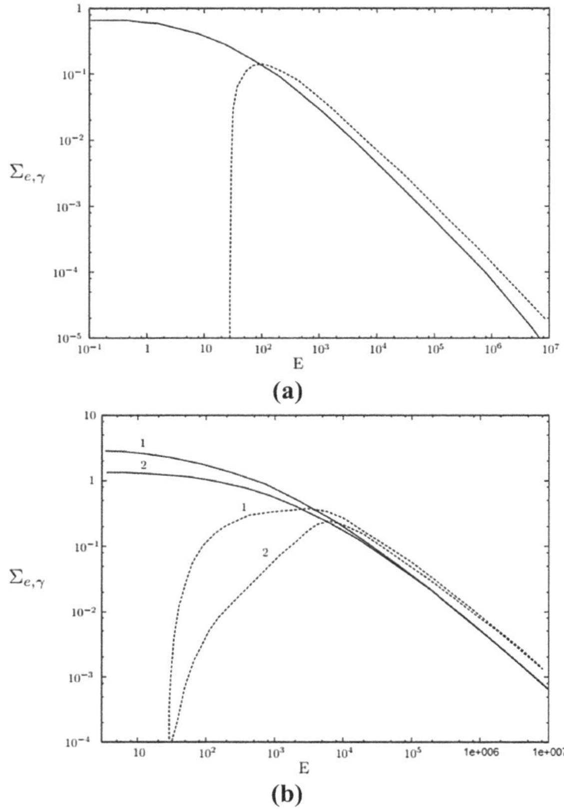

Fig. 13.3 Absorption cross sections for electrons (solid curves) and photons (dotted line): **a** monoenergetic photon field, $\omega = 10^{-2}$ MeV; **b** power-law distribution of field photons: $1 - \alpha_x = 1, 2 - \alpha_x = 2$ [6]

range $(2 \div 3)E_{th}(\omega_2) < E < (2 \div 3)E_{th}(\omega_1)$, the photon absorption cross section behaves like $E^{\alpha_x - 1}$. The limit value of $\Sigma_e(E)$ for small E is equal to

$$\Sigma_e(E) \sim (2/3)\alpha_x \omega_1^{\alpha_x - 1} \int\limits_{\omega_1}^{\omega_2} \frac{d\omega}{\omega^{\alpha_x}}.$$

13.4 EPC in Monochromatic Photon Field

The article [6] presents the EPC photon spectra in monoenergetic photon field calculated by using backward cascade equation approach. Relating to monoenergetic field with photon energy $\omega = 10^{-2}$ MeV, the numerical results can be used for an arbitrary value of ω (under condition that $\omega \ll E_t$).

Fig. 13.4 Differential cascade curves. $E = 10^4$ MeV. $E_t = 1(1)$; $6(2)$; $10.5(3)$; $26.1(4)$; $30(5)$; $104(6)$; $1040(7)$; $8257(8)$ MeV

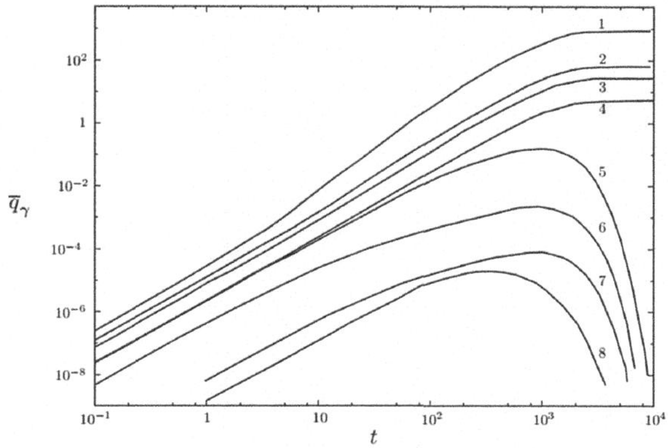

Cross sections in a monoenergetic photon field (see (13.9)–(13.14)) depend on products of the energies of interacting particles ωE and $\omega E'$ resulting in the quantity $\bar{q}_\alpha/\omega$, as it is easy to see from the Eq. (13.3), are also functions of ωE and ωE_t. Because of this values of EPC characteristics in the field of photons of arbitrary energy ω can be obtained from calculations for some basic photon field energy ω_0 according to the following formula:

$$\bar{q}_\alpha(t, E; E_t|\omega) = (\omega/\omega_0)\bar{q}_\alpha(t, E\omega/\omega_0|\omega_0).$$

Figures 13.4, 13.5 and 13.6 show the average characteristics of EPC photons, initiated by primary particles with energies greater than the maximum energy of the photon absorption cross section $E_{\max}$. As can be seen from the figures, the differential cascade curves for small t grow approximately in proportion to t^2, then the degree of growth of the curves for E_t from the interval $E_{\rm th} < E_t < E$ decreases and they have maxima at distances $t_{\max} \simeq (1-3)E/E_{\rm th}$—of the order or somewhat greater than the free path of the primary particle (see Fig. 13.3a). The degree of growth of the curves for $E_t < E_{\rm th}$, on the contrary, increases, but at distances $t > t_s \simeq 10E/E_{\rm th}$ they go out on the plateau.

In the region of small t, where, for all E_t, we have $\bar{q} \sim t^2$, the shape of the energy spectra is practically independent of t (see Fig. 13.5). When approaching $t_{\max}$, due to the indicated difference in the dependencies of q on t for different E_{th}, the subthreshold region of the spectrum ($E_t < E_{\rm th}$) grows faster than the above-threshold region ($E_t > E_{\rm th}$), as a result of which a step in the spectrum is formed in $E_t \sim E_{\rm th}$ region. At larger depths $t > t_s$, the view of the above-threshold area spectrum does not change, but $\bar{q}_{e,\gamma}$ decreases exponentially as t increases:

$$\bar{q}_{e,\gamma}(t, E) \sim (1/E_t)\exp(-\Sigma_{e,\gamma}(E)t).$$

In the subthreshold region for $t > t_s$, the spectrum ceases to depend on t and $\sim 1/E_t^{1.5}$. The invariability of the forms of suprathreshold and subthreshold areas

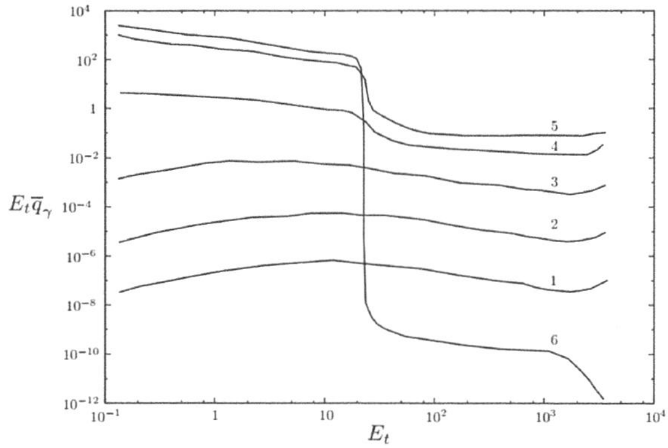

Fig. 13.5 Differential energy spectra. $E = 10^4$ MeV. $t = 0.1(1)$; $1(2)$; $10(3)$; $10^2(4)$; $10^3(5)$; $10^4(6)$

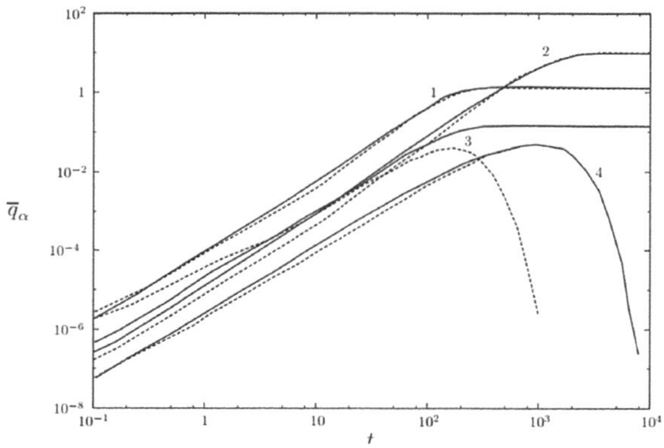

Fig. 13.6 Differential cascade curves. $E = 10^3$ MeV(1, 3), 10^4 MeV (2, 4); $E_t = 6$ MeV (1, 2), 30 MeV (3, 4), solid line for a primary photon, dotted line for a primary electron

spectra at a distance $t > t_s$ in [5, 7] is called the *standardization of spectra*. The difference between the suprathreshold and subthreshold regions of the spectra is explained by different mechanisms of their formation. Photons with energies $E_t \leq E_{\text{th}}$ are produced as a result of inverse Compton scattering of cascade electrons, while the spectrum of photons with energies $E_t \geq E_{\text{th}}$ is the result of competition between this process and absorption of photons due to pair photoproduction.

In the region $E_{\text{th}} < E_t < E$ at the maximum of the cascade curve, the value function $\bar{q}$ is practically independent of the primary energy E:

$$\bar{q}_{e,\gamma}(t_{\max}, E_t) \simeq 6.2\ 10^{-2}/E_t.$$

Figure 13.6 represents differential photon cascade curves $\bar{q}_{e,\gamma}$ in EFC, initiated by primary photons and electrons. A noticeable dependence on the type of primary particle, as can be seen from the figure, takes place only at the initial stage of cascade development, and the difference between $\bar{q}_e$ and $\bar{q}_\gamma$ is larger for large values of E_t/E. In the region of maxima cascade curves for $E_t > E_{\text{th}}$ and a plateau for $E_t < E_{\text{th}}$,

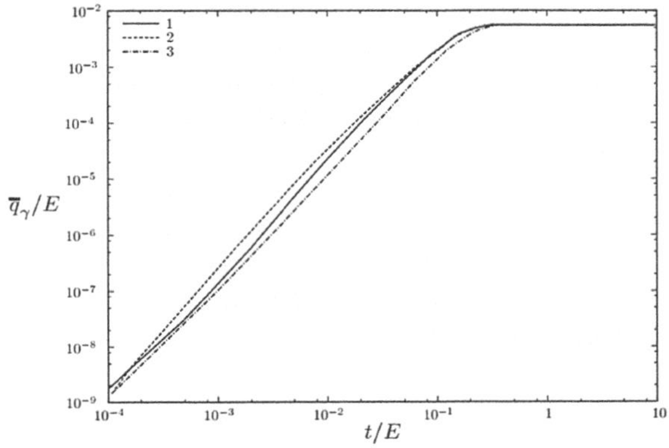

Fig. 13.7 Dependence $\bar{q}_\gamma/E$ on t/E. $E_t = 6$ MeV. $E = 10^3(1)$; $10^4(2)$; $10^5(3)$ MeV

the average EPC characteristics (not only $\bar{q}$) do not depend on the type of primary particle. On the distances $t > t_{\max}\bar{q}_e$ for $E_t > E_{\text{th}}$ again exceed $\bar{q}_\gamma$.

Calculations show that in EPCs initiated in a monoenergetic photon field, sufficiently energetic primary particles $E > E_m$ when $\Sigma_{e,\gamma} \sim (E_{\text{th}}/E)\ln(4E/E_{\text{th}})$, in certain ranges of t and $E_t\bar{q}\alpha(t, E_t; E)/E$ depends only on t/E, that is, the scaling behavior takes place. Calculations of average characteristics carried out in a wide range of primary particle energy allow one to study their dependencies on E, to establish the explicit form of scaling dependencies, and to determine the intervals where they are valid. Thus, analysis of the calculation results shows [7] that in the case of particle energies $E > E_m$ and $E < E_{\text{th}}$ the approximate representation

$$\bar{q}_{e,\gamma}(t; E; E_t) \simeq Ef(tE_{\text{th}}/E)g(E_t)$$

is valid, where the energy variables E and E_t are separated if to consider the form of the function f and g depending on the range of t.

Figure 13.7 $\bar{q}_\gamma/E$ demonstrates the *scaling form*, i.e., the quantity dependence on t/E. It follows from behavior of the function f, which first grows quadratically, then more slowly and at large distances is constant. On the first growth segment, the following approximation is valid:

$$\bar{q}_{e,\gamma}(t, E; E_t) \simeq E(tE_{\text{th}}/E)^2 Q_q(E_t).$$

In the plateau region $\bar{q}_{e,\gamma}$ are directly proportional to the energy of the primary particle,

$$\bar{q}_{e,\gamma}(t > t_s, E; E_t) \simeq EK_q(E_t), \tag{13.20}$$

and the coefficients K_q do not depend on the type of the primary particle. Note that its energy dependence is described as follows:

$$K_q \sim 1/E_t^{1.5},$$

which corresponds to the value of the index of the subthreshold sections of differential photon spectra (–1.5). Such a slope was also obtained in [7].

The independence of the standardized average characteristics of photons with energies below the photoproduction threshold of E_{th} pairs on t and a directly proportional dependence on the energy release of the source (in the case of sufficiently large size of its active region) makes these particles carriers of information about the luminosity of the source, which is its main characteristics.

If the high-energy photons generated by the source are not monoenergetic but are separated by the spectrum $S_\alpha(E)$, then the average EPC characteristics $J_\alpha(t;\, E_t)$ are given by expression

$$J_\alpha(t,\, E_t) = \int_{E_t}^{\infty} dE S_\alpha(E)\bar{q}_\alpha(t,\, E). \tag{13.21}$$

We consider the primary photon spectrum

$$S_\alpha(E) = \begin{cases} 1/E^\gamma, & E_{\text{min}} < E < E_{\text{max}}, \\ 0, & E < E_{\text{min}} \text{ and } E > E_{\text{max}}. \end{cases} \tag{13.22}$$

The value of the spectral index γ varied from 1 to 3, the upper limit spectrum cutoff was assumed to be 10^6 and 10^7 MeV, the lower limit E_{min} ranged from E_t to $E_{\text{max}}/10$.

Consider first the spectrum of high-energy primary photons, when E_{min} is greater than the maximum energy of the photon absorption cross section E_m and characteristics of each of the individual cascades initiated primary photons of a given energy E satisfy the scaling dependencies discussed above.

Let us preface the numerical results with some obvious relations following from the consideration of the EPC initiated by the spectrum primary photons as a superposition of individual cascades from individual photons of the spectrum.

Obviously, depending on the average characteristics of the distance three qualitatively different areas are distinguished. At short distances $T < 10E_{\text{min}}/E_{\text{th}}$, not exceeding the distance of the cascade maxima from of the least energetic photons, J_α will grow exponentially, the exponent will depend on E_t, as in the case of a single initiating particles. At large distances $t > 10E_{\text{max}}/E_{\text{th}}$, where all individual cascades are standardized, the EPC from the spectrum of primary photons will also be standardized. And there will be an intermediate area distances $10E_{\text{min}}/Ep < t < 10E_{\text{max}}/Ep$, where the individual cascades different energies are at qualitatively different stages of development, and the result of their superposition will essentially depend on the parameters spectrum of primary photons.

In the region of large distances, where for all $E \in [E_{\text{min}},\, E_{\text{max}}]$, $\bar{q}_\gamma$ satisfy the relation (13.20), the average characteristics of J_α also reach a plateau and their values are proportional to the average energy of the primary spectrum E_s

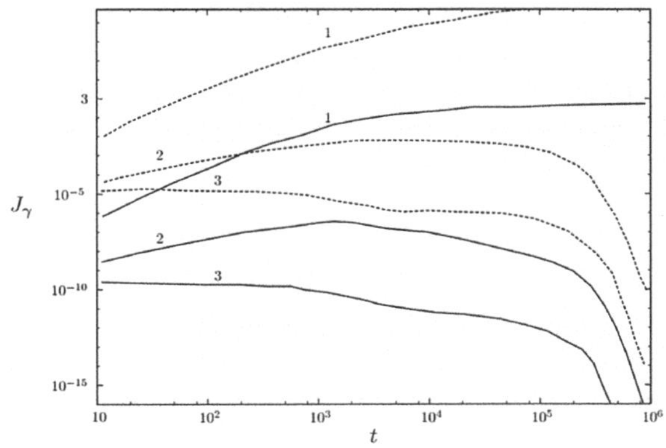

Fig. 13.8 The differential cascade curves. $E_{\min} = 10^4$ MeV, $E_{\max} = 10^6$ MeV, $E_t = 1(1)$, $10^4(2)$, $1.04 \cdot 10^5(3)$ MeV, $\omega = 10^{-2}$ MeV

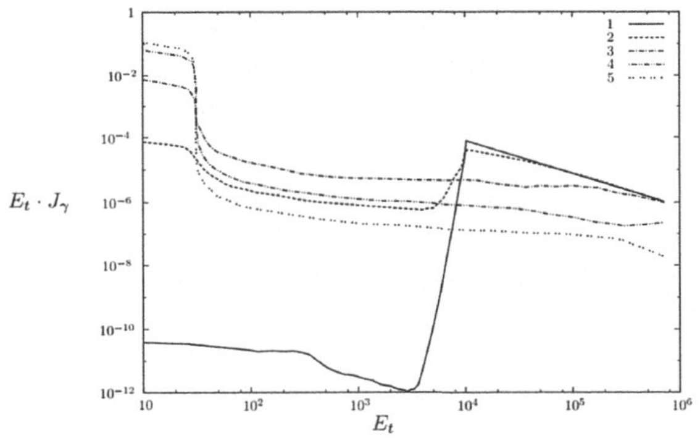

Fig. 13.9 Gamma quanta differential spectra. $E_{\min} = 10^4$ MeV, $E_{\max} = 10^6$ MeV. $t = 0$, $1(1)$; $10^2(2)$; $10^3(3)$; $10^4(4)$; $5 \cdot 10^4(5)$

$$J_\alpha(t, E_t) = K_\alpha(E_t)E_s, \quad E_s = \int\limits_{E_{\min}}^{E_{\max}} dE E S_\alpha(E).$$

On Figs. 13.8, 13.9 the value of $J_\alpha(t, E)$ in the EPC from the spectrum of primary photons is given. Other numerical results can be found in [6].

13.5 Variational Analysis

As a result of the development of the EPC in the active region of the NAG, the spectrum of particles S_a emerging from it differs from the primary spectrum $S_a(E)$. Changes in the spectrum of primary particles and the field of X-ray photons will lead to variations in the observed spectrum of γ-quanta.

In accordance with the apparatus described in Chap. 6, the sensitivity spectrum of γ-quanta to variations of the primary spectrum S_α is described by the variational derivative $\delta J_\alpha(t, E_t)/(\delta S_\alpha(E)dE)$. It gives the change in J_α caused by a unit change in the primary spectrum in a unit energy interval near the point E. From (13.21) it can be seen that this variational derivative is equal to $\bar{q}_a(t, E)$.

The change in J_α, caused by a unit change in the spectrum of X-ray photons at the point ω, is described in the first approximation of the perturbation theory by the variational derivative

$$\delta J_\alpha(t, E)/\delta n(\omega)d\omega = \int dE S_\alpha(E)\delta\bar{q}_\alpha(t, E)/(\delta n(\omega)d\omega). \tag{13.23}$$

For some perturbations of the primary spectrum δS_α and the X-ray photon field $\delta n(\omega)$, the variations of the flux emerging from the active region γ-quanta can be found using the sensitivity coefficients introduced above

$$\delta J_\alpha(t, E_t) = \int (\delta J_\alpha(t, E_t)/\delta S_\alpha(E)dE)\delta S_\alpha(E)dE$$
$$+ \int (\delta J_\alpha(t, E_t)/\delta n(\omega)d\omega)\delta n(\omega)d\omega. \tag{13.24}$$

The above expressions show that in order to study variations in the spectra of γ-quanta caused by changes in the primary spectrum and X-ray field, it is enough to know the value $\bar{q}_\alpha$ and its differential sensitivity $\delta\bar{q}_\alpha(t, E)/(\delta n(\omega)d\omega)$. Equations for the variational derivative

$$\delta\bar{q}_\alpha(t, E)/(\delta n(\omega)d\omega) \equiv \bar{q}_\alpha^{(1)}(t, E; \omega, n(\cdot))$$

are obtained acting on (13.3) by the operator $\delta/(\delta n(\omega)d\omega)$:

$$[\partial_t + \Sigma_e(E)]\bar{q}_e^{(1)}(t, E; \omega, n(\cdot)) - \int\limits_{E_t}^{E_e(\omega_2)} dE W_{e\gamma}(E \to E')\bar{q}_\gamma^{(1)}(t, E; \omega, n(\cdot))$$
$$- \int\limits_{\max\{E_t, E_e(\omega_2)\}}^{E_e(\omega_2)} dE' W_{e\gamma}(E \to E - E')\bar{q}_e^{(1)}(t, E'; \omega, n(\cdot))$$
$$= \frac{1}{n_0}\Bigg[\Sigma_e(E; \omega)\bar{q}_e(t, E) - \int\limits_{E_t}^{E_e(\omega)} dE' W_{e,\gamma}(E \to E'; \omega)\bar{q}_\gamma(t, E')$$
$$- \int\limits_{\max\{E_t, E_e(\omega)\}}^{E} dE' W_{e\gamma}(E \to E - E'; \omega)\bar{q}_e(t, E')\Bigg], \tag{13.25}$$

$$\partial_t \, \bar{q}_\gamma^{(1)}(t, E; \omega, n(\cdot)) + \Sigma_\gamma(E)\bar{q}_\gamma^{(1)}(t, E; \omega, n(\cdot))$$

$$- \int\limits_{\max\{E_t, E_{\gamma 1}(\omega_1)\}}^{E_{\gamma 2}(\omega_2)} dE' W_{\gamma e}(E \to E - E')\bar{q}_e^{(1)}(t, E'; \omega, n(\cdot))$$

$$- \frac{1}{n_0}\left[\Sigma_\gamma(E; \omega)\bar{q}_\gamma(t, E) - \int\limits_{\max\{E_t, E_{\gamma 1}(\omega_1)\}}^{E_{\gamma 2}(\omega_2)} dE' W_{\gamma e}(E \to E')\bar{q}_e(t, E')\right]. \quad (13.26)$$

It is easy to see that the equations for $q_e^{(1)}(t, E'; \omega, n(\cdot))$ coincide in form with the Eq. (13.3) for the value $\bar{q}_\alpha(t, E)$, but have other right-hand sides, which are expressed in terms of values q_α EPC developing in the unperturbed photon field $n(\omega)$, and the interaction cross section in monoenergetic field of photons with energy ω. In this regard, the scheme for calculating variations in the detector readings in a disturbed field is as follows. The Eq. (13.3) is preliminarily solved for the EPC value in the unperturbed field and the value of $J_\alpha(t, E_t)$ is calculated.

Then Eqs. (13.25)–(13.26) are solved for the differential sensitivity $q_\alpha^{(1)}(t, E; \omega, n(\cdot))$ at different points of the interval $[\omega_1, \omega_2]$ and the change in $\delta J_\alpha(t, E_t)$ caused by the perturbation $\delta n(\omega)$ is calculated. Equations (13.25)–(13.26) are solved by the numerical method. Calculations with different values of the energy intervals showed that for $\varepsilon = E_k/E_{k-1} = 10^{1/20} \div 10^{1/30}$ for $N = 3$ the relative differential sensitivity is calculated with an error of several percent.

The simplest case of photon field variability is a change in the photon density without changing the shape of the spectrum. Changes in the absolute intensity of X-rays with a constant slope of the spectrum have been observed in many nuclei of active galaxies. For calculation of variations in EPC characteristics caused by such variability photon field, it is enough to know only the important function at different depths of the development of a cascade in an unperturbed field. This is due to the use to measure the depth of development of the EPC of the cascade unit of length t_0, determined by the quantity n_0, the change of which is described by the variations of the field under consideration. The value of EFC characteristics at depth t_2 in photon field with density $n_2 = n_{02}/\omega^{\alpha_x}$ (cascade unit t_{02}) are equal to the values in the field with density $n_1 = n_{01}/\omega^{\alpha_x}$ (cascade unit t_{01}) at the depth t_1 defined by the relation

$$t_1 = t_2(t_{02}/t_{01}) = t_2(n_{01}/n_{02}).$$

Differential cascade curves of EPC gamma quanta developing in photon fields of different densities and the corresponding values of variations are shown in Figs. 13.10, 13.11. Here depth is measured in cascade units for a photon field of lower density. An increase in the photon density in the field leads to a more efficient development cascade, as a result, the cascade curve in a photon field of higher density turns out to be shifted toward shallower depths compared to a curve in a field of

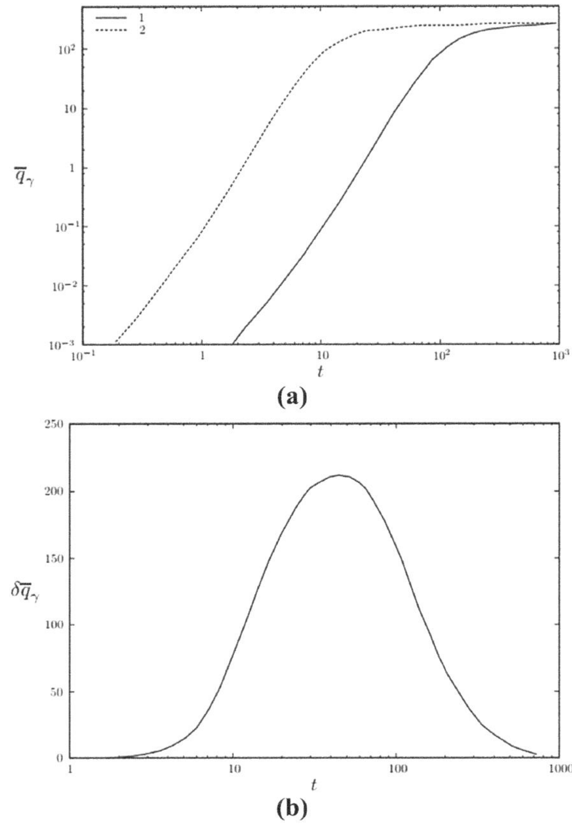

Fig. 13.10 **a** Differential cascade curves of EPC γ-quanta developing in photon fields of different density. $E = 10^3$ MeV, $E_t = 1$ MeV. 1—photon field with density n_1; 2—photon field with density $10n_1$; **b** Variations in the cascade curve caused by an increase in density photon field by 10 times

lower density. Dependence on the depth of variations spectra $\delta\bar{q}_\gamma$ of above-threshold $(E_t > E_{\text{th}})$ gamma quanta has a characteristic shape. As can be seen from the figure, even in the considered simplest case of variations in the density of the photon field, the variations in the spectra of gamma quanta depend significantly on the thickness of the compact region radiation in the source and the energy of primary particles, which determines the development of the cascade in depth.

Note that the data on variations in the gamma-ray spectra caused by changes in the density of X-ray photons of the field at a constant slope of the spectrum were also used to test the method that we developed for calculating variations. Comparison of the results of calculation of variations found using sensitivity coefficients $\bar{q}^{(1)}$ with variations obtained as a result of scaling showed that the error in calculating variations is several percent when the density of the photon field changes several times.

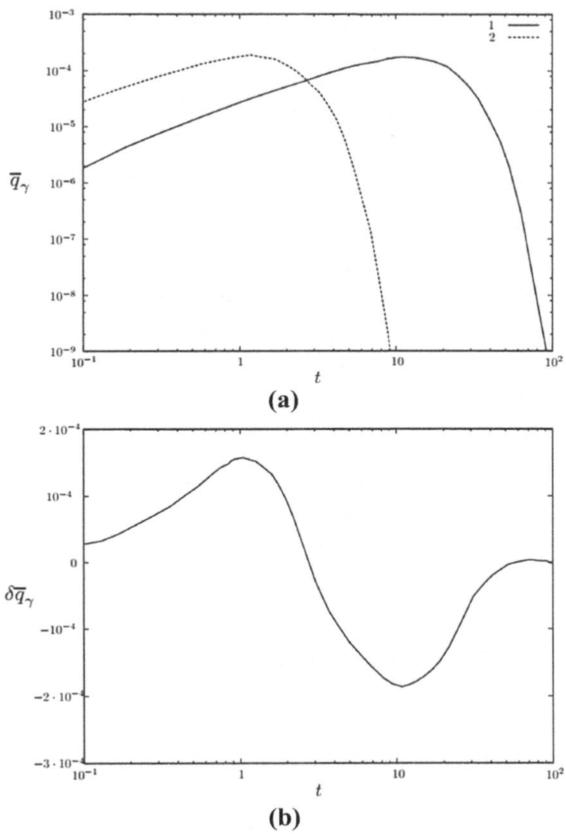

Fig. 13.11 **a** Differential cascade curves of EPC γ-quanta developing in photon fields of different density. $E = 10^5$ MeV, $E_t = 10^3$MeV. 1—photon field with density n_1; 2—photon field with density $10n_1$; **b** Variations in the cascade curve caused by an increase in density photon field by 10 times

13.6 EPC in Magnetic Fields

The EPC theory initially developed for high-energy cosmic rays passage through the atmosphere, nowadays, is significantly expanding its scope. The EPC in matter can be applied to some sources of high-energy cosmic radiation, in particular to the "hidden source" scenarios like massive black holes in centers of AGN or young pulsars inside the dense shells of recent supernovae explosions. Within another, so-called "beam dump" models applied to X-ray binaries, protons accelerated by the compact object (a neutron star or black hole), hit the atmosphere of the normal companion star, and thus result in the production of high-energy neutrinos and γ-rays.

F. Aharonian and A. Plyasheshnikov have published a very instructive in the pedagogical sense article "Similarities and differences between relativistic electron-photon cascades developed in matter, photon gas and magnetic field" [8].

For better understanding of the problem statement, we repeat the key features of the cascade model logics. The small-angle character of high-energy particle scattering makes the one-dimensional model to be quite sufficient for many aims. The small-angle nature of the scattering of high-energy particles makes the one-dimensional model quite sufficient for many purposes, and the homogeneity of the energy dependences of the inclusive cross sections gives the solutions of the kinetic equations a scaling form being suitable for the Mellin–Laplace transform method. As a result, the solutions are expressed through the set of the known functions $A(s)$, $B(s)$, $C(s)$ of the Mellin parameter s. Having fixed the energy of the primary particle that forms the cascade, we find a pair of particular solutions, the inverse Laplace transformation of which is carried out using the saddle point method. The real coordinate of the pass point, called the age of the cascade, establishes a relationship between the ratio E_0/E threshold and the depth of the cascade development. The dependence of the number of particles with such energies on the depth of observation is a cascade dependence with a characteristic maximum ($s = 1$), a steep increase in front of it ($s < 1$) and an exponential decrease after the maximum ($s > 1$). Quantitative behavior of the EPC in a specific medium is determined by the material parameters included in the system of cascade equations the main of which is the radiation length (or radiation unit) X_0, and the other is the particle registration threshold energy. One could expect, that the use of these parameters brings together the cascade curves in different materials, which gives grounds to speak on similarity. Nevertheless, this assumption needs to be verified by numerical calculations. Here are two papers comparing cascades in three different media that are relevant in astrophysical research [1, 2]. In the first of them, the ALTAI code and the SBCE methods were applied to investigate EPC properties in hydrogen gas and in ambient radiation and magnetic fields. For description of EPC in the monochromatic photon field with photon number density $n_0^{(G)}$ the radiation length was taken in the form

$$X_0^{(G)} = [4\pi n_0^{(G)} r_e^2]^{-1} k_0, \quad k_0 = \varepsilon_0 \omega_0,$$

and for the black-body radiation

$$X_0^{(BB)} \simeq 6.9 \times 10^{-7} (kT/m_e c^2)^{-2} \varepsilon \text{cm}.$$

Referring to the article [9], where EPC in a strong magnetic field was considered in the frame on the standard forward approach of the cascade theory, they applied the SBCE method adding two more environments to consideration. Let us dwell here on several interesting aspects of this work.

EPC development was considered in three various cosmic environments—gas (plasma), radiation, and magnetic fields. In these environments, γ-ray production may proceed with high efficiency through bremsstrahlung, inverse Compton scattering and synchrotron (and/or curvature) radiation, respectively. Generally, γ-ray production is effective when the cooling time that characterizes the rate of the process does not significantly exceed (i) the source (accelerator) age (ii) the characteristic time of non-radiative losses caused by adiabatic expansion of the source or particle escape, and (iii) the cooling time of competing radiation mechanisms that result

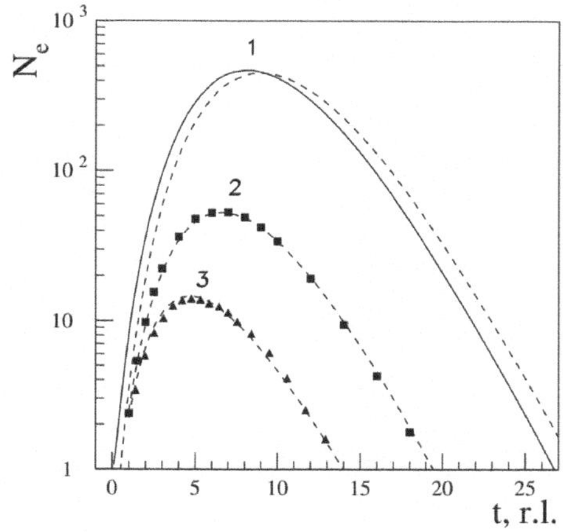

Fig. 13.12 Cascade curves of electrons for showers initiated by primary electrons (solid curves) and photons (dashed curves). The calculations are performed for the following primary energies $\varepsilon_0 = 2 \cdot 10^8$ (curve 1), $2 \cdot 10^7$ (curve 2), $2 \cdot 10^4$ (curve 3) and the ratio $\varepsilon_{\text{th}}/\varepsilon_{\text{cr}} = 125$ (curves 1 and 2), 0.05 (curve 3). For comparison, the results derived from the analytical cascade theory [11] (boxes) and by simulations with the ALTAI code [12] (triangles) are also shown

in low-energy photons outside the γ-ray domain. The differential cross sections for these processes were taken from [10].

The development of electromagnetic cascades in photon gas and magnetic fields is a more common phenomenon in astrophysics. In photon fields such cascades can be created on almost all astronomical scales, from compact objects like accreting black holes, fireballs in gamma-ray bursts and sub-pc jets of blazars to large-scale AGN jets clusters of galaxies (see the References in the article cited). As for the numerical calculations, the authors first of all compared the cascade curves obtained in the standard approximation of the cascade theory by two different methods SBCE and SAMC (ALTAI-version). The comparison revealed a good agreement (Fig. 13.12).

The description of EPC in the mono-energetic photon gas is more convenient to do introducing the radiation length $X_0^{(G)}$ in the following form

$$X_0^{(G)} = [4\pi n_0^{(G)} r_0^2]^{-1} \kappa_0,$$

where $n_0^{(G)}$ is the number density of photons. Apparently, $X_0^{(G)}$ corresponds, up to logarithmic terms, to the mean free path of γ-rays in the photon field at $\kappa_0 \gg \varepsilon_0 \omega_0$. Note that, unlike the radiation length in matter, the primary energy enters, through the parameter $\kappa_0 = \varepsilon_0 \omega_0$ in $X_0^{(G)}$. Obviously, for the ambient photon gas with a broadband energy distribution this parameter becomes meaningless.

In the magnetic fields, the cascade (radiation) unite is introduced by the formula

$$X_0^{(F)} = 0.207 \cdot 10^{-7} (H_{\text{cr}}/H) \chi_0^{1/3}$$

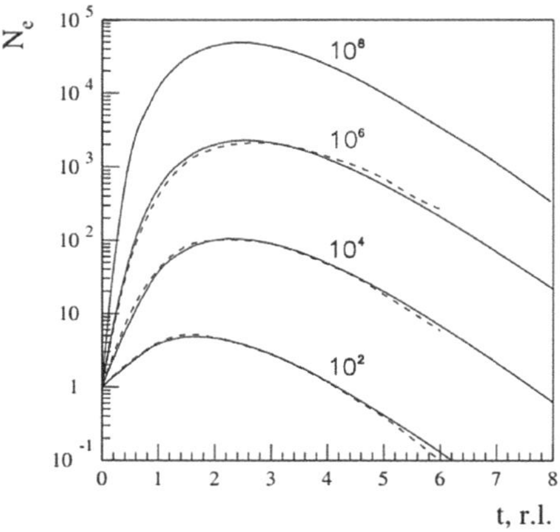

Fig. 13.13 Cascade curves of electrons for showers initiated by primary electrons in the magnetic field. Different values (indicated at the curves) of the ratio of primary and threshold energies are assumed for the fixed $\chi_{\text{th}} = \varepsilon_{\text{th}} H/Y_{\text{cr}} = 10^3$. For comparison, the results obtained in Ref. [13] are also shown (dashed curves)

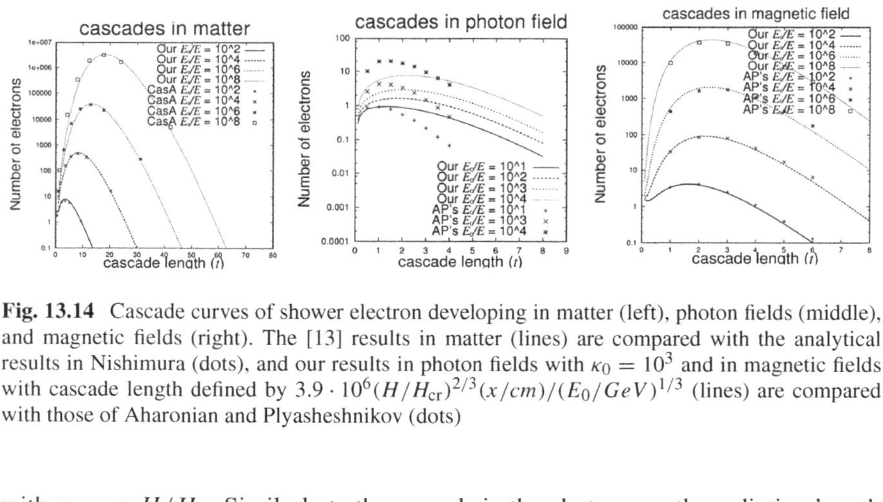

Fig. 13.14 Cascade curves of shower electron developing in matter (left), photon fields (middle), and magnetic fields (right). The [13] results in matter (lines) are compared with the analytical results in Nishimura (dots), and our results in photon fields with $\kappa_0 = 10^3$ and in magnetic fields with cascade length defined by $3.9 \cdot 10^6 (H/H_{\text{cr}})^{2/3}(x/cm)/(E_0/GeV)^{1/3}$ (lines) are compared with those of Aharonian and Plyasheshnikov (dots)

with $\chi_0 = \varepsilon_0 H/H_{\text{cr}}$. Similarly to the cascade in the photon gas, the radiation length in the magnetic field depends on the primary energy, but, in this case the dependence is slower, $X_0^{(F)} \sim \varepsilon_0^{1/3}$.

Nakatsuka et al [13] reevaluated electron–photon cascades developing in matter, photon fields, and magnetic fields by solving diffusion equations using a standard numerical integration method, and compared the results with the analytical results indicated in Nishimura (1967) and with those derived by the adjoint method of Aharonian and Plyasheshnikov (2003). These results turned out to be in agreement with those of Aharonian and Plyasheshnikov except the case of photon field (the central panel in Fig. 13.14). The cause for the difference is not discussed.

Finishing their article, the authors note that the cascade curves in the photon gas and magnetic field have features quite different from that in matter. The energy spectra of cascade particles are also considerably different from the conventional cascade spectra in matter. The spectra for the magnetic field have properties intermediate between those for cascade spectra in matter and in the photon field. Although for certain astrophysical scenarios the development of cascades in "pure" environments can be considered as an appropriate and fair approximation, at some conditions the interference of processes associated with interactions of cascade electrons and photons with the ambient photon gas and magnetic field (or matter) can significantly change the character of cascade development, and consequently the spectra of observed γ-rays. The impact is very complex and quite sensitive to the choice of specific parameters. Therefore, each practical case should be subject to independent studies.

References

1. Hayakawa S (1966) Electron-photon cascade process in intergalactic space. Suppl Progr Theor Phys 37–38:594–597
2. Ivanenko IP, Sizov VV (1991) Generation of gamma-rays in electron-photon cascades. Astrophys Space Sci 182:111–140
3. Aharonian FA, Plyasheshnikov AV (2003) Similarities and differences between relativistic electron-photon cascades developed in matter, photon gas and magnetic field. Astroparticle Phys 19:525–548
4. Nakatsuka A, Nishimura J (2011) An evaluation of electron-photon cascades developing in matter, photon and magnetic fields. Astrophys Space Sci Trans 7:345–350
5. Aharonian FA, Kirillov-Ugryumov VG, Vardanian VV (1985) Formation of relativistic electron-photon showers in compel X-ray sources. Astrophys Space Sci 115:201–225
6. Ivanenko IP, Lagutin AA, Linde IA (1991) Mean characteristics and fluctuations of electron-photon cascades developing in a photon field. Moscow State University, Nuclear Physics Institute, Preprint 91-24/228 (in Russian)
7. Ivanenko IP, Sizov VV, Trifonova SV (1988) Electron-photonic cascade in a photon field. I. Monoenergetic field. Moscow State University, Nuclear Physics Institute, Preprint 88-011/32 (in Russian)
8. Aharonian F, Plyasheshnikov A (2003) Similarities and differences between relativistic electron-photon cascades developed in matter, photon gas and magnetic field. Astroparticle Phys 19:525–548
9. Akhiezer AI, Merenkov NP, Rekalo AP (1994) On a kinetic theory of electromagnetic showers in strong magnetic fields. J Phys G: Nucl Particle Phys 20:1499–1514
10. Akhiezer AI, Berestetskii VB (1965) Quantum electrodynamics. Interscience, New York
11. Rossi B, Greisen K (1941) Cosmic-ray theory. Rev Mod Phys 13:240
12. Konopelko AK, Plyasheshnikov AV (2000) ALTAI: computational code for the simulations of TeV air showers as observed with the ground-based imaging atmospheric Cherenkov telescopes. Nucl Instr Meth 450:419–429
13. Nakatsuka T, Iyono A, Nishimura J (2011) An evaluation of electron-photon cascades developing in matter, photon and magnetic fields. Astrophys Space Sci Trans 7(3):345–350

For Further Reading

14. Anguelov V, Vankov H (1999) Electromagnetic showers in a strong magnetic field. J Phys G: Nucl Particle Phys 25(8):1755
15. Rao MV, Sreekantan BV (1998) Extensive air showers. World scientific
16. Grieder PK (2010) Extensive air showers: high energy phenomena and astrophysical aspects-A tutorial. Reference manual and data book
17. Heck D, Knapp J, Capdevielle JN, Schatz G, Thouw T (1998) CORSIKA: a Monte Carlo code to simulate extensive air showers. Report fzka 6019(11)
18. Alvarez-Muniz J, Engel R, Gaisser TK, Ortiz JA, Stanev T (2002) Hybrid simulations of extensive air showers. Phys Rev D 66(3):033011

Chapter 14
Cascade Ages, Similarity, Universality, and All That

14.1 Longitudinal Parameterizations and Uncertainties

A large number of particles continuously multiplying and changing their types form a complex stochastic process. Knowing the conservation laws and understanding the hierarchy of events allows to formulate the main characteristics of the process in terms of collective variables. The most important of them is the age parameter s, which relates the penetration depth of particles to the energy of the primary particle initiating the shower. The concept of the "age" of a shower has been in use in the cosmic ray community for more than half a century.

The most specific characteristic of a cascade process is the cascade curve. At a high energy E_0 of the primary particle, the deflection angles are small, and the particles move predominantly in one (initial) direction. The number of cascade particles crossing the plane perpendicular to this direction at a distance t is from the point of birth of the primary particle, and there is a cascade curve. Strictly speaking, this is a random curve, and (except some special problems, concerning, say, a measurement process) one usually deals with its mean value. The averaged cascade curve is a continuous curve with a maximum at some point $t_{\max}$. When solving the EPC development by using the Mellin–Laplace integral transformations, we obtain the cascade curve in an explicit form including the link between t and s. However, the primary particle, triggering the shower in atmosphere may be not only electron or photon but also proton or the iron nucleus.

The idea that all showers at the maximum, that is at the depth where the derivative of the shower size $N(t)$ vanishes, are "similar" independently from the energy and nature of the primary particle, can be naturally generalized, stating that all showers that have the same fractional rate of change with depth, that is the same "size slope" λ:

$$\lambda = \frac{1}{N(t)}\frac{dN(t)}{dt} \tag{14.1}$$

© Higher Education Press Limited Company 2024

V. V. Uchaikin, *Electron-Photon Cascades*, Nonlinear Physical Science,
https://doi.org/10.1007/978-981-99-7524-2_14

are also "similar". This means that to each value of the λ correspond well-determined shapes of the electron and photon spectra (again only valid for "most" particles) and a well-determined relative normalization for the two populations. It is intuitive (and will be later verified by detailed calculations) that the energy spectra of electrons and photons become progressively softer as λ decreases going from positive (when the shower size grows) to negative (when the shower size decreases) values. There is a one-to-one mapping between the values of the size slope λ and the values of the shower "age" s. This mapping is encoded by a function that, in the notation introduced by Rossi and Greisen in their classic paper, called $\lambda_1(s)$. The general definition of the shower age is therefore

$$s = \lambda_1^{-1}(\lambda) = \lambda_1^{-1}\left(\frac{1}{N(t)}\frac{dN(t)}{dt}\right), \tag{14.2}$$

where λ_1^{-1} is the inverse function to $\lambda_1(s)$.

The monotonically decreasing function $\lambda_1(s)$ has a single zero at $s = 1$, therefore according to Eq. (14.2) showers have age $s = 1$ at cascade maximum and age $s < 1 (s > 1)$ before (after) the maximum. The definition of the age parameter (14.2) may seem at first sight (and in some sense actually) arbitrary, since the size slope λ itself or any monotonic function of λ are also perfectly adequate to identify "similar" showers.

The cascade parameter s, also called the *shower age*, has played a very important role in the interpretation of the longitudinal development of cascade, highlighting the position of its maximum $t_{\max}$ by the condition $s = 1$, separating the cascade into the front (increasing) part and back (decreasing) parts. Despite the presence of a maximum, the cascade curve is far from symmetrical. The cause for this is easy to understand, considering the simplest model of particle reproduction: if at each step particle turns into two, distributing its mass (energy) in half, we will see a power-law increase in the number of particles corresponding to the initial stage of the cascade curve, and *counter reset* at the moment when the entire mass of the initial particle is used up. It is not difficult to obtain a smooth curve for the ensemble-averaged population of this process: it is enough to suggest a random spread of times between fission events. Introducing now an exponential (as in a Poisson process) distribution of intervals between events, we arrive at an exponentially increasing (now averaged) *population size–time* curve $N(t) = e^{\lambda t}$. This is a simple birth Furry process: if the initial mass is unlimited, this process continues indefinitely. It would seem that in order to obtain the decline in the average population size characteristic of cascade curves, it is enough to introduce absorption (in biological language, death).

This, however, was also not enough. In order to arrive at images like EPC cascade curves, *rapidly rising and then slowly falling*, it is necessary to introduce at least two types of particles with the possibility of converting them into each other. Biologists call such processes mutations, and from their point of view, electrons and photons are mutants with respect to each other.[1] Reasoning in terms of two-type (γ-e) particle cascade yields a sign-alternating slope of a shower curve

$$\frac{\partial \ln N}{\partial t} \simeq \lambda_1(s) \simeq \frac{1}{2}(s - 1 - 3\ln s), \quad s > 0.$$

The age parameter s illustrates the declining stage of a shower (growing for $s < 1$ and falling for $s > 1$) which is determined as the saddle point in the inverse Mellin transformation. In the one-dimensional cascade, it takes the very simple expression (using approximation B)

$$s = \frac{3t}{t + 2\ln(E_0/\beta)} = \frac{3t}{t + 2t_{\max}}.$$

It is easy to understand (thinking, for example, in terms of approximation B) the long tails of the cascade curves, formed by free paths of photons, having wider probability distributions both in longitudinal and lateral directions. This leads to an important conclusion: multiple Compton scattering is (at least, partially) responsible for the behavior of cascade curves at great depths, and what else more importantly, for the spatial distribution at large distances from the shower axis. Thus, one should expect refinement of those results of the cascade theory, which were obtained with neglecting Compton scattering and assuming $E_0 = \infty$. The direct experimental ways of reliably solving this problem were blocked: accelerators had not yet reached these energies, and in cosmic rays measurements there was no reliable way to separate showers created by gamma quanta from those created by protons.

The choice of the particular mapping of equation (14.2) is motivated by the fact that one can attach a direct physical meaning to the quantity s. The shapes of the electron and photon spectra for E above the critical energy ϵ (the average energy lost by an electron in a radiation length, which corresponds also to the electron energy for which radiative and collision losses are equal) and $E \ll E_0$ (with E_0 the primary particle energy) are well represented by a power law:

$$n_e(E) \sim n_\gamma(E) \sim E^{-(s+1)}. \tag{14.3}$$

These power-law behaviors stop when E approaches (from above) the electron critical energy.

First emerged in the study of the average longitudinal development of purely electromagnetic showers generated by photons or electrons, later it was applied to the lateral distribution of electrons around the EPC axis and then became useful for characterizing EMC families united into individual extensive atmospherical showers generated by hadronic primary particles such as protons or nuclei. Some recent works can refer to [1–3].

---

[1] Observe, that this is quite in the spirit of the modern theory of elementary particles.

It was then also applied [2, 3] to the lateral distribution of electrons around the shower axis. Soon, it was also understood that it is possible and useful to assign an "age" also to individual showers, and that the concept is applicable also to showers generated by hadronic primary particles such as protons or nuclei. We have rediscussed the concept of of shower age for the showers generated by ultrahigh-energy cosmic rays in the Earth's atmosphere. These results have been extended to the lateral distribution of electrons by Gora et al. [3]. This property of "universality" is obviously very important for the analysis and interpretation of high-energy cosmic ray observations, and it is therefore desirable to have a deeper understanding of its origin and of its limitations. In this work, we want to review critically the concepts of shower "age" and "universality". One of the main points we want to make is to argue that the definition of age of equation (14.1), while reasonably accurate in most cases, is in general not correct, and should be replaced by a better motivated and more accurate definition.

For energies below $E \sim \epsilon$ the electron spectrum has a sharp cutoff, while the photon spectrum has a "knee" and takes the form E^{-1}. The precise shapes of the cutoff of the electron spectrum, of the knee of the photon spectrum (that is the transition from the form $E^{-(s+1)}$ to the E^{-1} and the relative normalizations of the photon and electron spectra are all entirely determined by the age s (or equivalently by the size slope λ)) can be computed in detail.

The observations of the longitudinal profile of high energy showers obtained with the detection of fluorescence light, using the technique pioneered by Fly's Eye detector, and currently in use by the HiRes and Auger collaboration are commonly fitted using a four-parameter expression known as the "Gaisser–Hillas" profile: [4]:

$$N_{\rm GH}(t) = N_{\rm max}\left(\frac{t-\theta}{t_{\rm max}-\theta}\right)^{(t_{\rm max}-\theta)/\lambda} \exp\left((t_{\rm max}-t)/\lambda\right), \tag{14.4}$$

where t is the atmospheric depth (along the shower axis) in g cm^{-2}, $N_{\rm max} \equiv N(t_{\rm max})$, and θ and λ are two shape parameters relating to the cascade origin and the relaxation slope correspondingly.

A realistic estimate of the statistical uncertainties of important shower parameters is desired for many purposes, like data quality selection cuts, or the comparison between independent measurements like the surface and fluorescence detector measurements of the Pierre Auger Observatory or the Telescope Array. The uncertainties of $N_{\rm max}$, $t_{\rm max}$, θ, λ obtained after the minimization of Eq. (14.5), reflect only the statistical uncertainty of the light flux, which is referred to as "flux uncertainties" ($\sigma_{\rm flux}$). Additional uncertainties arise from the uncertainties on the shower geometry ($\sigma_{\rm geo}$), atmosphere ($\sigma_{\rm atm}$), and the correction for invisible energy ($\sigma_{\rm inv}$).

For the error propagation analysis, authors of [5], interpret the measured values of these parameters as random variables with expected values $\bar{\theta}$, $\bar{\lambda}$ coinciding with their true values, and variances σ_θ^2, σ_λ^2. Then, the parameter is determined by minimization of the function

$$\chi^2 = \chi^2_{\rm GH} + (\theta-\bar{\theta})^2/\sigma_\theta^2 + (\lambda-\bar{\lambda})^2/\sigma_\lambda^2. \tag{14.5}$$

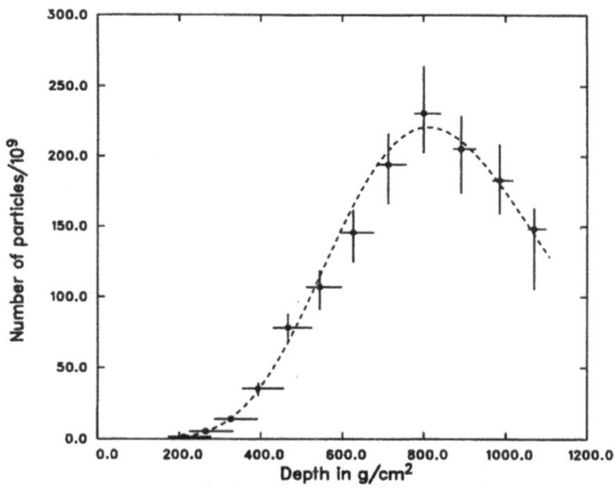

Fig. 14.1 The three-parameter best-fit shower profile is shown along with points obtained from the data in 5° intervals. The size at maximum is greater than 200 billion particles [6]

In this way, as the authors write, even if χ^2_{GH} is not sensitive to θ and λ, the minimization will still converge. On the other hand, if the measurements have small statistical uncertainties and/or cover a wide range in depth, the minimization function is flexible enough to allow for shape parameters differing from their mean values. These mean values can be determined from air shower simulations or, preferably, from high-quality data profiles which can be reconstructed without constraints.

Equation (14.5) can be easily extended to incorporate correlations between θ and λ and the energy dependence of their mean values. Air shower simulations indicate a small logarithmic energy dependence of the latter ($\leq 25\%$ and 5% per decade for θ and λ, respectively, as is stated in the article cited). In practice, it is sufficient to use energy-independent values determined at low energies, because at high energies the number of measured points is large and thus the constraints do not contribute significantly to the overall χ^2.

An example of GH-fitting is shown in Fig. 14.1 obtained for the light production on Fly's Eye from the intense Cherenkov beam along the shower axis ($E = 320$ EeV).

The authors of [5] give a very instructive separate analysis of the listed sources of uncertainties.

Flux uncertainty of the calorimetric energy. Even with the flux uncertainties of the Gaisser–Hillas parameters it is not straightforward to calculate the flux uncertainty of the calorimetric energy, which is given by the integral over the energy deposit profile:

$$Q = \int_0^\infty f_{\mathrm{GH}}(t)dt. \tag{14.6}$$

To solve this integral one can substitute

$$x = (t - \theta)/\lambda, \quad \xi = (t_{\max} - \theta)/\lambda \tag{14.7}$$

in the Gaisser-Hillas function to get

$$f_{\rm GH}(x) = N_{\max}(e/\xi)^{\xi} e^{-x} x^{\xi}, \tag{14.8}$$

so

$$Q = \lambda N_{\max}(e/\xi)^{\xi} \Gamma(\xi + 1) \tag{14.9}$$

and

$$\varphi_{\rm GH}(x) = \frac{Q}{\lambda(\xi + 1)} e^{-x} x^{\xi}. \tag{14.10}$$

In this way, $\sigma_{\rm flux}(Q)$ is obtained directly from the $\chi^2 + 1$ contour of Eq. (14.5).

Geometric uncertainties. Due to the uncertainties on the shower geometry, the distances r_i to each shower point are only known within a limited precision and correspondingly the energy deposit profile points are uncertain due to the transmission factors $T(r_i)$ and geometry factors $1/(4\pi r_i^2)$. Furthermore, the uncertainty of the shower direction, especially the zenith angle Θ, affects the slant depth calculation via $t_{\rm slant} = t_{\rm vert}/\cos\Theta$ and thus $t_{\max}$. Finally, the amount of direct and scattered Cherenkov light depends on the shower geometry, too, via the angles β_i.

The algorithms used to reconstruct the shower geometry from fluorescence detector data usually determine the following five parameters α_i [7], α_1, and α_2, denoting angles of the normal vector of a plane spanned by the shower axis and the detector (the so-called shower–detector–plane), α_3, denoting the angle of the shower within this plane, and α_4, α_5, being the time and distance of the shower at its point of closest approach to the detector. For any function $q(\alpha)$, standard error propagation yields the geometric uncertainty

$$\sigma^2_{\rm geom} = \sum_{i=1}^{5}\sum_{j=1}^{5} \frac{\partial q}{\partial \alpha_i}\frac{\partial q}{\partial \alpha_j}\sigma_{ij}(\alpha), \tag{14.11}$$

where $\sigma_{ij}(\alpha)$ denotes the covariance matrix of the axis parameters. As the calorimetric energy and $t_{\max}$ depend nontrivially on the shower geometry, the above derivatives need to be calculated numerically, i.e., by repeating the profile reconstruction and Gaisser-Hillas fitting for the ten new geometries given by $\alpha_i \pm \sqrt{\sigma_{ii}(\alpha)} \equiv \alpha_i \pm \sigma_i$ to obtain

$$\Delta_i = \frac{\partial q}{\partial \alpha_i}\sigma_i \approx \frac{q(\alpha_i + \sigma_i) - q(\alpha_i - \sigma_i)}{2}, \tag{14.12}$$

with which Eq. (14.11) reads as

$$\sigma^2_{\rm geom}(q) = \sum_{i=1}^{5}\sum_{j=1}^{5} \Delta_i \Delta_j \rho_{ij}(\alpha), \tag{14.13}$$

where

$$\rho_{ij} = \frac{\sigma_{ij}}{\sqrt{\sigma_{ii}\sigma_{jj}}} \tag{14.14}$$

denote the correlation coefficients of the geometry parameters α_i and α_j.

Atmospheric uncertainties. Whereas the Rayleigh attenuation is a theoretically well understood process, the molecular density profiles and aerosol content of the atmosphere vary due to environmental influences and need to be well monitored in order to determine the slant depth and transmission coefficients needed for the profile reconstruction. Uncertainties in these measured atmospheric properties can be propagated in the same way as the geometric uncertainties by determining the one sigma shower parameter deviations via Eq. (14.12).

Measured light yields at a receiver must be corrected for beam attenuation losses due to Rayleigh and Mie scattering. Let the location of source and receiver be t_1 and t_2 given in atmospheric slant depths, i.e., $t = t_0 e^{-h/H_0}$ sec, the vertical "depth" of the Fly's Eye detector. The Rayleigh transmission factor, T_R, or that fraction of photons not Rayleigh scattered from the beam can easily be calculated from Eq. (14.4).

Since $\rho dl = dt$, we have

$$T_R = \exp\left\{-\left(\frac{|t_1 - t_2|}{t_R}\right)\left(\frac{400\,\text{nm}}{\lambda}\right)^4\right\}. \tag{14.15}$$

For Mie scattering, we obtain a corresponding transmission factor, in this case, relating the two heights of source and receiver h_1 and h_2 by integrating Eq. (14.15) $(h_1 > h_2)$:

$$T_M = \exp\left\{\left(\exp\left(-\frac{h_1}{H_M}\right) - \exp\left(-\frac{h_2}{H_M}\right)\right) H_M \sec\Theta / L_M\right\}. \tag{14.16}$$

The overall transmission factor for a light beam propagating from source to receiver is thus

$$T = T_R T_M. \tag{14.17}$$

The energy uncertainty. Not all of the energy of a primary cosmic ray particle ends up in the electromagnetic part of an air shower. Neutrinos escape undetected and muons need long path lengths to fully release their energy. This is usually accounted by multiplying the calorimetric energy, Eq. (14.6), with a correction factor f_{enrg} linking the total primary energy of shower E_0 with its calorimetrical counterpart Q:

$$E_0 = f_{\text{enrg}} Q.$$

In EAS with neutrino and muon production, f_{enrg} depends on the primary energy. One of such parametrizations has a form

$$f_{\text{enrg}} = (a + bQ_c)^{-1}, \tag{14.18}$$

where a, b and c denote constants depending on the primary composition and interaction model assumed. This energy dependence needs to be taken into account when propagating the calorimetric energy uncertainty to the total energy uncertainty.

Due to the stochastic nature of air showers, the correction factor is subject to shower-to-shower fluctuations. The statistical uncertainty of f_{enrg} are parameterized as follows:

$$\sigma(f_{\text{enrg}}) \approx 1.66 \cdot 10^6 \lg(E_0/eV)^{-6.36}. \tag{14.19}$$

Their typical values are 2.5% at 10^{17} eV and 0.9% at 10^{20} eV.

The total statistical uncertainty. Summarizing the above considerations, the statistical variance of the total energy is

$$\sigma^2_{\text{stat}}(E_0) = E_0^2\sigma^2(f_{\text{enrg}}) + \left(\partial_Q f_{\text{enrg}} Q + f_{\text{enrg}}\right)^2 \sum_i \sigma_i^2(Q), \tag{14.20}$$

where i runs over the geometric, atmospheric, and flux uncertainties. Since the invisible energy correction does not affect the depth of shower maximum, its uncertainty is simply given by

$$\sigma_{\text{stat}}(t_{\max}) \sqrt{\sum_i \sigma_i^2(t_{\max})}. \tag{14.21}$$

The pull distributions of the reconstructed energy and shower maximum, shown in Fig. 14.2, both have a width of approximately one, which means that the total uncertainties from Eqs. (14.20) and (14.21) are good estimators for the actual event-by-event measurement uncertainties.

The Gaisser–Hillas profile implies the shower age:

$$s = \lambda_1^{(-1)}\left(\frac{1}{N_{\text{GH}}(t)}\frac{dN_{\text{GH}}(t)}{dt}\right) = \lambda_1^{(-1)}\left(-\frac{t - t_{\max}}{\lambda(t-\theta)}\right), \tag{14.22}$$

where $\lambda_1^{(-1)}(x)$ is the inverse function $\lambda_1(s)$. To study the age near shower maximum, one expands in a power series around the position of the maximum, using as expansion parameter the quantity

$$\delta = \frac{t - t_{\max}}{t_{\max} - \theta},$$

so that

$$s = 1 + \frac{\delta}{\Lambda} - \frac{4\Lambda - 3}{4\Lambda^2}\delta^2 + \frac{8\Lambda^2 - 12\Lambda + 5}{8\Lambda^3}\delta^3 + \ldots$$

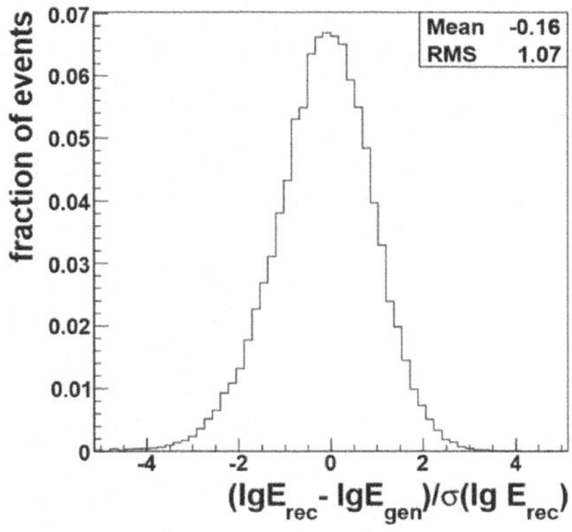

(a) Energy.

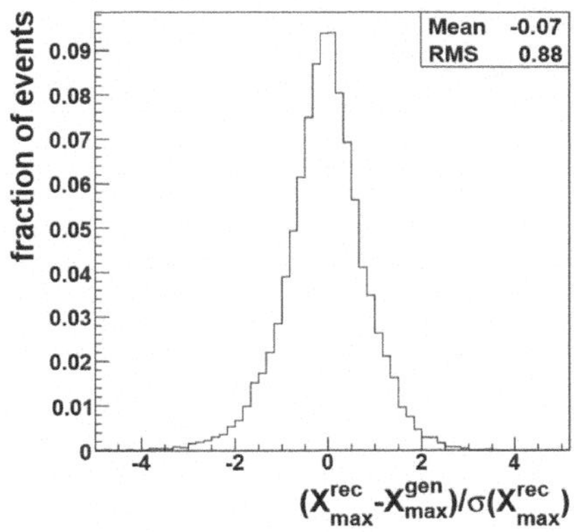

(b) $X_{\max}$.

Fig. 14.2 Energy and $X_{\max}$ reconstruction accuracy as function of the amount of detected Cherenkov light

The Gaisser–Hillas function can also be approximated by Gaussian in shower age

$$N(s) = N_{\max} e^{-(s-1)^2/2\sigma^2},$$

giving a good description of measured longitudinal profiles by means of only four ($N_{\max}$, $t_{\max}$, θ, Λ) fitting parameters, correlated because of the normalizing condition

$$s = \frac{3t}{t + 2t_{\max}}.$$

Talking about a cascade, in general, evokes first of all the image of a cascade curve describing a rapid increase in the number of particles in a cascade and a subsequent slower decrease in this number with the thickness of the layer passed by the cascade.

As an example, an EPC initiated by photons is shown in Fig. 14.3 taken from [8].

14.2 EAS with Respect to the Shower Age

One of the most popular codes for EAS calculations is the CORSIKA. We discuss here some results, obtained with this code and represented in [9]. Its authors note, that the article is motivated by an attempt to understand the EAS and to be able, despite their fluctuating development in the atmosphere, to reconstruct properly the primary energy on the shower-by-shower basis. Observations of the fluorescence light excited in the atmosphere by the charged particles of a shower, allow us to reconstruct its cascade curve $N(X)$, i.e., the particle number N as a function of the

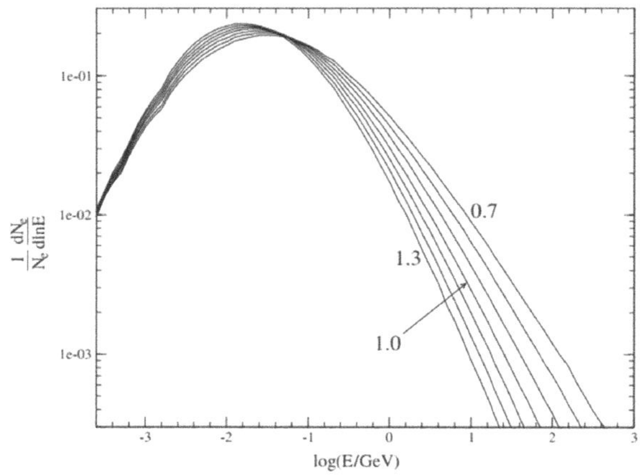

Fig. 14.3 Energy spectra of electrons at different shower ages s : 0.7, 0.8, ..., 1.3. Each curve is an average of 10 proton showers with $E_0 = 10^{19}$ eV [8]

slant depth X in the atmosphere. However, the transition from the observed light to the number of particles is not quite straightforward: the fluorescence light excited by an electron depends (although weakly) on its energy, also the light detected by a telescope contains a non-negligible fraction of Cherenkov photons. To take these effects into account one has to know the energy and angular distributions of the shower electrons at any (observed) level in the atmosphere. The fluorescence light depends on the shower age only, whereas the Cherenkov light also on the height in the atmosphere (as the threshold for this effect does depend on it). These yields do not fluctuate from shower to shower in such large showers, so that they can be calculated once forever and applied to any observed shower to deduce $N(X)$.

There were analyzed the energy spectra of electrons (normalized to 1) in proton-induced showers, for various ages s, each averaged over 10 showers, simulated with the CORSIKA program with the QGSJet model of nuclear interactions.

For each shower $X_{\max}$ has been determined separately. It has turned out that the age parameter s, defined for depth X of a hadronic shower in an analogous way describes very well the shower as a whole, and that for large showers the shape of the energy spectrum of electrons at a given level X in the atmosphere depends only on the shower development stage at this level, i.e. on its age s. The shape of the spectrum at a given age is the same independently of the nature of the primary particle (proton, or iron) and its primary energy. For big showers, where the number of particles in a given energy bin is large, the spectrum practically does not fluctuate from shower to shower. Thus, in the sense of the energy spectra of electrons, the authors conclude, *all large showers are similar*.

In conclusions to their article, the authors write: Based on shower simulations (with the CORSIKA program) we show that all large showers are similar when described by the shower age parameters. This means that electrons in all of them have the same shapes of the energy distributions at levels with the same ages. They also have the same shapes of the angular distribution, this being, however, a consequence of the fact that electrons of a given energy have an angular distribution independent of

s (i.e. the same everywhere in the shower). These distributions do not fluctuate from shower to shower as long as the number of particles in the energy- and angle-bins is large, what is well fulfilled in showers in the consideration here (above, say, 10^{18} eV). They do not depend on the primary particle, neither on its mass nor on energy.

The total number of particles (for a fixed primary energy) does, however, fluctuate to some extent from one shower to another, but when expressed as a function of s, these fluctuations are quite small. We have found that each shower curve $N(s)$ can be very well described by a four-parameter analytical curve—two halves of the Gaussian distributions with two different widths.

Thanks to the similarity of the showers, both the fluorescence and the direct and scattered Cherenkov light fluxes can be accurately predicted for any adopted $N(X(s))$ and compared with experimental results in order to find the best $N(X)$.

14.3 Lateral Distribution Fitting

The older generation of specialists in cosmic rays remember the end of the 1970s for a lively discussion about the lateral size of EPC given by the Nishimura–Kamata theory, and experimental data on EASs at the level of mountains. The author of this book (V.U.) was about 40 years old, and he managed to personally participate in two International Cosmic Ray conferences (1977, Plovdiv) and (1979, Kyoto), the agenda of which included the problem discussion. The situation that had developed by that time was concisely and adequately reflected in the Introduction to the article [10], referred to below as BCP (1980):

"Observation of extensive air showers shows that the age parameter describing the lateral profile is considerably younger (e.g., $0.6 \div 0.8$ for showers which have passed maximum size) than would be expected from the longitudinal development. This difficulty, if we consider the usual interaction models, cannot be explained by the complex association of parent photons issued from the hadronic cascade. It has therefore been proposed to separate the theoretical age parameter s_t from the local age parameter $s(r)$ (LAP) at any point using the NKG formula. The tendency of the LAP to increase at large distances from the axis has been noticed and a slight increase ($s(r) = \alpha \ln(\beta r) + s_t$) with axis distance r was deduced (Capdevielle et al. 1978, [12]) in modeling with the NKG formula and approximation B, after simulation of the hadronic cascade, but the amplitude of the variation of $s(r)$ was smaller than in experiment and the general values obtained for $s(r)$ were too large.

In the same period, some doubt has been cast on the validity of the NKG formula itself from Monte-Carlo calculations suggesting steeper lateral electron distributions (Allan et al. [29], Hillas and Lapikens [13]), but the primary photon energy (by reason of the time required for a complete Monte Carlo calculation of the electromagnetic cascade) was limited to near 100 GeV. This was not enough to assess the consequences for the usual extensive air shower energies of more than 10^6 GeV, as BCP wrote."

Our scientific group (Altai State University, Russia) accepted this challenge, and 2 years later, at the next conference in Kyoto, the result was already reported: the message published by Hillas about a significant underestimation of the size of EPC in the atmosphere by the NKG formula was confirmed. Moreover, this conclusion, made on the basis of the first Monte Carlo calculations at energies of 10–100 GeV, turned out to be relevant up to energies of 10^6 GeV and higher. The non-triviality of the problem was that the two main methods for calculating the EPC by that time (the Monte Carlo method and the Nishimura–Kamata integral transformations) had restrictions on the energy scale. "It was noticeable that the theorists were divided into two groups: one considered the MC results to be more reliable due to more scrupulous consideration of the cross sections of elementary processes, the other is that the NK theory is correct, and the MC calculations have not yet reached those high energies at which it becomes applicable.

Now, I insert my own reminiscence. Two years later, in Kyoto, in a narrow circle (J. Nishimura, L. Dedenko, V. Uchaikin), we continued the discussion. I made a cautious assumption that, due to the power-law decay of electron flux density with the distance akb the axis, the mean square radius is not a reliable measure of the width of this distribution in Monte Carlo results (as, indeed, in the analysis of experimental data), another kind of statistics is required here (Lévy statistics, which fascinated me at that time). An alternative explanation was advocated by Prof. Ivanenko group, indicating that the Nishimura-Kamata results were obtained in the $E_0 \to \infty$ approximation, and Monte Carlo calculations have not yet reached the required initial energies.

Our scientific team (Altai State University) accepted the challenge and thanks to SBCE-method easily raised the primary energy up to 10^6 GeV in parallel independent calculations both distribution function and its second moment. Results confirmed the validity of Ivanenko's point of view, or more exactly, agreement both methods—analytical (NK) and numerical (SBCE) ones.

Let me continue citing [10].

The radial electron distributions of EAS with size $N_e = 1.8 \cdot 10^5$ particles are shown in Fig. 14.1 for the different fitting functions of Uchaikin, NKG and Hillas. In spite of the large extrapolation in energy, the profile obtained in the assumption of Hillas' results is very similar to Uchaikin's with a slight underestimation. When compared with this last distribution, the distributions obtained via NKG look considerably flatter. At 5 m from the axis, the electron density is nearly two times larger with Uchaikin's assumption and at 200 m from the axis, it is two times smaller. Uchaikin's results always imply a steeper distribution for larger sizes, both at sea level and at 600 g/cm^2."

Figure 14.4 (taken from the BCP-paper) demonstrates the perfect agreement of results obtained with SBCE method, with other distribution, obtained by using Monte Carlo.

BCP has used the lateral distribution problem for the demonstration of comparative analysis of multi-parametric distributions obtained by different numerical methods and, for this reason, not allowing to make unambiguous conclusions about their compliance. Moreover, the comparison process involves experimental data for EAS, i.e., related not to pure EPC, but to their families generated by a nuclear cascade with incompletely defined interaction characteristics.

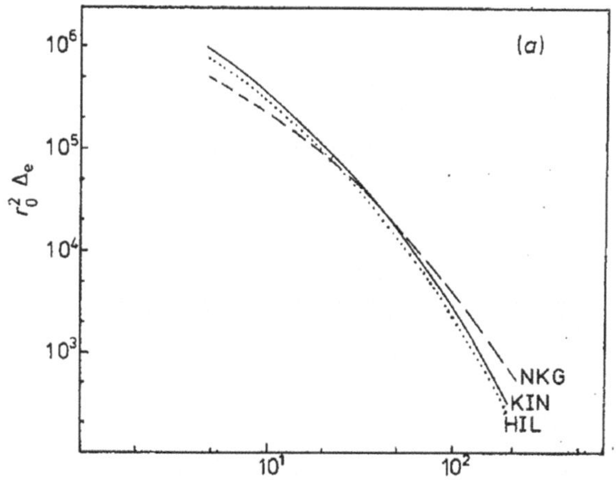

Fig. 14.4 Lateral electron density $r_0^2\Delta_\varepsilon$, as a function of axis distance r. Solid line–Uchaikin's assumption (KIN), dotted line—Hillas' assumption (HIL), dashed line—NKG assumption. $x = 1033$ g cm $^{-2}$. For the approximate description $\Delta_e \sim r^{-n}(5 \le r \le 10)$, the steepness n is given for the assumptions of NKG and Uchaikin (KIN)

These circumstances gave rise to an instructive discussion in the last quarter of the last century, an idea of which can give few quotes below from the article (Fig. 14.5).

It is easy to understand (thinking, for example, in terms of approximation B) the long tails of the cascade curves, formed by free paths of photons, having wider probability distributions both in longitudinal and lateral directions. This leads to an important conclusion: multiple Compton scattering is (at least, partially) responsible for the behavior of cascade curves at great depths, and what else more importantly, for the spatial distribution at large distances from the shower axis. Thus, one should expect refinement of those results of the cascade theory, which were obtained with neglecting Compton scattering and assuming $E_0 = \infty$. The direct experimental ways of reliably solving this problem were blocked: accelerators had not yet reached these energies, and in cosmic rays measurements there was no reliable way to separate showers created by gamma quanta from those created by protons.

Considering the function

$$f_{\rm NKG}(r) = C(s)(r/r_0)^{s-2}(1 + r/r_0)^{3-4.5},$$

CG takes two neighboring points, i and j, to involve a lateral age parameter (LAP) for any distribution $f(x)$ $(x = r/r_0)$ that characterizes the best fit by an NKG-type function in $[x_i, x_j]$:

$$s_{ij} = \ln(F_{ij} X_{ij}^2 X_{ij}^{4.5}) / \ln(X_{ij} Y_{ij}),$$

where $F_{ij} = f(r_i)/f(r_j)$, $X_{ij} = r_i/r_j$ and $Y_{ij} = (x_i + 1)/(x_j + 1)$. More generally, if $r_i \to r_j$, this suggests the definition of the LAP $s(r)$ at each point:

$$s(r) = \frac{1}{2x+1}\left((x+1)\partial_{\ln x} \ln f + (2 + \beta_0)x + 2\right).$$

J. Phys. G: Nucl. Phys. 8 (1982) 1317–1335. Printed in Great Britain

The radial electron distribution in extensive air showers

J N Capdevielle† and J Gawin‡

† Laboratoire de Physique Théorique§, Université de Bordeaux I, Chemin du Solarium, 33170 Gradignan, France

‡ Institute of Nuclear Research, ul. Uniwersitecka 5, 90-137 Lodz, Poland

Received 22 February 1982

Abstract. A three-dimensional simulation of EAS (extensive air showers) has been performed at different altitudes, including the new descriptions of electromagnetic cascades (steeper profiles than NKG). The various consequences are investigated: variation of age parameter with distance, effects of shower selection, size under-estimations induced by the treatment of experimental data assuming NKG and uniqueness of age parameter.

1. Introduction

It is now generally well known that the NKG (Nishimura–Kamata–Greisen) function does not give a good fit to the lateral distribution of electrons for a single value of the age parameter. This is admitted on the basis of analysis of experimental results (Tonwar 1981) over a large range of distances, from the shower core at mountain altitude as well as at sea level, and from sizes of 10^4 up to 10^8 particles.

In previous works (Capdevielle 1980, Bourdeau *et al* 1980) we considered the consequences of the conflicting propositions of solutions for three-dimensional transport equations of pure electromagnetic (EM) cascades (Uchaikin *et al* 1979a, b) in extensive air shower simulation and, in the same way, the effects of results of Monte-Carlo simulation (Hillas and Lapikens 1977). The remarkable agreement between Uchaikin's and Hillas' formulae was underlined and a typical variation of the local age parameter with distance with a general decrease was noticed.

The aim of this paper is to extend the effects of these new propositions, taking into account the complex situation resulting from the presence of the hadronic cascade, to obtain a better analysis of the general features of the radial electron distribution of EAS, since this is the fundamental basis of our understanding of ultra-high-energy phenomena.

2. The local age parameter (LAP)

2.1. Definition

In the longitudinal development of a pure EM cascade, the slope of a shower curve can be expressed approximately by a function of s alone as (Hayakawa 1970)

$$\partial N/\partial t \simeq \lambda_1(s) \simeq \tfrac{1}{2}(s-1-3\ln s) \qquad \text{if } s \text{ is not too small.}$$

§ Equipe de Recherche associée au CNRS.

Fig. 14.5 Beginning the debate on radial distribution

If $\beta_0 = 4.5$, $f_{\rm NKG}(r)$ with $s = s(r)$ can be used to fit f in the neighborhood of r.

"The similarity of Hillas' and Uchaikin's results, CG wrote, giving a steeper profile than NKG, is now well known: a typical effect in the new propositions is the very characteristic variation, for the same showers, of the LAP, $s(r)$. Compared with NKG, $s(r)$ decreases from the core to reach a minimum and increases far from the axis. In a hybrid simulation (Monte Carlo between $10^4 - 10^2$ GeV and use of the same formulae below 10^2 GeV), it was established that the age parameter can be lower by some 30%. In both cases, the effect of the fluctuations of the most energetic particles gives a larger increase of $s(r)$ at a large distance. Information from parallel work (Krys et al. 1980, [27]) indicates that the distribution could be steeper when the starting point of the cascade is deeper in the atmosphere. Such an effect of atmospheric modulation can be described by our parametrization, replacing the factor m in Uchaikin et al. formula by the factor

$$m_{\rm AM} = (0.78 - 0.21s)(a_0 + b_0 t),$$

where $a_0 = 1.54$ and $b_0 = 0.54 \cdot 10^{-3}$".

Returning to the time of the discussion, it is worth to mention that it was initiated with two papers of famous French astrophysicists. First of them, [12] reported on simulations of the electron lateral distribution for showers of sizes between 10^5 and 10^7 particles at mountain altitudes which cannot be described by a single age parameter and that its value increases with radial distance according to $s = \alpha' \ln r + \beta'$ (10 m $< r <$150 m). At the universal distance of 45 m (and not 70 m as has been considered previously), there exists a simple proportionality between shower size and density and here the age parameter can be considered to be equal to $s_{\rm t}$, the theoretical value. Comparison between these calculations and the Tien Shan experimental data shows a systematic size underestimation of 16%.

In Figs. 14.6–14.7, the authors compared various lateral distributions.[2]

"The approach of Uchaikin, write the authors further, is again analytic, starting from adjoint equations to (1): a different path is followed, involving the application of Bessel–Fourier transformations to r, numerical integration on t, and on E by partial polynomial functions. This method, which is also subject to several approximative steps and numerical inverse transformations, leads to a limiting radial distribution (as $E \to \infty$) differing from the NKG formula:

$$\rho_{\rm U}(r) = (mr_0)^{-2} \rho_{\rm NKG}(r/m),$$

where $m = 0.78 - 0.21s$ for $0.8 \leq s \leq 1.6$, r_0 is Moliere radius, $s = 3t/(t + 2\ln(E_0/\varepsilon_0))$ is the usual cascade parameter ($\varepsilon_0 = 84$ MeV) and E_0 is a primary photon energy. A lower energy threshold of 0.25 MeV in the calculations of Uchaikin is considered to give all the electrons. (p. 902).

[2] The symbol KIN arose because in the program of the Kyoto's Conference I was named in the Chine manner as *U Chai Kin*.

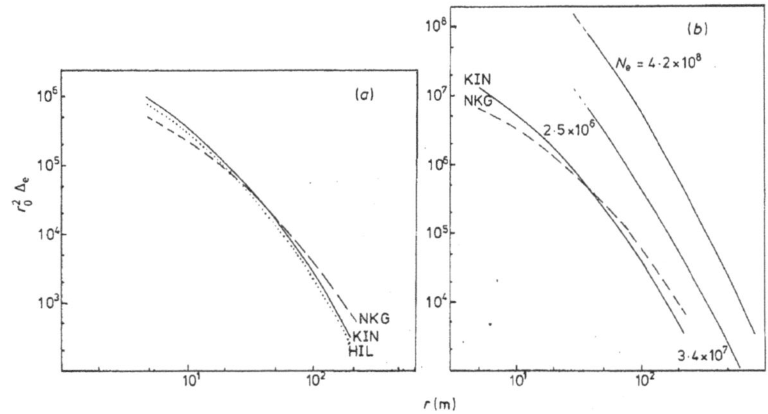

Fig. 14.6 Lateral electron density $r_0^2 \Delta_\varepsilon$, as a function of axis distance r. Solid line–Uchaikin's assumption (KIN), dotted line—Hillas' assumption (HIL), dashed line—NKG assumption. $x = 1033\ \mathrm{g\,cm}^{-2}$ (the see level). For the approximate description $\Delta_e \sim r^{-n}(5 \le r \le 10)$, the steepness n is given for the assumptions of NKG and Uchaikin (KIN)

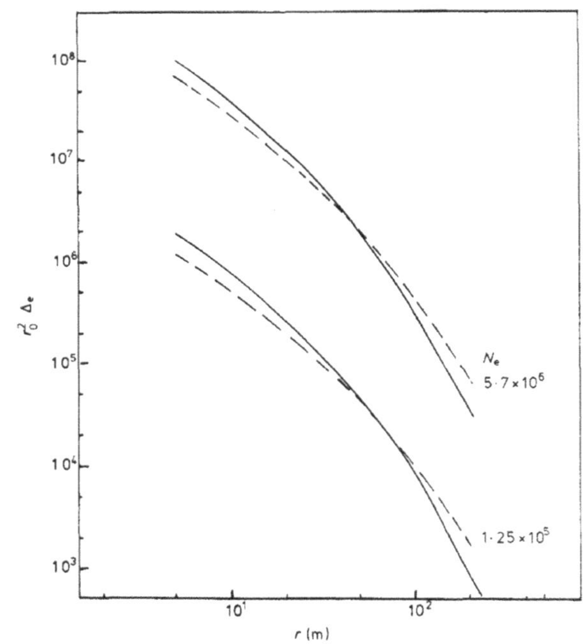

Fig. 14.7 Lateral electron distribution at high altitude ($x = 600\ \mathrm{g\,cm}^{-2}$ mountain level). Solid lines are KIN-results, dashed lines are NKG

This formula has been extrapolated according to the values of a_1, a_2, b_1, b_2 given at 10 and 100 GeV. In 1977 [13] Uchaikin's formula $\rho_U(r)$ allows us to calculate the total electron density at a given distance by superposition of the contribution of each individual EPC coming from the nuclear cascade. This hadronic cascade has been simulated by semi-Monte Carlo analytic procedure with the standard model (Capdevielle et al. [12]). The proton and pion mean free paths in air have been taken as 80 g cm^{-2} and 120 g cm^{-2}, respectively, and the electron radiation length to has been taken as 36.4 g cm^{-2}".

14.4 The TAP-LAP Analysis

The radial electron distributions of EAS with size $N_e = 1.8 \cdot 10^5$ particles are shown for the different assumptions of Uchaikin, NKG and Hillas. In spite of the large extrapolation in energy, the profile obtained in the assumption of Hillas' results is very similar to Uchaikin's with a slight underestimation. When compared with this last distribution, the distributions obtained via NKG look considerably flatter. At 5 m from the axis, the electron density is nearly two times larger with Uchaikin's assumption, and at 200 m from the axis, it is two times smaller.

The logarithmic derivative of density as a function of axis distance has been averaged by BCP between 5 and 10 m and the results are reproduced in Table 14.1 for different atmospheric depths x (g cm^{-2}) and sizes. For the approximate description $\Delta_e \sim r^{-n} (5 \le r \le 10)$ the steepness n is given for the assumptions of NKG and Uchaikin (KIN). Although the steepness is different with the two models, the densities are identical at a distance of about $0.3 \div 0.6$ Moliere radius shown for the same size.

Two new EPC parameters occur in the paper cited. One of them, the *theoretical age parameter* (TAP) s_t, following from the longitudinal EPC development, has been introduced in the paper [12] devoted to the lateral electron distribution at mountain altitudes (700 g cm^{-2}) for shower sizes between 10^5 and 10^7 particles and to the interpretation of its shape in terms of the age parameter and its fluctuations. The theoretical age parameter, s_t was introduced as averaged from the different ages obtained by the method of the moments in individual cascades and the local age

Table 14.1 The angular dependence of the effective registration area

$N_e/10^5$	$0 \cdot 257$	$1 \cdot 878$	$25 \cdot 41$	
n (NKG)	$1 \cdot 090$	$1 \cdot 111$	$1 \cdot 130$	$x = 1033$ g cm^{-2}
n (KIN)	$1 \cdot 231$	$1 \cdot 257$	$1 \cdot 283$	
$N_e/10^5$	$1 \cdot 246$	$6 \cdot 24$	$61 \cdot 9$	
n (NKG)	$1 \cdot 192$	$1 \cdot 227$	$1 \cdot 277$	$x = 600$ g cm^{-2}
n (KIN)	$1 \cdot 281$	$1 \cdot 321$	$1 \cdot 374$	

parameter, $s(r)$, defined from the logarithmic gradient of the total density, by identifying at this distance the NKG parameter having the same gradient. A statistical study of the crossing point of different normalized distributions of individual showers of fixed size has been carried out using two methods (minimization and a histogram of crossing points between structure functions in each interval of core distance).

A detailed analysis of Monte Carlo average lateral densities and of the age parameter in intervals of 10 m reveal the existence of three different size ranges, that the distribution is steeper at small core distances and flatters far from the shower axis than that for the theoretical age parameter. It looks as if it may be possible to describe the profile of the average lateral distribution by an age parameter which increases with radial distance. Such a dependence, established for three size N_e ranges ($1 \div 1.5 \cdot 10^5$; $3 \div 4.5 \cdot 10^5$; $3 \div 4.5 \cdot 10^6$) is approximated by the expression

$$\bar{s}(r) = \alpha \ln(\beta r / r_0) + \bar{s}_t,$$

where $\bar{s}_t$ is the average value of s_t at fixed shower size and $15 \le r \le 150$ m. The coefficients α and β are weakly model dependent, so it is convenient to express them in terms of s_t to get a pure phenomenological picture: for $\bar{s}_t \ge 0.9$,

$$\alpha(\bar{s}_t) = 0.126\bar{s}_t - 0.0993, \quad \beta(\bar{s}_t) = 6.99 - 4.2\bar{s}_t.$$

From Δ_{45} (the density at 45 m) it is possible to obtain the shower size and particle densities by replacing s in the NKG formula by $\bar{s}(r)$ for each value of r (or by determination of α', β' with $\bar{s}(r) = \alpha' \ln(r/r_0) + \beta'$ if $s(r)$ is measured at two points).

The authors conclude the cited paper with suggestion for a more accurate description of the lateral electron distribution at mountain altitudes:

$$\Delta(r) = \frac{N_e}{2\pi r_0^2} \frac{1}{\phi(r)} \frac{\Gamma(4.5 - s(r))}{\Gamma(s(r))\Gamma(4.5 - 2s(r))} \left(\frac{r}{r_0}\right)^{s(r)-2} \left(\frac{r}{r_0} + 1\right)^{s(r)-4.5} = N_e f(r) \tag{14.23}$$

for $10^5 \le N \le 10^7$ and $15 \le r \le 150$ m. This is the NKG formula including a logarithmic increase of age with axial distance:

$$s(r) = \alpha' \ln r + \beta'. \tag{14.24}$$

The coefficients α' and β' can be determined directly if s is measured at two points. The function $\phi(r)$, introduced for the renormalization of the NKG function (non-normalized with s independent of r), is expressed by

$$\phi(r) = 1.04 \left(\frac{0.7326}{r} + 0.937\right), \quad r \le 45 \text{ m},$$

$$\phi(r) = 1.04 + 0.0428 \tanh\left[0.52 \left(\frac{r - 45}{45}\right)\right], \quad r \ge 45 \text{ m}.$$

This formula (14.24) is as valid for individual showers as it is for the average. If the experimental situation is such that the shower age and particle densities are well estimated at 45 m from the axis, the shower size can be estimated directly from

$$N_e = 0.435 \cdot 10^5 \Delta(45 \text{ m}). \tag{14.25}$$

If this is not the case, $s(r)$ has to be estimated at two distances from the shower axis and $s(45)$ is estimated from Eq. (14.24). The density $\Delta(45)$ is estimated from (14.23) and one measured density at a given distance r_1:

$$\Delta(45) = \frac{f(45)}{f(r_1)}.$$

The shower size is given by (14.25). When $s(r_l)$ is measured to a high degree of accuracy, the formula (14.23) gives the shower size directly:

$$N_e = \frac{\Delta(r_1)}{f(r_1)}.$$

For the case of the Tien Shan experiment, the calculations show that measurements of the particle density near 45 m are a better measure of the shower size than the particle density at 70 m from the shower core. Measurements at 45 m from the axis yield significantly more accurate values for the shower parameters than the case at 70 m.

Numerical analysis of EAS data by using the theoretical age parameter s_t (TAP) directly obtained from the longitudinal electromagnetic component (Capdevielle et al. [12]) for estimation of the local age parameter (LAP), characterizing the logarithmic increase of age with axial distance and established in the form

$$s(r) = \alpha_1 \xi^2 - \beta_1 \xi - \delta_1 + s_t \tag{14.26}$$

with

$$\xi = \ln(\gamma_1 r/r_0),\ \alpha_1 = 0.0836,\ \beta_1 = 0.0216,\ \delta_1 = 0.348,\ \gamma = 2.633,$$

led BCP to the following conclusion.

The uncomfortable situation in which extensive air showers are assigned small values of the age parameter together with maximum longitudinal development at high altitude can be ruled out if Uchaikin's results are adopted. There need be no contradiction in the observation of steep profiles in the radial electron distribution at low altitude if relation (14.26) is used to estimate s.

The remarkable agreement between Uchaikin's and Hillas' formulae was underlined and a typical variation of the local age parameter with distance with a general decrease was noticed by Capdevielle and Gawin [12].

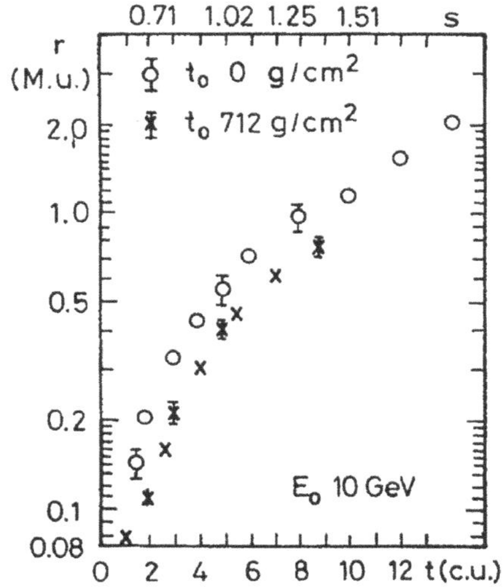

Fig. 14.8 Mean radii of the electromagnetic cascades

The general application of Uchaikin's distribution to EAS simulations appears to give a more coherent interpretation of experimental data on the lateral structure of EAS at all altitudes. If this distribution is admitted it is no longer necessary to introduce new exotic processes in the hadronic cascade. In particular, there would not appear to be any contradiction between the steepness observed in the lateral electron distribution and interaction models involving a high rate of multiple production.

I will end my discussion of lateral distribution by citing two papers published after our report in Kyoto and estimating its result.

One of them [27] demonstrates that, whereas the longitudinal development is well described by commonly used analytical formulae, the simulated lateral distributions are clearly narrower than those obtained from the so-called Nishimura–Kamaia–Greisen formulae. The deviations from those formulae are in fact more complicated. The lateral distributions of electrons in showers of a fixed age depend both on the level of shower starting point and on primary energy. Up to now, the author writes, it was always assumed that the lateral distribution of particles expressed in Molier units depends only on one parameter which describes longitudinal development of cascades, i.e., the age parameter. It means, that for the same stage of development of cascades the distributions of particles at different depths in the atmosphere should be identical. In Fig. 14.8, one can see the calculated values of the mean distances of secondary particles from the cascade axis versus depth of development of cascades. This dependence is shown for both above-mentioned cases, namely for the cascades initiated at the depth of 712 g/cm^2 and at the top of atmosphere for primary energies of 10 GeV. As one can see the mean distance of particles from the cascade axis (width of the lateral distributions) depends on the starting point of cascade in the atmosphere for the same stage of the development of the cascade (the density effect). The width of lateral distribution becomes larger with the larger density gradient in the

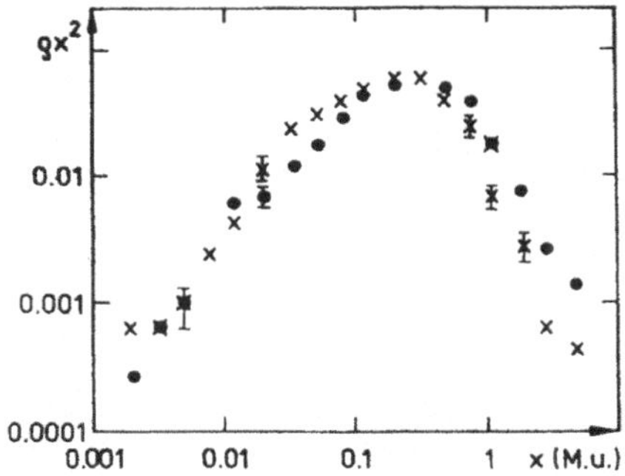

Fig. 14.9 Lateral distribution of all secondary electrons (above 1 MeV) in cascades initialed by primary photon of energy 50 GeV at the depth of 5 c.u. beyond the starting point. The two starting points of cascades are shown: crosses—the depth of 712 g/cm^2, circles—the top of the atmosphere. ρ is the density of electrons (including positrons). $X = \frac{r}{r_1} \cdot \rho x^2$ is normalized to one particle in the cascade on this level

atmosphere. It means that the larger height of production of particles (for cascades initiated higher in the atmosphere) relates to the wider lateral distribution. It is clearly seen in Fig. 14.9 where the lateral distributions of electrons in cascades initiated at different depths in the atmosphere are presented. These discrepancies may result from the fact that the Molier unit characteristic of the observation level was used, while actually the cascade develops in the same layer of air above this level, where the density of air is smaller and changes continuously.

It has pointed out by several authors, the author writes further, that the electron lateral distributions in electromagnetic cascades cannot be well described by the NKG formula. The Hillas approximation [13] of the lateral distributions of electrons obtained on the basis of simulation of the electromagnetic cascades in air and the theoretical approximation of Uchaikin [28] based on the numerical solution of the diffusion equation show that the lateral distributions are narrower than those predicted by the NKG formula.

"Comparison of MC results obtained for two different cascade starting points in atmosphere with the Hillas and the NKG curves is shown in Fig. 14.10 As it is visible in Fig. 14.9 MC lateral distributions are narrower when compared with the NKG formulae with the age parameter taken from the longitudinal development, especially for the cascades initiated deep in the atmosphere. The present situation as illustrated in Fig. 14.9 is such that at small distances from the cascade axis evident differences between the curves exist. The Hillas results and these simulations indicate that, at small distances from the core, the lateral distributions of particles are flatter than it was thought up to now and they are steeper at large distances. The Uchaikin formula was obtained for particles at distances greater than 2 m and cannot be used for the nearer ones. At large distances, they agree with the Hillas curves. It means that the number of particles near the axis as calculated from the NKG formula was overestimated.

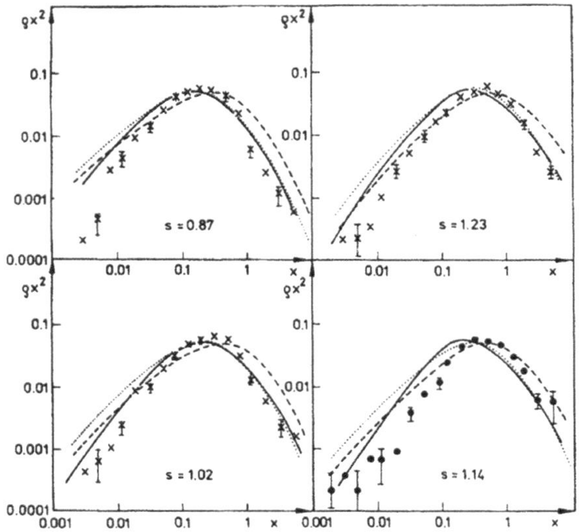

Fig. 14.10 Lateral distributions of electrons. The points are simulations, the curves are - - - NRG functions, —-Hillas formulae, and . . . Uchaikin formulae. The two starting points of the cascades: crosses—the depth 712 g/cm^2, circles—the top of atmosphere

These deviations have significant consequences for the experimental investigations of high-energy cosmic rays. For instance, the flat distributions of electrons at small distances from the axis would lead to underestimation of the energy in subcores in EAS" [29, p. 119].

I end this section with referring the relatively recent paper [33], showing that the theoretical investigation of EPC and the practical use of their results are still in the sphere of astrophysicists attention. The paper suggests the separate use of two cascade parameters, $s_\parallel$ and $s_\perp$ for fitting longitudinal and lateral sizes EPC: the cascade curve Greisen formula

$$N_e = \frac{0.31}{\sqrt{\ln(E_0/\epsilon)}} \exp[t(1 - 1.5\ln(s_\parallel)], \tag{1}$$

with

$$s_\parallel = \frac{3t}{t + 2\ln(E_0/\epsilon_0)}, \tag{2}$$

and the modified NKG structure function

$$f(r) = C(s_\perp)(r/r_m)^{s_\perp - 2}(1 + r/r_m)^{s_\perp - 4} \tag{3}$$

with

$$C(s_\perp) = \frac{\Gamma(4.5 - s_\perp)}{2\pi\Gamma(s_\perp\Gamma(4.5 - 2s_\perp))}.$$

$\rho(r) = (mr_{\rm M})^{-2}\rho_{\rm NKG}(r/m)$, being the mean electron density at r. The factor $m = 0.78 - 0.21 s_{\|}$ compensates the effect of NK asymptotic condition $E_0 \to \infty$.

14.5 Space–Time Similarity Relations in Inclined Air Showers

It is well known that radiation transport in an infinite homogeneous isotropic medium is invariant under shear and rotation transformation. This statement is formulated by E.Wigner in terms of group theory [21], and widely used by harmonic analysis of radiation flux [22]. The detailed and successive elucidation of the symmetry properties of transport equation solution can be found in the books [23, 24]. As an example one can point at the case of a chemically homogeneous medium with varying density ρ along one of cartesian coordinates z. The intensity of radiation from a unidirectional point source integrated over a plane being perpendicular to the axis depends only on mass distance

$$\tau = \int_{z_0}^{z} \rho(z')dz'$$

between source point z_0 and observation plane z and can be obtained from solution of transport equation for the medium with some constant density. Such kind of transformation was successfully used by investigation of transition effects in EPC and add (see Sects. 9.2–9.3). We consider a particular case of density dependence law (z-axis, the axis is directed downward (toward the Earth's surface):

$$\rho(z) = \rho_0 e^{z/h}, \quad h = \text{const} > 0). \tag{14.27}$$

Using the transformation technique, one can see that the z-shear transformation leads to change in ρ_0, when the z-scale transformation leads to change in h. The last one can be compensated by changing the cascade axis direction. This fact inspires the mathematical treatment of the problem, solution of which can be useful for experimental analysis and computer simulation EAS, if it is possible to neglect its curvature.

Let us denote the flux density of EPC a-particle number by $\phi_a(\mathbf{x}, t)$, $\mathbf{x} = (\mathbf{r}, \mathbf{p})$, $\mathbf{r}$ and $\mathbf{p}$ are the radius-vector and momentum of the particle, respectively, t denotes a moment of observation.

$$\hat{L}\phi \equiv [v^{-1}\partial_t + \Omega\nabla + M]\phi(\mathbf{x}, t) = S(\mathbf{x}, t), \tag{14.28}$$

where $v = v_a(\mathbf{p})$ is a speed of a-particle with the momentum $\mathbf{p}$, $\Omega = \mathbf{p}/p$, M is a linear operator describing the interaction of cascade particles with the atmosphere, and $S(\mathbf{x}, t)$ is a source term. The equation is a matrix one, but we have omitted the matrix unit symbol here because it does not matter for further consideration. Its Green function obeys the equation

$$\hat{L}G(\mathbf{x}, t; \mathbf{x}_0, t_0) = \delta(\mathbf{x} - \mathbf{x}_0)\delta(t - t_0). \tag{14.29}$$

We will further suppose that the medium is stationary one, i.e., its properties do not change during a cascade developing time. Introducing a time scale change operator $\Theta_t^{(\alpha)}$,

$$\Theta_t^{(\alpha)}G(\mathbf{x}, t; \mathbf{x}_0, t_0) = G(\mathbf{x}, \alpha t; \mathbf{x}_0, \alpha t_0), \quad \alpha > 0.$$

Applying it to both sides of Eq. (14.29) yields

$$[(\alpha v)^{-1}\partial_t + \Omega\nabla + \hat{M}]G(\mathbf{x}, \alpha t; \mathbf{x}_0, \alpha t_0) = \alpha^{-1}\delta(\mathbf{x} - \mathbf{x}_0)\delta(t - t_0).$$

Neglecting nonlinear density effects, we can represent the material operator $\hat{M}$ for the chemically homogeneous medium as the product

$$\hat{M} = \rho(\mathbf{r})\hat{K}, \tag{14.30}$$

where $\rho(\mathbf{r})$ is the mass density of the medium at the point $\mathbf{r}$. Let us introduce now a space scale change operator $\Theta_{\mathbf{r}}^{(\beta)}$,

$$\Theta_{\mathbf{r}}^{(\beta)}G(\mathbf{x}, t; \mathbf{x}_0, t_0) = G(\beta\mathbf{r}, \mathbf{p}, t; \beta\mathbf{r}_0, \mathbf{p}_0, t_0).$$

Applying it to both sides of Eq. (14.29) under condition (14.30), we arrive to the following equation:

$$[v^{-1}\partial_t + \beta^{-1}\Omega\nabla + \rho_\beta(\mathbf{r})\hat{K}]G(\beta\mathbf{r}, \mathbf{p}, t; \beta\mathbf{r}_0, \mathbf{p}_0, t_0) = \beta^{-3}\delta(\mathbf{x} - \mathbf{x}_0)\delta(t - t_0),$$

where

$$\rho_\beta(\mathbf{r}) = \Theta_{\mathbf{r}}^{(\beta)}\rho(\mathbf{r}) = \rho(\beta\mathbf{r}).$$

Because the Green function depends on $\rho(\mathbf{r})$ in a functional way, it is convenient to use the notation $G(\mathbf{x}, t; \mathbf{x}_0, t_0; \rho(\cdot))$ for the solution of the equation

$$[v^{-1}\partial_t + \Omega\nabla + \rho(\mathbf{r})\hat{K}]G(\mathbf{x}, t, \mathbf{x}_0, t_0; \rho(\cdot)) = \delta(\mathbf{x} - \mathbf{x}_0)\delta(t - t_0). \tag{14.31}$$

Now, applying the product of these operators to both sides of Eq. (14.31), and assuming $\alpha = \beta = k$, we represent the result in the form

$$[v^{-1}\partial_t + \Omega\nabla + k\rho_k(\mathbf{r})\hat{K}]G(k\mathbf{r}, \mathbf{p}, kt; k\mathbf{r}_0, \mathbf{p}_0, kt_0; \rho(\cdot)) = k^{-3}\delta(\mathbf{x} - \mathbf{x}_0)\delta(t - t_0). \tag{14.32}$$

Taking into account the linearity of the equation and comparing (14.31) with (14.32), we have got

$$G(k\mathbf{r}, \mathbf{p}, kt; k\mathbf{r}_0, \mathbf{p}_0, kt_0; \rho(\cdot)) = k^{-3}G(\mathbf{r}, \mathbf{p}, t; \mathbf{r}_0, \mathbf{p}_0, t_0; k\rho_k(\cdot)). \tag{14.33}$$

The equation connects the Green functions for two inhomogeneous media with mass densities $\rho(r)$ and $k\rho(kr)$. If the media are homogeneous,

$$\rho(\mathbf{r}) = \rho(k\mathbf{r}) = \text{const}$$

and we arrive at a well-known result [25]

$$G(k\mathbf{r}, \mathbf{p}, kt; k\mathbf{r}_0, \mathbf{p}_0, kt_0; \rho) = k^{-3}G(\mathbf{r}, \mathbf{p}, t; \mathbf{r}_0, \mathbf{p}_0, t_0; k\rho). \tag{14.34}$$

In the case of the exponential density (14.27), the medium realization becomes a function of two parameters, h and ρ_0.

$$\rho(x, y, z) = \rho_0 e^{z/h}, \tag{14.35}$$

so that

$$G(\mathbf{x}, t; \mathbf{x}_0, t_0; \rho(\cdot)) = G(\mathbf{x}, t; \mathbf{x}_0, t_0; \rho_0, h).$$

Because

$$k\rho_k(\mathbf{r}) = k\rho_0 e^{zk/h},$$

Eq. (14.34) becomes

$$G(k\mathbf{r}, \mathbf{p}, kt; k\mathbf{r}_0, \mathbf{p}_0, kt_0; \rho_0, h) = k^{-3}G(k\mathbf{r}, \mathbf{p}, kt; \mathbf{r}_0, \mathbf{p}_0, kt_0, h/k).$$

The values ρ_0 and $k\rho_0$ mean here the mass densities of the origin of coordinates. It is more convenient to use the density at one of the points given in the list of Green function arguments, say at the observation point $\mathbf{r}$:

$$G(\mathbf{r}, \mathbf{p}, t; \mathbf{r}_0, \mathbf{p}_0, t_0; \rho_0, h) = k^{-3}G(\mathbf{r}/k, \mathbf{p}, t/k; \mathbf{r}_0/k, \mathbf{p}_0, t_0/k, k\rho(\mathbf{r}/k), h/k). \tag{14.36}$$

This is a general similarity relation between two Green functions describing development of EPCs in two media with different values of parameter h.

Integrating (14.36) over time and lateral coordinates, we arrive at interrelation

$$G(z, \mathbf{p}; z_0, \mathbf{p}_0; \rho(z), h) = G(z/k, \mathbf{p}; z_0/k, \mathbf{p}_0; k\rho(z/k), h/k). \tag{14.37}$$

Calculating the mass distance between source and observation levels we are getting for left-hand side

$$\tau_{\text{lhs}}(z_0, z) = \rho(z)e^{-z/h}h\left[e^{z/h} - e^{z_0/h}\right]$$

and for the right-hand side

$$\tau_{\text{rhs}}(z_0/k, z/k) = \rho(z/k)e^{-z/h}h\left[e^{z/h} - e^{z_0/h}\right].$$

According to Eq. (14.35), we get $\tau_{\text{lhs}}(z_0, z) = \tau_{\text{rhs}}(z_0/k, z/k)$.

It is well known, that for chemically homogeneous media with mass density depending on one space coordinate only the solution of a plane problem depends on the mass distance, and the formula (14.37) obtained above obeys this condition.

The general similarity relation obtained above can be used for approximate investigation and simulation of space–time structure of EPC in atmosphere. This approximation is based on three assumptions: firstly, we suppose the atmosphere to be flat and chemically homogeneous; secondly, the density is falling with growing altitude in an exponential way, and thirdly, the lateral dimension of inclined EPC is sufficiently small to neglect the change of the air density in the lateral directions.

Let $\theta < \pi/2$ denote the arrival angle particle producing EPC and $\mu = \cos\theta$. Within the scope of the above assumption, one can regard that the cascade develops in the medium the density of which changes along EPC axis only, so that the parameter $h_\theta = h_0/\mu$, where h_0 corresponds to density dependence in vertical direction ($\theta = 0$). Denoting the Green function for a vertical EPC by G_0 and for inclined EPC by G_θ, after substitution $k = \mu$ into Eq. (14.37), we obtain

$$G_\theta(\mathbf{r}, \mathbf{p}, t; \mathbf{r}_0, \mathbf{p}_0, t_0; \rho(\mathbf{r})) = \mu^3 G_0(\mu\mathbf{r}, \mathbf{p}, \mu t; \mu\mathbf{r}_0, \mathbf{p}_0, \mu t_0; \rho(\mu\mathbf{r})/\mu).$$

Thus, if we have a Green function G_0 for vertical EPC and need corresponding function G_θ for inclined EPC at observation point $\mathbf{r}$, where the atmosphere density is ρ, we should take G_0 for such observation point $\mathbf{r}'$, where the density equals to ρ/μ, find space-time position of the new source with the aid of transformation

$$\mathbf{r}_0' - \mathbf{r}' = (\mathbf{r}_0 - \mathbf{r}_1)\mu,$$

$$t_0' - t' = (t_0 - t)\mu$$

and multiply result by μ^3. This generalizes the previous result obtained for the case when the cascade starts from infinity [26].

References

1. Giller M, Wieczorek G, Kacperczyk A et al (2004) Energy spectra of electrons in the extensive air showers of ultra-high energy. J Phys G: Nuclear Particle Phys 30(2):97–105
2. Nerling F, Blümer J, Engel R, Risse M (2006) Universality of electron distributions in high-energy air showers-Description of Cherenkov light production. Astroparticle Phys 24(6):421–437
3. Gora D, Engel R, Heck D et al (2006) Universal lateral distribution of energy deposit in air showers and its application to shower reconstruction. Astroparticle Phys 24(6):484–494
4. Gaisser T, Hillas AM (1977) Reliability of the method of constant intensity cuts for reconstructing the average development of vertical showers. In: Proceedings of the 15th ICRC, Plovdiv, vol 8, p 353

5. Unger M, Dason BR, Engel R (2008) Reconstruction of longitudinal profiles of ultra-high energy cosmic ray showers from fluorescence and Cherenkov light measurements. Nuclear Instrum Methods Phys Res Sect A: Accelerat, Spectrom, Detect Assoc Equip 588(3):433–441
6. Bird DJ, Corbato SC et al (1995) Detection of a cosmic ray with measured energy well beyond the expected spectral cutoff due to cosmic microwave radiation. Astrophys J 441:144–150
7. Baltrusaitis RM, Cady R, Cassiday GL et al (1985) The Utah Fly's Eye detector. Nuclear Instrum Methods Phys Res A 240(2):410–428
8. Giller M, Wieczorek G, Kacperczyk A, Stojek H, Tkaczyk W (2004) Energy spectra of electrons in the extensive air showers of ultra-high energy. J Phys G: Nucl Part Phys 30:97–105
9. Giller M, Kacperczyk A, Malinowski J, Tkaczyk W, Wieczorek G (2005) Similarity of extensive air showers with respect to the shower age. J Phys G: Nucl Part Phys 31:947–958
10. Bourdeau MF, Capdevielle JN, Procureur J (1980) The lateral age parameter in extensive air showers. J Phys G: Nucl Phys 6:901–908
11. Uchaikin VV (1979) The theory of cascade process nonlinear measurements and the statement of backward problems. In: Proceedings of 16-th ICRC, vol 7. Kyoto, pp 24–29
12. Capdevielle JN, Gawin J, Procureur J (1978) The lateral distribution of electrons in EAS at mountain altitude. J Phys G: Nucl Phys 4(2):303–313
13. Hillas A, Lapikens J (1977) Electron-photon cascades in the atmosphere and in detectors. In: Proceedings of the 15th of ICRC, Plovdiv, vol 8, p 460
14. Capdevielle JN, Gawin J (1982) The radial electron distribution in extensive air showers. J Phys G: Nuclear Phys 8(9):1317–1335
15. Aharonian FA, Plyasheshnikov AV (2003) Similarities and differences between relativistic electron-photon cascades developed in matter, photon gas and magnetic field. Astroparticle Phys 19:525–548
16. Aharonian FA, Coppi PS, Volk HJ (1994) Very high energy gamma-rays from agn: cascading on the cosmic background radiation fields and the formation of pair halos. Astrophys J 423:L5
17. Gaisser TK (1990) Cosmic rays and particle physics. Cambridge University Press, Cambridge
18. Aharonian FA, Atoyan AM (1985) Ultrahigh energy gamma-quanta - carriers of cosmological information. Zh Eksp Teor Fiz 62:189
19. Anguelov V, Vankov H (1999) Electromagnetic showers in a strong magnetic field. J Phys G: Nucl Part Phys 25:1755
20. Nakatsuka T, Iyono A, Nishimura J (2011) An evaluation of electron-photon cascades developing in matter, photon and magnetic fields. Astrophys Space Sci Trans 7:345–350
21. Wigner E (1954) The problem of multiple scattering. Phys Rev 94:17
22. Fano U, Spencer LV, Berger MJ (1959) Penetration and diffusion of X rays. Handbuch der Physik Bd XXXVIII/2 (1959)
23. Case KM, Zweifel PF (1967) Linear transport theory. Addison-Wesley Publishing Company
24. Kolchuzhkin AM, Uchaikin VV (1978) Introduction to the theory of penetration of particles through the matter. Atomizdat, Moscow (in Russian)
25. Zerby C (1967) Density transformations for time-dependent radiation transport problems. Nucl Sci Engin 29:151–153
26. Uchaikin VV (1989) Application of the theory of similarity to the analysis of spatial and temporal distributions in extensive air showers. Izv Vyssh Uchebn Zaved Fizika 7:119 (in Russian)
27. Krys E (1985) Longitudinal and lateral structure of electromagnetic cascades simulated by full Monte Carlo method. Acta Phys Polon B 16(2):109–123
28. Lagutin AA, Plyasheshnikov AV, Uchaikin VV (1979) The radial distribution of electromagnetic cascade particles in the air. In: Proceedings of the 16th ICRC, vol 7. Kyoto, pp 18–23
29. Allan HR, Crannell CJ, Hough JH et al (1975) The width of electron-photon cascades in air. In: Proceedings of the 14th ICRC, vol 6. Munich, pp 3071–3076
30. Procureur J, Stamenov JN, Ushev SZ (1988) On the lateral distribution of electrons in electromagnetic cascades in air. J Phys G: Nucl Phys. 14:807–810
31. Belenki SZ, Ivanenko IP (1958) Cascade theory of showers. Uspekhi Fiz Nauk 69:591–656

32. Ivanenko IP, Roganova TM (1983) Cascades generated by particles with ultra. High energy, Moscow, Nauka (in Russian)
33. Dey RK, Bhadra A, Capdevielle JN (2012) Scaling behaviour of lateral distribution of electrons in EAS. J. Phys. G: Nucl. Part. Phys. 39, 085201, 14pp

Appendix A
Method of Integral Transformations

One of the most fruitful and general approaches to the theoretical description of cascade processes in the mathematical language is the group of integral transformations.

The essence of integral transformations is that, replacing the desired functions with their mathematical images, they at the same time change the mathematical operations that reflect the physical understanding of the process. In this relation, the integral transformation is an analogue of the logarithm. It is worth to recall that in the pre-computer (more precisely, pre-calculator) era, tables of logarithms and the rule that the logarithm of a product is equal to the sum of logarithms were used to multiply multi-digit numbers. Thus, the multiplying problem was reduced to summation, and it remained only to solve the inverse problem: finding its original using the found "image" of the product (logarithm). For this, special numerical tables *antilogarithms* were compiled. For similar purposes, integral transformations serve as well. The most known of them—the *Fourier*, *Laplace transformations*, and *Mellin transformations*—can be met in any section of theoretical physics. Their applications convert the operations frequently encountered in analysis, called *convolutions of functions*, into products of their images, and then (often using special inverse tables) to restore the desired original.

A.1 The Laplace Transformation

Let a function $f(x)$ be defined on the whole x-axis and integral

$$J = \int_{-\infty}^{\infty} f(x)e^{-\lambda x}dx, \tag{A.1}$$

where λ is a complex-valued parameter, converges on at least one straight line $\text{Re}\,\lambda = \text{c}$. Then, the function

© Higher Education Press Limited Company 2024

V. V. Uchaikin, *Electron-Photon Cascades*, Nonlinear Physical Science,
https://doi.org/10.1007/978-981-99-7524-2

$$\check{f}(\lambda) = \int_{-\infty}^{\infty} f(x)e^{-\lambda x}dx$$

is called the *two-sided Laplace transform* of the original function $f(x)$. It is similar to the Laurent series, its convergence domain is the strip $a \leq \text{Re}\,\lambda \leq b$, which may degenerate into a straight line $a = b$.

If a function $f(x)$ is defined on the positive semiaxis and the integral

$$\hat{f}(\lambda) = \int_{0}^{\infty} f(x)e^{-\lambda x}dx \tag{A.2}$$

converges, the function (A.2) is called the *one-sided Laplace transform* of the function $f(x)$. We shall assume, that integral (A.2) converges at the upper limit. If it converges at the upper limit under a determined value of $\lambda = \lambda_a$, Re $\lambda_a = a$, then it converges under all other λ, for which Re $\lambda > a$, and vice versa, if integral (A.2) diverges under some $\lambda = \lambda_{a'}$, Re $\lambda_{a'} = a'$, then it diverges also under all values of λ, for which Re $\lambda < a'$. Hereby, in general, $\hat{f}(\lambda)$ is defined in a half-plane lying to the right of a straight line parallel to the imaginary axis. The one-sided transformation is similar to a power series, the concept of a half-plane of convergence of an integral is equivalent to the concept of a convergence circle for a power series. Of great importance is the procedure for restoring the original function from its Laplace image. The most popular inversion formula

$$f(x) = \frac{1}{2\pi i} \int_{c-i\infty}^{c+i\infty} \hat{f}(\lambda)e^{\lambda x}d\lambda, \quad a \leq c \leq b, \tag{A.3}$$

is valid for both (one-sided and two-sided) transformations, since the one-sided transformation can be considered as a two-sided one from a function, equal to zero on the negative semiaxis (recall, that formulas for coefficients in Taylors' and Laurent's series have also the same form). For its applicability, transform $\hat{f}(\lambda)$ must be regular in the strip $a \leq \text{Re}\,\lambda \leq b$ and satisfy the condition

$$\hat{f}(\lambda) = \mathcal{O}|\lambda|^{-\alpha-1}, \quad 0 < \alpha < 1, \quad \lambda \to \infty, \quad a \leq \text{Re}\,\lambda \leq b,$$

or (equivalently) the original must be continuous on continuous on the entire axis, except for a countable set of discontinuity points that have no limit points at a finite distance, and satisfy two conditions:

$$\int_{-\infty}^{\infty} |f(x)|(e^{-ax} + e^{-bx})dx < \infty, \tag{A.4}$$

$$|f(x') - f(x)| < |x' - x|^{\alpha}, \quad 0 < \alpha < 1$$

for all x' in the vicinity of point x. In case of a one-sided Laplace transformation $b = \infty$, i.e., the integration contour in Eq. (A.3) lies to the right of all poles and singular points of the transform $\hat{f}(\lambda)$.

Note the following properties of the two-sided Laplace transform:

(a) Let $F(x) = f'(x)$, then $\check{F}(\lambda) = \lambda \check{f}(\lambda)$.

(b) Let $F(x) = xf(x)$, then $\check{F}(\lambda) = -\check{f}'(\lambda)$.

(c) Let $F(x) = \int\limits_{-\infty}^{x} f(x')dx'$, then $\check{F}(\lambda) = \check{f}(\lambda)/\lambda$.

(d) Let $F(x) = \int\limits_{-\infty}^{x} x'^n f(x')dx'$, then $F(\infty) = (-1)^n \check{f}^{(n)}(0)$.

Similar properties are inherent to one-sided Laplace transforms:

(a) Let $F(x) = f'(x)$, then $\hat{F}(\lambda) = \lambda \hat{f}(\lambda) - f(0)$.

(b) Let $F(x) = xf(x)$, then $\hat{F}(\lambda) = -\hat{f}'(\lambda)$.

(c) Let $F(x) = \int\limits_{0}^{x} f(x')dx'$, then $\hat{F}(\lambda) = [\hat{f}(\lambda) - f(0)]/\lambda$.

(d) Let $F(x) = \int\limits_{0}^{x} x'^n f(x')dx'$, then $F(\infty) = (-1)^n \hat{f}^{(n)}(0)$.

A.2 The Inversion of Transformation

A powerful tool for calculating (A.3) is the residue method, however, it only applies to closed loops. To close the integration path in (A.4), **Jordan's lemma** conditions should be verified and then the **decomposition theorem** may be applied: *any (meromorphic and regular in some domain* $Re\lambda > a$*) function satisfying these conditions, and possessing for any* $c > a$ *an absolutely convergent integral* $\int_{c-i\infty}^{c-i\infty} \tilde{f}(\lambda)d\lambda$ *is the Laplace transform of original function*

$$f(x) = \sum_k \text{res} \tilde{f}(\lambda_k) e^{\lambda x},$$

where the sum of the residues is taken over all poles λ_k *of the function* $\tilde{f}(\lambda)$ *in non-decreasing order of their modules.*

The *residue* of a function $F(\lambda)$ in a simple pole k is determined as

$$\text{res}_{\lambda_k} F(\lambda) = [(\lambda - \lambda_k) F(\lambda)]_{\lambda=\lambda_k}. \tag{A.5}$$

In particular, if $p(\lambda_k) = 0$ and $p(\lambda_k) \neq 0$, then

$$\text{res}_{\lambda_k} = \frac{q(\lambda_k)}{p'(\lambda_k)}.$$

If λ_k is a n-th order pole of a function $F_(\lambda)$, then

$$\mathrm{res}_{\lambda_k} F(\lambda) = \frac{1}{(n-1)!} \left\{ \left(\frac{d}{d\lambda} \right)^{n-1)} [(\lambda - \lambda_k)^n F(\lambda)] \right\}_{\lambda=\lambda_k} .$$

From approximate ways of the inversion integral computing, the *saddle-point method* is applied.

Let integral (A.3) can be represented in the form

$$I = \frac{1}{2\pi i} \int_{c-i\infty}^{c+\infty} H(z) e^{f(z)} dz, \tag{A.6}$$

where $H(z)$, and $f(z) = u(z) + iv(z)$ are real functions on the real axis and $u(z)$ has a minimum at $\bar{z}$, so

$$\frac{\partial u}{\partial x} = \frac{\partial u}{\partial y} = 0. \tag{A.7}$$

Since the functions $u(x, y)$ and $v(x, y)$ are harmonic,

$$\frac{\partial^2 u}{\partial x^2} = -\frac{\partial^2 u}{\partial y^2}, \quad \frac{\partial^2 v}{\partial x^2} = -\frac{\partial^2 v}{\partial y^2},$$

then when passing through the point $\bar{z}$ parallel to the imaginary axis, the function $u(z)$ takes a maximum value. This is the *saddle-point*. According to the Cauchy–Riemann conditions,

$$\frac{\partial v}{\partial x} = -\frac{\partial u}{\partial y}, \quad \frac{\partial v}{\partial y} = -\frac{\partial u}{\partial x},$$

so the derivative of $v(z)$ at this point is also equal to 0. Since $v(z) = 0$ on the real axis is zero, then

$$\frac{\partial^2 v}{\partial x^2} = -\frac{\partial^2 v}{\partial y^2} = 0, \tag{A.8}$$

i.e., in a neighborhood of a point $\bar{z}$, the function $v(z)$ along a straight line parallel to the imaginary axis, almost constant and equal to zero.

Let us move the integration contour (A.6) so that it passes through the point $\bar{z}$ (remaining parallel to the imaginary axis). Then, in a neighborhood of this point, the function $e^{f(z)} = e^{u(z)}$ will increase rapidly, passing through the maximum value $e^{u(\bar{z})}$ at the point $\bar{z}$, and then decrease just as quickly. If $H(z)$ is a sufficiently slowly varying function, its change inside the narrow region around the saddle point can be neglected:

$$I \approx \frac{1}{2\pi i} H(\bar{z}) \int\limits_{\bar{z}-i\infty}^{\bar{z}+i\infty} e^{u(z)} dz. \tag{A.9}$$

Expanding $u(\bar{x}, y)$ into a series in y

$$u(\bar{x}, y) = u(\bar{x}, 0) + y\left(\frac{\partial u(\bar{x}, y)}{\partial y}\right)_{y=0} + \frac{y^2}{2}\left(\frac{\partial^2 u(\bar{x}, y)}{\partial y^2}\right)_{y=0} + \ldots,$$

and taking into account Eqs. (A.7)–(A.8), we get

$$u(\bar{x}, y) = f(\bar{z}) - \frac{y^2}{2}|f''(\bar{z})| + \ldots,$$

where $f(\bar{z}) = u(\bar{x}, 0)$ and $|f''(\bar{z})| = \partial^2 u(\bar{x}, 0)/\partial x^2$. Taking into account, that $dz = idy$, where y runs from $-\infty$ to ∞, rewrite (A.9) as

$$I \approx \frac{1}{2\pi} H(\bar{z}) \int\limits_{-\infty}^{\infty} e^{-(y^2/2)|f''(\bar{z})|} dy.$$

Finally, after replacing $y^2|f''(\bar{z})| = t^2$, leading to the Poisson integral

$$\int\limits_{-\infty}^{\infty} e^{-t^2/2} dt = \sqrt{2\pi},$$

we have

$$I \approx \frac{1}{\sqrt{2\pi|f''(\bar{z})|}} H(\bar{z}) e^{f(\bar{z})}, \tag{A.10}$$

where $\bar{z}$ is determined by the saddle-point condition

$$f'(\bar{z}) = 0. \tag{A.11}$$

The presentation of the saddle-point method in the general case can be found in [9, 11, 19] and others.

Sometimes, it is possible to perform the inverse transformation in other ways, for example, represent the result as a definite integral of a real function, which can be calculated numerically. In particular, if a transform decomposes into the product of "reversible" transforms, one can use the convolution theorem.

A.3 The Fourier Transformation

Let $f(x)$ be an absolutely integrable function, given on the whole real axis, and k be a real parameter. The function

$$\tilde{f}(k) = \int_{-\infty}^{\infty} f(x)e^{ikx}dx \tag{A.12}$$

is called the *Fourier transform* of the function $f(x)$ (provided, that the integral exists).

There is no fundamental difference between a Fourier transform and its inverse. Therefore the tilde in $\tilde{f}(k)$ is usually omitted, and it is the name of the variable that determines whether a function is considered in real or Fourier space. For the same reason, the choice of the sign in the exponent is arbitrary, except that the sign in the Fourier transform has to be the opposite of that in the inverse transform. Finally, there is some flexibility with regard to the arrangement of the factor $1/2\pi$. In quantum mechanics, it has become customary to let the Fourier transform and its inverse each have a factor $(2\pi)^{-1/2}$.

In physics problems, one frequently deals with the Fourier transform of an observable quantity. Let $f(x)$ be such a quantity which must then be a real function, then

$$\widetilde{f^*}(k) = \int_{-\infty}^{\infty} f(x)e^{-ikx}dx = \tilde{f}(-k),$$

where the asterisk denotes the complex conjugate.

If $\check{F}(\lambda)$ is a two-sided Laplace transform of $f(x)$, then obviously $\tilde{f}(k) = \check{F}(-ik)$. This interrelation underlays properties of the Fourier transform by using the Laplace transform counterparts and vice versa. Thus, the inversion formula becomes

$$f(x) = \frac{1}{2\pi}\int_{-\infty}^{\infty} \tilde{f}(k)e^{-ikx}dk. \tag{A.13}$$

Replacing x by x' in (A.12) and inserting obtained $\tilde{f}(k)$ into Eq. (A.13), after changing the integration order and using expression

$$f(x) = \int_{-\infty}^{\infty} \delta(x - x')f(x')dx', \tag{A.14}$$

we arrive at the integral representation of Dirac's delta function:

$$\delta(x) = \frac{1}{2\pi} \int_{-\infty}^{\infty} e^{ikx} dk. \quad \text{(A.15)}$$

Equation (A.15) says, that $\delta(x)$ is the Fourier transform of the constant function $\tilde{f}(k) = 1$. The strict theory of integral transformations says, that not every function $f(x)$ has a Fourier transform. A sufficient condition is that it be *square-integrable*; that is, so that the following integral converges:

$$||f||^2 \equiv \int_{-\infty}^{\infty} |f(x)|^2 dx.$$

More generally, there exists the *Fourier inversion theorem*, which states that if $f(x)$ is square-integrable, then the Fourier transform $\tilde{f}(k)$ exists and moreover

$$\frac{1}{2\pi} \int_{-\infty}^{\infty} \tilde{f}(k) e^{-ikx} dk = \begin{cases} f(x), & \text{if } f \text{ is continuous at } x \text{ and,} \\ (1/2)[f(x-0) + f(x+0)], & \text{otherwise.} \end{cases}$$

In other words, at a point of discontinuity, the inverse transform produces the average of the left and right limiting values of the function $f(x)$.

Let us make a few remarks concerning the Fourier transformation.

(a) Its first basic property is that if $\tilde{f}(k)$ is the Fourier transform of $f(x)$, then $\tilde{f}^*(-k)$ is the Fourier transform of $f^*(x)$. This follows simply by taking the complex conjugate of the Fourier integral (A.12):

$$\left[\int_{-\infty}^{\infty} f^*(x) e^{ikx} dx\right] = \left[\int_{-\infty}^{\infty} f(x) e^{-ikx} dx\right]^* = \tilde{f}(-k).$$

Therefore, we conclude that if $f(x)$ is real, then $\tilde{f}^*(k) = \tilde{f}(-k)$ and vice versa.

(b) Suppose that $f'(x) = df(x)/dx$ is also square-integrable, and use in its Fourier transform formula integration by parts, we find:

$$\tilde{f}'(k) = (-ik)\tilde{f}(k).$$

After repeating n times,

$$\tilde{f}^{(n)}(k) = (-ik)^n \tilde{f}(k).$$

It is worth noting, that the use of this formula to the function $f(x)$ assumes its *well-behavior nature* (technically, it is enough that $f(x)$ belongs to the *Schwarz class of distributions*, consisting of those infinitely differentiable functions which decay, together with all its derivatives, sufficiently fast at infinity).

(c) *The Parseval identity.*

One can show, that the Fourier transform of a square-integrable function is itself square-integrable:

$$||\tilde{f}||^2 = \int_{-\infty}^{\infty} f(x) f^*(x) dx = \frac{1}{2\pi} = \int_{-\infty}^{\infty} |f(x)|^2 dx = \frac{1}{2\pi} ||f||^2. \tag{A.16}$$

Therefore, since $||f||^2$ is finite, so is $||\tilde{f}||^2$ (the *Parseval identity*).

(d) Multivariate generalization. Formulas (A.12), (A.13) and (A.15) admit obvious generalizations to multivariate cases. So, for three-dimensional case, we, respectively, obtain

$$\tilde{f}(\mathbf{k}) = \int\limits_{\mathbb{R}^3} f(\mathbf{x}) e^{i\mathbf{kx}} d\mathbf{x}, \tag{A.17}$$

$$f(\mathbf{x}) = \frac{1}{(2\pi)^3} \int\limits_{\mathbb{R}^3} \tilde{f}(\mathbf{k}) e^{-i\mathbf{kx}} d\mathbf{k}, \tag{A.18}$$

$$\delta(\mathbf{x}) = \frac{1}{(2\pi)^3} \int\limits_{\mathbb{R}^3} e^{i\mathbf{kx}} d\mathbf{k}. \tag{A.19}$$

(e) Computing two-dimensional Fourier transform for axially symmetric planar distribution

$$f(\mathbf{r}) = f(r), \quad \mathbf{r} \in \mathbb{R}^2, \quad r = |\mathbf{r}|,$$

yields

$$\tilde{f}(\mathbf{k}) = 2\pi \int\limits_0^{\infty} f(r) J_0(kr) r dr, \tag{A.20}$$

and

$$f(r) = \frac{1}{2\pi} \int\limits_0^{\infty} \tilde{f}(k) J_0(kr) k dk. \tag{A.21}$$

Here,

$$J_0(kr) = \frac{1}{\pi} \int\limits_0^{\pi} e^{ikr\cos\varphi} d\varphi, \; k = |\mathbf{k}| \tag{A.22}$$

is the zero-order Bessel function. From Eqs. (A.20)–(A.21), the Bessel function orthogonality follows:

$$\int_0^\infty J_0(kr)J_0(kr')kdk = \delta(r - r')/r. \tag{A.23}$$

Equation (A.21) is referred to as *Fourier–Bessel* (*Hankel*) integral transformation.

A.4 Spherical Harmonics and Legendre Polynomials

Spherical harmonics form a complete orthonormal set of functions on the unit sphere. This implies that a *reasonably well-behaved* function $\psi(\Omega)$ of the unit vector $\Omega = \{\sin\theta\cos\phi, \sin\theta\sin\phi, \cos\theta\}$ can be expanded in terms of spherical harmonics $Y_{lm}(\Omega)$,

$$\psi(\Omega) = \sum_{l=0}^{\infty}\sum_{m=-l}^{l} a_{lm} Y_{lm}^*(\Omega)d\Omega$$

with

$$a_{lm} = \int_{4\pi} d\Omega Y_{lm}^*(\Omega)\psi(\Omega) \equiv \int_{-1}^{1} d\cos\theta \int_0^{2\pi} d\phi Y_{lm}^*(\theta,\phi)\psi(\theta,\phi).$$

In common notation,

$$Y_{lm}(\Omega) = N_{lm} P_l^m(\cos\theta)e^{im\phi},$$

where $P_l^m(\mu)$ represents an associated Legendre function,

$$P_l^m(\mu) = \frac{(-1)^m}{2^l l!}(1-\mu^2)^{m/2}\frac{d^{l+m}(\mu^2-1)l)}{d\mu^{l+m}}$$

and

$$N_{lm} = \sqrt{\frac{2l+1}{4\pi}\frac{(l-m)!}{(l+m)!}}$$

is a normalization constant, so that

$$\int_{4\pi} Y_{lm}^*(\Omega)Y_{l'm'}(\Omega)d\Omega = \delta_{ll'}\delta_{mm'}.$$

In the special but not infrequent case of vanishing azimuthal dependence all terms with $m \neq 0$ drop out, and normalizing condition reduces to

$$\int_{-1}^{1} P_l(\mu)P_{l'}(\mu)d\mu = \frac{2}{2l+1}\delta_{ll'},$$

where

$$P_l(\mu) = P_l^0(\mu),\ l = 0,\ 1,\ 2,\ \ldots$$

are the Legendre polynomials obeying the differential equation

$$(1-\mu^2)P_l'' - 2\mu P_l' + l(l+1)P_l = 0.$$

The lowest order Legendre polynomials read

$$P_0(\mu) = 1, \quad P_1(\mu) = \mu, \quad P_2(\mu) = \frac{1}{2}(3\mu^2 - 1), \quad \ldots$$

The following relations are most often used in practice of analytical calculations:

- the addition theorem

$$P_l(\Omega \cdot \Omega') = \frac{4\pi}{2l+1}\sum_{m=-l}^{l} Y_{lm}^*(\Omega)Y_{lm}(\Omega'),$$

- the recurrence relation

$$\mu P_l^m = \frac{1}{2l+1}[(l-m+1)P_{l+1}^m + (l+m)P_{l-1}^m],$$

- and the generating function

$$G(s) = \frac{1}{\sqrt{1-2\mu s+s^2}} = \sum_{l=0}^{\infty} P_l(\mu)s^l,$$

  which is convenient to determine explicit expressions and matrix elements.

A.5 The Mellin Transformation

Dealing with probabilistic moments in mathematical statistics shows that with an increase in the number of moments, the accuracy of distributions restoring increases, and this can be achieved not only by increasing the total number of moments but also by decreasing intervals between them on the order scale, that is, by introducing moments of fractional orders. The obvious possibility of passage from discrete orders to a continuous variable opened the way to improve the accuracy of recovery problem solving, and the introduction of complex orders put it into the system of efficient algorithms based on the complex variable function theory. Bearing this in mind, one would expect the introduction of non-integer (even complex) order s moments:

$$\mu_s \equiv \bar{f}(s) = \int_a^b f(x)x^s dx. \tag{A.24}$$

This is a special kind of so-called *Mellin's transformation*.

In its origin, the Mellin transform method is related to the well-known moment problem in mathematical statistics, the main task of which is to restore the probability distribution of a random variable over a finite number of its moments. This task arises in different scientific applications (e.g., physics, chemical engineering, economics) as well as in computational mathematics itself, when the passage to moments simplifies the solving procedure.

The classical moment problem for a finite interval $[a, b]$, also known as the *Hausdorff moment problem*, can be stated as follows. A set of moments

$$\mu_n = \int_a^b f(x)x^n dx, \quad n = 0, 1, \ldots, m, \quad n \leq m \tag{A.25}$$

of a function $f(x)$ are integrable over the interval with $\mu_n < \infty \ \forall x \in [a, b]$. The problem is to reconstruct the PDF function $f(x)$ from the knowledge of its moments. The necessary and sufficient conditions for a solution to exist were given by Hausdorff [13]. The moment problem and its variants have been studied extensively in the literature. For a finite number of moments, the problem is underdetermined and it is not possible to construct a unique solution from the moment sequence unless further assumptions about the function are made.

Let us give hereafter a succinct description of the often-used reconstruction techniques. As a common point, the bounded support $[a, b]$ is rescaled to [0,1] for a general formulation.

By historical and practical reasons, the index had been shifted by 1, and the Mellin transform of a function $f(t)$, defined on the positive real semiaxes $0 < t < \infty$, is determined as follows.

Definition Let $f(t)$ be a function defined on the positive real axis $0 < t < \infty$. The *Mellin transformation* $\mathscr{M}(f; \sigma)$ is the operation mapping the function f into the function $\bar{f}$ called the *Mellin transform* and defined on the complex plane by the relation:

$$\bar{f}(\sigma) = \int_0^\infty f(t)t^{\sigma-1} dt. \tag{A.26}$$

If it diverges at the low limit under some value $\sigma = \sigma_a(\operatorname{Re}\sigma_a = a)$, then it diverges under all σ with $\operatorname{Re}\sigma < a$. If it diverges at the upper limit under some value $\sigma = \sigma_b(\operatorname{Re}\sigma_b = b)$, then it diverges under all σ with $\operatorname{Re}\sigma > b$. Thus, if the Mellin integral somewhere converges, then its convergence domain is a strip on the complex plane bounded by two straight lines parallel to the imaginary axis. This introduces what is called the *strip of definition* of the Mellin transform that will be denoted by $\Sigma(a, b)$. In some cases, this strip may extend to a half-plane ($a = -\infty$ or $b = +\infty$) or to the whole complex Σ-plane ($a = -\infty$ and $b = +\infty$).

A.6 Relation to Laplace and Fourier Transformations

The change of variables defined by

$$t = e^{-x}, \quad dt = -e^{-x}dx$$

transforms the integral (A.26) into

$$F(\sigma) = \int_{-\infty}^{\infty} f(e^{-x})e^{-\sigma x}dx. \tag{A.27}$$

After the change of function

$$g(x) \equiv f(e^{-x}),$$

one recognizes in (2.13) the two-sided Laplace transform of $g(x)$:

$$\breve{g}(\sigma) = \int_{-\infty}^{\infty} g(x)e^{-\sigma x}dx.$$

Symbolically, this can be written as

$$\mathscr{M}[f(t); \sigma] = \mathscr{L}[f(e^{-x}); \sigma]. \tag{A.28}$$

To find a link with the Fourier transform, write $\sigma = a - ik$ in (A.27):

$$F(\sigma) = \int_{-\infty}^{\infty} f(e^{-x})e^{-ax}e^{ikx}dx. \tag{A.29}$$

Thus,

$$\mathscr{M}[f(t); a - ik] = \mathscr{F}[f(e^{-ikx})],$$

i.e., for a given value of Re(σ) =a belonging to the definition strip, the Mellin transform of a function can be expressed as a Fourier transform. Applying the inverse Fourier's formula to (A.29) and going back to variables t and σ, we arrive at the inverse Mellin's transform formula, which finally reads

$$f(t) = \frac{1}{2\pi i} \int_{a-i\infty}^{a+i\infty} \bar{f}(\sigma) t^{-\sigma} d\sigma, \tag{A.30}$$

where the integration is along a vertical line through Re(σ)= a. Observe, that in case of s parametrization (A.24), the inverse transform formula reads

$$f(t) = \frac{1}{2\pi i} \int_{a'-i\infty}^{a'+i\infty} \bar{f}(s) t^{-s-1} ds \quad (t > 0).$$

The Mellin convolution $g(x) * h(x)$ of two functions $g(x)$ and $h(x)$ given on the positive semiaxis x, is determined by integral

$$g(x) * h(x) = \int_0^\infty g(x/x') h(x') dx'/x'. \tag{A.31}$$

In terms of Mellin's transform, it converts into the product of Mellin's images

$$\mathscr{M}\{g(x) * h(x)\} = \bar{g}(s)\bar{h}(s). \tag{A.32}$$

Another kind of convolution,

$$g(x) \star h(x) = \int_0^\infty g(xx') h(x') x' dx', \tag{A.33}$$

is transformed into

$$\mathscr{M}\{g(x) \star h(x)\} = \bar{g}(s)\bar{h}(-s).$$

A.7 Integral Representation of the Series

Let us touch on the method of analytical continuation, effectively used in the EPC cascade theory, developed by Nishimura and Kamata. Assume that the solution $f(x)$ of some equation is obtained as a series, for example,

$$f(x) = \sum_{n=0}^{\infty} a_n (-x)^n \tag{A.34}$$

with convergence area $0 \le x \le a$, then there may be a problem of calculating $f(x)$ outside this area. One way to solve this problem is based on the use of *analytic continuation*. The series (A.34) can be be considered as the result of calculating complex integral of a function having simple poles at points $z = 0, 1, 2, \ldots$ by the residue method, and the residues, according to (A.34), respectively, equal

$$\frac{1}{2\pi i}a_0, \quad \frac{1}{2\pi i}a_1(-x), \quad \frac{1}{2\pi i}a_2, (-x)^2, \quad \ldots \tag{A.35}$$

Recall that the gamma-function $\Gamma(z)$ has simple poles at the points $z = -n(n = 0, 1, 2, \ldots)$ with residues equal to $(-1)^n/n!$, therefore, the function $\Gamma(-z)$ has poles at points $z = n$ with residues $(-1)^n/n!$, and the function $-\frac{1}{2\pi i}\Gamma(z + 1)\Gamma(-z)a(z)x^z$ has residues (A.35) at corresponding points. Drawing an integration loop between the poles $z = 0$, and $z = -1$ of this function and closing it to the right (clockwise), we get the series

$$-\frac{1}{2\pi i}\oint \Gamma(-z)\Gamma(z+1)a(z)x^z = \sum_{n=0}^{\infty} a(n)(-x)^n, \tag{A.36}$$

coinciding with (A.34) if $a(n) = a_n$. Since the integral over the closing loop is equal to zero, the left side can be replaced by an integral along a contour parallel to the imaginary axis

$$-\frac{1}{2\pi i}\int\limits_{c-i\infty}^{c+i\infty} \Gamma(z)\Gamma(z+1)a(z)x^z = \sum_{n=0}^{\infty} a(n)(-x)^n. \tag{A.37}$$

The resulting integral serves as an analytic continuation of the function (A.34) given in the form of a series, outside the circle of its convergence. To get the decomposition of $f(x)$ by degrees x^{-1}, we close the integration contour with a semicircle of infinitely large radius to the left (against hour hand). Then the considered integral will be equal to the sum of the residues integrand at the points $z = -1, -2, \ldots$ plus integral over a closed loop.

It can be shown that the latter has the order e^{-x} [6]; neglecting them, we get

$$-\frac{1}{2\pi i}\oint \Gamma(-z)\Gamma(z+1)a(z)x^z = -\sum_{n=0}^{\infty} a(-n)\left(-\frac{1}{x}\right)^n. \tag{A.38}$$

This formula plays an important role in computing of energy spectra in B-approximation of the Nishimura–Kamata cascade theory, where x means the ratio of the electron kinetic energy to the critical energy characterized by the equality of ionization and radiation losses.

Such analytical continuation is also used when calculating the angular distribution of charged particles. In the small-angle axial approximation, its Fourier transform is expressed through the Bessel function, and we encounter the conditionally convergent integral

$$I = \int_0^\infty J_0(kr)(k^2/4)^p k dk,$$

computed as a limit of integrals

$$I_a = \int_0^\infty J_0(kr)(k^2/4)^p e^{-ak} k dk$$

as $a \to 0$. Expanding the Bessel function into the series, we obtain

$$I_a = 4^{-p} \sum_{n=0}^\infty \left[-\left(\frac{z}{2}\right)^2 \right]^n \frac{\Gamma(2n+2p+2)}{(n!)^2 a 2n + 2p + 2}.$$

This series can be represented as complex integral

$$I_a = \frac{1}{2\pi i} 4^{-p} \int_{c-i\infty}^{c+i\infty} \left(\frac{z}{2}\right)^{2z} \frac{\Gamma(-z)(2z+2p+2)}{\Gamma(z+1)a^{2z+2p+2}} dz, \quad -1 < c < 0.$$

Of all the residues of the integrand, only the residue at point $z = p$ will not contain a and will not vanish as $a \to 0$, so

$$\lim_{a\to 0} I_a = \mathrm{res}_{z=-p-1}[4^{-p} \left(\frac{z}{2}\right)^{2z} \frac{\Gamma(-z)(2z+2p+2)}{\Gamma(z+1)a^{2z+2p+2}}]. \tag{A.39}$$

Computing the right-hand side yields

$$\int_0^\infty J_0(kr) \left(\frac{k^2}{4}\right)^p k dk = 2r^{-2p-2} \frac{\Gamma(p+1)}{\Gamma(-p)}.$$

Appendix B
An Excerpt From Ivanenko–Roganova Book[1]

Spatial distribution of EPC electrons with finite $E_0/\beta(E_0/E)$ in heterogeneous environment

Ivanenko I.P., Roganova T.M.

This is an excerpt from the book,[2] published with kind permission of Professor Tatiana Roganova (Moscow State University). In my opinion, it can be considered as the last chord in the resounding discussion of the late 70s—early 80s about the asymptotic behavior of the spatial distribution of particles in ultra-high energy showers.

Let us analyze the spatial distribution function (SDF) of cascade electrons in the atmosphere, taking into account the finiteness of the ratio $E_0/\beta(E_0/E)$ and atmospheric inhomogeneities calculated by different methods. In [1], there are summarized works in which the functions $N_p(E_0, E(0), t, r)$ were calculated analytically [21] or by the moment method [22] taking into account both effects under consideration. As already mentioned, in [23, 25, 105–107], the question of the accuracy of the commonly used Nishimura–Kamata–Greisen (NKG) formula [21] in experimental data on EAS analysis. In [23, 105], ordinary Monte Carlo schemes have been used for calculating the spacial distribution in light–matter (air) taking into account and without taking into account the heterogeneity of the atmosphere. In a heterogeneous atmosphere cascades begin at depths of 40, 80, 120, 160 g/cm^2 above level of observation. Particles were traced to energies of the order $\sim$1 MeV, at lower energies up to 50 keV and the initial energies 1, 10, 10^2 and 10^3 GeV. The code allows one to calculate the behavior of cascades in a scintillator and in a water Cherenkov counter. In some calculations, photo-effect were turned on, up to energies $\sim$50 keV only effect Compton-effect. Low-energy photons in this case manifest themselves noticeably at distances of more than 100 m from the axis.

---

[1] Published with kindly permission of Prof. Tatiana Roganova.

[2] I.P. Ivanenko, T.M.Roganova, Cascade showers initiated by superhigh energy particles. Moscow, Nauka, 1983 (in Russian).

© Higher Education Press Limited Company 2024

V. V. Uchaikin, *Electron-Photon Cascades*, Nonlinear Physical Science,
https://doi.org/10.1007/978-981-99-7524-2

In these studies, the results differed from each other by approximately 20–30% as statistics were collected and the algorithm and codes were improved. In [105], the statistics correspond to the energy flux to the boundary of 10 TeV and 30 TeV at $E_0 \approx 1$ TeV. The quantity $r^2\rho/N$ was calculated for the signal in the scintillator 10 cm thick in the distance range from $\sim$1 to $\sim$$10^3$ m from the axis of a shower developing in a heterogeneous atmosphere for all E_0. Approximation formulas are given for the same quantity. The author indicates that the approximations are not very accurate and will change. Let us note a characteristic feature of approximation formulas, their explicit dependence on E_0. In earlier works by the same authors, where similar data were given for the SDF of electrons (though in more narrow interval of E_0), a qualitatively different statement was made about the independence of the PDF form from E_0 at 10 GeV, i.e., that 10 GeV with regard to the spatial distribution function can be considered as a *representative of infinity*.

In [105], the real E_0-dependence was admitted (it is clearly visible in Figs. 1a and 1b of [105]), but, noting this fact for interval $1 \div 10^3$ GeV, the author does not discuss the question on accuracy of this replacement. Figure 2 in [105] presents $r^2\rho/N$ for electron density in the shower from the primary photon $E_0 = 10^3$ GeV for $s = 1.0$ and 1.4 in the interval $r \approx 1 \div 10^3$ m. The author notes good agreement of his results for $r \approx 1 \div 10^2$ m with numerical calculations [108] and discrepancies with data [21]. However, the validity of the statement about a smaller change in the SDF for $r > 1$ with the value s is not confirmed (see [106], Fig. 2). With some parallel shift along the z-axis the difference between the functions will apparently be within the computational limits errors.

The works [24, 106–108] describe a new approach based on numerical solving adjoint equations of cascade theory. It is different from numerical method [109], used to solve one-dimensional problems, and from multigroup method [1], used to solve problems of various dimensions. The works [106–108] do not contain a strict proof of the convergence of the method and error estimation. Instead of this, comparisons with other numerical calculations with declared accuracy were performed and demonstrated good agreement. The parameters essential for the result accuracy: the degree of fitting polynomials, the number of steps in t and E on the interval of change of variables were found empirically by observing the numerical process convergence. The solution method is used to calculate average one-dimensional characteristics and their fluctuations and correlations, SDF of electrons in a homogeneous atmosphere at $E_0 = 10 \div 10^6$ GeV, $r = 0.3 \div 2 \cdot 10^2$ m and $t = 10 \div 24$ cascade unite. The results at $s = 1 \div 1.5$ for various E_0 and at $s = 1$ for $E = 10$ and 10^2 GeV with the corresponding data from [110] are compared. The comparison shows a quite satisfactory agreement. The numerical difference between reference (test) data and results of the works [104–106, 110] is interpreted as the error measure. Figure 1 in [108] shows, that at $s = 1$ the computational error is close to the discrepancy between curve 1 ($E_0 = 10$ GeV) and curve 5 ($E_0 = 10^5$ GeV), i.e., within the limits of calculation accuracy, there appears to be no difference between all the curves. The study showed that the shape of the SDF within the accuracy calculations does not depend on E_0. Therefore, the conclusion that $E_0 \approx 10^3$ GeV is identical to infinitely large energy is beyond doubt. The dependence of SDF-form from E_0 has not been studied in detail. Authors of [110] are limited themselves to the statement that comparison of their results at $E_0 \approx 10^3$ GeV and NKG are valid, but NKG is incorrect in the whole studied intervals of s and r. Their SDF practically coincides with NKG when replacing $r_M = 80$ m by $r'_M = 40$ m.

After successful applications of the adjoint equations method to solving various problems in EPC theory [25, 111-114], its authors had elaborated a new version of the Monte Carlo method, similar to that used in extensive air showers calculations [40].

From initial energy E_0 to energy $E^A \approx 10^{10}$ eV, the development of showers is described in the approximation of the A-theory. From E^A to $E_0 = 10$ MeV, the Monte Carlo scheme is used. The authors of the work [113] linked this approximate scheme with the adjoint equations of the cascade theory, traced the coarsening of the equations at the first stage consideration (approximation A) it corresponds. Used (quite obvious) approximations are written down, but the estimate of the error of the results for a particular task has not been carried out, so the work cannot be considered as a justification for a well-known approximate model electron–photon cascade.

In [25], two methods, numerical and Monte Carlo, are applied to solving the problem of electron's SDF calculation for shower with $E_0 = 10^2 \div 10^7$ GeV and $E = 0.25 \div 25$ MeV. The root mean square radius $r_{\rm rms}$ is given for the entire interval E and E_0 as a function of $t/t_{\max} = 0 \div 2.5$. Strong changes in $r_{\rm rms}$ are observed from 10^3 to 10^6 GeV for all $t/t_{\max}$ and all $E_{\rm th}$. The authors note that when $E_0 \geq 10^6$ GeV rms radius does not depend on E_0. This conclusion does not follow from the data presented in the work, and its accuracy is not discussed. There is contained interesting statement (see [25], Fig. 2) that the SDF for identical values of s, determined from a condition close to one-dimensional saddle-point condition is practically the same for all E_0 and $E_{\rm th}$ in variables $x = r/r_{\rm rms}$. Estimating of the statement accuracy from this drawing is almost impossible. However, there is no doubt that in these variables x the difference between the functions under consideration in a certain interval x, depending mainly on s, will be smaller, than in traditional variables.

Investigations of SDF performed using the method of moments [1, 22], show that in a certain interval distances r, the SDF is mainly determined by the magnitude of its second moment. In [112], the same result was obtained for angular functions distributions. The t-dependence of the angle rms are shown, the results for $E_0 = 10^3$ and 10^5 GeV are noticeably different. Authors state that $\theta_{\rm rms}$, when expressed in the age-parameter s, (i.e., in $t/t_{\max}$) at $E_0 \geq 10^3$ GeV does not depend on E_0 and, therefore, the angular distribution at 10^3 GeV reaches its limit form. The last statement is of clearly approximate nature. Nevertheless, this series (Uchaikin–Lagutin–Plyasheshnikov) of works brought a new qualitative and quantitative understanding of EPC nature: the SDF changes with growing E_0 at least up to $E_0 \geq 10^7$ GeV.[3]

[3] I regard these words as worthy recognition of the role of our Altai group in resolving the problem of lateral distribution in EPC, which was the subject of heated debate in the 80s of the last century. Unfortunately, some of this team are no longer alive – Prof. Plyasheshnikov and Dr Konopelko. By working on this book, I am fulfilling my duty to them.

1. Belyaev AA, Ivanenko IP, Kanevsky BL et al (1980) Electron-photon cascades in cosmic rays at extra-high energies. Moscow, Nauka (in Russian).
21. Nishimura J (1967) Theory of cascade shower. In: Handbuch der Physik. B, vol 46, pt 2. Springer, pp 3–114
40. Gaisser TK, Yodh GB (1980) Particle collisions above 10 TeV as seen by cosmic rays. Ann Rev Nucl Part Sci 30:475–512
105. Hillas AM (1981) Results of Monte Carlo simulations of electron-photon cascades in the atmosphere and detectors. In: Proceedings of the 17th international cosmic ray conference, vol 6, pp 244–247
106. Lagutin AA, Plyasheshnikov AV, Uchaikin VV et al (1979) The calculations of the electromagnetic cascade electrons. In: Proceedings of the 17th international cosmic ray conference, vol 7, pp 7–12
107. Plyasheshnikov AV, Lagutin AA, Uchaikin VV (1979) The numerical method of radial distribution of electromagnetic cascade particles calculation. In: Proceedings of the 17th international cosmic ray conference, vol 7, pp 13–17
108. Lagutin AA, Plyasheshnikov AV, Uchaikin VV (1979) The radial distribution of electromagnetic cascade particles in the air. In: Proceedings of the 17th international cosmic ray conference, vol 7, pp 18–23
109. Tielheim KO, Zöllner R (1972) Multigroup method of solution of cascade equations. J Phys A: Math Gen 6:1054–1066
110. Hillas AM, Lapikens J (1977) Electron-photon cascades in the atmosphere and in detectors. In: Proceedings of the 17th international cosmic ray conference, vol 8, pp 460–465
111. Lagutin AA, Plyasheshnikov AV, Uchaikin VV et al (1981) The calculations of correlations in the electron-photon cascade electrons. In: Proceedings of the 17th international cosmic ray conference, vol 5, pp 198–201
112. Lagutin AA, Plyasheshnikov AV, Uchaikin VV et al (1981) The calculations of the angular distributions of electron-photon cascade electrons. In: Proceedings of the 17th international cosmic ray conference, vol 5, pp 202–205
113. Plyasheshnikov AV, Vorobyev KV (1981) The algorithm of the calculation of extreme high energy electromagnetic cascade parameters by the Monte Carlo method. In: Proceedings of the 17th international cosmic ray conference, vol 5, pp 206–209
114. Plyasheshnikov AV, Vorobyev KV, Chernyaev GV (1981) The calculation of dissipated energy depth distributions of the electron photon cascade in heavy substance. In: Proceedings of the 17th international cosmic ray conference, vol 5, pp 210–213
115. Ivanenko IP, Osipova LN, Roganova TM, Fedorova GF (1982) On spatial electron-photon shower particle distribution function for various values of E_0/β in the iso-thermic atmosphere. Izv AN USSR, Ser Phys 46(7):500–505

Appendix C
A Few Images From Prof. Ivanenko Scientific Group Archiv[4]

Prof. I.P.Ivanenko
19.01.1929–26.01.1993

[4] Published with kindly permission of Prof. Tatiana Roganova.

© Higher Education Press Limited Company 2024

V. V. Uchaikin, *Electron-Photon Cascades*, Nonlinear Physical Science,
https://doi.org/10.1007/978-981-99-7524-2

Prof. T.M. Roganova and Prof. I.P.Ivanenko

Prof. I.P. Ivanenko with colleagues

Discussion between Uchaikin's and Ivanenko's groups

Index

© Higher Education Press Limited Company 2024

V. V. Uchaikin, *Electron-Photon Cascades*, Nonlinear Physical Science,
https://doi.org/10.1007/978-981-99-7524-2

Author Index

© Higher Education Press Limited Company 2024

V. V. Uchaikin, *Electron-Photon Cascades*, Nonlinear Physical Science,
https://doi.org/10.1007/978-981-99-7524-2

SPRINGER NATURE

GPSR Compliance

The European Union's (EU) General Product Safety Regulation (GPSR) is a set of rules that requires consumer products to be safe and our obligations to ensure this.

If you have any concerns about our products, you can contact us on ProductSafety@springernature.com

In case Publisher is established outside the EU, the EU authorized representative is:

Springer Nature Customer Service Center GmbH
Europaplatz 3
69115 Heidelberg, Germany

The manufacturer's authorised representative in the EU is Springer Nature Customer Service Centre GmbH, Europaplatz 3, 69115 Heidelberg, Germany. If you have any concerns regarding our products, please contact ProductSafety@springernature.com

Printed and bound by CPI Group (UK) Ltd, Croydon, CR0 4YY

01/10/2026

02225849-0016